Stratigraphy and Paleoenvironments of the Cretaceous Western Interior Seaway, USA

Edited by:
Walter E. Dean, U.S. Geological Survey, Denver, Colorado
And
Michael A. Arthur, Pennsylvania State University, University Park, Pennsylvania

On the front cover: (lower, background figure) Northeast face of the Kaiparowits Plateau, south-central Utah. Cliffs are formed by the Coniacian/Santonian Straight Cliffs Formation. Gray shale at the base of the cliffs is the Cenomanian/Turonian Tropic Shale; (Middle, left) Bedding cycles in the Bridge Creek Limestone Member of the Cenomanian/Turonian Greenhorn Formation, Pueblo, Colorado; (Middle, right) Map of major basins and uplifts in Colorado and adjacent states.

On the back cover: Photograph of bioturbated limestone in the Fort Hays Limestone Member of the Niobrara Formation in the Amoco Bounds core showing large *Chondrites* burrows at the base and large *Teichichnus* burrows at the top.

Copyright 1998 by
SEPM (Society for Sedimentary Geology)

Robert W. Dalrymple, Editor of Special Publications

SEPM Concepts in Sedimentology and Paleontology No. 6

Tulsa, Oklahoma, USA   *March 1998*

SEPM thanks the following for their generous contributions to
Stratigraphy and Paleoenvironments of the
Cretaceous Western Interior Seaway, USA

Walter E. Dean
U.S. Geological Survey

Robert W. Scott
Precision Stratigraphy Associates
Cleveland, Oklahoma

Paul C. Franks, Geologist
Tulsa, Oklahoma

UNIVERSITY OF PLYMOUTH	
Item No.	900 4530136
Date	1 7 JAN 2001 S
Class No.	551.770973 STR
Contl. No.	✓
LIBRARY SERVICES	

ISBN 1-56576-044-1 ✓

© 1998 by
SEPM (Society for Sedimentary Geology)
1731 E. 71st Street
Tulsa, OK 7413-5108
1-800-865-9765

Printed in the United States

90 0453013 6

Charles Seale-Hayne Library
University of Plymouth
(01752) 588 588
LibraryandITenquiries@plymouth.ac.uk

PREFACE

This volume presents the results of a coordinated, multidisciplinary study of Cretaceous carbonate and clastic rocks in cores collected along a transect across the old Cretaceous seaway that extended from the Gulf Coast to the Arctic by a team of academic, industry, and U.S. Geological Survey (USGS) scientists. Our overall goal was to construct a subsurface transect of mid-Cretaceous strata that were deposited in the U.S. Western Interior Seaway (WIS), ranging from pelagic, organic-carbon rich, marine hydro-carbon source rocks in Kansas and eastern Colorado to nearshore, coal-bearing units in western Colorado and Utah. This transect of cores has provided the basis for paleoenvironmental interpretation of organic-carbon burial in an epicontinental, foreland basin setting. In part, the objectives of our study were motivated by the research emphases outlined by the Cretaceous Resources, Events and Rhythms (CRER) Project of the Global Sedimentary Geology Program.

In particular, the papers in this volume focus on the Graneros Shale, Greenhorn Formation, Carlile Shale, and Niobrara Formation and equivalents in cores from six drillholes from western Kansas, southeastern Colorado, and eastern Utah. This series of cores provides unweathered samples and continuous, smooth exposures required for geochemical studies, mineralogical investigations, and biostratigraphic studies. Major objectives of the project covered in the collected papers include: (1) establishing the precise timing of sea level change, rates of subsidence, and facies change; (2) determining of controls on the accumulation, burial, and diagenesis of organic matter; (3) calibrating of depositional cycles using high-resolution stratigraphy; and (4) determining the paleogeography, paleoclimatology, and paleoceanography of the Western Interior Seaway and immediately adjacent landmasses.

We gratefully acknowledge the many sources of funding and materials that supported these studies. Much of the research was funded by the Continental Scientific Drilling Program through the USGS and by the Department of Energy (DOE) through a grant to Penn State (DE-FG02-92ER14251). A core from an Amoco drillhole from western Kansas was released to the USGS in 1992, and description and analysis of this core plus that of a previously acquired well (Schock-Errington #1) in northwestern Kansas constitute the data base for the eastern end of the transect. Three holes that form the western end of the transect, funded by USGS energy programs, were drilled and continuously cored in June 1991 in the Kaiparowits basin near Escalante, Utah. In June 1992, a 700-foot hole, funded by DOE, was drilled and continuously cored near Portland, Colorado, east of the Florence oil field. This sequence, deposited in relatively deep water on the west side of the WIS, includes cycles of terrigenous-clastic and pelagic-marine sediments to contrast with the pelagic-carbonate-dominated cycles of Kansas and the clastic-dominated cycles of western Colorado and Utah. A second, 800-foot hole, also funded by DOE, was drilled in July 1992 about 10 miles southwest of the Portland hole in Pierre Shale, which serves as the reservoir for hydrocarbons in the Florence field. Formal descriptions of the cores are available in a computer data base published as USGS Open-File Reports. All tabulated data from the project are available through the National Geophysical Data Center (NGDC).

We sincerely thank the many reviewers of individual papers in the volume for their concentrated efforts to maintain high technical standards and Peter Sholle for encouraging us to publish this volume through SEPM, the Society for Sedimentary Geology.

Most of the data presented in the papers in this volume are available through:

World Data Center-A for Paleoclimatology
NOAA/NGDC
325 Broadway
Boulder, CO 80303 USA

Phone: 303-497-6280
Fax: 303-497-6513
Telex: 592811 NOAA MASC BDR
http://www.ngdc.noaa.gov/paleo/paleo.html

DEDICATION

   We dedicate this volume to our colleague and friend, Erle G. Kauffman. Erle was a collaborator in this project but was unable to complete his planned contribution to meet the deadline for the volume because of a major illness. However, Erle Kauffman's profound influence is evident in nearly every paper and we are indebted to him for helping us in so many ways. Armed with a strong intellect, boundless energy, and the gift of a silver tongue, Erle has captured the imagination of a horde of students and colleagues and ensnared many to study the spectacular but complicated Cretaceous geology of the Western Interior. Erle's own "holistic" efforts, aided by a cadre of talented students, combine grueling field work with classical principles of biozonation, radiometric dating, lithostratigraphy, and paleoecology to build a foundation for understanding Cretaceous events and basin evolution in the U.S. Western Interior. This substantial body of work includes early recognition of the utility of "Milankovitch" cyclicity in the rock record that provides an unprecedented quasi-20,000-year resolution of events preserved in sedimentary strata. Likewise, he has examined the imprints of anoxic events on faunal compositions and sedimentary structures of these strata, and has made a number of intriguing suggestions, including the hypothesis that anoxia typically occurred at the sediment/water interface rather than within the water column as commonly assumed. Erle was also among the first to use sequence stratigraphic concepts in his studies--long before the techniques became a hallmark of sedimentary geology. In 1977, a landmark paper summarized many of these concepts as a series of enduring insights into the history of sea level and lithostratigraphy of Cretaceous Western Interior basin. He also found evidence for stepwise biotic extinctions accompanying the Cenomanian/Turonian "event," among many other contributions.

   Erle Kauffman has pursued both science and life with great vigor. His detailed knowledge and experience, infectious enthusiasm, and endless store of intriguing speculation coupled with boundless optimism and humor and profound generosity has affected us all. We wish him all the best as he rebounds from his unfortunate illness and hope that he will soon rejoin us in our quest to understand the intriguing strata of the Cretaceous Western Interior Seaway, of course with a few bottles of wine and strum of the ol' banjo.

Walter E. Dean
Michael A. Arthur

TABLE OF CONTENTS

CRETACEOUS WESTERN INTERIOR SEAWAY DRILLING PROJECT: AN OVERVIEW

WALTER E. DEAN[1] AND MICHAEL A. ARTHUR[2]

[1]*U.S. Geological Survey, Federal Center, Denver, Colorado 80225*

[2]*Department of Geosciences, Pennsylvania State University, University Park, Pennsylvania 16802*

ABSTRACT: The Cretaceous Western Interior Seaway Drilling Project was begun in 1991 under the auspices of the U.S. Continental Scientific Drilling Program. It was intended to be a multidisciplinary study of Cretaceous carbonate and siliciclastic rocks in cores from bore holes along a transect across the Cretaceous Western Interior Seaway. The study focuses on middle Cretaceous (Cenomanian to Campanian) strata that include, in ascending order, Graneros Shale, Greenhorn Formation, Carlile Shale, and Niobrara Formation. The transect includes cores from western Kansas, eastern Colorado, and eastern Utah. The rocks grade from pelagic carbonates containing organic-carbon-rich source rocks at the eastern end of the transect to nearshore coal-bearing units at the western end. These cores provide unweathered samples and the continuous depositional record required for geochemical, mineralogical, and biostratigraphic studies. The project combines biostratigraphic, paleoecological, geochemical, mineralogical, and high-resolution geophysical logging studies conducted by scientists from the U.S. Geological Survey, Amoco Production Company, and six universities.

INTRODUCTION

What Is the Cretaceous Western Interior Seaway Drilling Project (WISDP)?

The U.S. Continental Scientific Drilling Program (CSDP) was established in 1988 when Congress passed the Continental Scientific Drilling and Exploration Act (P.L.100-441) mandating that three U.S. Government agencies, the Department of Energy (DOE), the U.S. Geological Survey (USGS), and the National Science Foundation (NSF), develop long- and short-term policy objectives and goals for scientific drilling in the United States. Representatives from these agencies comprise the CSDP Interagency Coordinating Group (ICG/CSDP), which provides multiyear, multihole scientific research strategies for the CSDP. In 1990, we proposed to the ICG/CSDP to drill a transect of cores across the Cretaceous Western Interior Seaway (WIS) that extended from the Gulf of Mexico to the Arctic during maximum marine transgressions (Fig. 1). The proposed transect was to consist of four holes or groups of holes from western Kansas, eastern Colorado, southwestern Colorado, and eastern Utah. We proposed to focus on the two most extensive transgressive episodes in the seaway during the middle Cretaceous that resulted in deposition of two important organic-carbon-rich pelagic limestone units, the Cenomanian-Turonian Greenhorn Formation and Santonian-Campanian Niobrara Formation. An interdisciplinary team of researchers from government, academia, and industry would conduct biostratigraphic studies, paleoecologic studies, inorganic, organic and stable isotopic geochemical studies, mineralogical investigations, and high-resolution geophysical logging. Cores would provide the unweathered samples and continuous smooth exposures required for these studies.

The needs for the eastern end of the transect were met by a hole that was drilled and continuously cored with better than 90% recovery by Amoco Production Company in western Kansas (Amoco No. 1 Rebecca K. Bounds, Greeley County, Kansas; Fig. 2). The Cretaceous part of this core is archived in the USGS Core Research Center (USGS-CRC) in Denver. The 450-m Cretaceous section recovered in the Bounds core extends from the top of the Upper Jurassic Morrison Formation to the middle of the Upper Cretaceous Niobrara Formation (Fig. 3) (Dean et al., 1995).

The needs for the western end of the transect were met in 1991 when the USGS drilled and continuously cored three holes, all about 300 m deep, with better than 98% recovery in the Kaiparowits Basin of south central Utah (Fig. 2). Two holes (USGS-CT1-91 and USGS-SMP1-91) were drilled on top of the Kaiparowits Plateau to collect the coal-bearing sequences of the Upper Cretaceous Straight Cliffs Formation (Hettinger, 1995). A third hole (USGS No. 1 Escalante) was drilled at the base of the Kaiparowits Plateau near the town of Escalante, Utah, and collected all of the marine Tropic Shale and the top of the Dakota Sandstone (Fig. 4). The cores from these three holes are archived in the USGS-CRC in Denver.

During June 1992, the USGS, with DOE funding, drilled and continuously cored a 213-m hole (USGS No. 1 Portland; Fig. 2) in Cretaceous strata east of the Florence oil field in the Cañon City Basin near Florence, Colorado. Core recovery was essentially 100% and includes the lower half of the Niobrara Formation, all of the Carlile Shale, Greenhorn Formation, and Graneros Shale, and the top of the Dakota Sandstone (Fig. 5). Of particular note was the excellent recovery of very distinct limestone-marlstone cycles of the Greenhorn and Niobrara. Both of these pelagic carbonate units contain abundant marine organic matter that may be sources of petroleum in the Florence field as well as in the Denver basin.

The fourth site of the original transect in the northern San Juan Basin of southwestern Colorado was not drilled because the section there was collected by M. R. Leckie and colleagues in a series of trenches along the northern boundary of Mesa Verde National Park (Fig. 2) (Leckie et al., this volume). The results presented in this volume are mainly based on cores from the other three points along the transect (Bounds, Portland, and Escalante cores; Fig. 2), but data are presented from other cores that are publicly available in the USGS-CRC, Denver, from several outcrop sections, and from the Leckie et al. trench (Fig. 2).

Why the Cretaceous?

The Cretaceous Period (ca. 142-65 Ma) of earth history offers a significant opportunity for understanding global processes and their variations. Cretaceous marine and terrestrial strata are extremely widespread in outcrop, subcrop, and in ocean basins. Many of the subcrop sections are recoverable by shallow to

Stratigraphy and Paleoenvironments of the Cretaceous Western Interior Seaway, USA, SEPM Concepts in Sedimentology and Paleontology No. 6
Copyright 1998 SEPM (Society for Sedimentary Geology), ISBN 1-56576-044-1, p. 1-10.

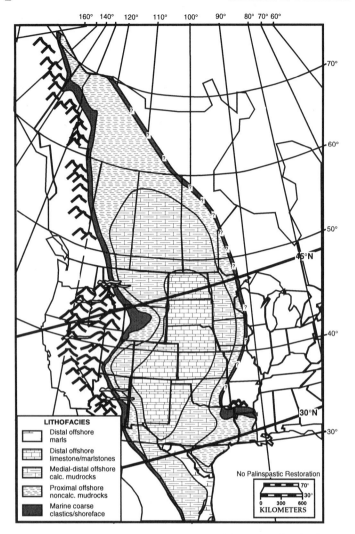

FIG. 1.–Lithofacies map for the early Turonian of the Western Interior Seaway. Paleolatitude lines at 30° and 45°N also are shown. Modified from Sageman and Arthur, 1994.

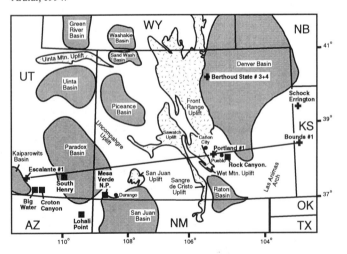

FIG. 2.–Map showing locations of major basins and uplifts in Colorado and adjacent states, locations of cores described in this volume (plus symbols), and locations of outcrop sections discussed in this volume (solid squares).

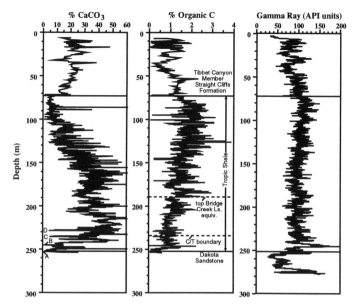

FIG. 3.–Plots of percentages of $CaCO_3$ and organic carbon in the Tropic Shale and lower part of the Tibbet Canyon Member of the Straight Cliffs Formation in the USGS No. 1 Escalante core. The position of the Cenomanian/Turonian (C/T) boundary and the top of strata equivalent to the Bridge Creek Limestone Member of the Greenhorn Formation also are shown.

intermediate (up to 1000 m) continental drilling and ocean-basin sites are accessible by deep-sea drilling. Variable combinations of tectonism, volcanism, atmospheric and ocean chemistry, climate, sea level, and sediment supply helped to produce some of the largest phosphorite deposits and hydrocarbon reserves known on earth (Fig. 6). It is estimated that Cretaceous source rocks are responsible for more than 70% of the world's reserves of crude oil (Tissot, 1979). Understanding the origin and distribution of Cretaceous, organic-rich sequences is tantamount to understanding the origin and distribution of much of the world's oil and gas. In addition, Cretaceous strata contain major reserves of coal, kaolinite, bauxite, and manganese. Accurate prediction of the availability of such resources, and an understanding of their distribution requires models based on a comprehensive knowledge of the Cretaceous world.

The Cretaceous is marked by substantial changes in the extent of shelf and epicontinental seas and in regional and global patterns of marine sedimentation. These changes were the result of major fluctuations in global eustatic sea level (Haq et al., 1987) (Fig. 7). Much of the Cretaceous is also recognized as having had a globally warm, equable, mostly ice-free climate that is about as far removed from our present glacially dominated climate as that of any other time in the Phanerozoic. The origins of this warm climate are not well understood, but are presumed to be related to a major "greenhouse" phenomenon, which possibly was the result of increased volcanic outgassing of carbon dioxide (e.g., Arthur et al., 1985b and 1991; Lasaga et al., 1985).

The middle Cretaceous between 120 Ma and 80 Ma (Aptian to Campanian) is characterized by several globally widespread episodes of organic-carbon burial in marine sequences (e.g., Schlanger and Jenkyns, 1976; Arthur and Schlanger, 1979; Jenkyns, 1980; Arthur et al., 1987; Schlanger et al., 1987; Arthur et al., 1990). These episodes represent periods of widespread

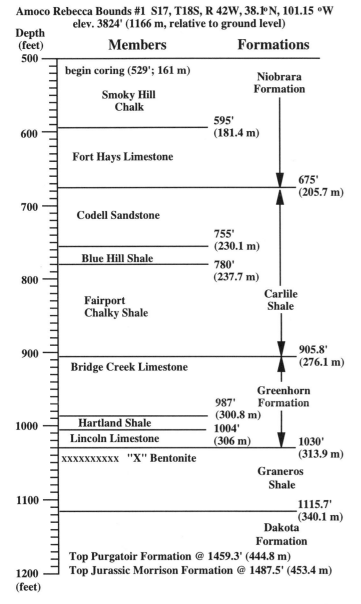

FIG. 4.–Depths to tops of formations and members recovered in the Bounds core.

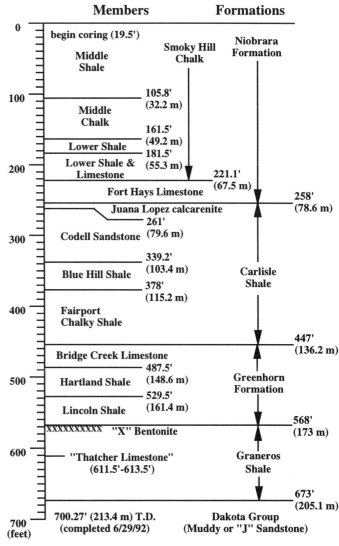

FIG. 5.–Depths to tops of formations and members recovered in the Portland core.

oxygen deficiency in oceanic mid- and deep-water masses that have been termed "Oceanic Anoxic Events" (OAEs; Fig. 7). The widespread occurrence of OAEs in time and space within the middle Cretaceous may imply fundamental changes in oceanic circulation and (or) the rate and mode of delivery of organic matter to the deep sea. The origins of the OAEs are not known for certain, but available data suggest that such events resulted from some combination of higher phytoplankton productivity and enhanced preservation under oxygen-depleted, deep-water masses. Arthur and Natland (1979) suggested that the injection of warm, saline water produced in marginal evaporite basins around the Atlantic during the middle Cretaceous may have created a stable salinity stratification and periodic anoxia in the deep Atlantic. Brass et al. (1982) concluded that the main effect of a warm, saline deep-water mass on oxygen concentration would be through reduced oxygen solubility in warmer water. Herbert and Sarmiento (1991) pointed out that a warm, saline deep-water mass would also increase the efficiency with which plankton

extract nutrients from convecting waters at low latitudes. Sarmiento et al. (1988) produced a more broadly applicable 3-box model for the widespread and rapid formation of deep-water oxygen deficits by varying the rate of supply of organic matter to the deep water relative to rate of supply of oxygen to deep water from high-latitude, oxygen-bearing surface water. Jewell (1996), using a three-dimensional ocean circulation model, determined that the surface-water residence time of the seaway was 0.6 to 2.5 years whereas that of deep water (<100 m) was 1.3 to 4.6 years. Because of the short deep-water residence time, apparent oxygen utilization was modest for all simulations that Jewell tried. Anoxic episodes that occurred in the Cenomanian/Turonian WIS apparently were the result of incursions of mid-depth, suboxic to anoxic waters from the open ocean, or restricted deep-water circulation due to a sill at the entrance to the seaway. Slingerland et al. (1996) used a three-dimensional, turbulent-flow, coastal-ocean model to estimate circulation and chemical evolution of the seaway. This model showed that drainages from the eastern margin were more important than previously suspected and created a strong counterclockwise gyre that was effective in mixing

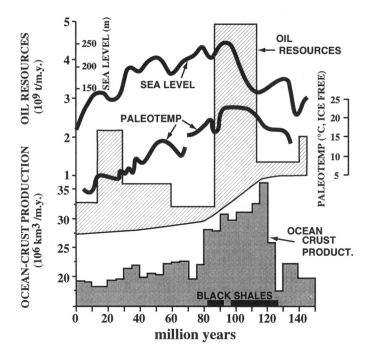

FIG. 6.–Combined plot of world ocean crust production (modified from Larson, 1991a), high-latitude sea-surface temperatures (Savin, 1977; Arthur et al., 1985b), long-term eustatic sea level (Haq et al., 1988), times of black shale deposition (Jenkyns, 1980), and world oil resources (Irving et al., 1974; Tissot, 1979) plotted on geologic time-scale of Harland et al. (1982). Modified from Larson, 1991b.

Tethyan and boreal waters. These studies, however, do not comprehensively explain the interplay between climate, sea level, surface- and deep-ocean circulation, and organic-carbon production and burial.

Why the Western Interior Seaway?

The U.S. Western Interior Seaway (WIS) extended from the Gulf Coast to the Arctic during maximum Cretaceous transgressions as a northwestern arm of the Tethys Ocean (Fig. 1). The numerous energy-rich Cretaceous foreland basins of the Rocky Mountain regions of North America are the remnants of this once extensive seaway. Cretaceous sequences in the WIS are an important part of the global expression of high eustatic sea levels, tectonism, warm climate, and oxygen depletion because of the relative geographic isolation of the seaway and, therefore, its strong susceptibility to paleoclimatic and paleoceanographic change (e.g., Kauffman, 1977 and 1984; Barron et al., 1985). The WIS sequences are renowned for their expression of small-scale cyclicity, with periods of tens to hundreds of thousands of years that commonly are attributed to Milankovitch orbital variations (Fischer, 1980 and 1993; Arthur et al., 1984; Fischer et al., 1985; Bottjer et al., 1986; ROCC Group, 1986; Pratt et al., 1993; Dean and Arthur, this volume; Sageman et al., this volume; Scott et al., this volume). However, the linkages of these presumed orbital variations to sedimentary processes in a supposed ice-free world are poorly understood in contrast to Quaternary cycles with similar periodicities. The WIS sequences provide a tremendous opportunity to examine the sedimentary expression of orbital cyclicity in a geographically restricted part of the Cretaceous ocean that apparently was finely tuned to these orbital variations.

Two important organic-carbon-rich pelagic limestone units

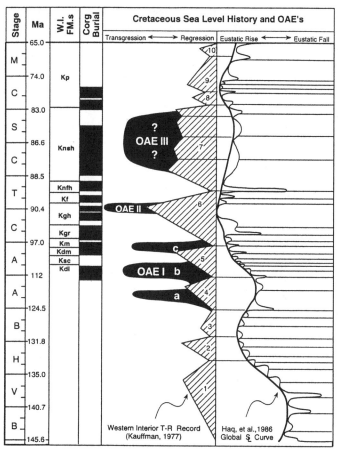

FIG. 7.–Global sea-level history compared with transgressions and regressions and development of organic-carbon-rich deposits in the Western Interior Seaway. The global sea-level curve of Haq et al. (1987) is correlated to transgressions and regressions of the Western Interior Seaway compiled by Kauffman (1977). Major sequence boundaries are correlated to regressive maxima by solid lines. Black bars mark stratigraphic intervals with high levels of preserved organic carbon (Corg>2%). Oceanic Anoxic Events (OAE) intervals are from Arthur et al. (1990). Western Interior Cretaceous formations (W.I. FMs) include: Kp=Pierre Shale; Knsh= Smoky Hill Chalk Member of the Niobrara Formation; Knfh= Ft. Hays Limestone Member of the Niobrara Formation; Kf= Fairport Shale; Kgh= Greenhorn Formation; Kgr= Graneros Shale; Km=Mowry Shale; Kdm=Muddy Sandstone Member of the Dakota Formation; Ksc=Skull Creek Shale Member of the Dakota Formation; Kdl=Lytle Sandstone Member of the Dakota Formation. Stage abbreviations, from top to bottom, are: M=Maastrichtian; C=Campanian; S=Santonian; C=Coniacian; T=Turonian; C=Cenomanian; A=Albian; A=Aprian; B=Barremian; H=Hauterivian; V=Valanginian; B=Berriasian. Ages of stage boundaries are from Harland et al. (1982).

were deposited in the eastern part of the WIS during the two most extensive transgressive episodes (e.g., Kauffman, 1977; Arthur et al., 1985a; Pratt et al., 1993). The Cenomanian-Turonian transgression resulted in deposition of the Bridge Creek Limestone Member of the Greenhorn Formation (OAE II, Fig. 7), and the Late Turonian through early Campanian transgression led to sedimentation of the Niobrara Formation (OAE III, Fig. 7). Both of these pelagic carbonate units were deposited in a rapidly subsiding basin (Bond, 1976; Cross and Pilger, 1978) that was characterized by substantial clastic sediment input (e.g., Ryer, 1977a and b) from uplifted tectonic terranes of the Sevier orogenic belt to the west (Armstrong, 1968; Weimer, 1970) as well as an abundant supply of windblown ash from volcanic activity in the same region (Kauffman, 1977). At maximum transgression, the shoreline of the WIS extended as far west as western Utah and western Arizona (Fig. 1). The predominantly pelagic carbonate units of Kansas and eastern Colorado are replaced by a more clastic-rich sequence in western Colorado and eastern Utah. Important coal-bearing sequences equivalent in age to the Niobrara were deposited along the western shoreline in Utah, and were cored as part of the WISDP (Hettinger, 1995). In the northern San Juan basin in southwestern Colorado, the Niobrara equivalents are represented by sandstones of the lower Mesa Verde Group, and the upper part of the Mancos Shale, and the time equivalent of the Greenhorn Formation is the lower Mancos Shale (Fig. 8). We have selected these time intervals, which represent extremes in global eustatic sea level and resulting continental flooding, to test the sensitivity of ocean circulation to global climatic forcing as the result of the changing configuration of continental topography and shallow seas.

The Cretaceous sequences of the Western Interior are among the most studied in the world. Knowledge of the Western Interior, built up over the years by industry, government, and academic scientists, provides exceptional temporal and spacial control for working out sequence stratigraphic relationships, and provides extensive knowledge on the sedimentology, geochemistry, paleontology, and paleobiology of these sequences. The temporal control is the most precise of any Cretaceous sequence in the world, both in terms of absolute radiometric ages (Obradovich, 1993) and high-resolution biostratigraphic zonation (Kauffman et al., 1993; Scott et al., this volume). The extensive paleogeographic, stratigraphic, biotic, and geochemical data available for Cretaceous strata in the WIS suggests that an understanding of the response of a large epicontinental sea to variations in global forcing is within our grasp. This knowledge can then be extended to less explored foreland basins of the world that have similar sequences. Our approach focuses on the middle Cretaceous marine record of the WIS and the extent to which marine sedimentation in the seaway reflects global events at various times. For example, with respect to so-called "anoxic events", many workers favor models that involve propagation of external marine environmental conditions, especially oxygen-deficient deeper water masses, into the WIS as the result of global eustatic sea level rise with a superimposed imprint of regional environmental conditions (e.g., Fischer and Arthur, 1977; Barron et al., 1985; Arthur et al., 1987; Glancy et al., 1993). Others hypothesize that the WIS and equivalent depositional environments on other continents are the sources of global oxygen-deficient mid- to deep-water masses (e.g., Eicher and Diner, 1989; Fisher et al., 1994). These contrasting hypotheses are amenable to testing by modeling efforts tied to comprehensive data syntheses and model validation.

Because the Cretaceous is so important to so many different disciplines of the geosciences, the Global Sedimentary Geology Program (GSGP) identified "Cretaceous Resources, Events and Rhythms" (CRER) as its first major international project (Ginsburg and Beaudoin, 1990). Under this project are five Working Groups: (1) Sequence stratigraphy and sea level change; (2) Black shales and organic-carbon burial; (3) Cyclostratigraphy; (4) Carbonate platform evolution; and (5) Paleogeography, paleoclimatology and sediment fluxes. Marine sedimentary units deposited in the Cretaceous WIS provide an exciting opportunity to develop and integrate the five working group themes listed above in one project. The results of east-west WISDP transect described in this volume address the objectives of all of the CRER working groups except those of the carbonate platform working group. We hope that a future WISDP north-south transect will include sections of carbonate platforms in New Mexico, Texas,

and Mexico. The strata of interest monitor global change during the Cretaceous and include a number of petroliferous units; are characterized by well-developed cyclicity in the Milankovitch band; pose a variety of interesting paleoclimatic, paleoceanographic and sediment flux problems; and are dominated by sequences produced by marked transgressive-regressive cycles. The following sections describe how the WISDP is addressing the objectives of the working groups of CRER.

OBJECTIVES OF WESTERN INTERIOR SEAWAY DRILLING PROJECT

Sea Level Change and Sequence Stratigraphy

The WISDP transect provides sampling of strata deposited during two major transgressive-regressive cycles of the WIS that reflect global changes in sea level (Figs. 6, 7). The origin of these sea-level changes is puzzling because much of the Cretaceous is thought to have been ice-free and characterized by warm, latitudinally equable climates and, therefore, ice-volume changes could not have induced much variation in sea level. On the other hand, the Cretaceous was a time of unusually active sea-floor spreading, with plate-margin and mid-plate volcanism so extensive that tectonically induced changes in sea level should be expected, possibly corresponding to thrust loading on the western edge of the basin.

Key objectives are: (1) to determine the precise timing and rates of sea-level change and their influence on lithology, sedimentation rate, and sediment geochemistry; (2) to determine the amount, type, and degree of preservation of organic matter and its relation to sea-level variations, whether tectonic or eustatic. (3) to examine rates of subsidence, apparent sea-level change, and facies migration in light of purported global tectonic-eustatic events.

Organic Matter Deposition and Burial

The cored sequences on the WISDP transect encompass several episodes of enhanced burial of organic carbon that are important both within the WIS and globally. These periods of widespread oxygen deficiency in oceanic deep-water masses (so-called "Oceanic Anoxic Events"; Fig. 7), marked by widespread organic-carbon-rich sequences loosely called "black-shales," include several hydrocarbon source-rock sequences as well as economically important coal sequences (Sageman and Arthur, 1994).

Key objectives are: (1) to determine onshore-offshore facies patterns and relationships of organic-matter accumulation, changes in organic-matter type and degree of preservation, and how these relate to transgressive episodes; (2) to use variations in faunal and floral components and sedimentary structures to evaluate the extent and intensity of oxygen deficiency during black-shale episodes and correspondence in timing to global events; (3) to determine the effects of organic-matter enrichment on the geochemical characteristics of the sediments, particularly as they relate to the cycling of Fe, Mn, S, P, and trace elements; and (4) to determine the effects of thermal metamorphism and other geochemical characteristics with increasing burial depth along the transect of cores using the fresh, unweathered samples that the cores provide.

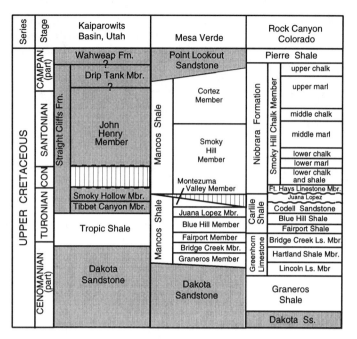

Series	Stage	Kaiparowits Basin, Utah	Mesa Verde	Rock Canyon Colorado

FIG. 8.–Stratigraphic columns showing correlation of stratigraphic units from Rock Canyon near Pueblo, Colorado, Mesa Verde, Colorado, and the Kaiparowits Basin, Utah. Modified from West and Leckie (this volume). White panels represent units of fine-grained carbonate or mud rocks; shaded panels represent sandstone units; panels with vertical stripes represent lacunas.

Cyclostratigraphy

The Cretaceous sequences of the WIS are renowned for their expression of small-scale cyclicity, with periods of tens of thousands of years that commonly are attributed to Milankovitch orbital variations (Gilbert, 1895; Hattin, 1971 and 1985; Kauffman, 1977; Fischer, 1980; Pratt, 1981; Arthur et al., 1984; Barron et al., 1985; Barlow and Kauffman, 1985; Bottjer et al., 1986; ROCC Group, 1986; Pratt et al., 1993). In addition, longer-term cycles with periodicities of up to several million years, are defined on geochemical and geophysical logs (Arthur and Dean, 1992; Pratt et al., 1993; Dean and Arthur, this volume) and also may be related to orbital cycles or possibly to sea level changes that are correlated with Milankovitch forcing. However, the linkages of these presumed orbital variations to sedimentary processes in a supposed ice-free world are poorly understood in contrast to ice-volume-dominated Quaternary cycles with similar periodicities. The continuously cored sequences and high quality geophysical logs of the WISDP transect provide a tremendous opportunity to examine the sedimentary expression of orbital cyclicity in a geographically restricted part of the Cretaceous ocean that apparently was finely tuned to these orbital variations. In addition, the sequences of the WIS contain the greatest diversity of faunal and floral assemblages of any Cretaceous sequence in the world.

Key objectives are: (1) to use the detailed biostratigraphies provided by the faunal and floral assemblages, with abundant volcanic ash layers, to provide a unique, high-resolution time sequence for precisely calibrating cyclicity; (2) to use these "absolute" age data to calculate important variables such as rates and timing of sediment supply, organic productivity, and changes in water-mass characteristics that contribute to the sedimentary expression of Milankovitch orbital cycles.

Paleogeography, Paleoclimatology, and Sediment Fluxes

The origins of the warm climate of the Cretaceous are poorly understood, but are presumed to be related to a major expression of a greenhouse phenomenon, such as might result from increased volcanic outgassing of carbon dioxide. However, even the warm, equable climate paradigm is being challenged – at least for some parts of the Cretaceous – as new data are obtained. The WIS sequences of the North America are an important part of the global expression of high eustatic sea levels, tectonism, warm climate, and oxygen depletion because of the relative geographic isolation of the seaway and, therefore, its strong susceptibility to paleoclimatic and paleoceanographic change.

Key objectives are: (1) to relate trends in geochemical properties and organic-carbon accumulation rates to global sea level, tectonism, warm climate, and oxygen depletion; (2) to provide validation of climate models (e.g. GCMs); (3) to provide explanations for the significance of bioevents and substantial changes in global biotic diversity during the Cretaceous; (4) to document rates and timing of biotic extinction and evolution in detail; (5) to investigate the possibility that shallow Cretaceous seas such as the Western Interior Seaway were the sources of oxygen-depleted oceanic deep-water masses and anoxia.

SUMMARY OF RESULTS

Scott et al. provide a chronostratigraphic reference section for the upper Albian (Purgatoire Formation) to Coniacian (Fort Hays Member of the Niobrara Formation) for western Kansas based on detailed graphic correlation of paleontological data (dinoflagellates, foraminifera, calcareous nannofossils, pollen, spores, and molluscs) from the Bounds core. The Bounds reference section is linked to key reference sections in Europe and North Africa by graphic correlation. Scott et al. then use the chronology produced by the Bounds reference section to define four orders of cycles. The longest cycles are parasequences of 2.0 to 3.4 My in duration represented by the Purgatoire Formation, the lower Dakota Formation, the upper Dakota and Graneros Formations, and the Greenhorn and Carlisle Formations. The shortest cycles with periodicities of tens of thousands of years occur in the Greenhorn Formation and are best displayed by the striking limestone/marlstone cycles in the Bridge Creek Limestone Member of the Greenhorn. Using their graphic correlation chronology, Scott et al. concluded that the 13 cycles in the Bridge Creek in the Bounds core, which average 1.5 m in thickness, have an average duration of 17,333 years, very close to the 19,000 and 21,000 Milankovitch orbital precession cycles.

Numerous investigators during the last century, beginning with G. K. Gilbert (1895), have tried to refine the timing of the carbonate cycles in the Greenhorn Formation, as well as those in the Niobrara Formation, and several papers in this volume, in addition to Scott et al., address this issue. Another hotly debated aspect of these cycles is their origin. Because of the shallow depths of the Western Interior Seaway, carbonate dissolution can be ruled out as the primary cause, but arguments have been made both for and against carbonate production and clastic dilution as the primary cause. Burns and Bralower examined assemblages of calcareous nannofossils and carbon and oxygen isotopic composition of the fine fractions (<38mm) of these two carbonate units in the Bounds and Portland cores to see if they could distinguish between carbonate production and clastic dilution. They found that despite the shallow water depths during deposition of these carbonate units, there is considerable variability in the preservation of calcareous nannofossils. The best preserved nannofossils are in the Niobrara Formation in the Bounds core, and the worst preserved are in the Niobrara Formation in the Portland core. Preservation of nannofossils in the Bridge Creek Limestone is moderate in both cores. Variations in abundances of nannofossil fertility indicators do not correlate with those in carbonate content, which suggests that the cyclic lithologic variations may not be associated with surface-water conditions. Values of $\delta^{18}O$ of the fine-fraction carbonate also do not correlate with carbonate content, but they are negatively correlated with abundances of species of these calcareous phytoplankton that are thought to indicate of high surface-water fertility. This suggested to Burns and Bralower that surface-water fertility was higher when runoff of freshwater from the western highlands was greater, which brought in more detrital clastic material, and freshwater with lower values of $\delta^{18}O$. They conclude, therefore, that the limestone/marlstone cycles in the Bridge Creek and Niobrara appear to be controlled by both carbonate production and clastic dilution.

Bralower and Bergen describe the calcareous nannofossil assemblages in 1100 samples from the Cenomanian to Santonian sections in the Bounds, Portland, and Escalante cores. Based on this biostratigraphy, they correlated the Cenomanian/Turonian boundary between cores and with an outcrop section in southern Utah. However, other stage boundaries could not be correlated. Also, there is a discrepancy in the stratigraphic position of the Cenomanian/Turonian boundary between western, marginal sections and basin-center sections, probably due to one or a combination of environmental factors such as salinity and fertility.

Biostratigraphy and paleoecologic interpretations based on planktonic and benthic foraminifera by West et al. provide an important link between the subsurface (USGS core from Escalante, Utah) and outcrop sections along the western margin of the seaway at maximum transgression in late Cenomanian and early Turonian. Cyclic changes in foraminifera assemblages in the Tropic Shale in the Escalante core and in outcrops of the Mancos Shale at Lohali Point, Arizona, and Mesa Verde National Park, southwestern Colorado (Fig. 2), record dynamic changes in paleoceanography linked to major transgressive-regressive episodes. Relative abundance of the benthic foraminifer *Gavelinella* clearly delimit several of the sequence stratigraphic boundaries of Leithold (1994) in the Tropic and Tununk Shales in southern Utah. Changes in species of benthic foraminifera occur in response to cyclic variations in food and bottom-water oxygenation associated with shifting water masses. At the time of maximum transgression (early Turonian), the seaway was filled with a warm, oxygen-depleted, southerly (Tethyan) water mass. However, the transition from calcareous to agglutinated benthic foraminifera in the middle Turonian suggests that the warm, southern water mass was replaced by a cold, boreal water mass.

Leckie et al. extend the foraminifera interpretations of West et al. for the western margin of the basin eastward toward the center of the basin at the classic section at Rock Canyon near Pueblo, Colorado (Figs. 2, 8). Leckie et al. confirm the findings of West et al.: that the benthic foraminifera assemblages in the Mancos Shale at Lohali Point, Arizona are very similar to those in the Mancos at Mesa Verde National Park, Colorado, and both are different than those in the time-equivalent Bridge Creek Limestone at Rock Canyon. The benthic foraminifera assemblages at

Mesa Verde and Lohali Point contain distinct arenaceous taxa with northern affinities. This suggests that a cool, northern bottom-water mass predominated along the western margin of the basin. In contrast, the planktonic foraminifera assemblages at Lohali Point are more similar to those at Rock Canyon than those at Mesa Verde, and suggest a Tethyan source. The clay mineral suites at each of the three sites are dominated by detrital illite and smectite derived from the Sevier orogenic belt to the west. However, the clay mineral suites at Lohali Point and Rock Canyon also contain substantial amounts of kaolinite derived from somewhere other than the Sevier orogenic belt. Because of the similarities in planktonic foraminifera and clay-mineral assemblages in the Cenomanian/Turonian rocks at Lohali Point and Rock Canyon, Leckie et al. conclude that these sites received a southern surface-water mass and a southern detrital source in addition to the dominant detrital influx from the western Sevier region.

Savrda presents the results of careful, detailed analyses of ichnofossil variations in both the Bridge Creek Limestone and Niobrara Formation in the Bounds and Portland cores. Analyses of individual burrowed intervals involved identification of burrow types (to the ichnogeneric level when possible) and assessment of burrow diameters and penetration depths. Within the Bridge Creek Limestone, Savrda identified four ichnocoenoses based on recurring ichnofossil associations. Vertical stacking patterns of ichnocoenoses and laminated intervals (laminites) were then used to reconstruct the histories of paleo-oxygenation. The analyses reveal that paleo-oxygenation progressively decreased during deposition of the Bridge Creek at both core sites. Comparison of interpreted oxygenation curves (IOC) indicates that the floor of the seaway was less oxygenated at the Bounds core site in western Kansas than at the Portland core site in central Colorado, at least during deposition of the clastic-rich parts of bedding couplets. These curves also define high-amplitude redox cycles that correspond to the limestone/marl couplets of the Bridge Creek as well as higher-frequency, lower-amplitude cycles within marlstone and shale units.

For the Niobrara, Savrda used the vertical disposition of six oxygen-related ichnocoenoses and associated laminites to reconstruct oxygenation histories at both sites. As for the Bridge Creek, oxygenation curves indicate a broad trend towards benthic deoxygenation during deposition of the Fort Hays and Smoky Hill Members. Oxygenation cycles of variable amplitude and frequency, which generally correspond to carbonate rhythms, are superimposed upon this trend. Decimeter-scale and lower-frequency cycles in both members may be linked to orbitally forced climate cycles, whereas higher-frequency cycles in the Smoky Hill record periodic or episodic processes operating at time scales shorter than Milankovitch orbital rhythms. Laminated fabrics are more common in the lower Smoky Hill in the Bounds core than they are in the Portland core. This suggests that differences in depositional conditions between the two sites during deposition of the lower Smoky Hill were similar to those during deposition of the Bridge Creek; i.e., benthic oxygenation of the seaway at the Bounds core site was significantly lower than at the Portland core site.

Sageman et al. applied quantitative spectral analyses to lithologic and paleoecologic cycles at the C/T boundary in the Bridge Creek Limestone in the Portland core to determine the dominant periodicities of the cycles. They used a combination of percent carbonate, percent organic carbon, grayscale pixel values from scanned photographs, ichnologic measurements from Savrda (this volume) and nannofossil data from Burns and Bralower (this volume). The analyses were aided by recently published radiometric (Obradovich, 1993) and biostratigraphic data (Kauffman et al., 1993). Spectra of percent $CaCO_3$, percent organic carbon, and grayscale pixel values produced significant periodicities that closely approximate the three major Milankovitch cycles of eccentricity, obliquity, and precession in the upper 6 m of the Bridge Creek Limestone. Spectra for maximum burrow diameter and oxygen-related ichnocoenosis rank are similar to those for percent organic carbon, which suggests a common control by bottom-water oxygenation. Spectra for some calcareous nannofossil taxa are less definitive due to poor preservation. Sageman et al. then developed a new model for causes of cyclicity in the Bridge Creek Limestone in which obliquity controlled dilution by terrigenous clastic material through its effects on precipitation at high latitudes. This dilution also brought in fresher water, resulting in stratification of the water column, lower benthic oxygen levels, better preservation of organic matter, and less bioturbation, the classic dilution-redox model for the WIS cycles. On the other hand, cyclic fluctuations in carbonate production are mainly controlled by evaporation and nutrient upwelling in the Tethyan realm south of the WIS. The two forcing mechanisms mix in the shallow seaway to produce the complex bedding patterns observed in the Bridge Creek Limestone.

Pancost et al. used hydrocarbon biomarkers and compound-specific carbon isotope analyses at the Cenomanian/Turonian boundary interval in the Bounds, Portland, and Escalante cores to try to detect trans-basinal variations in organic matter input and diagenesis. As expected, the influx of terrestrial organic matter was greatest along the western margin (Escalante core). Marine contributions were greatest in the central basin (Bounds and Portland cores). The organic matter in the Bounds core is enriched in ^{13}C relative to that in the Portland core, which Pancost et al. interpret as indicating that the seaway at the site of the Bounds core in Kansas was dominated by warm, CO_2-poor, Tethyan waters, whereas the seaway at the site of the Portland core in Colorado, the seaway was dominated by CO_2-rich, boreal waters. The isotopic composition of the isoprenoid hydrocarbons pristane and phytane suggests that greater selective degradation of marine organic matter and (or) lower marine productivity occurred during deposition of the limestone beds relative to the marlstone beds.

The upper Cenomanian to middle Turonian Tropic Shale, and the correlative Tununk Shale Member of the Mancos Shale, accumulated along the western margin of the Western Interior Seaway in Utah. From studies of these units in outcrops and southern Utah and in the Escalante core, Leithold and Dean conclude that these units record relatively rapid deposition in prodeltaic environments. Detailed examination of primary sedimentary structures, ichnofabric, and fecal pellets suggest that sediment accumulation in this setting was the result of both suspension fallout from river plumes and storm-induced turbidity currents. Carbon burial in the deltaic deposits was apparently affected by global, regional, and local processes. In the lower part of the Tropic Shale in the Escalante core, for example, a marked positive carbon isotopic excursion near the Cenomanian/Turonian boundary parallels that seen at many localities around the world. In the Tropic Shale, however, unlike in other Cenomanian and Turonian successions, peak concentrations of marine organic matter are not observed at the boundary but stratigraphically higher. In the Escalante core, an upward trend toward increasing concentration of hydrogen-rich, marine organic matter par-

allels evidence both for progressively increasing sediment accumulation rates and for decreasing oxygen levels.

Changes in paleoceanographic conditions across the Cenomanian/Turonian boundary in the Tropic Shale in the Escalante core were investigated by Pagani and Arthur, who measured major and trace elements and carbon and oxygen isotopic composition of the carbonate in bulk samples, inoceramids, and ammonites. Stable isotopic compositions of well-preserved shell carbonate derived from inoceramids and ammonites reflect temporal and spatial environmental variations within the water column. Inoceramid oxygen-isotopic compositions are probably influenced by "vital effects" and do not appear to directly record primary conditions. However, temporal trends in inoceramid isotopic values probably reflect relative environmental variations. Negative shifts in ammonite $\delta^{18}O$ coincide well with short-term, base-level cyclicity. The amplitude of these negative shifts is interpreted to reflect the influence of two distinct water masses. A freshened, northern component water with a depleted $\delta^{18}O$ signature is closely associated with the western boundary of the basin and dominates during regressive events and (or) progradation of the shoreline. This agrees with results of previous studies of planktonic organisms (e.g., West et al., this volume) as well as with recent numerical circulation models for the WIS.

Cyclic variations on scales of decimeters to tens of meters in concentrations of carbonate, detrital clastic material, organic-carbon, and in degree of bioturbation characterize the cyclic pelagic limestone sequences of the Niobrara Formation. Although the cycles in the Niobrara Formation on all scales can adequately be represented as a three-component system of carbonate, clay, and organic carbon, multivariate Q-mode factor analyses of elemental geochemical data by Dean and Arthur identified which other elements are associated with these three components, and also defined several other element associations that reflect more subtle geochemical differences in the Niobrara, due mainly to changes in redox conditions. Biogenic carbonate, windblown volcanic ash, and terrigenous detritus transported by winds, rivers, and surface currents competed against one another during deposition of the carbonate units of the Niobrara. Dean and Arthur attempted to use elemental ratios to try to detect relative changes in influx of detrital illite, volcanic smectite, and eolian quartz. Although they found marked changes in clastic sources with time, they found little difference in the composition of the clastic faction that can be related to individual limestone/marlstone cycles and, therefore, dry/wet climatic cycles. Dean and Arthur used the oxygen-isotope composition of bulk carbonate and concentrations of Mg and Sr to detect evidence for burial diagenesis in the Niobrara. Values of $\delta^{18}O$ in bulk carbonate are about 2‰ lower in the Smoky Hill Member in more deeply buried sections in eastern Colorado than in more shallow buried sections in western Kansas. Most of this difference is due to greater burial diagenesis in Colorado with formation of isotopically light carbonate cements. However, even when corrected for burial diagenesis, the values of $\delta^{18}O$ in the Niobrara are still lower than those in open marine carbonates of the same age (e.g., Austin Chalk), providing further evidence for seawater of lower than normal salinity in the Western Interior Seaway. If the carbonates in the Niobrara underwent burial diagenesis, as suggested by the oxygen-isotope data, then Sr and Mg should have been lost during recrystallization and cementation. However, the observed concentrations of Sr and Mg are remarkably close to those predicted by theoretical mixing lines between their concentrations in pure pelagic biogenic carbonate and Cretaceous clay, suggesting that the carbonates in the Niobrara did not lose Sr or Mg through diagenesis.

REFERENCES

ARMSTRONG, R. L., 1968, Sevier orogenic belt in Nevada and Utah: Geological Society of America Bulletin, v. 79, p. 429-458.

ARTHUR, M. A., AND NATLAND, J. H., 1979, Carbonaceous sediments in the North and South Atlantic: the role of salinity in stable stratification of Early Cretaceous basins, in Talwani, M., Hay, W. W., and Ryan, W. B. F., eds., Deep Drilling Results in the Atlantic Ocean: Continental Margins and Paleoenvironment: Washington, D.C., American Geophysical Union, Maurice Ewing Series, v. 3, p. 375-401.

ARTHUR, M. A., AND SCHLANGER, S. O., 1979, Cretaceous 'oceanic anoxic events' as causal factors in development of reef-reservoired giant oil fields: American Association of Petroleum Geologists Bulletin, v. 63, p. 870-885.

ARTHUR, M. A., AND DEAN, W. E., 1992, An holistic geochemical approach to cyclomania: Examples from Cretaceous pelagic limestone sequences, in Einsele, G., Ricken, G. W., and Seilacher, A., eds., Cyclic and Event Stratigraphy II: Heidelberg, Springer-Verlag, p. 126-166.

ARTHUR, M. A., DEAN, W. E., BOTTJER, D. A., AND SCHOLLE, P. A., 1984, Rhythmic bedding in Mesozoic-Cenozoic pelagic carbonate sequences: The primary and diagenetic origin of Milankovitch-like cycles, in Berger, A., Imbrie, J., Hays, J. D., Kukla, G., and Saltzman, B., eds., Milankovitch and Climate, pt. 1: Amsterdam, Reidel Publishing Co., p. 191-222.

ARTHUR, M. A., DEAN, W. E., POLLASTRO, R., SCHOLLE P. A., and Claypool, G. E., 1985a, A comparative geochemical study of two transgressive pelagic limestone units, Cretaceous Western Interior basin, U.S., in Pratt, L. M., Kauffman, E. G., and Zelt, F. B., eds., Fine-grained Deposits and Biofacies of the Cretaceous Western Interior Seaway: Evidence of Cyclic Sedimentary Processes, Fieldtrip Guidebook 4: Tulsa, Society of Economic Paleontologists and Mineralogists, p. 16-27.

ARTHUR, M. A., DEAN W. E., AND SCHLANGER, S. O., 1985b., Variations in the global carbon cycle during the Cretaceous related to climate, volcanism, and changes in atmospheric CO_2, in Sundquist, E. T., and Broecker, W. S., eds., The Carbon Cycle and Atmospheric CO_2: Natural Variations Archean to Present: Washington, D. C., American Geophysical Union Monograph 32, p. 504-529.

ARTHUR, M. A., SCHLANGER, S. O., AND JENKYNS, H. C., 1987, The Cenomanian-Turonian oceanic anoxic event, II. Paleoceanographic controls on organic matter production and preservation, in Brooks, J., and Fleet, A., eds., Marine Petroleum Source Rocks: London, Geological Society of London Special Publication 26, p. 401-420.

ARTHUR, M. A., BRUMSACK, H.-J., JENKYNS, H. C., AND SCHLANGER, S. O., 1990, Stratigraphy, geochemistry, and paleoceanography of organic carbon-rich Cretaceous sequences, in Ginsburg, R. N., and Beaudoin, B., eds., Cretaceous, Resources, Events, and Rhythms: Background and Plans for Research: Dordrecht, the Netherlands, Kluwer Academic Publishers, p. 75-119.

ARTHUR, M. A, KUMP, L. R., DEAN, W. E., AND LARSON, R. L., 1991, Superplume, supergreenhouse? (abs.): Eos, Transactions of the American Geophysical Union, v. 72, p. 301.

BARLOW, L.K., AND KAUFFMAN, E. G., 1985, Depositional cycles in the Niobrara Formation, Colorado Front Range, in Pratt, L. M., Kauffman, E. G., and Zelt, F. B., eds., Fine-grained Deposits and Biofacies of the Cretaceous Western Interior Seaway: Evidence of Cyclic Sedimentary Processes, Fieldtrip Guidebook 4: Tulsa, Society of Economic Paleontologists and Mineralogists, p. 199-208.

BARRON, E. J., ARTHUR, M. A., AND KAUFFMAN, E. G., 1985, Cretaceous rhythmic bedding, sequences: A plausible link between orbital variations and climate: Earth and Planetary Science Letters, v. 72, p. 327-340.

BOND, G., 1976, Evidence for continental subsidence in North America during the Late Cretaceous global submergence: Geology, v. 4, p. 557-560.

BOTTJER, D. J., ARTHUR, M. A., DEAN, W. E., HATTIN, D. E., AND SAVRDA, C. E., 1986, Rhythmic bedding produced in Cretaceous pelagic carbonate environments—sensitive recorders of climatic cycles: Paleoceanography, v. 1, p. 467-481

BRASS, G. W., SOUTHAM, J. R., AND PETERSON, W. H., 1982, Warm saline bottom water in the ancient ocean: Nature, v. 296, p. 620-623.

CROSS, T. A., AND PILGER, R. H., Jr., 1978, Tectonic controls of Late Cretaceous sedimentation, Western Interior, U.S.A.: Nature, v. 274, p. 653-675.

DEAN, W. E., ARTHUR, M. A., SAGEMAN, B. B., AND LEWAN, M. D., 1995, Core descriptions and preliminary geochemical data for the Amoco Production Company Rebecca K. Bounds # 1 well, Greeley County, Kansas: United States Geological Survey Open-File Report 95-209, 243 p.

EICHER, D. L., AND DINER, R., 1989, Origin of the Cretaceous Bridge Creek cycles in the Western Interior, United States: Palaeogeography, Palaeoclimatology, Palaeoecology, v. 74, p. 127-146.

FISCHER, A. G., 1980, Gilbert-bedding rhythms and geochronology, in Yochelson, E. I., ed., The Scientific Ideas of G.K. Gilbert: Boulder, Geological Society of America Special Paper 183, p. 93-104.

FISCHER, A. G., 1993, Cyclostratigraphy of Cretaceous chalk-marl sequences, in Caldwell, W. G. E. and Kauffman, E. G., eds., Evolution of the Western Interior Basin: St. John's, Geological Association of Canada, Special Paper 39, p. 283-296.

FISCHER, A. G., AND ARTHUR, M. A., 1977, Secular variations in the pelagic realm, in Cook, H.E., and Enos, P., eds., Deep Water Carbonate Environments: Tulsa, Society of Economic Paleontologists and Mineralogists, Special Publication 25, p. 19-50.

FISCHER, A. G., HERBERT, T. D., AND PREMOLI-SILVA, I., 1985, Carbonate bedding cycles in Cretaceous pelagic and hemipelagic sediments, in Pratt, L. M., E. G. Kauffman, E. G., and Zelt, F. B., eds., Fine-grained Deposits and Biofacies of the Cretaceous Western Interior Seaway: Evidence of Cyclic Sedimentary Processes, Fieldtrip Guidebook 4: Tulsa, Society of Economic Paleontologists and Mineralogists, p. 1-10.

FISHER, C. G., HAY, W. W., AND EICHER, D. L., 1994, Oceanic front in the Greenhorn sea (late middle through late Cenomanian): Paleoceanography, v. 9, p. 879-892.

GILBERT, G. K., 1895., Sedimentary measurement of geologic time: Journal of Geology, v. 3, p. 121-127.

GINSBURG, R. N., AND BEAUDOIN, B., eds., 1990., Cretaceous Resources, Events and Rhythms: Background and Plans for Research: Dordrecht, the Netherlands, Kluwer Academic Publishers, 352 p.

GLANCY, T. J., ARTHUR, M. A., BARRON, E. J., AND KAUFFMAN, E. G., 1993, A paleoclimate model for the North American Cretaceous (Cenomanian-Turonian) Epicontinental Sea, in Caldwell, W. E., and Kauffman, E. G., eds., Evolution of the Western Interior Basin: St. John's, Geological Association of Canada, Special Paper 39, p. 219-242.

HAQ, B., HARDENBOL, J., AND VAIL, P. R., 1987, Chronology of fluctuating sea levels since the Triassic (250 million years ago to present): Science, v. 235, p. 156-1167.

HARLAND, W. B., ARMSTRONG, R. L., COX, A. V., CRAIG, L. E., SMITH, A. G., AND SMITH, D. G., 1982, A Geologic Time Scale: London, Cambridge University Press, 263 pp.

HATTIN, D. E., 1971. Widespread synchronously deposited burrow-mottled limestone beds in the Greenhorn Limestone (Upper Cretaceous) of Kansas and southeastern Colorado, American Association of Petroleum Geologists Bulletin, v. 55, p. 412-431.

HATTIN, D. E., 1985, Distribution and significance of widespread, time-parallel pelagic limestone beds in Greenhorn Limestone (Upper Cretaceous) of the central Great Plains and southern Rocky Mountains, in Pratt, L. M., Kauffman, E. G., and Zelt, F. B., eds., Fine-grained Deposits and Biofacies of the Cretaceous Western Interior Seaway: Evidence of Cyclic Sedimentary Processes, Fieldtrip Guidebook 4: Tulsa, Society of Economic Paleontologists and Mineralogists, p. 28-37.

HERBERT, T. D., AND SARMIENTO, J., 1991., Ocean nutrient distribution and oxygenation: limits on the formation of warm saline bottom water over the past 91 My: Geology, v. 19, p. 702-705.

HETTINGER, R. D., 1995, Sedimentological descriptions and depositional interpretations, in sequence stratigraphic context, of two 300-meter cores from the Upper Cretaceous Straight Cliffs Formation, Kaiparowits Plateau, Kane County, Utah: U. S. Geological Survey Bulletin 2115-A, 32 p.

IRVING, E., NORTH, F. K, AND COUILLARD, R., 1974, Oil, climate, and tectonics: Canadian Journal of Earth Science, v. 11, p. 1-17.

JENKYNS, H. C., 1980, Cretaceous anoxic events—from continents to oceans: Journal of the Geological Society of London, v. 137, p. 171-188.

KAUFFMAN, E. G., 1977, Geological and biological overview: Western Interior Cretaceous Basin, in Kauffman, E. G., ed., Cretaceous Facies, Faunas and Paleoenvironments across the Western Interior Basin: Mountain Geologist, v. 14, p. 75-99.

KAUFFMAN, E. G., 1984, Paleobiogeography and evolutionary response dynamic in the Cretaceous Western Interior Seaway of North America, in Westermann, G. E. G., ed., Jurassic-Cretaceous Biochronology and Paleogeography of North America: St. John's, Geological Association of Canada, Special Paper 27, p. 273-306.

KAUFFMAN, E. G., SAGEMAN, B. B., KIRKLAND, J. I., ELDER, W., P., HARRIES, P. J., AND VILLAMIL, T., 1993, Molluscan biostratigraphy of the Cretaceous Western Interior Basin, North America, in Caldwell, W. G. E. and Kauffman, E. G., eds., Evolution of the Western Interior Basin: St. John's, Geological Association of Canada, Special Paper 39, p.397-434.

LARSON, R. L, 1991a, Latest pulse of Earth: Evidence for a mid-Cretaceous superplume: Geology, v. 19, p. 547-550.

LARSON, R. L., 1991b, Geological consequences of superplumes: Geology, v. 19, p. 963-966.

LASAGA, A. C., BERNER, R. A., AND GARRELS, R. M., 1985, An improved geochemical model of atmospheric CO_2 fluctuations over the past 100 million years, *in* Sundquist, E. T., and Broecker, W. S., eds., The Carbon Cycle and Atmospheric CO_2: Natural Variations Archean to Present: Washington, D. C., American Geophysical Union, Monograph 32, p. 397-411.

LEITHOLD, E. L., 1994, Stratigraphical architecture at the muddy margin of the Crertaceous Western Intrerior Seaway, southern Utah: Sedimentology, v. 41, p. 521-542.

OBRADOVICH, J. D., 1993, A Cretaceous time scale, *in* Caldwell, W. G. E. and Kauffman, E. G., eds., Evolution of the Western Interior Basin: St. John's, Geological Association of Canada, Special Paper 39, p. 379-396.

PRATT, L. M., 1981, A paleo-oceanographic interpretation of the sedimentary structures, clay minerals, and organic matter in a core of the middle Cretaceous Greenhorn Formation near Pueblo, Colorado: Unpublished Ph. D. Dissertation, Princeton University, Princeton, N.J., 176p.

PRATT, L. M., ARTHUR, M. A., DEAN, W. E., AND SCHOLLE, P. A., 1993, Paleoceanographic cycles and events during the Late Cretaceous in the Western Interior Seaway of North America, *in* Caldwell, W. E., and Kauffman, E. G., eds., Evolution of the Western Interior Basin: St. John's, Geological Association of Canada, Special Paper 39, p. 333-354.

RESEARCH ON CRETACEOUS CYCLES (ROCC) GROUP, 1986, Rhythmic bedding in Upper Cretaceous pelagic carbonate sequences: Varying sedimentary response to climatic forcing: Geology, v. 14, p. 153-156.

RYER, T. A., 1977a. Coalville and Rockport areas, Utah, *in* Kauffman, E. G., ed., Cretaceous Facies, Faunas, and Paleoenvironments Across the Western Interior Basin: Mountain Geologist, v. 14, p. 103-128.

RYER, T.A., 1977b, Patterns of Cretaceous shallow marine sedimentation, Coalville and Rockport areas, Utah: American Association of Petroleum Geologists Bulletin, v. 8, p 177-188.

SAGEMAN, B. B., AND ARTHUR, M. A., 1994, Early Turonian paleogeographic/ paleobathymetric map, Western Interior, U.S., *in* Caputo, M. V., Peterson, J. A., and Franczyk, K. J., eds., Mesozoic Systems of the Rocky Mountain Region, USA: Denver, Rocky Mountain Section, Society of Economic Paleontologists and Mineralogists, p. 457-469.

SARMIENTO, J. L., HERBERT, T. D., AND TOGGWEILER, J. R.,1988, Causes of anoxia in the world ocean: Global Biogeochemical Cycles, v. 2, p. 115-128

SAVIN, S. M., 1977, The history of the earth's surface temperature during the past 100 million years: Annual Review of Earth Planetary Sciences, v. 5, p. 319-355.

SCHLANGER, S. O. AND H. C. JENKYNS, 1976. Cretaceous oceanic anoxic events – Causes and consequences, Geologie en Mijnbouw, v. 55, p. 179-184.

SCHLANGER, S. O., ARTHUR, M. A., JENKYNS, H.C., AND SCHOLLE, P.A., 1987, The Cenomanian-Turonian oceanic anoxic event, 1. Stratigraphy and distribution of organic- carbon-rich beds and the marine $\delta^{13}C$ excursion, *in* Brooks, J., and Fleet, A., eds., Marine Petroleum Source Rocks: London, Geological Society of London Special Publication 26, p. 371-399.

SLINGERLAND, R., KUMP, L. R., ARTHUR, M. A., FAWCETT, P. J., SAGEMAN, B. B., AND BARRON, E. J., 1996, Estuarine circulation in the Turonian Western Interior Seaway of North America: Geological Society of America Bulletin, v. 108, p. 941-952.

TISSOT, B., 1979, Effects on prolific petroleum source rocks and major coal deposits caused by sea-level changes: Nature, v. 77, p. 463-465.

WEIMER, R. J., 1970, Rates of deltaic sedimentation and intrabasin deformation, Upper Cretaceous of Rocky Mountain region, *in* Morgan.J.P., ed., Deltaic Sedimentation, Modern and Ancient: Tulsa, Society of Economic Paleontologists and Mineralogists Special Publication 15, p. 211-222.

TIMING OF MID-CRETACEOUS RELATIVE SEA LEVEL CHANGES IN THE WESTERN INTERIOR: AMOCO NO. 1 BOUNDS CORE

ROBERT W. SCOTT[1], PAUL C. FRANKS[2], MICHAEL J. EVETTS[3], JAMES A. BERGEN[4], AND JEFFRY A. STEIN[5]

[1]*Precision Stratigraphy Associates, RR 3 Box 103-3, Cleveland, Oklahoma 74020*
[2]*2720 S. Cincinnati, Tulsa, Oklahoma 74114*
[3]*1227 Venice Street, Longmont, Colorado 80501*
[4, 5]*Amoco Corporation, 501 WestLake Park Boulevard, Houston, Texas 77079*

ABSTRACT: The Upper Albian-Coniacian section cored in the Amoco No. 1 Rebecca K. Bounds well in Greeley County western Kansas, serves as a reference section for the timing of depositional events in the Western Interior Seaway. Chronostratigraphy of this section was calibrated by a multidisciplinary study of nannofossils, dinoflagellates, spores, pollen, foraminifers, and mollusks. Range data of the biota in the Bounds core were compared by graphic correlation to a global composite standard that includes key reference sections in Europe and North Africa.

The basal Upper Albian sequence boundary is overlain by transgressive facies of the Purgatoire Formation dated as 102.8 Ma. The upper Upper Albian sequence boundary between the Purgatoire and Dakota Formations marks a hiatus in deposition from 99.4 to 98.2 Ma. The Albian-Cenomanian intra-Dakota sequence boundary spans from 96.0 to 94.1 Ma. The Turonian-Coniacian sequence boundary between the Carlile and Niobrara Formations spans from 89.9 to 88.3 Ma. Maximum flooding is documented within the Purgatoire at 101.4 Ma and in the Graneros Shale at 93.7-92.8 Ma. The Albian-Cenomanian boundary defined by European ammonites and correlated by dinoflagellates is placed at the intra-Dakota unconformity.

Graphic correlation is an independent method of measuring the durations of Milankovitch-scale depositional cycles and can separate climatic cycles from longer tectono-eustatic cycles. Four orders of depositional cycles are recorded by lithological changes, and their durations are constrained by graphic correlation. The longest cycles range from 2.0 to 3.4 My and are found in the sequences defined by the Purgatoire Formation, the lower part of the Dakota Formation, the upper Dakota and Graneros Formations, and the Greenhorn and Carlile Formations. The next lower order comprises transgressive-regressive subcycles of about 0.5 My long in the Purgatoire. The third-scale cycles include sandstone-mudrock cycles in the Dakota, limestone-marl cycles in the lower part of the Greenhorn, and cyclical strata in the Fort Hays Limestone Member of the Niobrara Formation that are about 100 ka long. The shortest cycles are limestone-marl couplets in the upper Greenhorn that are about 41 ka long.

INTRODUCTION

The methodology of sequence stratigraphy defines orders of magnitude of depositional cycles under the assumption that cycle scales have natural durations. Different scales of cycle durations, however, are defined by different authors (compare Vail et al., 1991 with Goldhammer et al., 1991). The question of which scale is better can be tested by techniques of high-precision quantitative stratigraphy, such as graphic correlation using high-resolution stratigraphic data. The Amoco No. 1 Rebecca K. Bounds core from Greeley County, Kansas, near the Colorado state line (Fig. 1) provides such data. It also serves as a key reference section for Cretaceous strata in the middle of the Western Interior Seaway.

The Bounds well is located on the southeast flank of the Las Animas Arch (Fig. 1), which functioned as a paleodrainage divide (Macfarlane et al., 1991; Hamilton, 1994). This section complements the Rock Canyon section west of Pueblo, Colorado (Kauffman and Pratt, 1985) and it provides complete paleontological control across the interval from the Upper Albian Purgatoire Formation to the Coniacian Fort Hays Limestone Member of the Niobrara Formation. The Dakota Formation is a significant aquifer in Kansas and Colorado and the Bounds core serves as an important data point in modeling the stratigraphy (Hamilton, 1994; Macfarlane et al., 1994). The core also provides a complete, unweathered Dakota section that yields palynomorphs for dating. The accurate age of the Dakota and placement of the Albian/Cenomanian boundary are difficult to decipher because of the poor fossil recovery from outcrops. The core provides excellent data for resolving these problems.

The Bounds well was cored continuously with virtually com-plete recovery within the Cretaceous interval (Fig. 2). The core was originally measured in feet, which are reported here to facilitate comparison of our descriptions with the core. Conversion to metric is made where appropriate. One half of the cored Cretaceous interval is archived at the U. S. Geological Survey Core Research Center in Denver, and the remaining part is at Amoco Corporation Technology Center in Tulsa, Oklahoma. Microfossil preparations used in this study are in the care of the

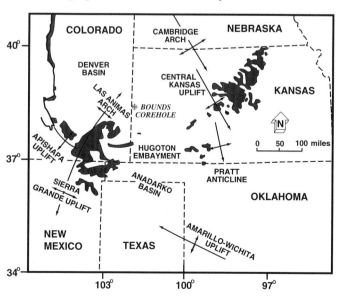

FIG. 1.–Map of Kansas and parts of surrounding states showing location of the Amoco No. 1 Bounds corehole, generalized outcrop distribution of Albian and lower Cenomanian strata (shaded), and major structural elements.

Stratigraphy and Paleoenvironments of the Cretaceous Western Interior Seaway, USA, SEPM Concepts in Sedimentology and Paleontology No. 6
Copyright 1998 SEPM (Society for Sedimentary Geology), ISBN 1-56576-044-1, p. 11-34.

TABLE 1.–STRATIGRAPHIC NOMENCLATURE IN CORE AND OUTCROP

Colorado Front Range[a]	Bounds Core[b]	Kansas Outcrop[c]

Niobrara Formation
    Smoky Hill Chalk Member
    Fort Hays Limestone Member
	Niobrara Formation
	    Smoky Hill Chalk Member
	    Fort Hays Limestone Member
		Niobrara Formation
		    Smoky Hill Chalk Member
		    Fort Hays Limestone Member

Carlile Shale
    Sage Breaks equiv. Sh. Mbr.
    Juana Lopez Member
    "upper shale member"
    Codell Sandstone Member
    Blue Hill Shale Member
    Fairport Chalky Shale Mbr.
	Carlile Shale
	
	    Codell Sandstone Member
	    Blue Hill Shale Member
	    Fairport Chalk Shale Mbr.
		Carlile Shale
		
		    Codell Sandstone Member
		    Blue Hill Shale Member
		    Fairport Chalk Shale Mbr.

Greenhorn Formation
    Bridge Creek Limestone Mbr.
 	
    Hartland Shale Member
    Lincoln Limestone Member
	Greenhorn Formation
	    Bridge Creek Limestone Mbr.
	
	    Hartland Shale Member
	    Lincoln Limestone Member
		Greenhorn Formation
		    Pfeifer Shale Member
		    Jetmore Chalk Member
		    Hartland Shale Member
		    Lincoln Limestone Member

Graneros Shale
    "upper Shale member"
    Thatcher Limestone Member
    "lower shale member"
	Graneros Shale
	
	    Thatcher Limestone Member
		Graneros Shale

Dakota Group
    Muddy Sandstone
    "upper transitional member"
    "lower channel ss. member"
	Dakota Formation
	    "upper member"
	    "lower member"
		Dakota Formation
		    Janssen Clay Member
		    Terra Cotta Clay Member

Glencairn Formation
Plainview Formation
Lytle Formation
	Purgatoire Formation
	    "upper part"
	    "lower part"
		Kiowa Formation
		    Longford Member
		    Cheyenne Sandstone

[a]Kauffman and Pratt, 1985	[b]this report	[c]Hattin, 1975; Franks, 1975

Amoco biostratigraphy group in Houston. Casts of the megafossils are curated by the Paleontology Museum at The University of Kansas in Lawrence.

Initial results of the Bounds core and Cretaceous depositional cycles of Kansas are the subject of a number of presentations, papers, and abstracts: Scott and Franks (1988), Bergen et al. (1990), Scott (1992a, b), Scott et al. (1993), Scott et al. (1994a, b), and Hamilton (1994).

LITHOSTRATIGRAPHY

Lithostratigraphic names are applied to the section in the Bounds core (Fig. 2, Table 1) following the North American Commission on Stratigraphic Nomenclature (1983). Most of the stratigraphic units are well known and can be traced from Kansas to Colorado. Depths to tops of key lithostratigraphic marker units in the Bounds core are given in Table 2.

Purgatoire Formation

The name Purgatoire Formation, rather than Kiowa Formation, is applied to the sandstone and shale interval above the Morrison Formation and below the Dakota Formation because the interval consists of 38% sandstone (Fig. 3) and is lithically like type Purgatoire in southeastern Colorado, about 170 km to the west (Fig. 3). In its type area, the Purgatoire consists of interbedded shale and sandstone and generally has a mappable sandstone unit at its base, the Lytle Formation. The Purgatoire is coeval with the Kiowa Formation, which, in its type area in south-central Kansas, about 230 km southeast of the Bounds well, is dominantly a dark-gray shale.

Clearly an arbitrary cutoff must be drawn between the type area of the Kiowa Formation in Kansas and the type area of the Purgatoire Formation in Colorado. We suggest that the lithologic similarity and percentage of interbedded sandstone should

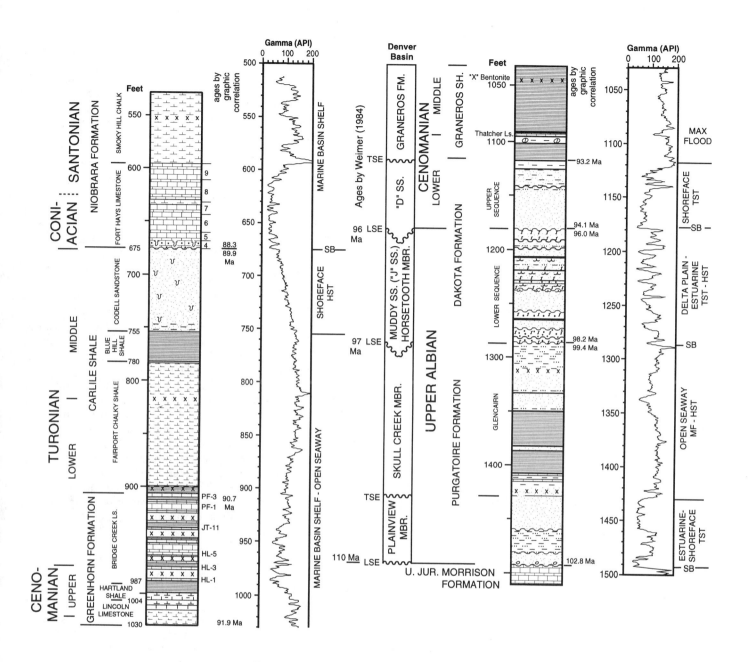

FIG. 2.–Gamma-ray log and lithologic section of mid-Cretaceous strata in the Amoco No. 1 Bounds core showing depositional cycles and subcycles and sequence interpretation. TST - transgressive systems tract; HST - highstand system tract; LSE - lowstand surface of erosion; SB - sequence boundary; TSE - transgressive surface of erosion. Ages of sequence boundaries derived by graphic correlation for the Bounds core are compared with ages of equivalent sequence boundaries in the Denver Basin by Weimer (1984).

be the primary criteria. For example, the two cross sections of Macfarlane et al. (1994, Figs. 5A, B) show that this unit in central Kansas contains a few thin sandstone beds whereas in western Kansas it contains thicker, interspersed sandstone beds. A practical cutoff would be in the subsurface of western Kansas.

A basal Cretaceous sandstone unit that crops out below the Kiowa Formation in south-central Kansas is mapped as the Cheyenne Sandstone, which pinches out a few kilometers west of its type area (Scott, 1970; Franks, 1975). The Kiowa/Cheyenne contact here is a transgressive disconformity. Sandstone intervals at the base of the Kiowa Formation elsewhere in Kansas are identified by Macfarlane et al. (1994, Fig. 5) and Hamilton (1994, Figs. 4, 5) as "Cheyenne Sandstone." Hamilton (1994) also applies the name to sandstone at the base of the Purgatoire Formation in the Bounds core. Wire-line log correlations by Scott (1970), Macfarlane et al. (1994), and Scott et al. (1994b), however, show that the basal Cretaceous sandstone bodies in Kansas are physically discontinuous. Application of the name, Cheyenne Sandstone, to the basal Cretaceous sandstone in the Bounds core treats discontinuous and lithically different sandstone bodies at the base

TABLE 2.–LITHOSTRATIGRAPHIC MARKER UNITS (FT)

STRATIGRAPHIC UNIT/ SURFACE		LOG DEPTH	CORE DEPTH
Core top in Smoky Hill Chalk Member		522	529.0
Fort Hays Limestone Member			
Hattin's units	9	595	594.7
	8	611	616
	7	631	631
	6	644	640
	5	660	660
	4	668	668
Carlile Shale/Codell Sandstone Mbr.		676	675.0
Blue Hill Shale Member		760	755.0
Fairport Chalk Member		788	780.0
Marl-shale cycles at 853, 856, 860, 868, 876, 881, 885, 893, 897, 900			
Greenhorn Fm./Bridge Creek Ls. Mbr./PF-3		905	905.8
Marl-limestone cycles		908	908.3
		912	912.3
Hattin's units	PF-2	919	916.0
	PF-1	925	921.8
	JT 12	930	930.7
	JT 11	938	938.0
		942	942.7
	JT-10	946	946.0
	JT-9	950	948.8
	JT 6	955	955.0
		958	957.7
	HL 5	962	962.2
	HL-4	968	967.6
		972	972.5
	HL-3	976	976.3
	HL-2	982	982.7
	HL-1	988	985.5
Base Bridge Creek/Top Hartland		989	987.0
Base Pfeifer-equivalent		936	938.0
Base Jetmore-equivalent		954	952.5
Base Hartland/Top Lincoln		1008	1003.9
Base Lincoln/Top Graneros		1034	1029.9
"X Bentonite"		1042	1046.0
Thatcher Limestone Mbr.		1091	1084.0
Dakota Formation		1123	1115.7
Ravinement surface		1180	1171.2
Purgatoire Formation		1287	1286.5
Transgressive contact		1332	1333.3
Transgressive contact		1430	1421.3
Ravinment surface		1462	1459.3
Morrison Formation		1495	1487.5

of the Cretaceous as a physically continuous lithostratigraphic unit. The name, Cheyenne Sandstone, should not be extended beyond its pinch-out margins in south-central Kansas.

Hamilton (1994) also applies the name, Longford Member, to bioturbated sandstones and marine shales above the basal sandstone bed and below the thick, dark-gray marine shale in the Bounds core. In his cross sections he extends the Longford from its type area in north-central Kansas to western Kansas as a physically continuous unit. Along the Kiowa Formation outcrop belt, the Longford is an aerially restricted lithosome that forms an assemblage of red-mottled and carbonaceous clay rocks, mudstones, and siltstones that is capped by a distinctive, laminated siltstone (Franks, 1979, 1980). The Longford is a marginal paralic facies of the Kiowa marine shales that disappears south-westward along the Kiowa outcrop belt. The name, Longford, should not be applied to rocks in the Bounds core that are lithically distinct and physically separated from the type Longford.

The Purgatoire Formation represents an unconformity-bounded transgressive-regressive depositional cycle, the Kiowa-Skull Creek T5 cycle of Kauffman (1984), the Kiowa-Skull Creek-Purgatoire-Tucumcari sequence of Scott et al. (1994b), or the Cheyenne/Kiowa Sequence of Hamilton (1994). The Purgatoire in the Bounds core can be divided into genetic depositional subcycles bounded by transgressive contacts (Fig. 3). The lower part of the Purgatoire, from 1487.5 to 1428.6 ft, consists of three subcycles of bioturbated sandstone and dark-gray shale that represent a shallow marine transgressive systems tract. The upper part of the Purgatoire consists of two sandstone subcycles separated by a transgressive contact at 1333.3 ft.

The contact of the Purgatoire Formation with the underlying Morrison Formation is a regional unconformity representing about 35 My. It is mapped as a regional sequence boundary (Scott, 1970; Weimer, 1984; Scott and Franks, 1988; Hamilton, 1994; Macfarlane et al., 1994; and Scott et al., 1994b). The precise contact is not preserved in the Bounds core (Fig. 4) because the topmost Morrison core segment was worn during coring. The uppermost part of the Morrison is a light-gray marlstone that is weathered greenish gray in the top 20 cm. The basal Purgatoire is a 6-cm-thick, dark-gray shale showing discontinuous, sandy ripple laminae. It is overlain by a 2-m-thick, light-gray sandstone with clasts of olive gray to brownish-gray clay rock similar to the Morrison as well as siltstone clasts, chert pebbles, and quartz granules at its base. The shale represents the first Cretaceous marine or marginal-marine deposit in the core and the sandstone overlies it with an erosional, transgressive contact.

Lower part of the Purgatoire Formation.—

The 6-cm-thick, dark-gray shale at the base of the Purgatoire Formation is a marine or marginal-marine deposit. The interval above from 1487.5 to 1428 ft consists of two bioturbated sandstones separated by medium-gray shale. Contacts between each sand and shale unit are sharp and irregular and suggest transgressive and locally erosive conditions. These contacts may represent ravinement surfaces that formed during transgressive retrogradational processes resulting in backstepped marine or marginal-marine muds and nearshore or shoreface sands. Except for the conglomeratic interval at the base of the lower sandstone, the sandstones are fine- to medium-grained and well sorted. Bioturbation and contorted clay laminae are common. The shales contain ripple-laminated sand laminae and trace fossils. Marine dinoflagellate specimens make up more than 50% of the palynomorph assemblages and indicate a nearshore marine environment.

Upper part of the Purgatoire Formation.—

The Purgatoire Formation from 1428 to 1286.5 ft represents a succession from maximum flooding to progradation of nearshore sandstone and shale. The top of the Purgatoire was truncated by erosion. Dark-colored shale between 1428 to 1400 ft contains calcareous shell beds, cone-in-cone concretions, inoceramids, am-

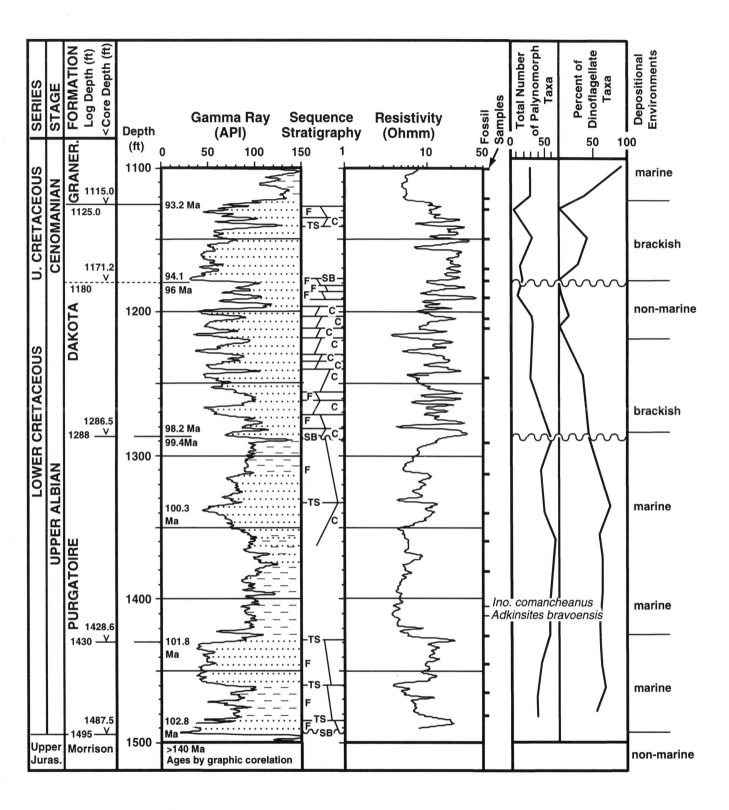

FIG. 3.–Integrated well logs (gamma ray and resistivity), lithology, and palynology data for the Purgatoire/Dakota section in Bounds core. Interpreted depositional environments and sequence stratigraphy also are shown. Ages of key sequence boundaries determined by graphic correlation are given in millions of years ago (Ma). Smaller-scale cycles are delineated within sequence stratigraphic units by fining-up (F) and coarsening-up (C) lithologic changes, the extent of which are shown by inclined lines. SB - sequence boundary; TS - transgressive surface. Percent of dinoflagellate taxa is relative to total palynomorph taxa.

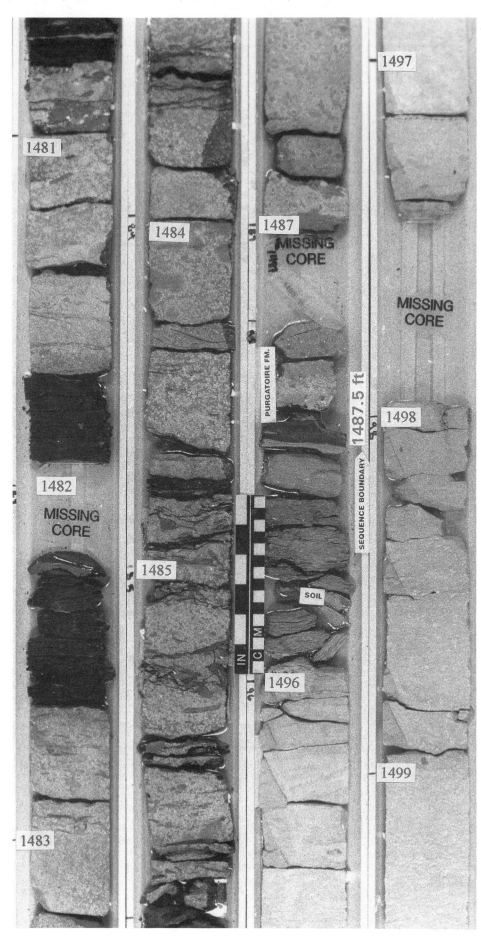

FIG 4.–Core photograph of Purgatoire/ Morrison boundary at 1487.5 ft and low- ermost Purgatoire transgressive contact at 1487.3 ft. Photo spans depth interval of 1480.8 to 1499.4 ft.

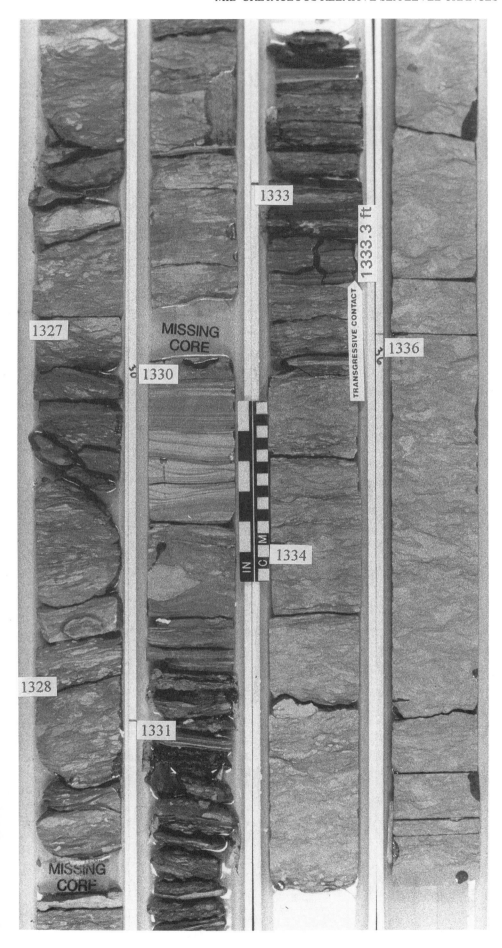

Fig. 5.–Core photograph of transgressive contact in sandstone in upper part of Purgatoire Formation at 1333.3 ft between bioturbated sandstone below and shale above, grading upward into bioturbated sandstone. Photo spans depth interval of 1326.5 to 1337 ft.

FIG. 6.–Core photograph of Purgatoire/Dakota contact at 1286.5 ft, marked by inclined color change between light-colored, very fine-grained sandstone below and darker colored mudstone above. Photo spans depth of 1280 to 1291.5 ft.

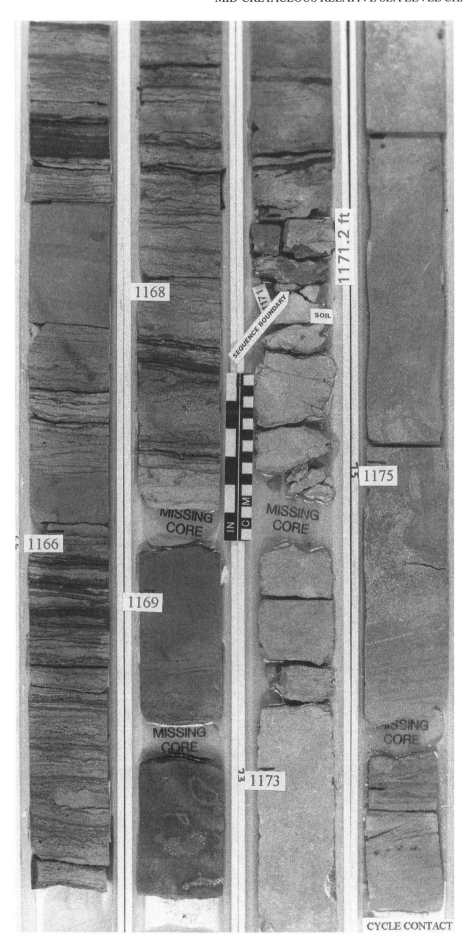

FIG. 7.–Core photograph of intra-Dakota sequence boundary at 1171.2 ft between mudstone below and shaly sandstone above. A fining-upward cycle is seen from 1176.3 to 1171.2 ft at right base of core. Top of core photo at 1165 ft.

A. Upward coarsening cycle

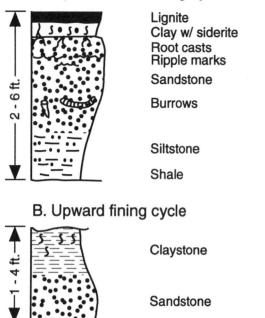

Lignite
Clay w/ siderite
Root casts
Ripple marks

Sandstone

Burrows

Siltstone

Shale

2 - 6 ft.

B. Upward fining cycle

1 - 4 ft.

Claystone

Sandstone

FIG. 8.–Models of an upward-coarsening depositional cycle (A) and an upward-fining depositional cycle (B) in the Dakota Formation in the Bounds core.

monites, and marine dinoflagellates. It represents maximum flooding of the open seaway although condensed sedimentation is not evident. Beginning at 101.4 Ma the Tethyan and Boreal Realms were connected, as indicated by regional correlations (Williams and Stelck, 1975). The seaway was closed later during deposition of the Dakota Formation.

The sandstone interval from 1352 to 1312 ft represents progradation of nearshore conditions. This interval is divided by a sharp contact at 1333.3 ft (Fig. 5) that matches a kick on the gamma log and represents a transgressive surface. The lower sandstone unit is glauconitic and bioturbated and coarsens upward from shale to very fine-to medium-grained sand. The upper sandstone unit is clayey, rippled, and bioturbated and coarsens upward from shale and silt to very fine-grained sand. The abundance of dinoflagellate specimens decreases near the top, suggesting onset of nearshore conditions as the seaway filled and contracted.

The uppermost contact with the overlying Dakota Formation is contained in the core (Fig. 6). The topmost Purgatoire Formation consists of interlaminated sandstone and shale that becomes lighter colored just below the contact and is difficult to distinguish from the Dakota rocks. However, a sharp, wavy contact separates this siltstone from 1 m of mudstone and silty shale grading up to sandstone, which are allocated to the Dakota (Fig. 6). Within the mudstone are deformed, 2-cm-long, rip-up clasts of sandstone evidently from the Purgatoire and early diagenetic siderite spherules. The uppermost Purgatoire mainly consists of quartz with some kaolinite and illite (G. Powers, pers. commun., 1988). The lithologic succession in the Bounds core is directly analogous to the contact at type Kiowa Formation where gray Dakota clay rocks enclose zones of spherulitic siderite and rest on topmost fine-grained Kiowa sandstone (Franks, 1975). This contact also represents a regional unconformity across the basin (Scott, 1970).

Dakota Formation

Outcrops of the Dakota Formation in Kansas consist of kaolinitic clay rocks, siltstone, and lenticular sandstone (Plummer and Romary, 1942; Franks, 1975). The lower two-thirds of the Dakota is characterized by light-gray claystone and mudstone with red hematitic mottles; this interval is the Terra Cotta Clay Member. The upper one-third is the Janssen Clay Member, which is characterized by gray and dark-gray claystone, lignite, and sandstone. The Terra Cotta represents lowstand alluvial-plain environments and the Janssen represents delta-plain environments in a transgressive systems tract (Franks, 1975; 1979). This change in depositional regime suggests that the boundary between the Terra Cotta and Janssen is a sequence boundary. However, a physical stratal contact between Terra Cotta and Janssen has not yet been identified in the Kansas outcrop belt. Nonetheless, the environmental shift represented by the change from the Terra Cotta to the Janssen agrees with a regional base-level change that corresponds to the sequence boundary in the Bounds core at 1171.2 ft. This boundary, however, differs from the boundary chosen by Hamilton (1994). He places the intra-Dakota sequence boundary in the Bounds core at a lignite bed above a weathered clay rock at the log depth of 1204 ft (core depth of 1204.8 ft). However, several lignite beds and weathered clay beds are developed within the interval below the 1171.2 ft core depth. We prefer to use a lithically distinct and traceable stratal contact to divide the Dakota.

The top of the Janssen Clay Member at the outcrop is a transgressive contact marked by locally developed ravinement surfaces. Locally, relict, brackish to nearshore fossil assemblages are found in the uppermost Janssen strata (Hattin, 1967). The Dakota Formation in Kansas is described in more detail by Franks (1975), and usage of the name Dakota in Colorado and New Mexico is reviewed by Mateer (1987) and Holbrook and Dunbar (1992).

Lower Part of the Dakota Formation—

The lower two-thirds of the Dakota Formation in the Bounds core is a depositional sequence bounded by erosional unconformities at 1286.5 and 1171.2 ft (Figs. 3, 6, 7). This interval consists of interbedded sandstone, shale, mudstone, and lignite. The diversity and abundance of marine dinoflagellates decrease up the section, suggesting that the depositional environment became less marine. This sequence consists of 13 depositional units or cycles (Fig. 3) defined by vertical successions of facies bounded by sharp contacts. Each unit represents a genetically related set of environmental changes. These units may also be parasequences. Most of the units in the lower part of the sequence coarsen upward. The units in the upper part fine upward and are thinner and less bioturbated (Fig. 8). The coarsening upward units represent deltaic distributaries that terminated at sea level with exposure to soil-forming processes. Siderite nodules in the base of the units may have been reworked from soils formed under wet conditions (Stoops, 1983; Stoops and Eswaran, 1985; Landuydt, 1990). The lowstand incised valley-fill at the base of the "J Sandstone" in the Denver Basin (Fig. 2) (Dolson et al., 1991) is absent in this core. Units 1 to 3 (1272.7 to 1230.1 ft) in the core, which are capped by thick lignite and weathered mud rocks, form the transgressive system tract. The remaining cycles, which thin up section and become more terrestrial, compose the highstand systems tract.

The typical coarsening upward unit begins with siltstone or

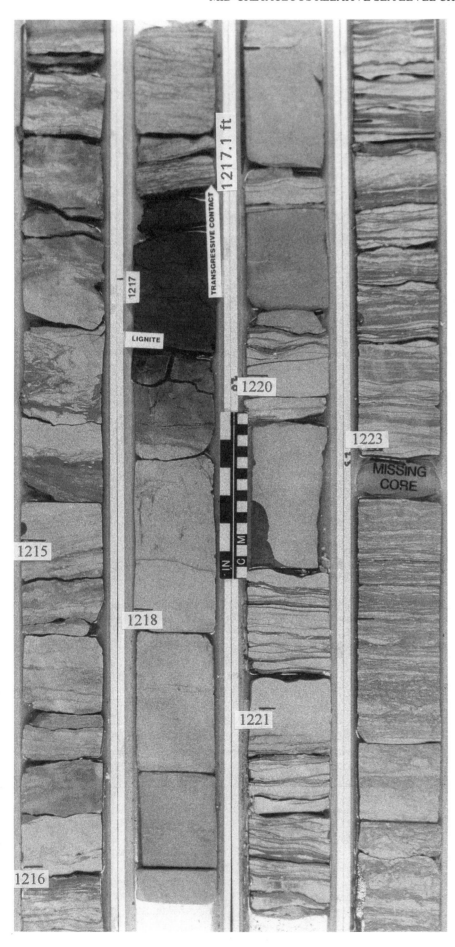

1217.1 ft

1217

TRANSGRESSIVE CONTACT

LIGNITE

1220

1223

MISSING CORE

1215

1218

1221

1216

Fig. 9.–Core photograph of coarsening-upward, shale-sandstone-lignite cycle in lower part of Dakota; transgressive cycle contact at 1217.1 ft (label has slipped); note root casts below lignite. Photo spans core depth of 1214 to 1224 ft.

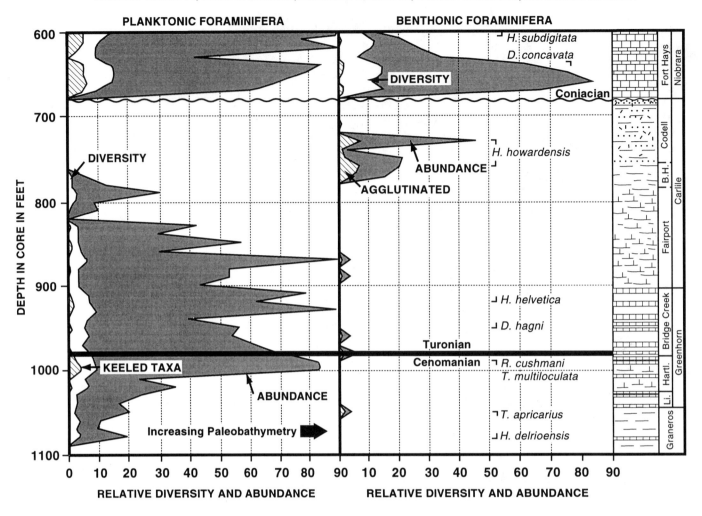

FIG. 10.–Diversity and abundance of planktonic and benthonic foraminifera through the Graneros to Niobrara interval in Bounds core. Diversity is numbers of taxa; note percentage of keeled and arenaceous taxa. Diversity of benthonic taxa is very low below the Niobrara Formation. Paleobathymetry is inferred to increase with increasing abundance of planktonic taxa. (First occurrence datums and last occurrence datums of key marker species indicated by brackets.

mudstone and grades upward to fine-grained sandstone (Figs. 8A, 9). The basal mud rocks may be yellowish-gray to dark-brownish-gray shale or claystone with siderite spherules and carbon flecks. The basal contact is sharp and wavy. The siltstone to sandstone interval is brownish-gray and well sorted. Scattered wavy ripple laminae or cross laminae are locally intersected by distinct burrows of *Skolithos* or *Teichichnus,* among others. Many cycles are capped by dark claystone or lignite. Root casts are common in sandstone beneath the lignite. The claystone consists of mainly quartz, illite, and kaolinite, as well as some siderite, feldspar, and, in one sample at the top of unit 2, mixed layer illite-smectite (G. Powers, pers. commun., 1988). This type of cycle is represented by units 1 to 9. Unit 10 (1185.8-1189.8 ft), which caps this interval and is 1.2 m (4 ft) thick, is composed of claystone and mudstone. The cyclic units average 3.2 m (10.4 ft) thick.

Units 11, 12, and 13 form fining upward cycles and average 1.5 m (4.9 ft) in thickness. The typical fining upward cycle begins with yellowish-gray, fine- to very fine-grained sandstone with ripple and contorted laminae above a sharp base (Figs. 8B, 9). The sandstone grades upward to yellowish-gray claystone consisting of quartz, kaolinite, and illite with sparse carbon flecks,

pyrite, and in unit 13, siderite (identification by x-ray diffraction, G. Powers, pers. commun., 1988). Dinoflagellates are absent.

Upper Part of the Dakota Formation.–

The top of the Dakota Formation can be placed at the base of a thin sandstone bed at 1115.7 ft (Fig. 3). The bed is bioturbated and has contorted shale laminae like outcrop beds that are allocated to the Graneros Shale. The upper part of the Dakota extends from 1171.2 to 1115.7 ft and is dominantly sandstone. It can be divided into seven units or cycles marked by sharp basal contacts between the underlying sandstone and the overlying shale. These shale and sandstone units represent the transgressive systems tracts of the upper Dakota-Graneros sequence. The lowstand incised valley fill, which is generally identified as the "D Sandstone" in the Denver Basin (Fig. 2) (Dolson et al., 1991), is absent in this core.

The first unit in this facies set (Fig. 3, cycle 14) is a bioturbated sandstone from 1171.2 to 1128.6 ft (Fig. 7). The base is glauconitic and consists of illite and kaolinite clays with small amounts of feldspar (G. Powers, pers. commun., 1988). In the upper part of this unit, siltstone and mudstone become more preva-

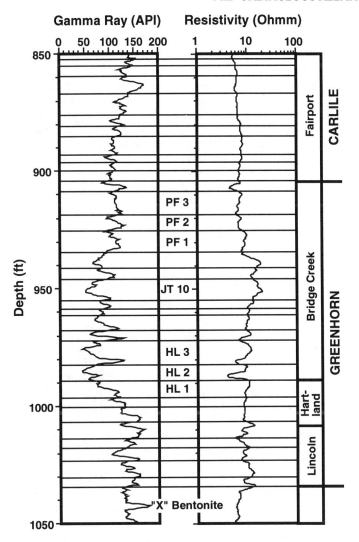

FIG. 11.—Gamma-ray (left) and resistivity logs (right) of the Bounds well showing depositional cycles in the Greenhorn Formation marked off by thin horizontal lines. Selected units are labeled (e.g., "HL 1"; "JT 10"; See Table 2 for a listing of all units).

lent and marine dinoflagellates are also present. The overlying six units are shale-sandstone cycles coarsening upward. Their sharp tops suggest retrogradational transgressive processes. They are 0.5-1 m (1.5-3 ft) thick and bioturbated.

Graneros Shale

The Graneros Shale is a dark-gray shale between the Dakota Formation and the Greenhorn Formation. In Kansas and eastern Colorado it ranges from 7.3 to 32 m (24 to 105 ft) thick (Hattin, 1965; Kauffman, 1985). In the Bounds core it is 25.6 m (84.1 ft) thick. Nomenclatural history of the Graneros was reviewed by Hattin (1965), who also described it in Kansas. Near Pueblo, Colorado, Kauffman (1985) applied an informal three-part subdivision of the Graneros that is useful in the Bounds core (Table 1).

The "Lower Shale Member" is 9.3 m (30.4 ft) thick in the Bounds core and consists of clay shale and thin beds of ripple-laminated sandstone and siltstone (1115.7-1085.3 ft). In Kansas outcrops the contact is transitional in some places, but the top of

the Dakota locally is marked by clay-ironstone concretions (Hattin, 1965).

The Thatcher Limestone Member (Fig. 2) is a distinct marker bed that is less than 30 cm (1 ft) thick in Colorado and pinches out in Kansas. In the Bounds core, the Thatcher is identified as a 0.5-m (1.7-ft) thick interval of calcarenite and shale laminae having a sharp basal contact. Just above this unit in the core is a flood of microfossils that have Tethyan affinities (Fig. 10). The Thatcher in outcrop also is associated with a diverse southern biota that indicates opening of the Western Interior Seaway to the south, which closed at the end of Purgatoire deposition (Williams and Stelck, 1975). A bentonite 20 cm below the Thatcher at Pueblo, Colorado, is dated as 95.78 +/- 0.61 Ma by ^{40}Ar/^{39}Ar methods (Obradovich, 1993).

The "Upper Shale Member" in Colorado is a 16.4-m (53.7-ft) thick, dark-gray shale (Kauffman, 1985). The top of the Graneros Shale is placed at the base of the first coquinoid limestone or calcarenite (1029.9 ft) above the "X Bentonite Marker Bed" (1046.4-1046.0 ft) (Fig. 11). The Graneros Shale records the transgression and maximum flooding of the Western Interior Seaway during Middle Cenomanian time. The Thatcher Limestone Member represents the maximum flooding of this cycle.

Greenhorn Formation

In the Kansas outcrops, the Greenhorn Formation is an interbedded succession of dark-colored, shaly chalk, chalky limestone, calcarenite, and calcareous shale 21 to 41 m (69 to 134 ft) thick (Table 2). In the Bounds core the Greenhorn is 37.8 m (124.1 ft) thick (1029.9-905.8 ft) (Table 2). The Greenhorn in the Bounds core (Fig. 11) is divided into three members used by Kauffman and Pratt (1985) in the section at Pueblo, Colorado; in ascending order these are the Lincoln Limestone, the Hartland Shale, and the Bridge Creek Limestone Members (Table 1). This nomenclature is used here because subdividing the upper part of the Greenhorn into the Jetmore Chalk and Pfeifer Shale Members, as in central Kansas, is not practical. Moreover, the Bounds well was drilled only 50 km north of the type locality of the Bridge Creek Limestone Member in Hamilton County, Kansas.

The basal contact with the Graneros Shale in central Kansas is a sharp, bedding-plane disconformity between shale and limestone (Hattin, 1975). However, in western Kansas this contact is conformable, and nearly 10 ft of calcareous and chalky shale lie between the "X Bentonite Marker Bed" and the lowest calcarenite (Hattin, 1975). In the Bounds core, the contact with the Graneros is conformable just as it is in nearby outcrops. The top of the Greenhorn Formation is also conformable and placed at the top of a distinct limestone marker bed, but lithology and biota change little into the basal Fairport Chalk Member of the Carlile Shale (Hattin, 1975). In the Bounds core, a pair of 2-3 cm thick limestone beds at 905.8 ft mark the top of the Greenhorn. This corresponds to log kicks at 905 ft (Fig. 11).

The Lincoln Limestone Member in the core is a 7.9 m (26.0 ft) section of dark-gray, calcareous shale, calcisiltite, and bentonite capped by a 15-cm-(0.6-ft-) thick limestone bed (Table 1). The Lincoln in the core contains more shale than in central Kansas outcrops. The overlying Hartland Shale Member is a 5.2-m-(16.9-ft-) thick, calcareous shale interval between two prominent limestone beds that mark the top of the Lincoln and the base of the overlying Bridge Creek Limestone Member. The Hartland is an average of 8.1 m (26.6 ft) thick in the outcrops (Hattin, 1975). The Bridge Creek consists of an interbedded interval of limestone and shale that is 25 m thick (81.2 ft) in the Bounds core; in the outcrop near Pueblo, Colorado, it is 15.2 m (50 ft) thick. Ten distinctive, synchronous limestone marker beds defined and traced into Colorado by Hattin (1971, 1985) are identified in the Bounds core by their thickness and spacing within the section, and by their molluscan guide fossils, and their positions relative to correlatable bentonite beds. However, identifying them in the Bounds core is complicated by the development of additional limestone beds (Fig. 11; Appendix 2).

The Greenhorn Formation, and particularly the Bridge Creek Limestone Member, is characterized by decimeter-meter-scale cycles of limestone-shale or chalk-marl (Hattin, 1985) that are assumed to be periodic at an estimated frequency of 41 ky (Fischer et al., 1985; Pratt et al., 1993). In the Bounds core, cycle boundaries were picked at the top of limestone beds where contacts with overlying shale beds are generally sharp and irregular. Some contacts are burrowed and appear gradational. The shale or marlstone beds may be laminated by thin layers of foraminiferal tests or inoceramid prisms. Some shale intervals are barren of fossil laminations. The shale and marlstone units grade up into the limestone beds. In some cycles a thinner limestone bed underlies the uppermost limestone bed. These 1-3 m-thick cycles are readily detected on the well logs (Fig. 11). The basal shaly intervals have high gamma-ray intensities and low electrical resistivities, and the limestone beds have the opposite signature. Bentonite beds thicker than 2 cm correspond with very high gamma-ray peaks and moderate electrical resistivity; they modify the typical character of the shale-limestone cycles. The cycle definition, then, is based on log features calibrated to the core lithology (Table 2).

The Cenomanian/Turonian boundary is placed between 980 and 978.2 ft in the Bounds core based on various biostratigraphic data to be discussed in the next section. The global carbon isotope event, OAE2, is an additional marker for placement of the Cenomanian/Turonian boundary (Pratt, 1985; Pratt et al., 1993). This event is defined by higher values of $\delta^{13}C$ in both organic and carbonate carbon (Fig. 12). M. D. Lewan analyzed 55 samples from the Greenhorn and Graneros Formations for $\delta^{13}C$ on isolated kerogen at the Amoco Tulsa Research Center. Whole-rock samples of carbonates collected at 10-cm intervals from the Bridge Creek Limestone Member of the Greenhorn were analyzed for carbon and oxygen isotopes by M. A. Arthur at Penn State University (Fig. 12) (Dean et al., 1995). The Amoco data show a shift in values of $\delta^{13}C$ in kerogen from -25.1 ‰ to -23.2 ‰ between 983.92 and 979.73 ft. The upper shift in kerogen from -24.6 ‰ to -25.7 ‰ is between 970.58 and 968.83 ft. The Penn State values of $\delta^{13}C$ on whole-rock carbonates shifts from 0.92 ‰ to 3.61 ‰ between 980.06 and 979.45 ft below and from 2.65 ‰ to 1.82 ‰ between 969.55 and 968.88 ft at the top. Although the curves are based on different data, boundaries can be placed in about the same place.

Carlile Shale

The Carlile Shale is conformable with the underlying Greenhorn Formation and is unconformably overlain by the Niobrara Formation. It represents the regressive part of Kauffman's (1984) Cycle R6. In Kansas the Carlile is from 79-100 m (260 to 330 ft) thick (Hattin, 1962) and near Pueblo, Colorado, it is about 70 m (230 ft) thick (Glenister and Kauffman, 1985). In the Bounds core the Carlile Shale is 70.3 m (230.8 ft) thick (Table 2). The Carlile consists of three members in ascending order: the Fairport Chalk Member, the Blue Hill Shale Member, and the Codell Sand-

stone Member (Table 1) (Hattin, 1962). The Fairport in the Bounds core is a calcareous shale with calcisiltite laminae, inoceramids, and bentonites; it grades with the Bridge Creek Limestone Member below and the Blue Hill Shale Member above. Meter-scale cycles of calcareous shale and shaly chalk are indicated by the gamma log (Fig. 2), but they are not obvious in the core. The distribution of calcisiltite laminae is greater in the lower 12 m (40 ft) of the member. The Bluehill is a dark-gray, noncalcareous shale with sparse fossils that grades upward into the Codell. The Codell consists of an upward-coarsening succession of silty shale or siltstone at the base. It becomes a very fine- to medium-grained, highly bioturbated, quartz sandstone near the unconformity at the top. Bioturbation reworked sand into the basal Niobrara Formation.

Niobrara Formation

The Niobrara Formation in Kansas unconformably overlies the Carlile Shale and conformably underlies the Pierre Shale, or it is truncated by Cenozoic erosion and, as in the Bounds core, is overlain by alluvium. The Niobrara ranges from 122 to 198 m (400 to 650 ft) thick in western Kansas (Hattin, 1982), but only the lower 44.5 m (146 ft) are contained in the Bounds core (Table 2). The Fort Hays Limestone Member comprises the basal 24 m (80 ft) in the core. Its base is sandy and appears to grade down into the Codell Sandstone Member because of bioturbation. However, a log kick at 676 ft is a useful marker for separating these lithostratigraphic units (Fig. 13). In the core, the appearance of a calcareous matrix above 675 ft distinguishes the Fort Hays from the underlying Codell. This unconformity is a regional contact traceable from Kansas to Colorado (Laferriere and Hattin, 1989). Below the contact in Colorado, a separate depositional sequence is comprised of the Juana Lopez and Sage Breaks Shale Members of the Carlile Shale (Fisher et al., 1985). The base of the Smoky Hill Member of the Niobrara Formation is picked at 594.7 ft in the core, between a consistent interval of limestone below and marl above; it is also marked by a strong gamma-ray log kick (Fig. 13).

Based on a study of outcrop and well log data, Laferriere and Hattin (1989) define seven bundles of limestone and shale marker beds in the Fort Hays Limestone Member. They traced the bed sets across the Western Interior Seaway using distinct low-resistivity kicks representing thick shale intervals overlain by limestone-shale couplets. Each of the units is about 1.5 to 3 m (5 to 10 ft) thick and consists of thinner limestone-shale couplets. Laferriere and Hattin attribute these bed sets to Milankovitch-type processes and Fischer (1993) concludes that the thinner of the 20-cm-thick couplets represent the precessional 21-ky frequency and the 1-3-m-thick marker bundles are the 100-ky eccentricity signal.

In the Bounds gamma log, Laferriere and Hattin's (1989) units 4 through 9 were identified by picking the base of the prominent shale markers in the Fort Hays Limestone Member (Fig. 13). The depths of these units are given in Table 2; they range in thickness from 2 to 6.4 m (6.8 to 21.1 ft). The time span for units 4 through 9 measured by graphic correlation is 480,000 yrs (Fig. 14), which gives an average duration of 80,000 yrs per cycle. This duration is within the range estimated by Fischer (1993), and a small decrease in the slope of the line of correlation (LOC; dashed line on Fig. 14) matches a gradient for 100,000-yr cycles. This gradient is defined by the base and top of *Valvulineria plummerae* in the Fort Hays Limestone Member (Fig. 14). So,

TABLE 3.–KEY FOSSIL RANGES IN CORE DEPTHS

TAXA	BASE	TOP
Ammonites		
Adkinsites bravoensis	1412.9	
Collignoniceras woollgari	790.0	
Prionocyclus hyatti	790.0	761.8
Bivalves		
Inoceramus arvanus	1085.2	1071.3
Inoceramus comancheanus	1406.5	
Inoceramus cuvieri	834.8	
Inoceramus pictus	999.6	
Inoceramus prefragilis	1034.4	1008.5
Inoceramus rutherfordi	1067.6	1053.8
Mytiloides costellatus?	761.8	
Mytiloides hercynicus?	838.2	
Mytiloides labiatus	851.6	
Mytiloides mytiloides	950.5	943.9
Mytiloides opalensis	973.4	963.4
Mytiloides subhercynicus	943.5	
Ostrea beloiti	1057.8	1048.0
Platyceramus platinus	592.0	551.0
Pseudoperna bentonensis	838.2	
Pseudoperna congesta	592.0	578.5

Planktonic Foraminifera	Base	Top
Archaeoglobigerina cretacea	610	600
Clavihedbergella moremani	870	
C. simplex	1020	600
C. subcretacea	960	870
Dicarinella concavata	660	640
D. hagni	950	
D. imbricata	610	
D. primitiva	670	610
Globigerinelloides bentonensis	1020	1000
G. ultramicra	1030	600
Globotruncana sp. cf. G. lapparenti	660	600
G. linneiana	660	600
G. renzi	670	650
Hasterigerinoides subdigitata	600	
Hedbergella delrioensis	1080	630
H. loetterlei	1000	600
H. planispira	1080	600
H. simplicissima	1020	600
Helvetoglobotruncana helvetica	920	910
Heterohelix globulosa	1080	600
H. moremani	1080	620
H. pulchra	930	600
Marginotruncana coronata	640	600
M. marginata	670	600
M. schneegansi	930	670
Praeglobotruncana stephani	1000	
Pseudotextularia browni	670	630
Rotalipora cushmani	1000	990
R. greenhornensis	1020	1000
Schackoina cenomana	990	
Ticinella multiloculata	990	
Ventilabrella austinana	990	640
Whiteinella paradubia	1020	600

Benthonic Foraminifera		
Anomalina nelsoni	650	610
Ammomarginulina perimpexus	760	
Bifarina geneae	670	620
Bulimina fabiliis	980	
Eouvigerina aculeata	620	610
Frondicularia archiaciana	640	
F. bidenta	650	
F. undulosa	670	650
Gaudryina bentonensis	730	
G. nebracensis	670	650
G. pupoides	640	
G. rugosa	670	640
G. spiritensis	770	730
Gavelinella kansasensis	770	640
Globorotalites subconicus	660	600
G. umbilicatus	640	620
Goesella rubulosa	610	
Gyroidinoides nitidus	670	600
Haplophragmoides howardense	760	730
H. rota	760	
Lenticulina isidis	650	
L. kansasensis	670	660
L. muensteri	610	
L. sublaevis	670	610
Marginulina austinana	660	
Miliammina ischnia	760	
Neobulimina albertensis	980	
N. canadaensis	640	600
N. irregularis	620	610
Neofabellina cushmani	670	
N. suturalis	630	620
Nodosaria affinis	670	650
Planularia dissona	640	
Pleurostomella austinana	670	620
Praebulimina kickapooensis	660	
P. reussi	660	600
Pseudoclavulina hastata	750	730
Quadrimorphina minuta	610	600
Reophax pepperensis	1050	
Saccamina alexanderi	760	750
Tappanina costifera	670	620
Trochammina wickendeni	770	730
Trochamminoides apricarius	1050	
Valvulineria infrequens	670	
V. loetterlei	980	770
V. plummerae	670	600
Verneuilinoides bearpawensis	730	710
Virgulina tegulata	670	610

Nannofossils		
Acaenolithus galloisii	1080	970
Ahmuellerella octoradiata	970	600
Amphizygus brooksii	1005.2	600
Broinsonia enormis	1070	890
Assipetra terebrodentarius	1040	610
Axopodorhabdus albianus	1080	983
Bidiscus rotatorius	1074.8	600
Biscutum constans	1080	600
Braarudosphaera bigelowi	1030	640
Broinsonia dentata	1074.8	600

Taxon		
Bukrylithus ambiguus	1080	600
Calculites obscurus	930	850
Calculites axosuturalis	860	800
Chiastozygus fessus	994	600
Chiastozygus litterarius	1084.7	600
Chiastozygus playrhethus	1050	920
Chiastozygus spissus	1080	880
Corollithion achylosum	1060	880
Corollithion exiguum	920	790
Corollithion kennedyi	1080	978.16
Corollithion madagascarensis	1074.8	800
Corollithion signum	1080	600
Cretrhabdus conicus	1080	600
Cretrhabdus surirellus	1084.7	600
Cretrhabdus loriei	1080	980
Cretrhabdus octofenestratus	1080	640
Cretrhabdus schizobrachiatus	1045	620
Cribrosphaerella ehrenbergii	1070	600
Cyclagelosphaera margerelii	1080	600
Cylindralithus asymmetricus	960	940
Cylndralithusi biarcus	670	600
Cylindralithus coronatus	660	600
Cylindralithus nudus	1065	610
Darwinilithus pentarhethum	1070	994
Discorhabdus watkinsii	1010	994
Eiffellithus eximius	670	600
Eiffellithus turriseiffelii	1084.7	600
Ellipsagelosphaera britannica	1070	1065
Eprolithus eptapetalus	962	780
Eprolithus floralis	1084.7	600
Eprolithus octopetalus	978.2	920
Eprolithus rarus	910	860
Flabellites oblongus	1065	600
Gartnerago nanum	1084.75	1025.75
Gartnerago obliquum	1000	600
Grantarhabdus coronadventis	1080	600
Helicolithus anceps	1015.6	600
Helicolithus compactus	1074.83	600
Kamptnerius magnificus	670	620
Liliasterites angularis	850	800
Lithastrinus moratus	670	600
Lithraphidites acutum	1050	980
Lithraphidites alatus	1045	984.5
Lithraphidites carniolensis	1055.16	610
Lordia xenotus	1080	1065
Loxolithus sp. 1	1074.83	600
Loxolithus sp. 2	860	840
Lucianorhabdus cayeuxi	640	610
Lucianorhabdus compactus	1070	979
Lucianorhabdus maleformis	920	630
Manvitiella pemmatoidea	1084.75	600
Markalius circumradiatus	1030	600
Marthasterites furcatus	670	600
Microrhabdulus decoratus	610	600
Microstaurus chiastius	1074.83	975.55
Micula staurophora	630	600
Miravetesina ficula	890	640
Octocyclus magnus	1025.75	1025.75
Percivalia fenestrata	1070	920
Percivalia hauxtonensis	1070	994
Placozygus sp. aff. P. spiralis	1074.83	610
Prediscosphaera cretacea	1084.75	600
Prediscosphaera spinosa	1080	610
Quadrum gartneri	977.08	600
Radiolithus planus	1074.8	850
Radiolithus undosus	860	860
Reinhardtites biperforatus	660	600
Reinhardtites scitula	1084.75	600
Reinhardtites sisyphus	670	600
Rhagodiscus achlyostaurion	1070	610
Rhagodiscus angustus	1074.83	610
Rhagodiscus asper	1080	974.25
Rhagodiscus splendens	1070	610
Rhomboaster svabenickae	920	650
Rotelapillus crenulatus	1080	600
Scampanella cornuta	1080	670
Sollasites horticus	1010	600
Tegumentum stradneri	1080	600
Tetrapodorhabdus coptensis	1070	976.5
Tetrapodorhabdus decorus	1065	610
Tranolithus exiguus	1084.75	610
Tranolithus gabalus	1025.75	600
Tranolithus minimus	983	600
Tranolithus phacelosus	1060	610
Tranolithus bitraversus	1016.25	790
Tubodiscus jurapelagicus	1040	1020
Vagalapilla dibrachiata	1080	600
Vagalapilla dicandidula	940	870
Watznaueria barnesae	1084.75	600
Watznaueria biporta	1055.16	600
Watznaueria fossacincta	1080	640
Watznaueria ovata	1080	600
Zeugrhabdotus embergeri	1084.75	600
Zeugrhabdotus theta	1065	1035
Zeugrhabdotus trivectis	1065	1035
Zeugrhabdotus sp. A	1080	600
Zygodiscus bicrescenticus	1084.75	600
Zygodiscus diplogrammus	1084.75	600
Zygodiscus elegans	1080	610

Palynomorphs
[depth in ( ) is abundant occurrence]

Taxon		
Afropollis jardinus	1212	1128
Aldorfia deflandrei	1090	670
Appendicsporites unicus	1312.4	
Aptea polymorpha	1380	1360
Aptea cf. A. anaphrissa	1465	1340
Apteodinium grande	1380	1100
Atopodinium haromense	1380	670
Batioladinium jaegeri	1380	1340
Chatangiella ditissima	800	680
Chatangiella granulifera	860	720
	(800)	
Chatangiella spectabilis	960	670
Chatangiella verrucosa	780	702
Chichaquadinium vestitum	1482	1150
Chlamydophorella huguonoti	1000	970
		(940)
Cicatricosisporites crassiterminatus	1212	680
Classopollis echinatus	1425	1400
Complexiopollis funiculus	970	702
Cyclonephelium membraniphorum	980	740
Dapsilidinium laminaspinosum	1400	702
Dingodinium albertii	670	

Dingodinium cerviculum	1340	1312.4	Odontochitina rhakodes	1482	1312.4
Dingodinium? cretaceum	960	670	Odontochitina singhi	1380	
Dingodinium euclaense	1000	860	Oligosphaeridium prolixispinosum	980	702
Dinopterygium cladoides	1482	680	Oligosphaeridium totum minor	1482	1340
Dorocysta litotes	970	740 (860)	Ornamentifera distalgranulata	1425	1090
Ellipsodinium rugulosum	980	670	Ovoidinium verrucosum	1168.9	1119.5/(1090)
Elytrocysta druggii	800	680	Palaeohystrichophora infusorioides	1445	670
Endocrinium campanula	950	740	Pervosphaeridium truncigerum		670
Epelidosphaeridia spinosa	1000	970	Pilosisporites trichopappilosus	1482	1119.5
Eurydinium glomeratum	960	740	Pilosisporites verus	1292	1212
Foraminisporis asymmetricus	1340	1245	Plicapollis serta	680	
Florentinia deanei	1050	680	Prolixosphaeridium conulum	1050	840/720
Florentinia mantelli	950	680	Protoellipsodinium spinosum	1425	1380
Florentinia resex	1482	680	Psaligonyaulax deflandrei	840	760
Florentinia verdieri	1466	670	Pseudoceratium expolitum	1482	1100
Ginginodinium evittii	1150	1119.5	Rhiptocorys veligera	960	702
Hystrichosphaeridium difficle	920	670	Rugubivesiculites rugosus	1312.4	670
Isabelidinium? amphiatum	740	670	Senoniasphaera protrusa	960	670
Isabelidinium magnum	1000	702	Scopusporis lautus	1425	1212
Ischyosporites crateris	1292	1179	Stephodinium coronatum	1400	970
Januaspoites spiniferus	1482		Stellatopollis barghoorni	1202	1150
Kiokansium perprolatum	1482	1050	Subtilisphaera cheit	1482	680
Litosphaeridium conispinum	1400	1380	Taurocuspoites segmentatus	1425	1179
Litosphaeridium siphonophorum	1100	980	Tigrisporites reticulatus	1465	1212
Luxadinium primulum	1150	1100	Trilobosporites apiverrucatus	1482	1150
Luxadinium propatulum	1340	1150	Trithyrodinium vermiculatum	840	740
Maghrebinia perforata	1380	1360	Trithyrodinium sp.	1050	670
Muderongia asymmetrica	1400	1119.5	Xenascus ceratioides	1090	670
Neoraistrickia robusta	1360	1179			

the constraint of graphic correlation supports Fischer's conclusions derived by other means. In this basin, the marker beds defined by Laferriere and Hattin (1989) seemingly are products of allocyclic processes and can be used as synchronous units.

BIOSTRATIGRAPHY

The Bounds core reference section is used to measure the durations of different scales or orders of depositional cycles. Biostratigraphic data from the core constrain the timing of transgressive and regressive events within the central part of the seaway and measure the durations of hiatuses at unconformities. The technique utilized to measure precisely the ages of fossil tops and marker horizons in the Bounds core is graphic correlation. The depths to tops and bases of key fossil zones used in preparing the graphic correlation are given in Table 3.

Graphic correlation is a quantitative, but nonstatistical, technique for determining the coeval relationships between two sections by comparing the ranges of taxa in both sections. By plotting the tops and bases of taxa found in both sections on an X/Y graph, the rate of sediment accumulation in one section can be compared with that in the other. Rates may be equal, or they may change through time, or an unconformity may develop in one section and not the other. Graphic correlation (Mann and Lane, 1995) enables the stratigrapher to compare the sedimentological events with the biotic events so that conclusions based on one can be tested by the other.

The elements of a graph are the scaled X/Y axes, the tops

(LAD-last appearance datum) and bases (FAD-first appearance datum) of the fossils, and the line of correlation. Where two drilled sections are cross plotted, the scale of both axes is in measured thickness. Fossil tops are plotted as + and bases are o (Fig. 14). The line of correlation between the two sections is placed by the stratigrapher to indicate the most constrained hypothesis of synchroneity between the two sections. The line of correlation is placed through the maximum number of tops and bases while taking into account depositional hiatuses indicated by the lithostratigraphic record.

A composite standard (CS), the integrated record of fossil ranges from widespread stratigraphic sections, is built by iterative graphic correlation of numerous sections. The CS establishes the virtual total range and the precise succession of fossil tops and bases. A plot of a new section to the CS shows most tops to the right of the line of correlation and most bases to the left. However, some ranges may be extended or even new taxa added. The scale of the CS is proportional to the thickness of the first reference section on the X-axis (standard reference section, SRS). The SRS must have no hiatuses and a uniform rate of sediment accumulation. The CS scale can be converted to a time scale either by graphing dated horizons into the CS or by graphing the SRS to a suitable time scale. A converted scale will not have uniform time increments, as in Fig. 14, because the original CS scale was not a time scale.

The Amoco composite standard data base was utilized to measure the ages of biotic and depositional events recorded in the Bounds core (Scott et al., 1994b); refer to Figure 6 in that

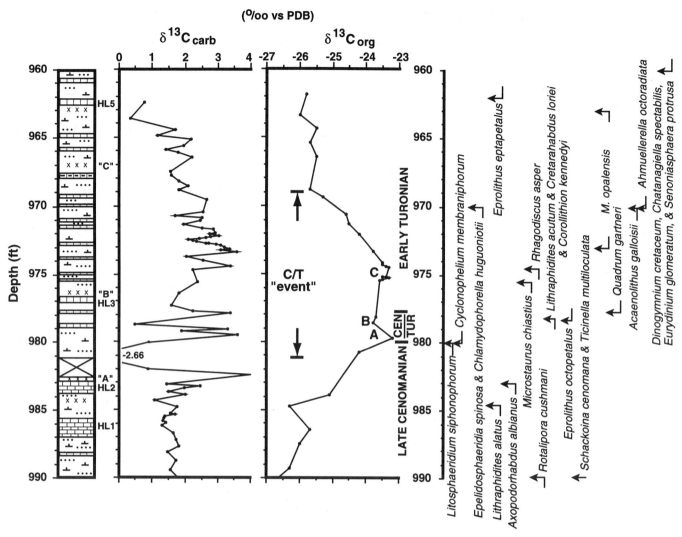

FIG. 12.—Carbon-isotope and paleontological data from the Bridge Creek Member, Greenhorn Formation across Cenomanian/Turonian boundary in Bounds core. Carbon-isotope data are from Dean et al. (1995). HL marker beds of Hattin (1971) and Elder (1985), and Pratt's (1985) sub-peaks of the carbon-isotope excursion (A, B, C) also are shown. The C/T boundary is bracketed by the top of *L. acutum* and the base of *Q. gartneri* zones.

report for a graph of the Bounds core section from Graneros to Purgatoire Formations. All CS are initially built from sections in local basins; then sections from other basins are added to determine the virtual total ranges and correct successions. This extension allows hiatuses in local sections to be recognized and measured that may have geologic significance. Amoco has developed a mature composite standard for the Cretaceous based on over 100 sections in all known biogeographic provinces, and it includes most reference sections of the stage boundaries.

Megafossils

Ammonites, inoceramids, and oysters are sparsely distributed throughout the Bounds core and are useful for identifying the stages. Because megafossil distribution is spotty, zonal boundaries are approximate. The distribution is related to the zonal schemes and concepts summarized by Kauffman et al. (1993); their biozone codes are indicated in parentheses with their interpolated radiometric dates. The megafossils used in preparing the graphic correlation interpretation (Fig. 14) (Fig. 6, Scott et al., 1994b) are listed in Table 3.

Two Upper Albian zones can be recognized in the Purgatoire Formation. The *Adkinsites bravoensis* range zone (AL-8, 102.8-102 ma) is represented at 1412.9 ft by this species. *Inoceramus comancheanus* is at 1406.5 ft; its FAD is in the *bravoensis* zone and its LAD is in the overlying biozone AL-9 defined by the ranges of *Boeseites romeri* and *Craginites serratescens*. The former defines a widespread zone in the Kiamichi Formation in Texas, and the Kiowa Formation in southeastern Kansas. The latter zone is known from the Duck Creek Formation above the Kiamichi in Texas and also from the upper Kiowa Formation in north-central and south-central Kansas (Scott, 1970).

In the Graneros Shale and lower Greenhorn Formation , three inoceramids occur in succession: *Inoceramus arvanus* (CE-3, 96-95 ma) at 1085.2-1071.3 ft, *Inoceramus rutherfordi* (CE-4, 95-94.7 ma) at 1067.6-1053.8 ft, and *Inoceramus prefragilis* (CE-5, 94.7-94.6 ma) at 1034.4-1008.5 ft. The first species extends through the upper Graneros with acanthocerid ammonites (Kauffman et al., 1993, Fig. 7). *I. rutherfordi* extends from the Thatcher Limestone Member of the Graneros through the Lincoln Limestone Member of the Greenhorn (Kauffman et al., 1977a). *I. prefragilis* also is known in the Lincoln as it is in this core. The oyster, *Ostrea beloiti* at 1057.8-1048.0 ft just below the Thatcher in the core, represents only the basal part of its occurrence, which extends to near the top of the Lincoln. These bivalves represent mainly the Middle Cenomanian, but the tops of *I. prefragilis* and *O. beloiti* may extend into the basal Upper Cenomanian.

Inoceramus pictus (94.6-93.4 ma) in the Hartland Shale Member of the Greenhorn Formation at 999.6 ft is near the base of its occurrence, which extends to the top of the Cenomanian in the Bridge Creek Limestone Member of the Greenhorn. *Mytiloides opalensis* (93.35-93.30 ma) at 973.4-963.4 ft in the lower part of the Bridge Creek is Lower Turonian. *Mytiloides mytiloides* (T2C, 93.3-92.2 Ma) at 950.5-943.9 ft is within its known range of Lower Turonian. *Mytiloides subhercynicus* (T4-5, 92.15-91.9 Ma) at 943.5 ft defines a total range zone spanning the Lower-Middle Turonian boundary and the Greenhorn-Carlile Shale contact. *Mytiloides labiatus* (T5, 92.10-91.9 Ma) at 851.6 ft in the Fairport Chalk Member of the Carlile also spans the same boundaries according to Kauffman et al. (1993). *Mytiloides hercynicus?* (T5) at 838.2 ft in the Fairport represents a range zone in the lower Middle Turonian. *Inoceramus cuvieri* (T5) at 834.8 ft in the Fairport, with the preceding species, characterize the Middle Turonian.

The lower Middle Turonian range zone of *Collignoniceras woollgari* (T5-6A, 92.1-91.5 Ma) is represented at 790.0 ft at the top of the Fairport Chalk Member of the Carlile Shale. The upper Middle Turonian range zone of *Prionocyclus hyatti* (T8-9, 90.6-90.45 Ma) is represented in the Blue Hill Shale Member of the Carlile. The occurrence of *Mytiloides costellatus?* (T10B, 90.4 Ma) at 761.8 ft at the top of the Fairport in this core is lower than reported by Kauffman et al. (1993); they found it in the Blue Hill and the Codell Members of the Carlile.

In the Niobrara Formation of the Bounds core, *Platyceramus platinus* (base of CO-5, 87.6 Ma) ranges from 592.0-551.0 ft in the lower Smoky Hill Shale Member of the Niobrara. This interval is dated by nannofossils as Santonian (Fig. 12), which is consistent with its reported Middle Coniacian to Lower Campanian range in the Smoky Hill (Kauffman et al., 1993).

Foraminifers

Foraminifers in washed residues of 51 core samples from the interval 1100 to 600 ft in the Bounds core were collected at a 10 ft spacing. The sampled interval extends from the Graneros Shale to the Fort Hays Limestone Member of the Niobrara Formation. Recovery was good to excellent for most of the samples. Benthonic species were abundant in samples from the Fort Hays (670-600 ft), but few in the underlying strata. Planktonic species were fairly abundant from the Niobrara, Fairport Member of the Carlile Shale, and the Greenhorn Formation (Fig. 13). Planktonic species were absent in the Codell Member of the Carlile (760-680 ft). Fifteen planktonic species recovered from the Bounds core have biostratigraphically important ranges (Leckie, 1985; Caldwell et al., 1993). The foraminifers used in the graphic correlation interpretation (Fig. 14) are listed in Table 3.

Biostratigraphic ages of the sampled interval are middle Cenomanian to upper Coniacian. The Fort Hays Limestone Member of the Niobrara Formation is entirely Coniacian in age in this core and is bounded by an unconformity at its base, where Upper Turonian to Lower Coniacian strata appear to be missing. The underlying Carlile Shale and the Bridge Creek Limestone Member of the Greenhorn Formation are Turonian in age.

Graphic correlation analysis (Scott et al., 1994b, Fig. 6) places the Cenomanian/Turonian boundary just above 980 ft within the lower part of the Bridge Creek and slightly above the highest occurrences of *Rotalipora cushmani* and *Ticinella multiloculata*, consistent with traditional approaches (Leckie, 1985; Caldwell et al., 1993).

The diversity and abundance of planktonic foraminifera decline sharply from the Greenhorn Formation into the Graneros Shale (Fig. 10). However, a calcareous microfauna characteristic of the Greenhorn — *Hedbergella delrioensis*, *Hedbergella planispira*, *Heterohelix globulosa*, and *Heterohelix moremani* — persists downward into the upper Graneros to 1080 ft. The highest occurrence of two key species of agglutinated foraminifera in the Graneros is at 1050 ft: *Trochamminoides apricarius* and *Reophax pepperensis*. This assemblage represents Eicher's (1965) *Trochamminoides apricarius* Assemblage Zone in the upper Graneros. Samples between 1100-1090 ft are barren of foraminifera.

The top of the Greenhorn Formation is placed at the top of the highest limestone bed (905.8 ft). The basal contact at 1029.9 ft is gradational and near the top of the agglutinated foraminiferal assemblage that characterizes the Graneros Shale. The Greenhorn contains a diverse assemblage of keeled planktonic species, including *Ticinella* and *Rotalipora* at 990-980 ft. The base of the Bridge Creek Limestone Member of the Greenhorn is at 987 ft, just below the Cenomanian/Turonian boundary. The Turonian species, *Dicarinella hagni* at 950 ft and *Helvetoglobotruncana helvetica* at 920-910 ft, were found only in the Bridge Creek. Below the Bridge Creek the following six Cenomanian planktonic species were found: *Globigerinelloides bentonensis*, *Praeglobotruncana stephani*, *Rotalipora cushmani*, *Rotalipora greenhornensis*, *Schackoina cenomana*, and *Ticinella multiloculata*. Benthonic foraminifera are generally absent in the Greenhorn, except between 980 to 960 ft, where *Buliminella fabilis*, *Neobulimina albertensis*, and *Valvulineria loetterlei* occur.

The microfauna of the Fairport Shale Member of the Carlile Shale is characterized by a relatively diverse and well preserved assemblage of non-keeled planktonic foraminifera. Large individuals of *Hedbergella* spp. and *Whiteinella* spp. dominate the association and range into the underlying Greenhorn Formation. *Neobulimina albertensis* is the only benthonic species recovered from this interval. Seven planktonic species from the Fairport are listed in Table 3.

The Blue Hill Shale Member of the Carlile Shale is dark gray and non-calcareous between 785 and 755 ft and displays gradational contacts with overlying and underlying units. It contains six benthonic foraminifers in addition to those in the Codell Sandstone Member (Table 2). Two planktonic species are found in the Blue Hill below 770 ft: *Hedbergella delrioensis* and *Heterohelix globulosa*.

Planktonic foraminifers were not recovered from the Codell Sandstone Member of the Carlile Shale. However, a low diversity assemblage of agglutinated species was found in the interval at 750-730 ft (Table 3). This assemblage is not age-diagnostic, but it is characteristic of the lower part of the Codell and the underlying Blue Hill Shale Member. The assemblage is assigned to the *Haplophragmoides howardense* Assemblage Zone (760-730 ft) (Kauffman, et al., 1977b). Associated taxa in the core include *Pseudoclavulina hastata* and *Gaudryina bentonensis*.

Foraminiferal recovery and preservation are excellent throughout the Fort Hays Limestone Member of the Niobrara

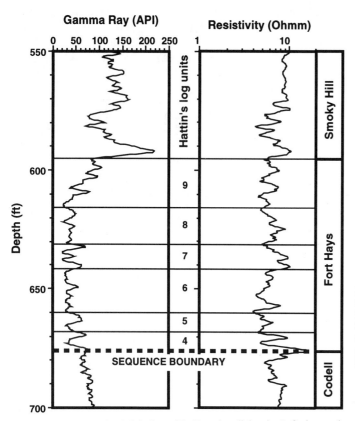

FIG. 13.–Gamma-ray and resistivity logs of the Bounds well showing Laferriere and Hattin's (1989) depositional cycles in the Fort Hays Limestone Member of the Niobrara Formation. The sequence boundary is the unconformable contact between the Codell and Fort Hays.

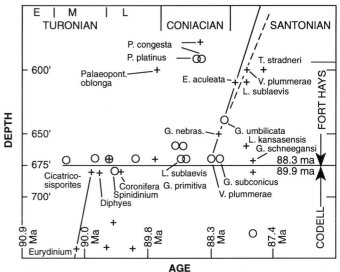

FIG. 14.–Graphic correlation solution of age constraints on Hattin's marker beds in the Niobrara Formation in Bounds core (y-axis is depth in feet). Fossil tops indicated by + and fossil base by o. Only the key fossils are named on this graph, although bases and tops of other taxa constraining the line of correlation are shown. The x-axis is age in the Amoco composite standard at a variable scale. The solid line is the preferred line of correlation; the dashed line is the alternative if cycles are at a 100 ka frequency.

Nannofossils

Nannofossils were processed from the Thatcher Limestone Member of the Graneros Shale to the top of the Fort Hays Limestone Member of the Niobrara Formation (1084.75-600 ft) because it is the most marine interval in the Bounds core. The initial sample spacing was 10 ft, but very closely spaced samples were collected across the Cenomanian-Turonian boundary. Recovery was good to excellent. The nannofossils used in the graphic correlation interpretation (Fig. 14) are listed in Table 3. For more detailed analysis of the nannofossil data, see Bralower and Bergen (this volume).

A number of nannofossil datums used in global biostratigraphy are identified in the Western Interior section (Watkins et al., 1993). The relative positions of many of these datums is confirmed in the Bounds core. The base (FAD) of *Lithraphidites acutum* at 1050 ft is in the upper part of the Graneros Shale. The top (LAD) of *Gartnerago nanum* at 1025.75 ft is in the upper Lincoln Limestone Member of the Greenhorn Formation, Upper Cenomanian. The base of *Gartnerago obliquum* at 1000 ft in the Hartland is uppermost Cenomanian. However, the bases of *Microrhabdulus decoratus* and *Lithastrinus moratus* are reported to be close to the base of *G. obliquum*; but in the Bounds core they were first found in the Niobrara Formation at 610 ft and 670 ft, respectively. The base of *Eiffellithus eximius* in outcrop is also reported from the uppermost Cenomanian *Neocardioceras juddii* zone. However, in the Bounds core, *E. eximius* is first found in the basal Niobrara at 670 ft.

In the Bounds core, the Cenomanian-Turonian boundary at the top of the *Neocardioceras juddii* zone in the lower Bridge Creek Limestone Member of the Greenhorn Formation is bracketed by the top of *Lithraphidites acutum* at 980 ft and by the base of *Quadrum gartneri* at 977.08 ft. Key tops in the underlying interval are *Axopodorhabdus albianus* at 983 ft, *Corollithion kennedyi* at 978.16 ft, and *Rhagodiscus asper* at 974.25 ft. The

Formation. Planktonic and benthonic species are diverse, well known, and diagnostic of the lower part of the Niobrara and its equivalents in the Western Interior. The *Hastigerinoides subdigitata* Total Range Zone of latest Coniacian to early Santonian age (Frerichs et al., 1977) is represented by the nominative species at 600 ft in the upper Fort Hays. The presence of *Dicarinella primitiva* between 670 and 610 ft indicates an early Coniacian to early Santonian age; and *Dicarinella concavata* from 660 and 640 ft is late Coniacian to early Santonian. Twenty-one other planktonic species are associated with these key taxa 34 benthonic species are present (Table 3).

The patterns of foraminiferal abundance and diversity in Figure 10 represent two episodes of maximum marine transgression. The first occurs in the Bridge Creek Limestone Member of the Greenhorn Formation and the second in the Fort Hays Limestone Member of the Niobrara Formation. The regressive phase between these two transgressions is represented by the Carlile Shale. The sequence boundary between these cycles is the unconformable Carlile-Niobrara contact. The initial transgressive phase of the Niobrara cycle appears to be absent in the Bounds core. Farther north in the Black Hills region, this transgressive portion of the Niobrara cycle is represented by an expanded section of the upper Carlile Shale that includes the Turner Sandy Member and the lower part of the Sage Breaks Member of the Carlile (Evetts, 1976). In the Black Hills region, the upper part of the Sage Breaks is equivalent to the Fort Hays Limestone Member of the Niobrara in the Bounds core.

base of *Corolithion exiguum* is reported from the uppermost Cenomanian part of the lower Bridge Creek (Watkins et al., 1993), but in the Bounds core it occurs in the Lower Turonian middle Bridge Creek (Hattin's marker bed PF-1; 1971) at 920 ft. The top of *Percivalia fenestrata* is reported to approximate that of *R. asper* but, in the Bounds core, *P. fenestrata* occurs at 920 ft in the upper Bridge Creek, 54 ft higher than *R. asper*.

In the Niobrara Formation the base of *Marthasterites furcatus* is reported from the top of the Fort Hays Limestone Member in the western Kansas outcrops, but in the Bounds core it occurs near its base at 670 ft. This taxon indicates that the rocks are Upper Coniacian and suggests that the base of the Fort Hays is younger here than in nearby outcrops. The Upper Turonian-lower Upper Coniacian *Lucianorhabdus maleformis* overlaps the range of *M. furcatus* in the Bounds core (920-630 ft). In this core its base in the upper Greenhorn Formation is lower than Upper Turonian. Other markers of the Smoky Hill Chalk Member of the Niobrara have different ranges in the Bounds core than reported from outcrop by Watkins et al. (1993). The base of *Lucianorhabdus cayeuxii* is reported from the Middle Santonian lower Smoky Hill, but here it occurs in the middle Fort Hays at 640 ft. The base of *Calculites obscurus* is reported from the basal Campanian part of the Smoky Hill, but in the Bounds core its base is in the Bridge Creek Member of the Greenhorn Formation at 930 ft.

Palynomorphs

Palynomorph samples were collected at about 10-foot intervals from 1480 ft in the basal Purgatoire Formation to 670 ft in the basal Niobrara Formation. Recovery was good to excellent and preservation of most specimens was very good. A number of fossil tops (LAD) and bases (FAD) (Table 3) are used here for the first time to identify European stages in the Western Interior and refine the assemblages described by Norris et al. (1975). The palynomorphs used in the graphic correlation interpretation (Fig. 14) are listed in Table 3.

A single sample from the Morrison Formation at 1539.3 ft yielded Upper Jurassic spores: *Callielisporites dampieri, Callielisporites turbatus, Cerebropollenites mesozoicus, Classopollis torosus,* and *Vitreisporites pallidus* (E. J. Kidson, pers. commun., 1988; R. W. Hedlund, pers. commun., 1992).

Six Upper Albian taxa have tops in the Purgatoire Formation in the Bounds core: *Appendicisporites unicus, Classopollis echinatus, Janusporites spiniferus, Litosphaeridium conispinum, Oligosphaeridium totum minor,* and *Trilobosporites apiverrucatus.* Seven taxa have bases in the Purgatoire: *Apteodinium grande, Florentinia verdieri, Odontochitina singhii, Ovoidinium verrucosum, Palaeohystrichophora infusorioides, Rugubivesiculites rugosus,* and *Subtilisphaera cheit.* Therefore, the Purgatoire is Upper Albian.

The Lower Cenomanian is indicated between 1150 and 1090 ft by the tops of *Afropollis jardinus, Apteodinium grande, Ginginodinium evitti, Luxadinium primulum, Luxadinium propatulum, Muderongia asymmetrica, Neoraistrickia robusta, Ovoidinium verrucosum, Pilosisporites trichopappilosus, Pseudocertium expolitum,* and *Stellatopollis barghoornii.* This interval includes the uppermost Dakota and lowermost Graneros Formations (Fig. 2).

The Middle Cenomanian is characterized by the base of *Isabelidinium magnum* at 1000 ft, and the top of *Kiokansium perprolatum* at 1050 ft, which spans the contact of the Graneros

and Greenhorn formations. The Upper Cenomanian between 1000 and 970 ft in the Lincoln Limestone and Bridge Creek Limestone Members of the Greenhorn Formation contains the tops of *Chlamydophorella huguoniotii, Epelidosphaeridia spinosa,* and *Litosphaeridium siphonorphorum,* and the bases of *Cyclonophelium membraniphorum* and *Dinogymnium euclaensis.*

The Cenomanian-Turonian boundary is placed at or above 980 ft by the top and the bases of *Ellipsodinium rugulosum, Oligosphaeridium prolixispinosum, Complexiopollis funiculus,* and *Dorocysta litotes.* This boundary is bracketed by nannofossils between 980 and 977.08 ft. The other fossils also agree based on the preliminary coarse sample spacing.

Fossil tops characteristic of the Lower Turonian are *Chatangiella granulifera, Chatangiella spectabilis, Dinogymnium cretaceum, Eurydinium glomeratum, Hystrichosphaeridium difficile,* and *Senoniasphaera protrusa,* which bracket the interval between 740 and 670 ft in the Codell Sandstone Member of the Carlile Shale. These tops in the Bounds core are higher than expected and some may have been reworked into the basal Niobrara Formation. Key taxa for the Middle Turonian are the bases of *Chatangiella ditissima, Chatangiella verrucosa,* and *Elytrocysta druggii,* and the tops of *Dorocysta litotes, Eurydinium glomeratum,* and *Prolixosphaeridium conulum,* which bracket the interval from 800 and 740 ft in the Blue Hill Shale Member and Codell. Characteristic Upper Turonian markers are the base of *Isabelidinium? amphiatum* and the top of *Cyclonophelium membraniphorum* at 740 ft in the Codell. However, ammonites in the Carlile outcrop and core and graphic correlation of the Bounds core indicate that this interval is Middle Turonian. The Coniacian marker bases are *Dinogymnium albertii* and *Pervosphaeridium truncigerum* at 670 ft and marker tops are *Atopodinium haromense, Cicatricosisporites carassiterminatus, Dinopterygium cladoides, Florentinia mantelli,* and *Florentinia resex* at 680-670 ft in basal Niobrara.

TIMING AND SCALE OF DEPOSITIONAL CYCLES

The sequence-stratigraphy methodology assumes that depositional cycles represent geologic time and that natural orders or scales of duration are represented by successively thinner scales of depositional cycles. Four scales or orders of depositional cycles are developed within the mid-Cretaceous section cored in the Amoco No. 1 Bounds well (Fig. 2). The Purgatoire Formation and the lower part of the Dakota Formation are the two thickest depositional cycles. These sequences, which are bounded by unconformities, are from 2.2 to 3.4 My long. Additional sequences may be present in the 5.1 My-stratigraphic interval represented by the upper Dakota transgressive systems tracts to the Carlile Shale progradational systems tracts. The Hartland Shale Member of the Greenhorn Formation is a candidate for the interval that may contain a conformable basinal contact that is equivalent to the unconformity in the Greenhorn on the western shelf (Sageman, 1985). The Bridge Creek Limestone Member may be the basinal equivalent of a maximum transgressive interval on the shelf. Another unconformity-bounded sequence, the Juana Lopez Member of the Carlile Shale and equivalents, lies within the 1.6-My hiatus between the Carlile and Niobrara formations in the Bounds core. The Niobrara also represents at least one cycle 1-3 My long.

The second scale of depositional cycles averages about 0.5 My in duration. It is represented by the two shale-sandstone intervals that comprise the basal transgressive systems tracts of the Purgatoire Formation (1487.5-1428.6, depositional units 1-4, Table 2). These transgressive events have been identified by numerous authors as representing global eustatic events.

The third scale of depositional cycles is represented by the 13 coarsening up and fining up cycles in the lower part of the Dakota Formation. If deposition were continuous and no cycles were missing, each would average 169,000 years. Even though these assumptions cannot be tested, these Dakota cycles are candidates for the 100,000 yr eccentricity climatic cycle. A second set of eccentricity cycles is suggested for the bundled marl-limestone marker beds in the Fort Hays Limestone Member of the Niobrara Formation. Laferriere and Hattin (1989) traced these marker beds across the basin using well logs.

The fourth scale of depositional cycle is the marl-limestone, 1-3 meter-thick cycles in the Bridge Creek Limestone Member of the Greenhorn Formation. These cycles are similar in scale to either the precession or obliquity climatic cycles of 23,000 or 41,000 yr, respectively (Pratt et al., 1993). To constrain the periodicity of the cycles in the Greenhorn and Carlile formations, the rate of rock accumulation was determined by graphic correlation of the fossils with an Amoco composite standard (Fig. 14) (Scott et al., 1994b, Fig. 6). The slope of the line of correlation changes significantly between 963 and 970 ft, near the base of the Turonian and just above the top of the $\delta^{13}C$ shift. For calculations, a cycle top at 968 ft was used as the depth at which the change in slope occurred. The gradient above this point is 9.15 cm/ky (3,333 yrs/ft) and below it is 2.44 cm/ky (12,500 yrs/ft). The 10 cycles in the Fairport Chalk Member average 1.59 m (5.2 ft) in thickness for about 17,000 yrs/cycle. The 13 Bridge Creek Limestone cycles average 1.48 m (4.85 ft) in thickness for about 16,000 yrs/cycle. The 11 cycles in the Greenhorn Formation below 968 ft average 1.83 m (6 ft) thickness for about 75,000 yrs/cycle.

Although this exercise does not prove that these depositional cycles have a Milankovitch frequency, it suggests that the cycles above 968 ft are in the range of the 21-ky precessional frequency and that the cycles below are in the range of the 41-ky eccentricity frequency. Interestingly, Fischer et al. (1985) and Pratt et al. (1993) suggest that Bridge Creek cycles are at the obliquity frequency of 41 ky. If the 34 cycles in the Fairport and Greenhorn interval of 179 ft are averaged over the time span of 1.7 My, the duration of each averages about 50,000 yrs/cycle. So, graphic correlation analysis suggests that a different frequency of cycle was recorded in the sedimentary column after the Cenomanian/Turonian $\delta^{13}C$ event. This hypothesis may provide a clue about climatic and oceanographic processes operating within this basin. Conversely, if these cycles are of a Milankovitch frequency, that would constrain the slope of the line of correlation.

The slopes of the lines of correlation for the Greenhorn and Carlile formations can be converted to rates of sediment accumulation per thousand years to compare with rates calculated by Elder and Kirkland (1985) for the Bridge Creek Limestone Member of the Greenhorn Formation. They estimated a rate on the order of 0.5-1.0 cm/ky. In the Bounds core, our data suggest a rate of 9.0 cm/ky in the upper part of the Greenhorn and Carlile section and 2.4 cm/ky in the lower part of the Greenhorn. The coeval Second White Specks Formation in western Canada has an accumulation rate of 2.5 cm/ky (Schröder-Adams et al., 1996). The 5-10x difference between the two parts of the section seems significant and merits additional testing.

The Greenhorn Formation was deposited during the longer-term T6 transgressive/regressive cycle in the Western Interior scheme of Kauffman (1984) and represents maximum transgression. Shorter term cycles of sequence scale (so-called third-order) were suggested by Sageman (1985) based on correlations between the basin center near Pueblo, Colorado, and the western margin. The Graneros-Greenhorn formational contact may represent a maximum flooding interval. Other intervals of transgression may be represented by the upper part of the Lincoln Limestone Member, by the middle Hartland Shale Member and by most of the Bridge Creek Limestone Member. Basinal type sequence boundaries were placed within the upper Graneros member, the lower part of the Lincoln Member and the lower and upper parts of the Hartland Member (Sageman, 1995). The 2-3 My represented by the interval between the Thatcher Limestone Member and the Bridge Creek Limestone may possibly contain four sequences averaging 500-750 ky long.

Clearly, climatic processes greatly affected deposition in the Western Interior basin, but eustasy and tectonics also affected the depositional record (Weimer, 1984). Only with this type of data, which uses depositional cycles that have constrained ages, can a statistical data base be acquired that will test the concept of natural orders of cycle scales and provide data for determining the class boundaries.

ACKNOWLEDGMENTS

The Bounds well was drilled and cored jointly by Amoco Research Center and Amoco Exploration in Denver. The core-design engineering team included K. Millheim, S. Randolph, D. Skidmore, H. Mount, C. Rai, and S. Walker. The core and log data were released for publication by Amoco. Numerous Amoco staff supported this project by core and sample preparation and by drafting, photography, and word processing. Amoco management recognized the technical value of the Cretaceous study.

REFERENCES

BERGEN, J. A., EVETTS, M. J., FRANKS, P. C., LEWAN, M. D., SCOTT, R. W., AND STEIN, J. A., 1990, Mid-Cretaceous sequences and cyclostratigraphy, Western Kansas (abst.): Society of Economic Paleontologists and Mineralogists Research Conference on Cretaceous Resources, Events, and Rhythms, August 20-24, 1990, Denver, Colorado, Program Abstracts.

CALDWELL, W. G. E., DINER, R., EICHER, D. L., FOWLER, S. P., NORTH, B. R., STELCK, C. R., AND VON HOLDT, L., 1993, Foraminiferal biostratigraphy of Cretaceous marine cyclothems, in Caldwell, W. G. E. and Kauffman, E. G., eds., Evolution of the Western Interior Basin: Waterloo, Ontario, Geological Association of Canada Special Paper 39, p. 77-520.

DEAN, W. E., M. A. ARTHUR, B. B. SAGEMAN, AND M. D. LEWAN, 1995, Core descriptions and preliminary geochemical data for the Amoco Production Company Rebecca K. Bounds #1 Well, Greeley County, Kansas: USGS Open-file Report 95-209, 243 p.

DOLSON, J., MULLER, D., EVETTS, M. J., AND STEIN, J. A., 1991, Regional paleotopographic trends and production, Muddy Sandstone (Lower Cretaceous), Central and Northern Rocky Mountains: American Association of Petroleum Geologists Bulletin, v. 75, p. 409-435.

EICHER, D. L., 1965, Foraminifera and biostratigraphy of the Graneros Shale: Journal of Paleontology, v. 39, p. 875-909.

ELDER, W. P., 1985, Biotic patterns across the Cenomanian-Turonian extinction boundary near Pueblo, Colorado, in Pratt, L.M., Kauffman, E.G., and Zelt, F.B., eds., Fine-grained Deposits and Biofacies of the Western Interior Seaway: Evidence of Cyclic Sedimentary Processes, Field Trip Guidebook No. 4: Tulsa, Society of Economic Paleontologists and Mineralogists, p. 157-169.

ELDER, W. P., AND KIRKLAND, J. I., 1985, Stratigraphy and depositional environments of the Bridge Creek Limestone Member of the Greenhorn Limestone at Rock Canyon Anticline near Pueblo, Colorado, in Pratt, L.M., Kauffman, E.G., and Zelt, F.B., eds., Fine-grained Deposits and Biofacies of the Western Interior Seaway: Evidence of Cyclic Sedimentary Processes, Field Trip Guidebook No. 4: Tulsa, Society of Economic Paleontologists and Mineralogists, p. 122-134.

EVETTS, M. J., 1976, Microfossil biostratigraphy of the Sage Breaks Shale (Upper Cretaceous) in northeastern Wyoming: The Mountain Geologist, v. 13, p. 115-134.

FISHER, C. G., KAUFFMAN, E. G., AND VON HOLDT, L., 1985, The Niobrara transgressive hemicycle in central and eastern Colorado: The anatomy of a multiple disconformity, in Pratt, L.M., Kauffman, E.G., and Zelt, F.B., eds., Fine-grained Deposits and Biofacies of the Western Interior Seaway: Evidence of Cyclic Sedimentary Processes, Field Trip Guidebook No. 4: Tulsa, Society of Economic Paleontologists and Mineralogists, p. 184-198.

FISCHER, A. G., 1993, Cyclostratigraphy of Cretaceous chalk-marl sequences in Caldwell, W. G. E. and Kauffman, E. G., eds., Evolution of the Western Interior Basin: Waterloo, Ontario, Geological Association of Canada, Special Paper 39, p. 283-295.

FISCHER, A.G., HERBERT, T., AND PRIMOLI-SILVA, I., 1985, Carbonate bedding cycles in Cretaceous pelagic and hemipelagic sediment, in Pratt, L.M., Kauffman, E.G., and Zelt, F.B., eds., Fine-grained Deposits and Biofacies of the Western Interior Seaway: Evidence of Cyclic Sediemntary Processes, Field Trip Guidebook No. 4: Tulsa, Society of Economic Paleontologists and Mineralogists, p. 1-10.

FRANKS, P. C., 1975, The transgressive-regressive sequence of the Cretaceous Cheyenne, Kiowa, and Dakota Formations of Kansas: Waterloo, Geological Association of Canada Special Paper No. 13, p. 469-521.

FRANKS, P. C., 1979, Record of an Early Cretaceous marine transgression — Longford Member, Kiowa Formation: Kansas Geological Survey Bulletin 219, p. 1-55.

FRANKS, P. C., 1980, Models of marine transgression-Example from Lower Cretaceous fluvial and paralic deposits, north-central Kansas: Geology, v. 8, p. 56-61.

FRERICHS, W. E., POKRAS, E. M., AND EVETTS, M. J., 1977, The genus Hastigerinoides and its significance in the biostratigraphy of the Western Interior: Journal of Foraminiferal Research, v. 7, p. 149-156.

GLENISTER, L. M. AND KAUFFMAN, E. G., 1985, High resolution stratigraphy and depositional history of the Greenhorn regressive hemicyclothem, Rock Canyon Anticline, in Pratt, L.M., Kauffman, E.G., and Zelt, F.B., eds., Fine-grained Deposits and Biofacies of the Western Interior Seaway: Evidence of Cyclic Sedimentary Processes, Field Trip Guidebook No. 4: Tulsa, Society of Economic Paleontologists and Mineralogists, p. 170-183.

GOLDHAMMER, R. K., OSWALD, E. J., AND DUNN, P. A., 1991, Hierarchy of stratigraphic forcing: Example from Middle Pennsylvanian shelf carbonates of the Paradox Basin in Franseen, E. K., Watney, W. L., Kendall, C. G. St. C., and Ross, W., eds., Sedimentary modeling: Computer simulations and methods for improved parameter definition: Kansas Geological Survey Bulletin 233, p. 361-413.

HAMILTON, V. J., 1994, Sequence stratigraphy of Cretaceous Albian and Cenomanian strata in Kansas, in Shurr, G. W., Ludvigsen, G. A., and Hamilton,R. H., eds., "Perspectives on the Eastern Margin of the Cretaceous Western Interior Basin": Boulder, Geological Society of America Special Paper 287, p. 79-96.

HATTIN, D. E. 1962, Stratigraphy of the Carlile Shale (Upper Cretaceous) in Kansas: Kansas Geological Survey Bulletin 156, 155 p.

HATTIN D. E., 1965, Stratigraphy of the Graneros Shale (Upper Cretaceous) in central Kansas: Kansas Geological Survey Bulletin 178, p. 1-83.

HATTIN, D. E., 1967, Stratigraphic and paleoecologic significance of macro-invertebrate fossils in the Dakota Formation (Upper Cretaceous) of Kansas, in Essays in Paleontology and Stratigraphy, Raymond C. Moore Commemorative Volume, Lawrence, Univ. Kansas Department Geology Special Publication 2, p. 570-589.

HATTIN, D.E., 1971, Widespread, synchronously deposited, burrow-mottled limestone beds in the Greenhorn Limestone (Upper Cretaceous) of Kansas and southeastern Colorado: American Association of Petroleum Geologists Bulletin v. 55, p. 412-431.

HATTIN, D. E., 1975, Stratigraphy and depositional environment of Greenhorn Limestone (Upper Cretaceous) of Kansas: Kansas Geological Survey Bulletin 209, p. 1-128.

HATTIN, D. E., 1982, Stratigraphy and depositional environment of Smoky Hill Chalk Member, Niobrara Chalk, (Upper Cretaceous) of the type area, western Kansas: Kansas Geological Survey Bulletin 225, p. 1-108.

HATTIN, D. E., 1985, Distribution and significance of widespread, time-parallel pelagic limestone beds in Greenhorn Limestone (Upper Cretaceous) of the central great Plains and southern Rocky Mountains, in Pratt, L.M., Kauffman, E.G., and Zelt, F.B., eds., Fine-grained Deposits and Biofacies of the Western Interior Seaway: Evidence of Cyclic Sedimentary Processes, Field Trip Guidebook No. 4: Tulsa, Society of Economic Paleontologists and Mineralogists, p. 28-37.

HOLBROOK, J. M. AND DUNBAR, R. W., 1992, Depositional history of lower Cretaceous strata in northeastern New Mexico: Implications for regional tectonics and depositional sequences: Geological Society of America Bulletin, v. 104, p. 802-813.

KAUFFMAN, E. G., 1984, Paleobiogeography and evolutionary response dynamic in the Cretaceous Western Interior seaway of North America, in Westermann,

G. E. G., ed., Jurassic-Cretaceous biochronology and Paleogeography of North America: St. John's, Newfoundland, Geological Association of Canada Special Paper 27, p. 273-306.

KAUFFMAN, E. G., 1985, Depositional history of the Graneros Shale (Cenomanian), Rock Canyon Anticline, in Pratt, L.M., Kauffman, E.G., and Zelt, F.B., eds., Fine-grained Deposits and Biofacies of the Western Interior Seaway: Evidence of Cyclic Sedimentary Processes, Field Trip Guidebook No. 4: Tulsa, Society of Economic Paleontologists and Mineralogists, p. 90-99.

KAUFFMAN, E. G., Hattin, D. E., and Powell, J. D., 1977a, Stratigraphic, paleontologic, and paleoenvironmental analysis of the Upper Cretaceous rocks of Cimarron County, northwestern Oklahoma: Geological Society of America Memoir 149, 150 p.

KAUFFMAN, E. G., COBBAN, W. A., AND EICHER, D. L., 1977b, Albian through Lower Coniacian strata and biostratigraphy, Western Interior United States, in Proceedings Second International Conference on Mid-Cretaceous Events, Special Volume: Nice, France, Annals Museum of Natural History, p. 21-1 to 21-33.

KAUFFMAN, E. G., AND PRATT, L. J., 1985, Field reference section, in Pratt, L.M., Kauffman, E.G., and Zelt, F.B., eds., Fine-grained Deposits and Biofacies of the Western Interior Seaway: Evidence of Cyclic Sedimentary Processes, Field Trip Guidebook No. 4: Tulsa, Society of Economic Paleontologists and Mineralogists, p. FRS/1-FRS/26.

KAUFFMAN, E. G., SAGEMAN, B. B., KIRKLAND, J. I., ELDER, W. P., HARRIES, P. J., AND VILLAMIL, T., 1993, Molluscan biostratigraphy of the Cretaceous Western Interior Basin, North America in Caldwell, W. G. E. and Kauffman, E. G., eds., Evolution of the Western Interior Basin: Waterloo, Geological Association of Canada Special Paper 39, p. 397-434.

LAFERRIERE, A. P., AND HATTIN, D. E., 1989, Use of rhythmic bedding patterns for locating structural features, Niobrara Formation, United States Western Interior: American Association Petroleum Geologists Bulletin v. 73, p. 630-640.

LANDUYDT, C. J., 1990, Micromorphology of iron minerals from bog ores of the Belgian Campine area, in Douglas, C. A., ed., Soil Micromorphology: A Basic and Applied Science: Developments in Soil Science 19: Elsevier, Amsterdam, p. 289-294.

LECKIE, R. M., 1985, Foraminifera of the Cenomanian-Turonian boundary interval, Greenhorn Formation, Rock Canyon Anticline, Pueblo, Colorado, in Pratt, L.M., Kauffman, E.G., and Zelt, F.B., eds., Fine-grained Deposits and Biofacies of the Western Interior Seaway: Evidence of Cyclic Sedimentary Processes, Field Trip Guidebook No. 4: Tulsa, Society of Economic Paleontologists and Mineralogists, p. 139-149.

MACFARLANE, P. A., WADE, A., DOVETON, J. H., AND HAMILTON, V. J., 1991, Revised stratigra- phic interpretation and implications for pre-Graneros paleogeography from test-hole drilling in central Kansas: Lawrence, Kansas Geological Survey Open-file Report 91-1A, 73 p.

MACFARLANE, P. A., DOVETON, J. H., FELDMAN, H. R., BUTLER, J. J., JR., COMBES, J. M., AND COLLINS, D. R., 1994, Aquifer/aquitard units of the Dakota aquifer system in Kansas: Methods of delineation and sedimentary architecture effects on ground-water flow and flow properties: Journal of Sedimentary Research, v. B64, p. 464-480.

MANN, K. O. AND LANE, H. R., 1995, Graphic correlation: Tulsa, SEPM Society for Sedimentary Geology Special Publication No. 53, 263 p.

MATEER, N. J., 1987, The Dakota Group of northeastern New Mexico, in Guidebook, 38th Field Conference, Northeastern New Mexico: Albuquerque, New Mexico Geological Society, p. 223-236.

NORRIS, G., JARZEN, D. M., AND AWAI-THORNE, B. V. 1975, Evolution of the Cretaceous terrestrial palynofora in western Canada, in W. G. E. Caldwell, ed., The Cretaceous System in the Western Interior of North America: Waterloo, Geological Association of Canada Special Paper 13, p. 333-364.

NORTH AMERICAN COMMISSION ON STRATIGRAPHIC NOMENCLATURE, 1983, North American stratigraphic code: American Association of Petroleum Geologists Bulletin, v. 67, p. 841-875.

OBRADOVICH, J. D., 1993, A Cretaceous time scale, in Caldwell, W. G. E. and Kauffman, E. G., eds., Evolution of the Western Interior Basin: Waterloo, Geological Association of Canada Special Paper 39, p. 379-396.

PLUMMER, N., AND ROMARY,J. F., 1942, Stratigraphy of the pre-Greenhorn Cretaceous beds of Kansas: Kansas Geological Survey Bulletin 41, pt. 9, p. 313-348.

PRATT, L. M., 1985, Isotopic studies of organic matter and carbonate in rocks of the Greenhorn marine cycle, in Pratt, L.M., Kauffman, E.G., and Zelt, F.B., eds., Fine-grained Deposits and Biofacies of the Western Interior Seaway: Evidence of Cyclic Sedimentary Processes, Field Trip Guidebook No. 4: Tulsa, Society of Economic Paleontologists and Mineralogists, p. 38-48.

PRATT, L. M., ARTHUR, M. A., DEAN, W. E., AND SCHOLLE, P. A., 1993, Paleo-oceanographic cycles and events during the Late Cretaceous in the Western Interior Seaway of North America in Caldwell, W. G. E. and Kauffman, E. G., eds., Evolution of the Western Interior Basin: Waterloo, Geological Association of Canada Special Paper 39, p. 333-353.

SAGEMAN, B. B., 1985, High-resolution stratigraphy and paleobiology of the Hartland Shale Member: Analysis of an oxygen-deficient epicontinental sea, *in* Pratt, L.M., Kauffman, E.G., and Zelt, F.B., eds., Fine-grained Deposits and Biofacies of the Western Interior Seaway: Evidence of Cyclic Sedimentary Processes, Field Trip Guidebook No. 4: Tulsa, Society of Economic Paleontologists and Mineralogists, p. 110-121.

SCHRODER-ADAMS, C. J., LECKIE, D. A., BLOCH, J., CRAIG, J., MCINTYRE, D. J. AND ADAMS, P. J., 1996, Paleoenvironmental changes in the Cretaceous (Albian to Turonian) Colorado Group of western Canada: Microfossil, sedimentological and geochemical evidence: Cretaceous Research, v. 17, p. 311-366.

SCOTT, R. W., 1970, Stratigraphy and sedimentary environments of Lower Cretaceous rocks, southern Western Interior: American Association Petroleum Geologists Bulletin, v. 54, p. 1225-1244.

SCOTT, R. W., 1992a, High-resolution correlation of Mid-Cretaceous sequences in the Gulf Coast and Western Interior (abst.): American Association of Petroleum Geologists Annual Convention Program, p. 116.

SCOTT, R. W., 1992b, Are Seismic/depositional sequences chronostratigraphic units? (abst.): Fifth North American Paleontological Convention, Abstracts and Program, The Paleontological Society Special Publication No. 6, p. 264.

SCOTT, R. W. AND FRANKS, P. C., 1988, The Kiowa-Dakota transgressive-regressive cycle (Late Albian to Early Cenomanian), Kansas (abst.): Program and Abstracts, American Association for the Advancement of Science, Southwestern and Rocky Mountain Division, 64th Annual Meeting, p. 17.

SCOTT, R. W., EVETTS, M. J., AND STEIN, J. A., 1993, Are seismic/depositional sequences time units? Testing by SHADS cores and graphic correlation: 25th Annual Offshore Technology Conference, OTC 7110, p. 269-276.

SCOTT, R. W., BERGEN, J. A., EVETTS, M. J., STEIN, J. A., KIDSON, E. J., AND FRANKS, P. C., 1994a, Chronostratigraphic calibration of Mid-Cretaceous depositional cycles in the Western Interior by graphic correlation (abst.): American Association of Petroleum Geologists Annual Convention Program, p. 254-256.

SCOTT, R. W., FRANKS, P. C., STEIN, J. A., BERGEN, J. A., AND EVETTS, M. J., 1994b, Graphic correlation tests the synchronous Mid-Cretaceous depositional cycles: Western Interior and Gulf Coast *in* Dolson, J. C. et al., eds., Unconformity-related hydrocarbons in sedimentary sequences: Denver, Rocky Mountain Association of Geologists, p. 89-98.

STOOPS, G., 1983, SEM and light microscopic observations of minerals in bog-ores of the Belgian Campine: Geoderma, V. 30, p. 179-186.

STOOPS, G., AND ESWARAN, H., 1985, Morphological characteristics of wet soils, *in* Wetland Soils: Characterization, Classification, and Utilization: Proceedings of a Workshop held under the joint sponsorship of the International Rice Research Institute, U. S. Dept. Agriculture, and Philippine Ministry of Agriculture, Manila, Philippines, p. 177-189.

VAIL, P. R., AUDEMARD, F., BOWMAN, S. A., EISNER, P. N., AND PEREZ-CRUZ, C., 1991, The stratigraphic signatures of tectonics, eustasy and sedimentology - an overview *in* Einsele, G. et al., eds., Cycles and Events in Stratigraphy: Springer-Verlag, Berlin, p. 617-659.

WATKINS, D. K., BRALOWER, T. J., COVINGTON, J. M., AND FISHER, C. G., 1993, Biostratigraphy and paleoecology of the Upper Cretaceous calcareous nannofossils in the Western Interior Basin, North America *in* Caldwell, W. G. E., and Kauffman, E. G. eds., Evolution of the Western Interior Basin: Waterloo, Geological Association of Canada Special Paper 39, p. 521-537.

WEIMER, R. J., 1984, Relation of unconformities, tectonics, and sea-level changes, Cretaceous of Western Interior, U.S.A. *in* Schlee, J. S., ed., Interregional Unconformities and Hydrocarbon Accumulation: American Association of Petroleum Geologists, Memoir 36, p.7-35.

WILLIAMS, G. D. AND STELCK, C. R., 1975, Speculations on the Cretaceous palaeogeography of North America *in* Caldwell, W. G. E., ed., The Cretaceous System in the Western Interior of North America: Waterloo, Geological Assocaiation of Canada Special Paper No. 13, p. 1-20.

UPPER CRETACEOUS NANNOFOSSIL ASSEMBLAGES ACROSS THE WESTERN INTERIOR SEAWAY: IMPLICATIONS FOR THE ORIGINS OF LITHOLOGIC CYCLES IN THE GREENHORN AND NIOBRARA FORMATIONS

CELESTE E. BURNS AND TIMOTHY J. BRALOWER

Department of Geology, University of North Carolina, Chapel Hill, North Carolina 27599-3315

ABSTRACT: Calcareous nannofossil assemblages were investigated in two cyclic stratigraphic intervals from the Late Cretaceous Western Interior Seaway to determine the causes of lithologic variations. Relative abundance data were collected from the Bridge Creek Limestone Member (Cenomanian-Turonian) of the Greenhorn Formation and the transition between the Fort Hays Limestone and Smoky Hill Chalk Members (Coniacian-Santonian) of the Niobrara Formation from two cores, the No. 1 Amoco Rebecca Bounds Core, western Kansas, and the USGS No. 1 Portland Core, central Colorado. Stable-carbon and -oxygen isotopic analyses of fine ($< 38 \mu m$) fractions were carried out on samples from the interval with the best preserved nannofossils, the Fort Hays Limestone-Smoky Hill Chalk transition in the Bounds core.

Preservation of nannofossil assemblages varies in the two cores. Correlations between $CaCO_3$ and the abundance of nannofossil species, which are used as proxies for organic productivity, are rarely significant, which indicates an inconsistent relationship between fertility and lithology. Fine-fraction, oxygen-isotopic values from the Fort Hays Limestone-Smoky Hill Chalk transition of the Bounds core correlate highly with nannofossil fertility markers, indicating a close relationship between organic productivity and the amount of run-off into the basin. However, the lack of consistent correlation between organic productivity (as seen in nannofossil assemblages) and lithology suggests that lithology was influenced by a variety of complex processes including variations in carbonate productivity and dilution with clastic material. Carbonate productivity was likely influenced by multiple water masses, including run-off from mountainous regions to the west, warm waters from the Tethys, and cooler waters from the Arctic. The interplay of the water masses and the additional dilution signal render the lithologic cycles out of phase with the periodicities that control organic productivity.

INTRODUCTION

The Origin of Upper Cretaceous Lithologic Cycles

Well-known for its rhythmically bedded sedimentary deposits, the Western Interior Seaway is one of the most intensively studied ancient marine basins (e.g., Hattin, 1971; Kauffman, 1977; Pratt et al., 1985; Caldwell and Kauffman, 1993). As an epicontinental arm of the Tethys Ocean, it extended from the Gulf of Mexico to the Arctic during maximum transgressions. The Bridge Creek Limestone Member of the Greenhorn Formation (Cenomanian-Turonian) and the Fort Hays Limestone and Smoky Hill Chalk Members of the Niobrara Formation (Coniacian-Santonian) were deposited during major transgressive events (Fig. 1) and contain some of the most marked cyclicities. The limestone-marlstone cycles of the Bridge Creek Limestone were extensively studied (e.g., Pratt et al., 1985), but less is known of lithologically similar cycles in the Fort Hays Limestone and Smoky Hill Chalk. In both units, cycles reflect orbitally controlled climatic fluctuations (e.g., Fischer, 1980; Pratt et al., 1993; Dean and Arthur, this volume; Sageman et al., this volume). However, the exact mechanisms through which climate cycles affected sedimentation in both the Greenhorn and Niobrara Formations remain debatable.

Two models are proposed for the origin of Bridge Creek Limestone bedding rhythms in the Western Interior Seaway. One explains the cyclicities as a result of fluctuations of river input to the seaway that diluted the sea with fresh water (Fischer, 1980; Pratt 1981, 1984; Arthur et al., 1985; Barron et al., 1985; Sageman, 1985; ROCC Group, 1986). During wet climatic periods, increased terrigenous run-off brought nutrient-rich fresh waters to the seaway that caused higher productivity of organic material (hereafter termed fertility), water column stratification, and oxygen-depleted deep waters. The result was accumulation of relatively organic-rich shales or marlstones in these intervals. Drier periods, on the other hand, led to decreased run-off, a well-mixed water column, and deposition of carbonate-rich sediments (limestone precursors) in an oxygenated environment.

The other model suggests that fluctuations in carbonate productivity, caused by variations in mixing intensity regulated by the oceanography of the Tethys, led to the origin of the limestone-marlstone cycles (Eicher and Diner, 1985, 1989). Studies of planktonic foraminiferal assemblages and other data suggest that marlstones accumulated during periods of water column stratification that decreased surface-water carbonate productivity. In contrast, limestones formed during drier climatic periods when ocean circulation improved and vertical mixing was vigorous.

In both models different lithologies reflect variable surface-water productivity. However, the processes that control variations in productivity and the effects on that productivity are different. In the dilution model, limestones form because of a decrease in fertility. For the productivity model, on the other hand, limestones are deposited during periods of higher carbonate productivity (i.e., calcareous nannofossil and foraminifera) resulting from increased vertical mixing. Thus, the effect of the two models on organic productivity may not be that different. High carbonate productivity does not necessarily indicate high organic productivity in surface water.

Calcareous Nannofossils

Research on the ecology and biogeography of modern calcareous nannoplankton shows that most species thrive in oligotrophic conditions (e.g., Honjo and Okada, 1974). Particular nannoplankton taxa are associated with latitudinal bands in the South Pacific (Burns, 1973) and Atlantic (McIntyre and Bé, 1967) Oceans and are predominantly controlled by temperature. Some modern nannoplankton species exist in low- or high-salinity environments such as estuaries and lagoons, but most species live in open oceans where salinity is lower (e.g., Bukry, 1974).

Research on nannofossil paleobiogeography over the past 20 years provides a strong basis for interpreting nannofossil assemblages as a record of paleoceanographic and/or paleoecologic changes (e.g., Roth and Bowdler, 1981; Roth and Krumbach, 1986; Erba et al., 1992). These studies show that it is possible

Stratigraphy and Paleoenvironments of the Cretaceous Western Interior Seaway, USA, SEPM Concepts in Sedimentology and Paleontology No. 6
Copyright 1998 SEPM (Society for Sedimentary Geology), ISBN 1-56576-044-1, p. 35-58.

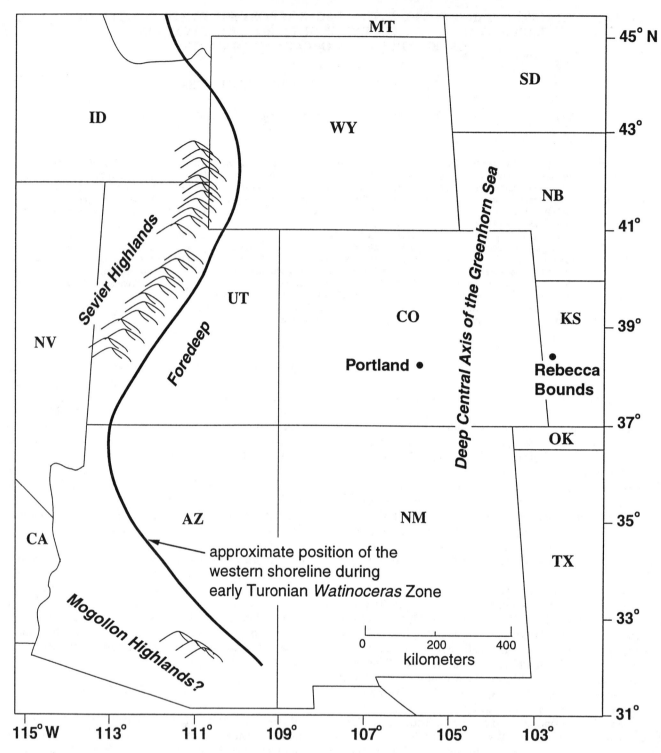

Fig. 1—Map of the Western Interior Seaway in the Late Cretaceous showing the locations of the Portland and Bounds cores.

to link a particular nannofossil species with specific conditions of fertility, temperature, and salinity. High abundances of *Biscutum constans* and *Zygodiscus erectus* are associated with areas of high surface-water fertility (Roth and Bowdler, 1981; Roth and Krumbach, 1986; Premoli-Silva et al., 1989; Watkins, 1989; Erba et al., 1992). *Watznaueria barnesae*, the most abundant species in most Cretaceous sections, is controlled by preservation (e.g., Roth and Bowdler, 1981), but more recently is also affiliated with low-fertility settings (Roth and Krumbach, 1986;

Roth, 1989; Premoli Silva et al., 1989). In moderately to well-preserved assemblages *W. barnesae* often negatively correlates with *B. constans* (Erba et al., 1992; Williams and Bralower, 1995).

Nannofossil assemblages are used to interpret fluctuations in surface-water fertility in the Bridge Creek Limestone. In one cycle unit of the Bridge Creek, Watkins (1989) found that taxa indicative of high surface-water fertility (e.g., *Zygodiscus erectus* and *Biscutum constans*) are more abundant in the marlstone than

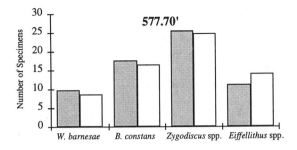

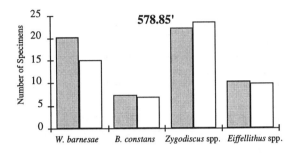

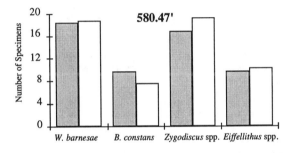

Fig. 2.—Comparison of counts of 200 (gray) and 300 (white) specimens of four species or groups of species in three different samples from the Fort Hays Limestone-Smoky Hill Chalk transition in the Bounds core.

in the limestone member. This conclusion tends to agree better with the dilution model than the productivity model. However, the dilution model does not account for near normal surface-water salinities that are also indicated by nannofossil assemblages in both limestones and marlstones (Watkins, 1989). Limestone beds were found to have higher Shannon-Weaver diversities than marlstone beds, showing that assemblages in marlstone units may have been dominated by a few taxa (Watkins, 1989). For this one studied cycle correlations suggested that nannofossil assemblages and lithology respond to the same mechanism: changes in fertility.

As part of the Cretaceous Western Interior Seaway Scientific Drilling Project, this investigation of calcareous nannofossil assemblages examines correlations between nannofossil preservation, abundances of fertility markers, and lithology in multiple cycles of the Greenhorn and Niobrara Formations.

Stable Isotopes

As indicators of temperature and salinity ($\delta^{18}O$ values) and fertility changes ($\delta^{13}C$ values), stable isotopic studies, in conjunction with assemblage data, may potentially shed light on the paleoceanography of the Western Interior Seaway. Stable iso-

topes provide a record of likely changes in circulation, freshwater influx, and nutrients. Fine-fraction (< 38 μm, dominantly the remains of nannoplankton) isotopic values can be compared to values from whole rock and benthic and planktonic fossils to determine how fertility, temperature, and/or salinity varied through the water column. Potential problems with using fine fractions include the polyspecific nature of the sample such that vital effects may obscure paleoceanographic trends (e.g., Paull and Thierstein, 1987). Diagenetic alteration can also change the original isotopic ratios.

In this investigation quantitative studies of nannofossil assemblages are complemented by measurements of fine-fraction oxygen and carbon isotopes from the Fort Hays-Smoky Hill transition to further explain the relationship between lithology and surface-water environments.

MATERIALS

Samples were obtained from two cores with well-defined lithologic variations, the Amoco Bounds core (western Kansas) and the Portland core (near Cañon City, Colorado) (Fig. 1). In both cores nannofossil assemblages were investigated in the Bridge Creek Limestone Member (Cenomanian-Turonian) of the Greenhorn Formation and the transition between the Fort Hays Limestone and Smoky Hill Chalk Members (Coniacian-Santonian) of the Niobrara Formation. The lithologies were described in detail (Dean et al., 1995; Savrda, this volume) and consist predominantly of well-consolidated marlstones and limestones. In this discussion the term marlstone to refers to fine-grained rocks with 30-60% CaCO₃, limestone refers to fine-grained rocks with greater that 60% CaCO₃, and shale describes rocks with less than 30% CaCO₃. Individual cycles generally range from several centimeters to several decimeters thick. Carbonate percentages of samples from which smear slides were made are available for all but the Fort Hays Limestone-Smoky Hill Chalk transition in the Portland core (Dean et al., 1995; Sageman et al., 1997). Carbonate data for the Bounds core were obtained by coulometry and appear in Dean et al. (1995). For the Bridge Creek interval in Portland core, digital carbonate data are courtesy of Sageman (unpublished).

METHODS

Assemblages of calcareous nannofossil were studied in 320 samples from the Fort Hays Limestone-Smoky Hill Chalk transition in the Bounds and Portland cores. Assemblages in 290 samples were studied from the Bridge Creek Limestone in the Bounds and Portland cores. Sampling intervals are varied to accommodate both large- and small-scale lithologic cyclicities. To insure that coccoliths are not damaged during slide preparation, samples are disaggregated in water whenever possible instead of being broken up by mechanical means. In a few limestone samples we used the technique of Monechi and Thierstein (1985) to disaggregate samples. Smear slides are made from well-mixed, unsettled sediment solution so that the nannofossils mounted on the slides accurately represent the samples.

Nannofossil specimens were observed under a light microscope at a magnification of 1250x and identified using standard taxonomy as described in Perch-Nielsen (1985). All specimens one-half the size of a coccolith or larger are identified: an average of approximately 220 specimens are counted in each sample. Counts of 200 and 300 specimens on separate smear slides from

TABLE 1.—NANNOFOSSIL TAXONOMY

Watznaueria barnesae (Black, 1959) Perch-Nielson, 1968

Sollasites barringtonensis Black, 1967

Markalius circumradiatus (Stover, 1966) Perch-Nielsen, 1968

Biscutum constans (Gorka, 1957) Black, 1959

Serisbiscutum ehrenbergii (Arkhangelsky, 1912) Deflandre, 1952

Scapholithus fossilis Deflandre in Deflandre & Fert, 1954

Marthasterites furcatus (Deflandre in Deflandre & Fert, 1954) Deflandre, 1959

Cylindralithus gallicus Strandner, 1963

Lithrastrinus grillii Stradner, 1962

Reinhardtites levis Prins & Sissingh in Sissingh, 1977

Rotellapillus laffiitei Noël, 1957

Chiastozygus litterarius (Gorka, 1957) Manivit, 1971

Kamptnerius magnificus Deflandre, 1959

Cyclagelosphaera margerelii Noël, 1965

Manivitella pemmatoidea (Manivit, 1965) Thierstein, 1971

Discorhabdus rotatorius (Bukry, 1969) Theirstein, 1973

Micula staurophora (Gardet, 1955) Stradner, 1963

Tegumentum stradneri Thierstein, 1972

Broinsonia spp.
    Bronsonia enormis (Shumenko, 1968) Manivit, 1971
    Broinsonia signata (Noël, 1969) Noël, 1970

Corollithion spp.
    Corollithion acutum Thierstein in Roth and Thierstein, 1972
    Corollithion ellipticum Bukry, 1969
    Corollithion exiguum Stradner, 1961
    Corollithion kennedyi Crux, 1981
    Corollithion signum Stradner, 1963

Cretarhabdus spp.
    Cretarhabdus angustiforatus (Black, 1971) Bukry, 1973
    Cretarhabdus conicus Bramlette & Martini, 1964
    Cretarhabdus surirellius (Deflandre, 1954) Reinhardt, 1970

Eiffellithus spp.

Eiffellithus eximius (Stover, 1966) Perch-Nielsen, 1968
Eiffellithus primus Applegate and Bergen, 1989
Eiffellithus trabeculatus (Gorka, 1957) Reinhardt and Gorka, 1967
Eiffellithus turriseiffelii (Deflandre, 1954) Reinhardt, 1965

Eprolithus spp.
    Eprolithus floralis (Stradner, 1962) Stover, 1966
    Eprolithus moratus Stover, 1966

Lithraphidites spp.
    Lithraphidites acutum Manivit et al., 1977
    Lithraphidites carniolensis Deflandre, 1963

Parhabdolithus spp.
    Parhabdolithus achlyostaurion Hill, 1976
    Parhabdolithus embergeri (Noël, 1958) Bralower, 1989
    Parhabdolithus regularis Gartner, 1968

Axopodorhabdus spp.
    Axopodorhabdus albianus (Black, 1967) Wind & Wise in Wise & Wind, 1977
    Axopodorhabdus dietzamannii (Reinhardt, 1965) Wind & Wise in Wise, 1983

Prediscosphaera spp.
    Prediscosphaera cretacea (Arkhangelsky, 1912) Gartner, 1968
    Prediscosphaera spinosa (Bramlette & Martini, 1964) Gartner, 1968

Rhagodiscus spp.
    Rhagodiscus angustus (Stradner, 1963) Stradner et al., 1968
    Rhagodiscus splendens (Deflandre, 1953) Noël, 1969

Tranolithus spp.
    Tranolithus gabalus Stover, 1966
    Tranolithus orionatus (Reinhardt, 1966) Perch-Nielsen, 1968

Vagalapilla spp.
    Vagalapilla mutterlosei Crux, 1989
    Vagalapilla octoradiata (Gorka, 1957) Bukry, 1969
    Vagalapilla stradneri (Rood et al., 1971) Thierstein, 1970

Zygodiscus spp.
    Zygodiscus diplogrammus (Deflandre and Fert, 1954) Gartner, 1968
    Zygodiscus elegans (Gartner, 1968) Bukry, 1969
    Zygodiscus erectus (Deflandre, 1959) Bralower, Monechi, & Theirstein, 1989
    Zygodiscus spiralis Bramlette and Martini, 1964
    all other species with a long diameter less than 5 μm

the same sample were made for three samples and show little difference (Fig. 2), indicating that 200 specimens provide a statistically representative sample of the assemblage. The species richness of a sample as a measure of diversity was determined by counting the number of species identified in each sample.

Though 55 species were identified and recorded, only seven species or groups of species were chosen for detailed paleoecological analysis. Species were grouped on the basis of taxonomic and paleoecological affinities to enhance trends in the data. For example, *Zygodiscus* spp. includes *Z. diplogrammus, Z. elegans, Z. erectus, Z. spiralis,* and all other species with a long diameter less than 5 μm. Other groupings include *Eiffellithus* spp., *Cretarhabdus* spp., *Prediscosphaera* spp., and *Tranolithus* spp. (Table 1).

Fine fraction oxygen and carbon isotopic analyses were carried out on 158 samples with well-preserved nannofossils from the Fort Hays Limestone-Smoky Hill Chalk transition from the Bounds core. The 38 μm fraction was separated by sieving with pH-buffered water and then dried in a 40°C oven overnight. Samples were then roasted in a vacuum at 320°C for 1 hour to remove volatile organic carbon. Samples were reacted with phosphoric acid in a Kiel automated carbonate extraction device at 90°C and analyzed with a Finnegan-Mat 251 isotope-ratio mass spectrometer at North Carolina State University. Results are expressed as delta per mil (δ, ‰) deviation from the University of Chicago Pee Dee Belemnite (PDB) marine carbonate standard. The estimated analytical precision is ±0.10‰ for $\delta^{18}O$ and ±0.05‰ for $\delta^{13}C$ based on replicate analysis of a Carrara Marble internal standard and the NBS-20 calcite standard. Internal sampling precision for $\delta^{18}O$ and $\delta^{13}C$ was estimated from 18 duplicate measurements (approximately 11% of the samples run); the estimated precision for $\delta^{18}O$ is ±0.07‰ and for $\delta^{13}C$ it is ±0.03‰. Abundance and percentage data for the four investigated sequences are archived at the National Geophysical Data Center.

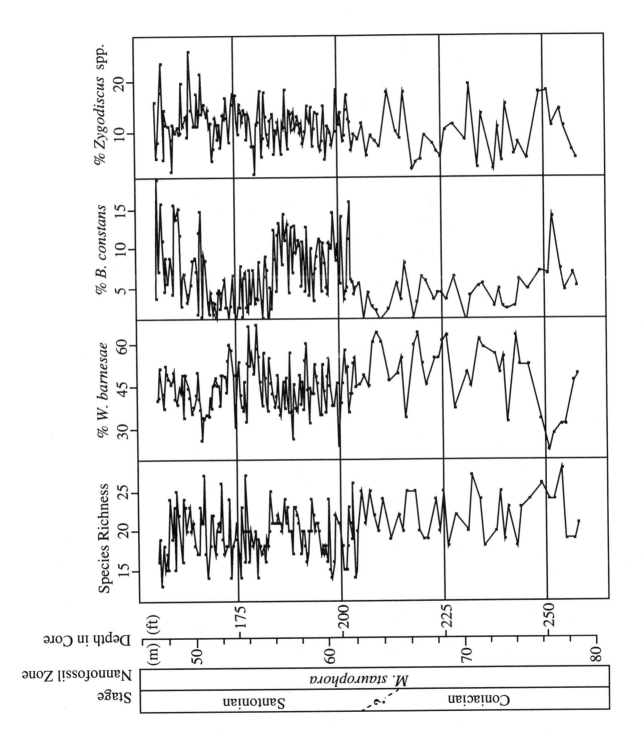

FIG. 3.—Percentages of nannofossil taxa and species richness in samples from the Fort Hays Limestone-Smoky Hill Chalk transition in the Portland core for (A) %*W. barnesae*, %*B. constans*, %*Zygodiscus* spp.; (B) (see following page) %*Eiffellithus* spp., %*Cretarhabdus* spp., %*Prediscosphaera* spp., %*Tranolithus* spp.

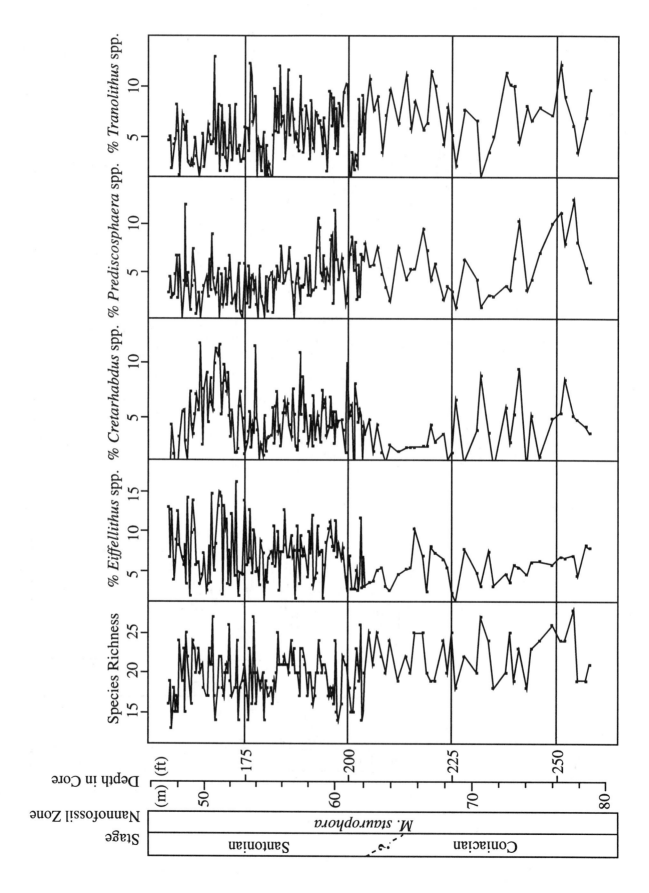

FIG. 3. (Continued)

A.

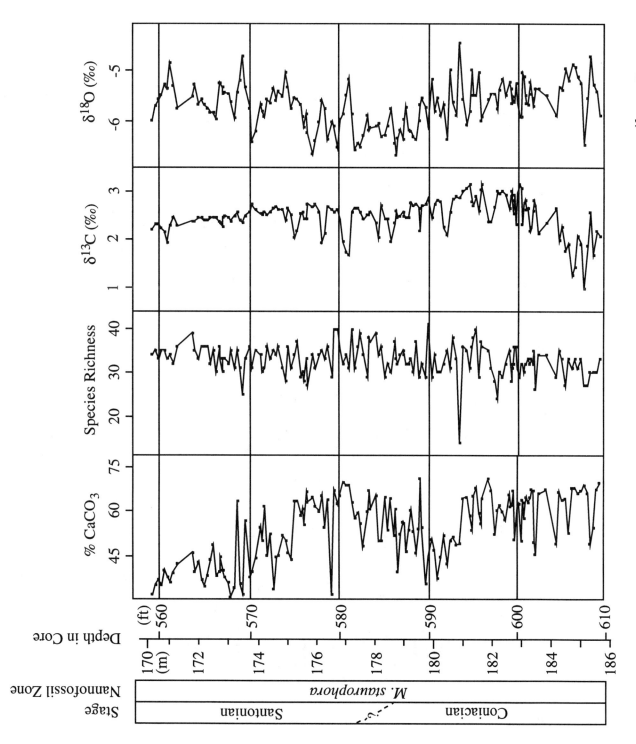

FIG. 4.—Carbonate percentages, species richness, and isotope data in samples from the Fort Hays Limestone-Smoky Hill Chalk transition in the Bounds core for (A) δ^{13}C(‰), δ^{18}O(‰); (B) %W. barnesae, %B. constans, %Zygodiscus spp.; (C) %Eiffellithus spp., %Cretarhabdus spp., %Prediscosphaera spp., %Tranolithus spp.

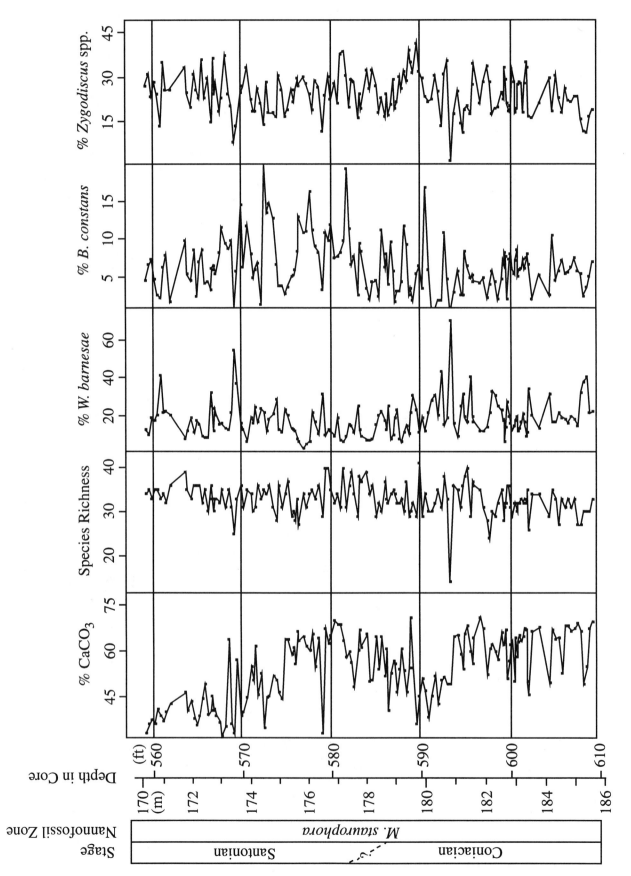

Fig. 4. (Continued)

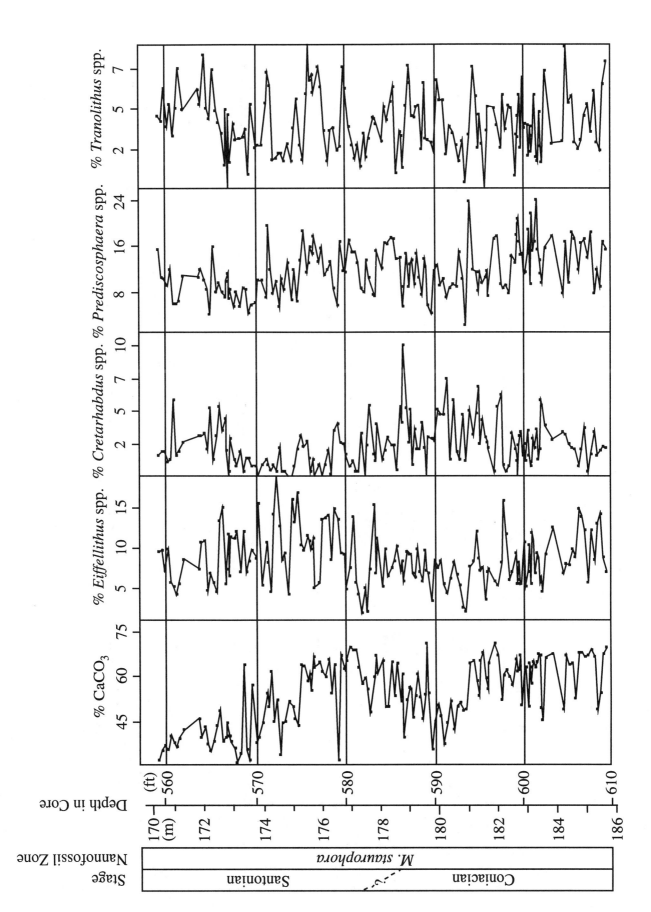

Fig. 4. (Continued)

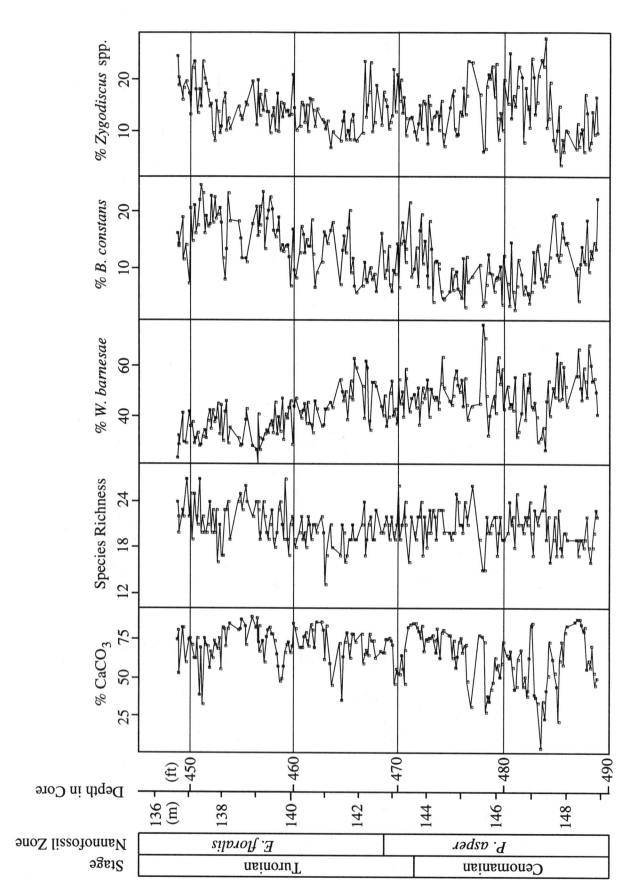

Fig. 5.—(A) Carbonate percentages, species richness, and percentages of nannofossil taxa in samples from the Bridge Creek Limestone in the Portland core for (A) %W. barnesae, %B. constans, %Zygodiscus spp.; (B) %Eiffellithus spp., %Cretarhabdus spp., %Prediscosphaera spp., %Tranolithus spp.

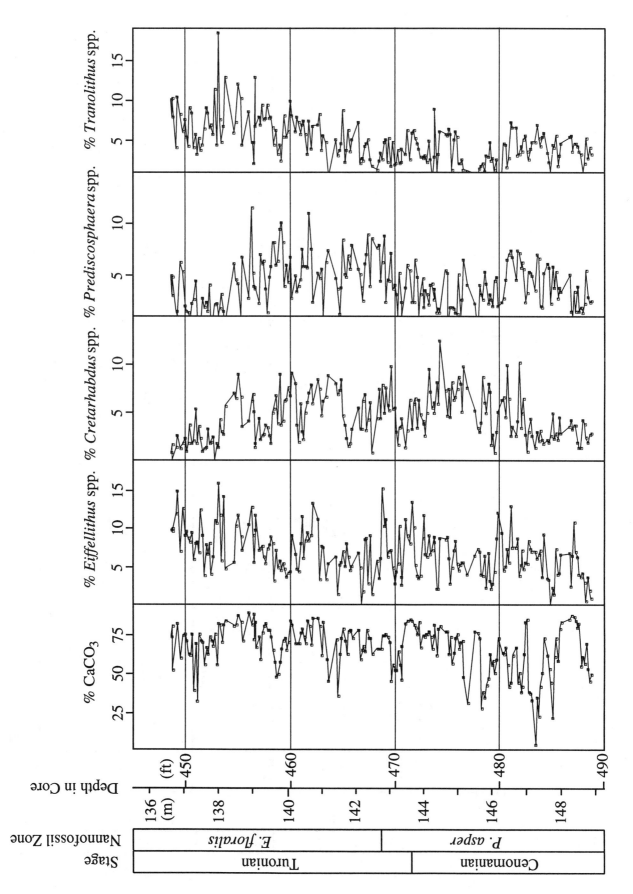

B.

Fig. 5. (Continued)

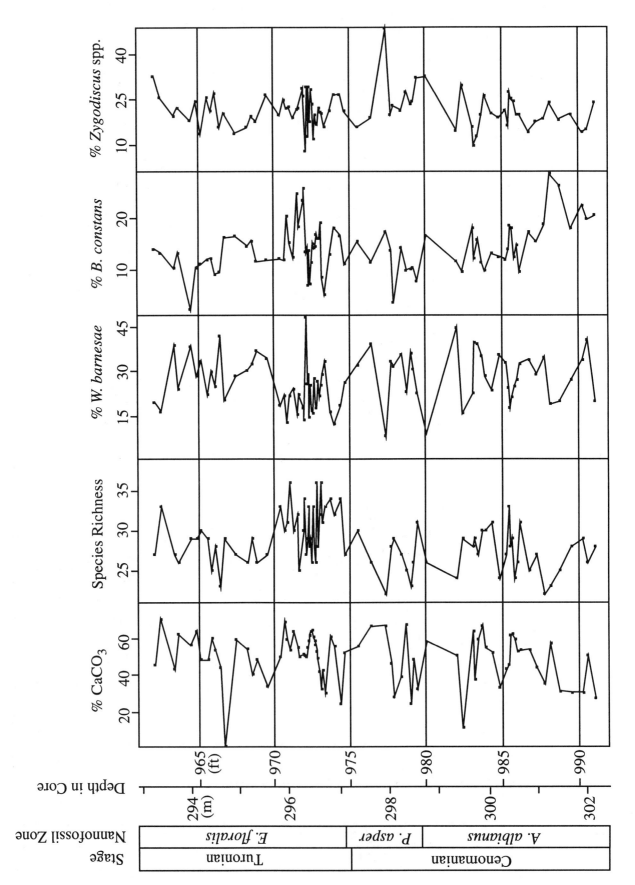

Fig. 6.—Carbonate percentages, species richness, and percentages of nannofossil taxa in samples from the Bridge Creek Limestone in the Bounds core for (A) %*W. barnesae*, %*B. constans*, %*Zygodiscus* spp.; (B) %*Eiffellithus* spp., %*Cretarhabdus* spp., %*Prediscosphaera* spp., %*Tranolithus* spp.

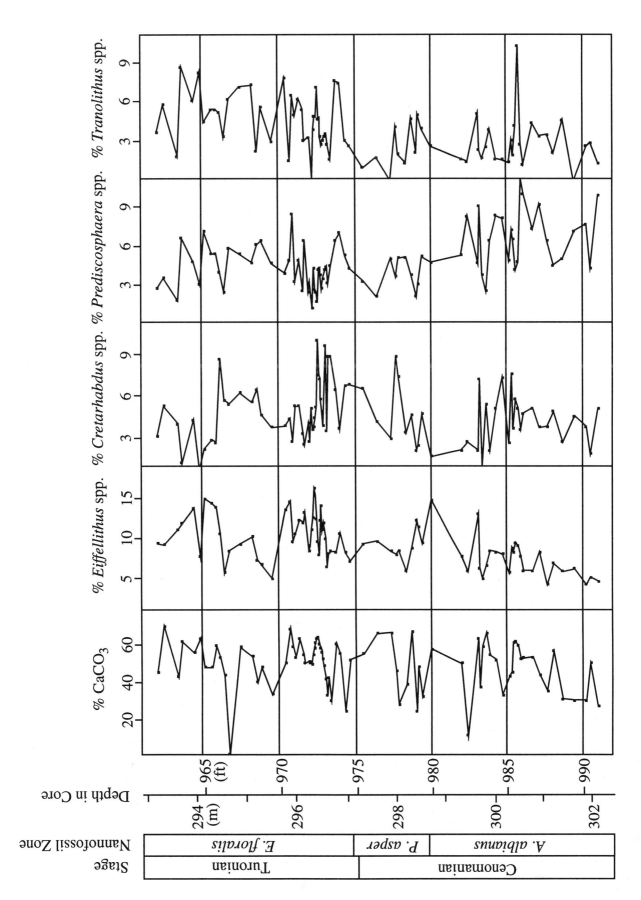

Fig. 6. (Continued)

CELESTE E. BURNS AND TIMOTHY J. BRALOWER

TABLE 2.-FORT HAYS LIMESTONE-SMOKY HILL CHALK TRANSITION IN PORTLAND CORE[1]

	Species Richness	*W. barnesae*	*B. constans*	*Zygodiscus* spp.	*Eiffellithus* spp.	*Cretarhabdus* spp.	*Prediscosphaera* spp.
Species Richness	1.00						
W. barnesae	*-0.18*	1.00					
B. constans	-0.05	**-0.50**	1.00				
Zygodiscus spp.	0.06	**-0.57**	0.17	1.00			
Eiffellithus spp.	0.10	**-0.34**	0.03	0.00	1.00		
Cretarhabdus spp.	0.02	**-0.34**	0.04	0.11	0.02	1.00	
Prediscosphaera spp.	*0.16*	**-0.29**	0.03	-0.02	0.10	*-0.13*	1.00
Tranolithus spp.	**0.20**	**-0.20**	0.12	*-0.14*	-0.03	-0.06	*0.17*

Italic type indicates values that are significant at the 95% confidence level.
Bold type indicates values that are significant at the 99% confidence level.
[1]Values given are Pearson coefficients.

TABLE 3.-FORT HAYS LIMESTONE-SMOKY HILL CHALK TRANSITION IN BOUNDS CORE[1]

	Species Richness	$CaCO_3$	$\delta^{13}C$	$\delta^{18}O$	*W. barnesae*	*B. constans*	*Zygodiscus* spp.	*Eiffellithus* spp.	*Cretarhabdus* spp.	*Prediscosphaera* spp.
Species Richness	1.00									
$CaCO_3$	0.00	1.00								
$\delta^{13}C$	0.04	-0.12	1.00							
$\delta^{18}O$	*-0.28*	-0.07	*-0.21*	1.00						
W. barnesae	**-0.46**	-0.19	-0.04	**0.57**	1.00					
B. constans	**0.24**	0.04	0.05	*-0.23*	**-0.43**	1.00				
Zygodiscus spp.	**0.21**	-0.01	0.11	**-0.37**	**-0.65**	*0.15*	1.00			
Eiffellithus spp.	-0.05	0.05	*-0.21*	0.06	-0.08	-0.06	*-0.20*	1.00		
Cretarhabdus spp.	0.10	*-0.15*	0.11	*-0.13*	0.08	**-0.27**	-0.11	*-0.16*	1.00	
Prediscosphaera spp.	0.09	**0.56**	*-0.18*	-0.05	**-0.38**	0.00	0.04	0.02	-0.11	1.00
Tranolithus spp.	0.03	*0.15*	0.01	**-0.22**	**-0.30**	-0.03	0.07	-0.02	-0.04	*0.19*

Italic type indicates values that are significant at the 95% confidence level.
Bold type indicates values that are significant at the 99% confidence level.
[1]Values given are Pearson coefficients.

RESULTS

Percentages of different nannofossil taxa, carbonate percentages, and species richness values for the four studied intervals are shown in Figures 3-6. There are considerably different *W. barnesae* percentages in the Fort Hays Limestone-Smoky Hill Chalk transition in the two studied cores. In samples from the Portland core, *W. barnesae* ranges from 23% to 67% with a mean of 46% [SD=9%, n=162] (Fig. 3A). In samples from the Bounds core *W. barnesae* ranges from 2% to 70% with a mean of 18% [SD=9%, n=158] (Fig. 4A). *W. barnesae* in the Bridge Creek Limestone of the Rebecca Bounds ranges from 9% to 48% with a mean of 27% [SD=8%, n=81] (Fig. 6A). Nannofossil assemblages in this unit in the Portland core are poorly preserved and have higher percentages of *W. barnesae*, ranging from 21% to

77% with a mean of 44% [SD=9%, n=209] (Fig. 5A). No other taxon shows such systematic differences between the two cores.

To subtract any possible preservational bias associated with high abundances of *W. barnesae*, percentages of all other species were calculated as percentages of the total counted specimens excluding *W. barnesae*. The sum of the percentages of these taxa without percentages of *W. barnesae* are also examined to determine the dominance of taxa in the overall assemblage. These "non-*W. barnesae*" percentages do not show any trends significantly different from the total percentages and thus are not presented. Nannofossil percentages are values from the total number of specimens counted in a particular sample.

Pearson correlation coefficients between taxon pairs and taxa and other variables were calculated for each of the two sections from both cores and range from 0.57 to -0.65 (Tables 2-5). We

TABLE 4.-BRIDGE CREEK IN PORTLAND CORE[1]

	Species Richness	CaCO$_3$	W. barnesae	B. constans	Zygodiscus spp.	Eiffellithus spp.	Cretarhabdus spp.	Prediscosphaera spp.
Species Richness	1.00							
CaCO$_3$	*-0.22*	1.00						
W. barnesae	*-0.37*	-0.03	1.00					
B. constans	0.06	*0.13*	**-0.56**	1.00				
Zygodiscus spp.	*0.35*	*-0.28*	**-0.58**	-0.02	1.00			
Eiffellithus spp.	0.02	*0.35*	**-0.46**	0.10	0.12	1.00		
Cretarhabdus spp.	0.05	0.04	0.05	*-0.37*	-0.05	-0.06	1.00	
Prediscosphaera spp.	0.06	*-0.12*	**-0.20**	-0.05	0.05	-0.01	*0.13*	1.00
Tranolithus spp.	0.07	*0.39*	**-0.52**	**0.29**	0.05	*0.39*	*-0.19*	-0.06

Italic type indicates values that are significant at the 95% confidence level.
Bold type indicates values that are significant at the 99% confidence level.
[1]Values given are Pearson coefficients.

TABLE 5.-BRIDGE CREEK IN BOUNDS CORE[1]

	Species Richness	CaCO$_3$	W. barnesae	B. constans	Zygodiscus spp.	Eiffellithus spp.	Cretrahabdus spp.	Prediscosphaera spp.
Species Richness	1.00							
CaCO$_3$	-0.02	1.00						
W. barnesae	*-0.33*	-0.17	1.00					
B. constans	-0.02	-0.02	**-0.39**	1.00				
Zygodiscus spp.	-0.01	0.08	**-0.65**	0.06	1.00			
Eiffellithus spp.	0.14	*0.40*	**-0.27**	-0.23	0.07	1.00		
Cretarhabdus spp.	*0.23*	-0.09	0.01	-0.22	-0.13	-0.11	1.00	
Prediscosphaera spp.	-0.03	*-0.28*	-0.02	0.10	-0.12	*-0.32*	-0.07	1.00
Tranolithus spp.	0.10	*0.30*	*0.26*	0.06	-0.03	*0.30*	-0.01	-0.08

Italic type indicates values that are significant at the 95% confidence level.
Bold type indicates values that are significant at the 99% confidence level.
[1]Values given are Pearson coefficients.

use the term "highly significant" in this discussion to refer to a correlation coefficient that is significant at the 99% confidence interval. The term "significant" refers to correlation coefficients that are significant at the 95% interval of confidence. Because correlations change markedly over a section, mean values are rarely representative. For example, percentages of *W. barnesae* and CaCO$_3$ appear to correlate well in the lower 3.1 m (10 ft) of the Bridge Creek Limestone of the Bounds core (Fig. 6A), although the correlation coefficient for the entire section is only -0.17.

Moving correlation curves provide a means of identifying specific intervals in the studied sections where two variables are significantly correlated at the 95% confidence interval. Plots of moving correlation curves for all data are illustrated in Burns (1995). A summary of all of the intervals in which the moving correlations are significant is given in Tables 6-9. For example, *W. barnesae* and *B. constans* have significant correlation coefficients in several intervals in the Fort Hays Limestone-Smoky

Hill Chalk transition of the Bounds core (Table 6), although the overall correlation coefficient is 0.43 (Table 3). Another example of the effect of moving correlation analysis is in the Bridge Creek Limestone in the Bounds core: there is a sharp change in the correlation between *Zygodiscus* spp. and *W. barnesae* in the top 4.6 m (15 ft) of the section (Table 7).

Because carbonate data are not available for the Fort Hays Limestone-Smoky Hill Chalk transition of the Portland core, color photographs of the core intervals were examined for possible qualitative correlations between nannofossil taxa and lithology. There is no noticeable correlation between lithology and *W. barnesae* or *Zygodiscus* spp. However, there may be a correlation between lithology and *B. constans* such that higher values of this taxon correlate preferentially with limestone.

To investigate relationships between variables that cannot be seen in plots of nannofossil assemblage data versus depth or in correlation analysis, Principal Components Analysis (PCA) was carried out. We utilize SYSTAT statistical software to ana-

TABLE 6.-FORT HAYS LIMESTONE-SMOKY HILL CHALK TRANSITION[1]

PORTLAND CORE	BOUNDS CORE
W. barnesae vs. *B. constans*	
48.5 m - 49.1 m (159 ft - 161 ft) (-)	172.8 m (567 ft) (-)
51.8 m (170 ft) (-)	173.1 m - 174.3 m (568 ft - 572 ft) (-)
54.9 m (180 ft) (-)	175.6 m (575 ft) (-)
56.1 m - 58.2 m (184 ft - 191 ft) (-)	179.8 m - 175.6 m (590 ft - 592 ft) (-)
59.1 m - 61.0 m (194 ft - 197 ft) (-)	182.0 m (597 ft) (+)
61.9 m - 63.1 m (203 ft - 207 ft) (-)	182.9 m (600 ft) (-)
64.3 m - 68.0 m (211 ft - 222 ft) (-)	183.5 m (602 ft)
74.4 m (244 ft)	
W. barnesae vs. *Zygodiscus* spp.	
48.5 m - 48.8 m (159 ft - 160 ft) (-)	172.2 m - 173.7 m (565 ft - 570 ft) (-)
49.1 m - 50.3 m (161 ft - 165 ft) (-)	174.0 m - 174.3 m (571 ft - 572 ft) (-)
51.8 m (170 ft) (-)	174.7 m - 174.9 m (573 ft - 574 ft) (-)
54.6 m - 56.1 m (179 ft - 184 ft) (-)	176.2 m - 177.4 m (578 ft - 582 ft) (-)
57.9 m - 59.5 m (190 ft - 195 ft) (-)	178.0 m - 178.6 m (584 ft - 586 ft) (-)
60.1 m - 61.6 m (199 ft - 202 ft) (-)	179.5 m (589 ft) (+)
62.2 m - 62.8 m (204 ft - 206 ft) (-)	182.9 m (600 ft) (-)
64.0 m - 69.5 m (210 ft - 228 ft) (-)	184.4 m (605 ft) (+)
B. constans vs. *Zygodiscus* spp.	
48.8 m (160 ft) (+)	*172.2 m (565 ft) (+)*
49.9 m (164 ft) (-)	*175.3 m (575 ft) (+)*
57.3 m - 58.5 m (188 ft - 192 ft) (+)	*179.8 m (590 ft) (-)*
61.3 m - 61.9 m (201 ft - 203 ft) (+)	*182.9 m (600 ft) (-)*
66.8 m - 67.4 m (219 ft - 221 ft) (+)	*184.4 m (605 ft) (-)*

[1]Intervals having significant correlations at the 95% interval of confidence
from moving correlation curves.

TABLE 7.-BRIDGE CREEK LIMESTONE[1]

PORTLAND CORE	BOUNDS CORE
W. barnesae vs. *B. constans*	
137.2 m - 137.8 m (450 ft - 452 ft) (-)	965' - 966' (-)
138.7 m (455 ft) (-)	295.1 m (968 ft) (-)
139.9 m - 140.5 m (459 ft - 461 ft) (-)	295.7 m - 296.3 m (970 ft - 972 ft) (-)
141.4 m - 142.3 m (464 ft - 467 ft) (-)	296.9 m - 297.8 m (974 ft - 977 ft) (-)
142.9 m - 143.3 m (469 ft - 470 ft) (-)	300.2 m (985 ft) (-)
143.9 m - 144.5 m (472 ft - 474 ft) (-)	
145.1 m - 146.0 m (476 ft - 479 ft) (-)	
146.6 m - 146.9 m (481 ft - 482 ft) (-)	
147.8 m - 148.4 m (485 ft - 487 ft) (-)	
W. barnesae vs. *Zygodiscus* spp.	
139.3 m - 141.1 m (457 ft - 463 ft) (-)	296.3 m - 301.5m (965 ft - 967 ft) (-)
141.7 m -142.3 m (465 ft - 467 ft) (-)	
143.3 m - 147.8 m (470 ft - 485 ft) (-)	
B. constans vs. *Zygodiscus* spp.	
139.6 m (457 ft) (-)	300.2 m (985 ft) (-)
140.2 m - 140.8 m (460 ft - 462 ft) (+)	
142.0 m (466 ft) (+)	
145.1 m - 146.0 m (477 ft - 479 ft) (+)	
146.9 m (482 ft) (+)	

[1]Intervals having significant correlations at the 95% interval of confidence
from moving correlation curves.

TABLE 8.-FORT HAYS LIMESTONE-SMOKY HILL CHALK TRANSITION[1]

PORTLAND CORE	BOUNDS CORE
B. constans vs. CaCO$_3$	
CaCO$_3$ not available	172.8 m (567 ft) (-)
	174.0 m - 175.3 m (571 ft - 575 ft) (-)
	176.2 m - 176.8 m (578 ft - 580 ft) (+)
	182.3 m - 182.9 m (598 ft - 600 ft) (+)
Zygodiscus spp. *vs.* CaCO$_3$	
CaCO$_3$ not available	176.2 m- 177.1 m (578 ft - 581 ft) (+)
W. barnesae vs. CaCO$_3$	
CaCO$_3$ not available	176.1 m - 177.1 m (578 ft - 581 ft) (-)
	179.8 m (590 ft) (-)
	181.4 m (595 ft) (-)
	183.5 m - 184.1 m (602 ft - 604 ft) (-)

TABLE 9.-BRIDGE CREEK LIMESTONE[1]

PORTLAND CORE	BOUNDS CORE
B. constans vs. CaCO$_3$	
139.3 m - 139.9 m (457 ft - 459 ft) (+)	296.6 m - 296.9 m (973 ft - 974 ft) (-)
140.2 m (460 ft) (-)	299.3 m - 299.9 m (982 ft - 984 ft) (+)
146.3 m - 146.6 m (480 ft - 481 ft) (+)	
Zygodiscus spp. vs. CaCO$_3$	
139.3 m (457 ft) (-)	300.2 m - 301.1 m (985 ft - 988 ft) (+)
140.8 m (462 ft) (+)	
141.7 m (465 ft) (-)	
142.6 m (468 ft) (-)	
144.5 m - 144.8 m (474 ft - 475 ft) (-)	
145.1 m - 146.3 m (476 ft - 480 ft) (-)	
146.9 m (482 ft) (+)	
148.1 m (486 ft) (+)	
W. barnesae vs. CaCO$_3$	
137.7 m - 138.7 m (452 ft - 455 ft) (-)	294.1 m (965 ft) (-)
145.4 m - 145.7 m (477 ft - 478 ft) (+)	295.4 m - 296.0 m (969 ft - 971 ft) (-)
146.9 m - 147.2 m (482 ft - 483 ft) (+)	296.9 m (974 ft) (-)
	299.9 m - 300.8 m (984 ft - 987 ft) (-)

[1]Intervals having significant correlations at the 95% interval of confidence from moving correlation curves.

lyze components of the assemblages and geochemical data from the two stratigraphic intervals. Variables are close to normally distributed and of similar magnitude but show a wide range in variance. For this reason, we ran PCA on the correlation matrix.

Loadings of variables on the first three principal components are shown in Table 10. The principal components account for about 60% of the variance in all of the data sets. Values are not highly significant which illustrates a lack of strong association between variables plotted in multidimensional space; nevertheless some information emerges. The results of PCA are similar for the Bridge Creek Limestone in the two cores. The first principal component shows high loadings for *W. barnesae*, with high inverse loadings for *B. constans*, *Zygodiscus* spp., *Eiffellithus* spp. and *Tranolithus* spp. The second principal component is different in the two cores: high loadings for *W. barnesae* and high inverse loadings for *B. constans* and *Zygodiscus* spp. in the Bounds core, and high loadings for CaCO$_3$ with high inverse loadings for species richness and *Zygodiscus* spp. in the Portland core.

The two Fort Hays Limestone-Smoky Hill Chalk transition intervals show different PCA results because CaCO$_3$ data are not available for the Portland core. We ran a separate PCA without CaCO$_3$ for the Bounds data, and the loadings for the other variables show values similar to when CaCO$_3$ is included. We include values for δ^{18}O and δ^{13}C from a separate PCA run, in which the loadings of the other variables also showed only minor changes. In the Bounds core, the first principal component shows high values for *W. barnesae* and δ^{18}O with inverse high loadings for *Zygodiscus* spp. The second principal component shows high loadings for CaCO$_3$, *Prediscosphaera* spp., and δ^{13}C. In the Fort Hays Limestone-Smoky Hill Chalk transition of the Portland core, *W. barnesae* dominates the first principal component, with *Zygodiscus* spp. and *B. constans* having smaller inverse loadings. *Prediscosphaera* spp. and *Tranolithus* spp. show high loadings for the second principal component.

Results of stable isotopic analysis performed on samples from the Fort Hays Limestone-Smoky Hill Chalk transition from the Bounds core are shown in Figure 4A. From light and scanning electron microscope (SEM) observation, approximately 80-90% of the fine fractions consists of coccoliths (Fig. 7). The remaining 10-20% consists predominately of micrite. Values of $\delta^{18}O$ range from -6.66‰ to -4.47‰ with a mean of -5.67‰ [SD=0.42, n=158], which is similar to bulk $\delta^{18}O$ values collected from Western Interior sediments of various ages (e.g., Pratt and Barlow, 1985; Pratt et al., 1993; Arthur et al., 1985; Barron et al., 1985). The variation in $\delta^{13}C$ ranges from 0.99‰ to 3.15‰ with a mean of 2.48‰ [SD=0.35, n=158].

DISCUSSION

Preservation

Diagenesis can alter the composition of nannofossil assemblages and isotopic ratios considerably, affecting their use as paleoecological and paleoceanographic indicators. Light microscope observations show considerable variation in nannofossil preservation throughout the cores. The abundance of *W. barnesae*, CaCO3 content, and scanning electron microscopic observations are included in the analysis of preservation.

One difficulty with interpreting relative abundance data is that high percentages of one species may cause what appears to be low percentages of another taxon. Low percentages of *B. constans*, for example, could result from an increase in *W. barnesae* due to preferential dissolution of the more solution-susceptible species. In contrast, low percentages of *B. constans* may be caused by paleoecologic changes independent of changes in another taxon. This "closed-sum" problem creates a seemingly insurmountable obstacle in using relative abundance data to determine paleoecological changes. Good coccolith preservation in a sample can improve paleoecological interpretations, although the "closed-sum" problem cannot be ruled out.

W. barnesae, a solution-resistant nannofossil species, is often used to indicate preservational variations within Cretaceous sediments (e.g., Roth and Bowdler, 1981). Assemblages with high percentages of *W. barnesae* (>40%, Thierstein, 1981; Roth and Krumbach, 1986; >70%, Williams and Bralower, 1995) are considered to be notably altered by dissolution of more fragile taxa. Experiments on Cenomanian assemblages by Hill (1975) show that during severe dissolution, *W. barnesae* increases while most other taxa decrease. Therefore, variations in other taxa should remain evident even if a moderate amount of dissolution increases *W. barnesae* percentages.

W. barnesae is more likely a better representative of preservational changes in the Portland core where mean values are over 40%. For the Bounds core we find that *W. barnesae* percentages are more paleoecologically significant. All sections studied have highly significant inverse correlations between *W. barnesae* and *B. constans* that may, as in other studies (e.g., Premoli-Silva et al., 1989), suggest that high *W. barnesae* contents are associated with low fertility environments. This is most likely to be true in intervals with good nannofossil preservation such as the Fort Hays Limestone-Smoky Hill Chalk transition from the Bounds core.

The relationship between CaCO3 and nannofossil preservation is not clearly defined. It is generally thought that samples with higher than 60% CaCO3 may have overgrowths on nannofossils while samples with less than 40% CaCO3 may have etched nannofossils (Thierstein and Roth, 1991). In either case, some original assemblage information may remain. Carbonate contents in the studied material vary between sections but are rarely over 70% or below 30%. *W. barnesae* shows no significant correlation with CaCO3 in either section in the Portland core (Tables 4-5). There is a low correlation, although significant, between CaCO3 and *W. barnesae* in the Fort Hays Limestone-Smoky Hill Chalk transition of the Bounds core (Table 3). Scanning electron- and light-microscopic observations indicate that there is little correlation between carbonate content and preservation in any of the sections.

Fort Hays Limestone-Smoky Hill Chalk transition, Portland core.—

Light microscope and SEM observations coupled with *W. barnesae* percentages (mean = 46%) indicate that nannofossil preservation in samples of the Fort Hays Limestone-Smoky Hill Chalk is poorer in the Portland core than in the Bounds core. It is difficult to isolate intervals in this section with sufficient preservation to warrant paleoecological interpretations, therefore taxonomic correlations are considered to be preservational products.

There are significant negative moving correlations (i.e., between *W. barnesae* and both *B. constans* and *Zygodiscus* spp.), but because nannofossil preservation in this interval is poor, we interpret these trends to result from preservational and "closed-sum" factors (Table 6). Because *B. constans* and *Zygodiscus* spp. are both considered to be indicators of high surface-water fertility, fluctuations in these taxa may be expected to mirror each other. Four intervals show significant positive correlation coefficients consistently between these taxa (Table 6), however we interpret these correlations to be largely the result of preservational and "closed-sum" factors. In addition, the moving correlation curves between *W. barnesae* and both *B. constans* and *Zygodiscus* spp. are similar (Burns, 1995), indicating that the data may be affected by the "closed-sum" factor. The entire interval has highly significant negative correlations between *W. barnesae* and *B. constans*, *Zygodiscus* spp., *Eiffellithus* spp., *Cretarhabdus* spp., *Prediscosphaera* spp., and *Tranolithus* spp. (Figs. 3A, B and Table 2). Preservational factors are thought to control the abundance of *W. barnesae*; "closed-sum" is thought to control the abundance of the other taxa.

Overall Pearson correlations coefficients of species richness and assemblage data support the importance of preservation in this interval. There are highly significant positive correlations between species richness and the abundances of both *W. barnesae* and *Tranolithus* spp. These two taxa tend to dominate samples with poorly preserved assemblages.

PCA also indicates the preservational control in this interval. The first principal component is dominated by *W. barnesae* (Table 10) and all of the other loadings are inverse, which suggests the closed-sum effect. We believe that nannofossil preservation in this interval of the Portland core is sufficiently poor that paleoecological interpretations of the data are unwarranted.

Fort Hays Limestone-Smoky Hill Chalk Transition, Bounds core.—

This interval has the best-preserved nannofossil assemblages. There are few broken or overgrown specimens and *W. barnesae* percentages are lower than 40% (mean = 18%) in all but three samples (Fig. 4A). Moving correlations between *B. constans*, *Zygodiscus* spp., and *W. barnesae* show intervals with significant positive and negative correlations (Table 7). Less signifi-

TABLE 10.-PRINCIPAL COMPONENTS ANALYSIS
OF ASSEMBLAGE DATA

BRIDGE CREEK LIMESTONE

Rebecca Bounds Core	1st PC	2nd PC	3rd PC
Species richness	0.29	0.02	-0.76
% CaCO$_3$	0.60	0.32	0.32
% *W. barnesae*	-0.75	0.56	0.26
% *B. constans*	0.09	-0.70	0.12
% *Zygodiscus* spp.	0.53	-0.51	0.06
% *Eiffellithus* spp.	0.67	0.43	0.11
% *Cretarhabdus* spp.	-0.11	0.27	-0.73
% *Prediscosphaera* spp.	-0.40	-0.42	-0.13
% *Tranolithus* spp.	0.52	0.25	-0.04
Percent variance	24.20	18.60	14.80

Portland Core	1st PC	2nd PC	3rd PC
Species richness	0.35	-0.60	0.05
% CaCO$_3$	0.24	0.76	0.34
% *W. barnesae*	-0.92	0.26	-0.09
% *B. constans*	0.60	0.18	-0.51
% *Zygodiscus* spp.	0.46	-0.65	0.02
% *Eiffellithus* spp.	0.60	0.29	0.40
% *Cretarhabdus* spp.	-0.28	-0.19	0.76
% *Prediscosphaera* spp.	0.05	-0.31	0.36
% *Tranolithus* spp.	0.70	0.37	0.11
Percent variance	28.00	20.00	14.00

FORT HAYS LIMESTONE -SMOKY HILL CHALK TRANSITION

Rebecca Bounds Core	1st PC	2nd PC	3rd PC
Species richness	0.51	-0.39	0.15
% CaCO$_3$	0.42	0.71	0.10
% *W. barnesae*	-0.91	0.17	-0.03
% *B. constans*	0.50	-0.29	-0.51
% *Zygodiscus* spp.	0.62	-0.43	0.08
% *Eiffellithus* spp.	-0.03	0.32	-0.48
% *Cretarhabdus* spp.	-0.26	-0.25	0.74
% *Prediscosphaera* spp.	0.54	0.63	0.23
% *Tranolithus* spp.	0.35	0.24	0.31
$\delta^{13}C$	(0.11)	(0.54)	(0.20)
$\delta^{18}O$	(-0.68)	(-0.29)	(-0.17)
Percent variance	26.40	17.40	13.60

Portland Core	1st PC	2nd PC	3rd PC
Species richness	0.29	0.49	0.36
% *W. barnesae*	-0.96	0.06	0.00
% *B. constans*	0.57	-0.12	-0.59
% *Zygodiscus* spp.	0.60	-0.43	0.01
% *Eiffellithus* spp.	0.37	0.14	0.61
% *Cretarhabdus* spp.	0.34	-0.43	0.23
% *Prediscosphaera* spp.	0.33	0.59	0.04
% *Tranolithus* spp.	0.23	0.63	-0.40
Percent variance	26.30	17.50	13.20

cant negative correlations suggest that original paleoecologic signals are preserved in the nannofossil assemblages from this interval. Although the first principal component has high negative loadings for *W. barnesae, B. constans,* and *Zygodiscus* spp. (Table 10) which suggests a possible role of preservation and/or "closed-sum," we believe this is the result of fertility control.

Bridge Creek Limestone, Portland core.—
Observations under the light microscope and the SEM and higher *W. barnesae* percentages (mean = 44%) indicate poorer

nannofossil preservation. Nannofossils are overgrown with micrite, and samples contain an abundance of large (approximately 10-15 μm) micrite crystals. All significant moving correlations between *W. barnesae* and taxa that represent higher fertility (*B. constans* and *Zygodiscus* spp.) are negative (Table 7) and, because nannofossil preservation is poor, are likely the result from preservational control and the "closed-sum" problem. This trend is also seen in significant loadings of *W. barnesae* on the first principal component and high negative loadings for several other species which indicates the "closed-sum" effect (Table 10).

Assemblage data from this interval have limited paleoecological and paleoceanographic significance. There is, however, a general decrease in the percentage of *W. barnesae* to approximately 40% along with a slight increase in *B. constans* above 141.7 m (465 ft), which suggests a slight improvement in nannofossil preservation (Fig. 5A). Discussion of paleoecological implications for the Bridge Creek Limestone from the Portland core is therefore limited to the section above 141.7 m (465 ft).

Bridge Creek Limestone, Bounds core.—
Preservation in the Bridge Creek Limestone in the Bounds core is fair. Nannofossils are slightly overgrown, but there is much less micrite than in the same interval in the Portland core. The lower mean percentage of *W. barnesae* (27%) also supports the visual estimate of fair preservation. There are only a few intervals with significant moving correlations between *W. barnesae* and *B. constans* (Table 7). On the other hand, *Zygodiscus* spp. is almost entirely negatively correlated with *W. barnesae,* especially below approximately 295.7 m (970 ft) (Table 7). If correlations of these two taxa with *W. barnesae* are controlled by preservation and "closed-sum" factors alone, moving correlation curves should be similar. Because this is not the case, we suggest that these data are not controlled by "closed-sum" factors and contain paleoecological information.

Species richness is significantly correlated with *W. barnesae.* We believe that preservational factors dominate this correlation, but in the best preserved intervals it may have paleoecological significance. Similarly, although the first principal component is likely controlled by preservation, the high loadings of only fertility markers on the second principal component and the inverse loadings for *W. barnesae* and *B. constans* and *Zygodiscus* spp. suggest that this principal component is controlled by fertility.

The Origin of Lithologic Cycles

Nannofossil data.—
We focus on *W. barnesae, B. constans,* and *Zygodiscus* spp. in this discussion because these taxa have well-known paleoecologic significance. Four other groups of species (*Eiffellithus* spp., *Cretarhabdus* spp., *Prediscosphaera* spp., and *Tranolithus* spp.) have unknown paleoecological affinities but are presented here because they make up a considerable percentage of the assemblage. We are also interested in any paleoecological significance for the groups that might be determined from relationships with the other well-established taxa.

Abundance data for the four sequences investigated are shown in Figures 3-6. Correlations between CaCO$_3$, *W. barnesae,* and the major fertility indicators (*B. constans* and *Zygodiscus* spp.) vary considerably in all sections which suggests that the mecha-

54   CELESTE E. BURNS AND TIMOTHY J. BRALOWER

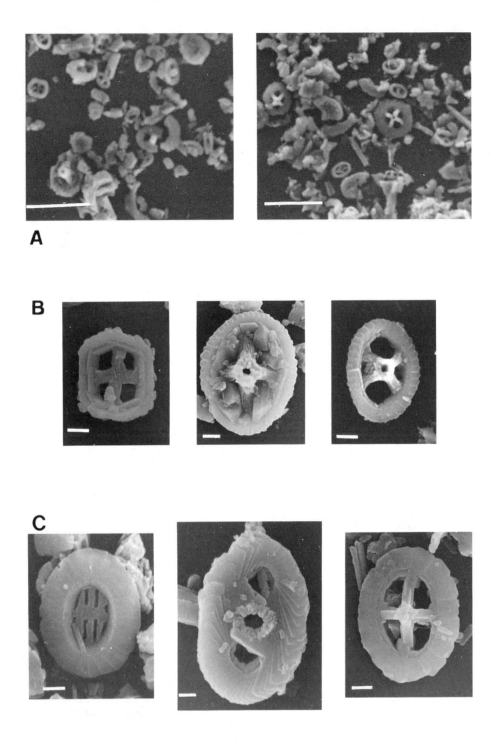

FIG. 7.—(A) SEM images from samples 567.19' (left) and 584.67' (right) from Fort Hays Limestone-Smoky Hill Chalk transition in the Bounds core. Scale bar = 10μ. (B) SEM images of *Corollithion signum* (left), *Eiffellithus anthophorus* (center), and *Chiastozygus* sp. (right) from sample 567.19' (Fort Hays Limestone-Smoky Hill Chalk transition in the Bounds Core). Scale bar = 1μ. (C) SEM images of *Sollasites barringtonensis* (left), *Reinhardtites levis* (center), and *Prediscosphaera spinosa* (right) from sample 584.67' (Fort Hays Limestone-Smoky Hill Chalk transition in the Bounds core).

nisms causing lithologic cycles are not just the result of changes in surface-water fertility (e.g., Arthur et al., 1985; Barron et al., 1985; Pratt and Barlow, 1985; Eicher and Diner, 1989; Watkins, 1989).

Fort Hays Limestone-Smoky Hill Chalk Transition, Bounds core.—
The good preservation of nannofossil assemblages allows us to interpret overall changes in nannofossil taxa in this interval in terms of surface-water fertility (Figs. 4B, C). Significant moving correlations between fertility indicators (*W. barnesae*, *B. constans*, and *Zygodiscus* spp.) and $CaCO_3$ are both positive and negative which indicates that in some intervals marlstones are associated with higher surface-water fertility (Fig. 4B, Table 8). In other intervals, higher fertility corresponds to limestones. Thus, there must be a complex relationship between surface-water conditions and lithology.

The moving correlation coefficients between *Zygodiscus* spp. and *B. constans* are not consistently positive (Table 6) which supports the conclusion of Erba et al. (1992) that these taxa indicate different levels of surface-water fertility. In general, *B. constans* has a stronger correlation with *W. barnesae* and $CaCO_3$ than with *Zygodiscus* spp. which suggests that for this section, particularly in the upper part (above 178.3 m (585 ft)), *B. constans* is a better paleoecological indicator.

There are some clear rhythms in the *B. constans* record between 172.2 m and 176.8 m (565 ft to 580 ft) that may indicate significant changes in fertility (Fig. 4A). Significant negative moving correlation coefficients between $CaCO_3$ and *B. constans* in the two intervals also suggest that marlstones were deposited during times of higher fertility (Fig. 4B). This, however, does not appear to be the case across the entire Fort Hays Limestone-Smoky Hill Chalk transition in the Bounds core; there are two intervals where moving correlation coefficients between $CaCO_3$ content and *B. constans* are positive (Fig. 4B, Table 8).

Eiffellithus spp. has barely significant negative correlations with *W. barnesae* and *Zygodiscus* spp. but no significant relation with $CaCO_3$ content (Table 3). *Prediscosphaera* spp. has a significant positive correlation with $CaCO_3$ content and a negative correlation with *W. barnesae* (Fig. 4C, Table 3). It is not clear whether these taxa have any paleoecological affinities in this section. A highly positive significant correlation between species richness and *W. barnesae* and significant negative correlations between species richness and *Zygodiscus* spp. and *B. constans* may reflect a decrease in diversity in times of low fertility contrary to the conclusions of Watkins (1989) for a single Bridge Creek cycle (Fig. 4B, Table 3). However, opposite correlations are observed between loadings for both sets of variables on the first principal component, which we also believe to be controlled by fertility (Table 10).

Bridge Creek Limestone, Portland Core.—
Interpretations of surface-water conditions in this interval are limited due to poor preservation. We interpret assemblages from above 141.7 m (465 ft) where percentages of *W. barnesae* are considerably lower (mean = 34.7%) to have some paleoecological significance (Figs. 5A, B). Table 9 shows intervals above this level with both negative and positive moving correlations between $CaCO_3$ content and relative abundance of fertility indicators, which suggests that marlstones could be associated with either higher or lower surface-water fertility.

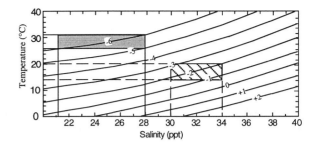

FIG. 8.—Plot of variation in oxygen isotopic composition of calcite as a function of temperature and salinity of the seawater from the model of Railsback et al. (1989) adapted from Woo et al. (1992). The solid, curved lines are isopleths of calcite $\delta^{18}O$ values. Estimated temperature and salinity ranges for values between -5 and -6 per mil are shown in the shaded area. The cross-hatched area represents a more reasonable temperature-salinity window for surface waters in the Cretaceous Western Interior Seaway.

Bridge Creek Limestone, Bounds Core.—
This section shows few significant correlations between $CaCO_3$ content and nannofossil taxa (Fig. 6A, Table 9) which suggests that there is little direct association between lithology and surface-water fertility. *Zygodiscus* spp. and *W. barnesae* are negatively correlated for much of the interval, indicating that a fertility signal is preserved (Figs. 6A, B). Percentages of *B. constans* and *Zygodiscus* spp. have almost no significant correlations (Fig. 6A, Table 7), further supporting the interpretation that these two taxa respond to different levels of high surface-water fertility. However, these taxa both possess high loadings (and inverse to *W. barnesae*) on the second principal component (Table 10), which we believe is associated with fertility.

The interval in the Bounds core with detailed sampling (28 samples between 296.7 m and 297.2 m (970 ft and 975 ft)) supports the interpretations of preservation and paleoecology discussed above. The correlation between *Zygodiscus* spp. and *B. constans* is significantly positive, which shows that these taxa respond similarly to changes in fertility. There are also significant negative correlations between *W. barnesae* and both *B. constans* and *Zygodiscus* spp. None of the correlations between $CaCO_3$ content and nannofossil taxa are significant for this detailed interval, which illustrates the absence of simple relationships between lithology and surface-water conditions.

Stable Isotopic Data.—
Stable isotopic analyses were conducted on fine fractions of samples from the Fort Hays Limestone-Smoky Hill Chalk transition in the Bounds core (Fig. 4A). This interval is characterized by moderate to good nannofossil preservation; SEM observations and assemblage analyses indicate that the abundance of micrite is low (< 10% of the sample) and dissolution and overgrowth of nannofossils are minor.

Most stable-isotopic research of Western Interior sediments was conducted on whole-rock samples, bulk organic-carbon fractions, or macrofossils (e.g., Pratt and Barlow, 1985; Pratt and Threlkeld, 1985; Arthur et al., 1985; Barron et al., 1985; Pratt et al., 1993; Dean and Arthur, this volume), and has shown that isotope ratios tend to have considerable diagenetic overprints (e.g., Pratt and Barlow, 1985; Dean and Arthur, this volume). Typical $\delta^{18}O$ values of sediments from the Niobrara Formation are generally lower than -7‰ and are thought to be affected by burial diagenesis (Laferriere, 1992). Arthur and Dean (1991) and Dean and Arthur (this volume) found that values of $\delta^{18}O$ in the Smoky Hill Chalk Member of the Niobrara Formation are about 2‰ lower in a core from eastern Colorado than those in a core from western Kansas, and they attribute this to greater burial diagenesis of the Smoky Hill in the Colorado core. Relationships between isotopic values and $CaCO_3$ content were also studied in Cretaceous cyclic sedimentary rocks and were correlated to primary oceanographic changes (e.g., Barron et al., 1985; Arthur and Dean, 1991; Pratt et al., 1993) and diagenetic factors (Thierstein and Roth, 1991; Laferriere, 1992). Thierstein and Roth (1991) found $CaCO_3$ and $\delta^{13}C$ to be negatively correlated when nannofossils were etched presumably during early methanogenic precipitation of $CaCO_3$.

As sediments are buried temperature increases, causing lower $\delta^{18}O$ values in re-precipitated calcite (Anderson and Arthur, 1983). Therefore, we would expect that if diagenesis has altered the original isotopic ratios, samples with lower $\delta^{18}O$ values should be more altered. Using the SEM we examined preservation in groups of samples with the lowest and highest $\delta^{18}O$ values and found that there is no correlation between lower $\delta^{18}O$ values and preservation. Although we cannot completely rule out some diagenetic alteration, in the discussion that follows we assume that a significant element of the primary fine-fraction isotopic signature remains.

Oxygen Isotope Data.—

It is well-accepted that there was considerable fresh water run-off into the Western Interior Seaway and that these waters had very low values of $\delta^{18}O$ (e.g., Barron et al, 1985; Pratt et al., 1993). Oxygen-isotope values are, therefore, often interpreted to reflect changes in input of these waters. Previously this influx was thought to be derived mostly from the Sevier Highlands to the west (e.g., Pratt, 1984), but it appears that a substantial volume may have come from highlands further south, at least during the Cenomanian-Turonian boundary interval (Leckie et al., this volume).

To assess water mass temperature and salinity during deposition of the Fort Hays Limestone-Smoky Hill Chalk transition, we compare our isotopic data to a model developed by Woo et. al (1992) based on values for an evaporative ocean basin, ice-free Cretaceous earth ($\delta_w = -1‰$), and typical Cretaceous oxygen-isotopic values. Figure 8 shows the estimated window of temperature and salinity ranges based on the majority of $\delta^{18}O$ values measured here (-5‰ to -6‰). This model indicates a temperature range of 26-31°C and a salinity range of 21-28 ppt. These temperatures are too high and salinities too low for the Western Interior Seaway considering its temperate latitude and the diverse nature of nannofossil assemblages (although the modern species, *Emiliania huxleyi* lives in waters as fresh as 18 ppt, diverse living nannofloral assemblages are not found in salinities under 30 ppt (e.g., Bukry, 1974). Figure 8 also shows the window of isotopic values based on more reasonable temperatures (14-20°C) and salinities (30-34 ppt); the difference between expected and measured $\delta^{18}O$ values is approximately 2‰ to 5.5‰. The amplitude of oxygen isotope values in cyclic parts of the record from one unit to another is about 1.5‰ which is equivalent to a change of 7 ppt of salinity or 5°C of temperature, both of which are unlikely magnitudes for short-term oscillations.

We believe that the amplitude of $\delta^{18}O$ cycles and the magnitude of $\delta^{18}O$ values is more likely controlled by periodic run-off of isotopically light waters from the west. The range of δ_w values draining from the west into the Western Interior Seaway dur-

ing the Cretaceous are thought to be about 20‰ (Glancy et al., 1993). Mixing of Tethys and Arctic waters and a small proportion of fresh water would produce values comparable to those presented here while maintaining a reasonable salinity.

There is little to no significant correlation between percent $CaCO_3$ and $\delta^{18}O$ in the Fort Hays-Smoky Hill Chalk transition, which indicates that there is no consistent relationship between the freshwater dilution and lithology. However, nannofossil assemblages are generally more closely correlated with $\delta^{18}O$ than with percent $CaCO_3$. Nannofossil taxa with known paleoecological associations correlate significantly with $\delta^{18}O$ (Table 3). There is a significant positive correlation between *W. barnesae* and $\delta^{18}O$ for much of the section (and high loadings for both variables on the first principal component (Table 10)) and negative moving correlations between $\delta^{18}O$ and *Zygodiscus* spp. and *B. constans* for several intervals (Burns, 1995). The trends between the abundance of *B. constans*, *Zygodiscus* spp., and *W. barnesae* and $\delta^{18}O$ may indicate an increase in surface-water fertility during intervals of increased run-off for parts of the Fort Hays Limestone-Smoky Hill Chalk transition.

Carbon Isotope Data.—

Correlations between $\delta^{13}C$ and the abundance of nannofossil taxa are less significant than correlations with $\delta^{18}O$; *W. barnesae* and $\delta^{13}C$ are positively correlated at the base of the section and gradually become negatively correlated upwards. There are almost no intervals with significant correlations between $\delta^{13}C$ and *B. constans* and *Zygodiscus* spp.

The Fort Hays Limestone-Smoky Hill Chalk transition carbon-isotopic profile (Fig. 4A) has potential for Upper Coniacian-Santonian stratigraphic correlation. There are no distinct excursions in the carbon-isotopic data: only at the base of the studied interval is there a negative shift between 185.6 m and 184.1 m (609 ft and 604 ft) (Fig. 4A). Isotopic values for whole-rock carbonate samples from this section indicate a sharp decrease in $\delta^{13}C$ in the Upper Coniacian (Pratt et al., 1993). This may correlate to the low carbon-isotope values at the base of the studied interval (Fig. 4a), but these correlations are somewhat tentative because different faunal groups have been used to date the section.

Comparison with Previous Studies.—

The dilution (e.g., Pratt, 1984) and productivity (Eicher and Diner, 1985) models were derived from analysis of Bridge Creek Limestone cycles. In a study of one Bridge Creek cycle, Watkins (1989) found a significant correlation between fertility indicators, species richness, and lithology, which supports the dilution model (e.g., Pratt, 1984) for the origin of sedimentary cycles. Here, we analyze data from both the Bridge Creek Limestone and the Fort Hays-Smoky Hill Chalk transition from approximately 45 lithologic cycles. The results are much more complex than those of Watkins (1989). In all of the intervals where paleoecologic interpretations are possible—the Fort Hays Limestone-Smoky Hill Chalk transition in both cores and the Bridge Creek Limestone in the Bounds core—there are no simple trends between abundance of nannofossil taxa (e.g., *B. constans* and *Zygodiscus* spp.) that are believed to indicate high fertility and lithology. Marlstones do not always have high percentages of fertility markers, and limestones do not always correlate with abundance of nannofossils associated with less productive environments.

Our data indicate that the connection between organic productivity, dilution, and lithology is complex, and that the dilution and productivity models are too simplistic to explain the origin of all Upper Cretaceous lithologic cycles. Because assemblage data are significantly correlated with $\delta^{18}O$, we believe that during deposition of the Fort Hays Limestone-Smoky Hill Chalk transition, organic productivity was closely related to the amount of run-off into the basin. Abundances of nannofossil fertility indicators are negatively correlated with $\delta^{18}O$, which suggests that organic productivity increased during periods of increased run-off of nutrient-rich, fresh waters.

A distinct difference exists between organic and carbonate productivity that, along with the flux of clastic material, determines $CaCO_3$ percentages. Calcareous microplankton groups, such as planktic foraminifera and nannoplankton, tend to thrive in oligotrophic conditions (e.g., Hallock et al., 1991). Thus it is quite possible that an interval of high organic productivity corresponds to low carbonate productivity and vice versa. The dilution model (e.g., Pratt, 1984) relates lithology to organic productivity, while the productivity model (e.g., Eicher and Diner, 1989) relates lithology to carbonate productivity. The former model's prediction that there was *higher organic* productivity during intervals of marlstone deposition may, therefore, not be at odds with the latter model, which considers that marlstones were the result of *lower carbonate* productivity.

The data collected here suggest that carbonate productivity was influenced not by a single watermass, but by complex interactions of different watermasses including run-off from the hinterland to the west, influx of Tethyan waters from the south, and Arctic waters from the north. Previous models (e.g., Barron et al., 1985; Eicher and Diner, 1989) suggest that surface fertility was modulated by a single water mass. The lack of consistent trends between $CaCO_3$ content and the fertility indicators implies that a dilution rhythm was superimposed out-of-phase on a carbonate productivity rhythm.

CONCLUSIONS

1. Analysis of calcareous nannofossil assemblages from the Fort Hays Limestone-Smoky Hill Chalk transition and the Bridge Creek Limestone from two cores indicate variable nannofossil preservation across the basin. The Fort Hays Limestone-Smoky Hill Chalk transition in the Bounds core has the best nannofossil preservation and its assemblages are the most paleoecologically significant. Samples from this interval in the Portland core are poorly preserved such that nannofossil percentages have been substantially altered by dissolution and "closed-sum" factors. The Bridge Creek Limestone intervals from both cores are moderately preserved and allow for paleoecological interpretations from assemblage data.

2. In some intervals nannofossil fertility indicators have increased abundances in marlstones; other intervals show increased abundances of these taxa in limestones. Correlations between nannofossil fertility markers and percent $CaCO_3$ are consistently insignificant which indicates that lithologic variations may not be associated with changes in surface-water conditions as seen in nannofossil assemblages.

3. Oxygen isotopic values of fine fraction carbonate from the Fort Hays Limestone-Smoky Hill Chalk transition of the Rebecca Bounds core are comparable to whole-rock values from this interval and most likely reflect input of fresh water into the seaway. The correlations between nannofossil fertility indicators and $\delta^{18}O$ values indicate significant correlation between organic

productivity and fresh-water input into the seaway. There is little relationship between $CaCO_3$ content (lithology) and organic productivity (as seen in high abundances of nannofossil fertility indicators). Lithologic cycles appear to be controlled both by variations in carbonate productivity and dilution, and thus are out of phase with variations in organic productivity. Therefore the interpretation of lithologic cycles as representative of organic productivity must be reevaluated.

ACKNOWLEDGMENTS

The authors gratefully acknowledge and thank Michael Arthur and Walter Dean for inviting us to participate in the WIKS project. We thank the staff at the USGS Core Laboratory for their efficient sampling of the two cores. We are grateful to Bernie Genna and Bill Showers for assistance at the NCSU stable isotope laboratory. We are grateful to Cindy Fisher, Charlie Paull, Joe Carter, and Walt Dean for their thoughtful reviews. And we thank Greg Spaniolo and Mal Foley for their help and advice with statistical analyses. Research was funded by a grant from the Department of Energy, the Geological Society of America, the American Chemical Society (Grant 24242-AC8), and the UNC-CH Martin Fund.

REFERENCES

ANDERSON, T. F., AND ARTHUR, M. A., 1983, Stable isotopes of oxygen and carbon and their application to paleoenvironmental problems, *in* Arthur, M. A., Anderson, T. F., Kaplan, I. R., Veizer, J., and Land, L. S., eds., Stable Isotopes in Sedimentary Geology: Tulsa, Society of Economic Paleontologists and Mineralogists Short Course No. 10, p. 1-151.

ARTHUR, M. A., AND DEAN, W. E., 1991, A holistic geochemical approach to cyclomania: Examples from Cretaceous pelagic limestone sequences, *in* Einsele, G., Ricken, W., and Seilacher, A., eds., Cycles and Events in Stratigraphy: New York, Springer-Verlag, p. 126-166.

ARTHUR, M. A., DEAN, W. E., AND SCHLANGER, S. O., 1985, Variations in the global carbon cycle during the Cretaceous related to climate, volcanism, and changes in atmospheric CO_2, *in* Sudquist, E. T., and Broeker, W. S., eds., The Carbon Cycle and Atmospheric CO_2: Natural Variations Archean to Present: Washington, D.C., American Geophysical Union Monographs, v. 32, p. 504-529.

BARRON, E. J., ARTHUR, M. A., AND KAUFFMAN, E. G., 1985, Cretaceous rhythmic bedding sequences: A plausible link between orbital variations and climate: Earth and Planetary Science Letters, v. 72, p. 327-340.

BUKRY, D., 1974, Coccoliths as paleosalintiy indicators—Evidence from Black Sea, *in* Degens, E. T. and Ross, D., eds., The Black Sea—Geology, chemistry, and biology: Tulsa, American Association of Petroleum Geologists Memoir 20, p. 353-363.

BURNS, D. A., 1973, The latitudinal distribution and significance of calcareous nannofossils in the bottom sediments of the south-west Pacific Ocean (Lat. 15-55° S) around New Zealand, *in* Fraser, R., ed., Oceanography of the South Pacific: Wellington, National Commission for UNESCO, p. 221-228.

BURNS, C. E., 1995, Upper Cretaceous nannofossil assemblages across the Western Interior Seaway: Implications for the origins of lithologic cycles in the Greenhorn and Niobrara Formations, Unpublished M.S. Thesis, University of North Carolina at Chapel Hill, 72 p.

CALDWELL, W. G. E., AND KAUFFMAN, E. G., 1993, Evolution of the Western Interior Basin: St. John's, Geological Association of Canada Special Paper 39, 680 p.

DEAN, W. E., ARTHUR, M. A., SAGEMAN, B. B., AND LEWAN, M. D., 1995, Core descriptions and preliminary geochemical data for the Amoco Production Company Rebecca Bounds Core #1 well, Greely County, Kansas: Washington, D.C., U.S. Geological Survey Open-File Report 95-209, 243 p.

EICHER, D. L. , AND DINER, R., 1985, Foraminifera as indicators of water mass in the Cretaceous Greenhorn Sea, Western Interior, *in* Pratt, L. M., Kauffman, E. G., and Zelt, F. B., eds., Fine-grained Deposits and Biofacies of the Cretaceous Western Interior Seaway: Evidence of Cyclic Sedimentary Processes, Fieldtrip Guidebook No. 4: Tulsa, Society of Economic Paleontologists and Mineralogists, p. 57-169.

EICHER, D. L., AND DINER, R., 1989, Origin of the Cretaceous Bridge Creek cycles in the Western Interior, United States: Palaeogeography, Palaeoclimatology, Palaeoecology, v. 74, p. 127-146.

ERBA, E., CASTRADORI, D., GUASTI, G., AND RIPEPE, M., 1992, Calcareous nannofossils and Milankovitch cycles: The example of the Albian Gault Clay Formation (southern England): Palaeogeography, Palaeoclimatology, Palaeoecology, v. 93, p. 47-69.

FISCHER, A. G., 1980, Gilbert-bedding rhythms and geochronology, in Yochelson, E. I., ed., The Scientific Ideas of G. K. Gilbert: Boulder, Geological Society of America Special Paper 183, p. 93-104.

GLANCY, T. J., ARTHUR, M. A., BARRON, E. J., AND KAUFFMAN, E. G., 1993, A paleoclimate model for the North American Cretaceous (Cenomanian-Turonian) epicontinental sea, in Caldwell, W. G. E., and Kauffman, E. G., Evolution of the Western Interior Basin: St. John's, Geological Association of Canada Special Paper No. 39, p. 219-241.

HALLOCK, P., PREMOLI-SILVA, I., AND BOERSMA, A., 1991, Similarities between planktonic and larger foraminiferal evolutionary trends through Paleogene paleoceanographic changes: Palaeogeography, Palaeoclimatology, Palaeoecology, v. 83, p. 49-64.

HATTIN, D. E., 1971, Widespread, synchronously deposited, burrow-mottled limestone beds in Greenhorn Limestone (Upper Cretaceous) of Kansas and Central Colorado: American Association of Petroleum Geologists Bulletin, v. 55, p. 412-431.

HILL, M. E., 1975, Selective dissolution of mid-Cretaceous (Cenomanian) calcareous nannofossils: Micropaleontology, v. 21, p. 227-235.

HONJO, S. AND OKADA, H., 1974, Coccoliths: Production, transportation and sedimentation: Micropaleontology, v. 20, p. 209-230.

KAUFFMAN, E. G., 1977, Geological and biological overview: Western Interior Cretaceous Basin, The Mountain Geologist, v. 14, nos. 3, 4, p. 75-99.

LAFERRIERE, L., 1992, Regional isotopic variations in the Fort Hays Member of the Niobrara Formation, United States Western Interior: Primary signals and diagenetic overprinting in a Cretaceous pelagic rhythmite: Geologic Society of America Bulletin, v. 104, p. 980-992.

McINTYRE, A., AND BÉ, A. W. H., 1967, Modern Coccolithophoridae of the Atlantic Ocean: Placoliths and cyrotoliths: Deep Sea Research, v. 14, p. 561-597.

MONECHI, S., AND THEIRSTEIN, 1985, Late Cretaceous Eocene nannofossil and magnetostratigraphic correlation near Gubbio, Italy: Marine Micropaleontology, v. 9, p. 419-440.

PAULL, C. P., and THIERSTEIN, H. R., 1987, Stable isotopic fractionation among particles in Quaternary coccolith-size deep-sea sediments: Paleoceanography, v. 2, p. 423-429.

PERCH-NIELSEN, K., 1985, Mesozoic calcareous nannofossils, in Bolli, H. M., Saunders, J. B., and Perch-Nielsen, K., eds., Plankton Stratigraphy, v. 1., p. 329-426.

PRATT, L. M., 1981, A paleo-oceanographic interpretation of the sedimentary structures, clay minerals, and organic matter in a core of the Middle Cretaceous Greenhorn Formation near Pueblo, Colorado: Ph.D. Dissertation, Princeton University, Princeton, N.J., 176 p.

PRATT, L. M., 1984, Influence of paleoenvironment factors on preservation of organic matter in middle Cretaceous Greenhorn Formation, Pueblo, Colorado: American Association of Petroleum Geologists Bulletin, v. 68, p. 1146-1159.

PRATT, L. M., AND THRELKELD, C. N., 1984, Stratigraphic significance of $^{13}C/^{12}C$ ratios in mid-Cretaceous rock of the western interior, U.S.A., in Stott, D. F. and Glass, D. J., eds., The Mesozoic of Middle North America: Calgary, Canadian Society of Petroleum Geologists, Memoir No. 9, p. 305-312.

PRATT, L. M., AND BARLOW, L. K., 1985, Isotopic and sedimentological study of the Lower Niobrara Formation, Lyons, Colorado, in Pratt, L. M., Kauffman, E. G., and Zelt, F. B., eds., Fine-grained Deposits and Biofacies of the Cretaceous Western Interior Seaway: Evidence of Cyclic Sedimentary Processes, Field Trip Guidebook No. 4: Tulsa, Society of Economic Paleontologists and Mineralogists, p. 209-214.

PRATT, L. M., KAUFFMAN, E. G., AND ZELT, F. B., eds., 1985, Fine-grained Deposits and Biofacies of the Cretaceous Western Interior Seaway: Evidence of Cyclic Sedimentary Processes, Fieldtrip Guidebook No. 4: Tulsa, Society of Economic Paleontologists and Mineralogists, 249 p.

PRATT, L. M., ARTHUR, M. A., DEAN, W. E., AND SCHOLLE, P. A., 1993, Paleo-oceanographic cycles and events during the Late Cretaceous in the Western Interior Seaway of North America, in Caldwell, W. G. E., and Kauffman, E. G., Evolution of the Western Interior Basin: St. John's Geological Association of Canada Special Paper No. 39, p. 333-353.

PREMOLI-SILVA, I., ERBA, E., AND TORNAGI, M. E., 1989, Paleoenvironmental signals and changes in surface fertility in Mid-Cretaceous Corg-rich pelagic facies of Fucoid Marls (Central Italy): Geobios, v. 11, p. 225-236.

Research on Cretaceous Cycles (R.O.C.C.) Group, 1986, Rhythmic bedding in Upper Cretaceous pelagic carbonate sequences—Varying sedimentary response to climatic forcing: Geology, v. 14, p. 153-156.

ROTH, P. H., 1981, Mid-Cretaceous calcareous nannoplankton from the central Pacific: Implications for paleoceanography, in Thiede, J., Vallier, T. L., et al., eds., Initial Reports Deep Sea Drilling Project, v. 62, p. 471-489.

ROTH, P. H., 1986, Mesozoic paleoceanography of the North Atlantic and Tethys Oceans, in Summerhayes, C. P., and Shackleton, N. J., eds., North Atlantic Palaeoceanography: London, Geological Society of London Special Publication No. 21, p. 299-320.

ROTH, P. H., 1989, Ocean circulation and calcareous nannoplankton evolution during the Jurassic and Cretaceous: Palaeogeography, Palaeoclimatology, Palaeoecology, v. 74, p. 111-126.

ROTH, P. H., AND BOWDLER, J. L., 1981, Middle Cretaceous Calcareous Nannoplankton Biogeography and Oceanography of the Atlantic Ocean: Tulsa, Society of Economic Paleontologists and Mineralogists Special Publication No. 32, p. 517-546.

ROTH, P. H., AND KRUMBACH, K. R., 1986, Middle Cretaceous calcareous nannofossil biogeography and preservation in the Atlantic and Indian Oceans: Implications for paleoceanography: Marine Micropaleontology, v. 10, p. 235-266.

SAGEMAN, B. B., 1985, High-resolution stratigraphy and paleobiology of the Hartland Shale Member: Analysis of an oxygen-deficient epicontinental sea, in Pratt, L. M., Kauffman, E. G., and Zelt, F. B., eds., Fine-grained Deposits and Biofacies of the Cretaceous Western Interior Seaway: Evidence of Cyclic Sedimentary Processes, Fieldtrip Guidebook No. 4: Tulsa, Society of Economic Paleontologists and Mineralogists, p. 110-121.

SAGEMAN, B. B., RICH, J., ARTHUR, M. A., BIRCHFIELD, G. E., AND DEAN, W. E., 1997, Evidence for Milankovich periodicities in Cenomanian-Turonian lithologic and geochemical cycles, Western Interior U.S.A.: Journal of Sedimentary Research, v. 67, p. 286-302.

THIERSTEIN, H. R., 1981, Late Cretaceous nannoplankton and the change at the Cretaceous-Tertiary boundary, in Warme, J. E., Douglas, R. G., and Winterer, E. L., eds., The Deep Sea Drilling Project—A decade of progress: Tulsa, Society of Economic Paleontologists and Mineralogists Special Publication No. 32, p. 355-394.

THIERSTEIN, H. R., AND ROTH, P. H., 1991, Stable isotopic and carbonate cyclicity in Lower Cretaceous deep-sea sediments; dominance of diagenetic effects: Marine Geology, v. 97(1-2), p. 1-34.

WATKINS, D. K., 1989, Nannoplankton productivity fluctuations and rhythmically-bedded pelagic carbonates of the Greenhorn Limestone (Upper Cretaceous): Palaeogeography, Palaeoclimatology, Palaeoecology, v. 74. p. 75-86.

WILLIAMS, J. R., AND BRALOWER, T. J., 1995, Nannofossil assemblages, fine fraction stable isotopes, and the paleoceanography of the Valanginian-Barremian (Early Cretaceous) North Sea Basin: Paleoceanography, v. 10, p. 815-839.

WOO, K. S., ANDERSON, T. F., RAILSBACK, L. B., AND SANDBERG, P. A., 1992, Oxygen isotope evidence for high-salinity surface sea water in the mid-Cretaceous Gulf of Mexico: Implications for warm, saline, deep water formation: Paleoceanography, v. 7, p. 673-685.

CENOMANIAN-SANTONIAN CALCAREOUS NANNOFOSSIL BIOSTRATIGRAPHY OF A TRANSECT OF CORES DRILLED ACROSS THE WESTERN INTERIOR SEAWAY

TIMOTHY J. BRALOWER[1] AND JAMES A. BERGEN[2]

[1]*Department of Geology, University of North Carolina, Chapel Hill, North Carolina 27599-3315*
[2]*Amoco, 501 West Lake Park Blvd., PO Box 3092, Houston, Texas 77253-3092*

ABSTRACT: The Cenomanian-Santonian calcareous nannofossil biostratigraphy of the Rebecca Bounds, Portland, and Escalante cores was investigated. Zonal markers were used to correlate the Cenomanian-Turonian boundary interval in the cores with outcrop sections in the Western Interior basin. However, it is difficult to apply existing zonations in the remainder of the section. We determined several new biohorizons that are useful for drawing correlations between the cores. These biohorizons are combined with some of the standard zonal markers in defining informal zonal units. Environmental factors appear to have reduced nannofossil diversity along the western margin of the basin and led to the premature extinction of several nannofossil markers in the Cenomanian-Turonian boundary interval.

INTRODUCTION

At its maximum extent the Western Interior Seaway stretched northwards from the Gulf of Mexico to the Arctic Ocean and westwards from Iowa to Utah (Fig. 1). Excellent exposure of sediments deposited in the center and western margin of the seaway enabled a host of paleontological investigations over the last 50 years (e.g., Cobban, 1951; Cobban and Scott, 1972; Elder, 1985). The macrofossil record of the basin is spectacular and allows detailed subdivision of the stratigraphic column. Recent paleoceanographic investigations show that the Western Interior Seaway has no modern analog; surface waters in the seaway were derived from the Gulf of Mexico, the Arctic, and from large river-delta systems draining mountainous terrane to the west (e.g., Hay et al., 1993; Kauffman and Caldwell, 1993).

The Upper Cretaceous section contains the record of five second-order sea level cycles and multiple shorter-order fluctuations (e.g., Kauffman, 1977; Hancock and Kauffman, 1979). The column has been divided into members and formations that are recognizable across the seaway (Fig. 2). Although most work was conducted on outcrop sections, coring is critical for studying the Upper Cretaceous section in much of the eastern part of the seaway where exposure diminishes and for obtaining materials less affected by surficial weathering. The Cretaceous Western Interior Seaway Drilling Project was designed to recover a transect of cores from Kansas in the east to Utah in the west (Dean and Arthur, this volume). Although macropaleontology was quite successful in dating the Escalante core (E. Kauffman, pers. commun., 1994), the other cores can only be dated using microfossil biostratigraphy. The use of planktonic foraminiferal biostratigraphy is limited by low diversity in the Escalante core and lithification in the Portland core (West and Leckie, this volume).

Nannofossils have long been known to be common in Upper Cretaceous sedimentary units of the Western Interior basin (e.g., Dawson, 1874). However, nannofossil biostratigraphic investigations of sedimentary rocks deposited in the Western Interior Seaway are limited. Much more attention has been paid to the application of nannofossil paleoecology in recreating surface-water circulation patterns (e.g., Watkins, 1986; Watkins, 1989; Fisher et al., 1994). Most biostratigraphic studies were directed at the Cenomanian-Turonian Greenhorn Cycle. This interval was first investigated by Cepek and Hay (1969) in a study of the Bunker Hill section in Kansas. Watkins (1985) described the Cenomanian-Turonian boundary nannofossil bios-

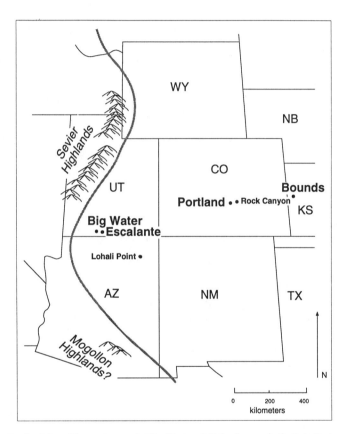

FIG. 1—Map of the western part of the Western Interior Seaway in the Late Cretaceous showing the locations of the sections investigated (in large type) and other locations discussed (in small type). Thick, gray line shows the approximate position of the western shoreline during the early Turonian *Watinoceras* Zone.

tratigraphy of the classic Rock Canyon section in central Colorado. Study of this same interval was expanded by Bralower (1988) to include sections from the Manitoba Escarpment, Canada, and Black Mesa, Arizona. The nannofossil biostratigraphy of the Coniacian-Campanian Niobrara Cycle was studied by Covington (1986) in outcrop sections of Kansas (see summary in Watkins et al., 1993). However, there are no other published studies of this interval from other sections of the Western Interior basin. The coeval section in Texas was investigated by numerous authors (e.g., Gartner, 1968; Bukry, 1969; Hill, 1976; Jiang, 1989).

The current investigation is based on observations of over

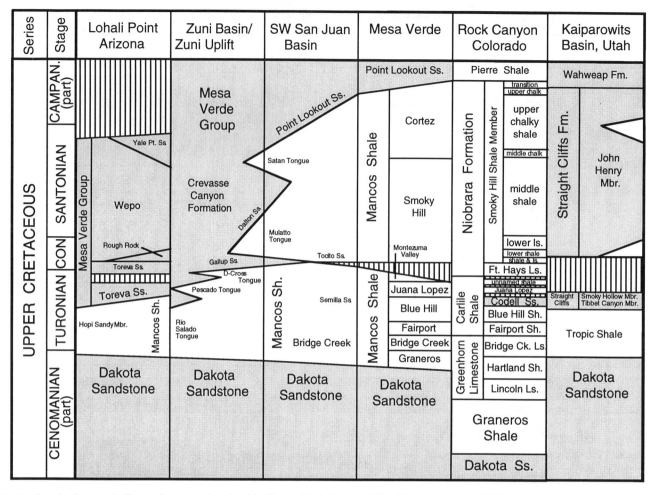

Fig. 2—Stratigraphic framework of Upper Cretaceous deposits of the Western Interior Seaway (M. Leckie, pers. commun., 1994).

1100 samples taken from the Escalante core (southern Utah), Portland core (central Colorado), Bounds core (western Kansas), and the Big Water Section (southern Utah) (Fig. 1).

METHODS

We carried out a detailed sampling of all cores, collecting lithologies including marlstone, shale, and chalk. Smear slides were prepared using the technique of Monechi and Thierstein (1985) devised for hard lithologies. All slides were observed in a transmitted light microscope at magnifications of 600x and 1000x. The spacing of samples investigated varied between 1ft and 10ft depending on the proximity to important biostratigraphic events. Relative abundances were assigned to each occurrence based on the number of specimens observed per field of view: an occurrence was described as abundant if more than 10 specimens were observed in each field, common if one to 10 specimens were observed in each field, few if one to nine specimens were observed in 10 fields of view, and rare if less than one specimen was observed in ten fields. Total nannofossil abundance in each sample was described using the same scheme. The nannofossil taxonomy applied in this investigation is standard (e.g., Perch-Nielsen, 1985; Bralower and Siesser, 1992; Varol, 1992; Bergen, 1994). All taxa identified are listed and problematic taxa are discussed in the Appendix. Five new species are described and four species recombined. Plates 1 and 2 show rep-

resentative taxa from the Bounds section.

The biostratigraphy of the Escalante and Portland cores and the Big Water Section was performed by Bralower. That of the Bounds core was performed by Bergen. Care was taken to maintain taxonomic consistency between the two authors. Because the preservation in the Bounds core is better than the other sections, considerably more taxa were observed within it.

Traditional Upper Cretaceous nannofossil biostratigraphy is based on a series of zonal and subzonal units. Zonations were established mostly in European and North African land sections (e.g., Sissingh, 1977) and DSDP sites from various ocean basins (e.g., Roth, 1978). These zonations are based partly on diagnostic taxa that can be observed in sequences from a variety of paleogeographic locations. However, the zonations are by no means applicable everywhere, and several studies were forced to apply parts of both zonations to subdivide the column (e.g., Bralower and Siesser, 1992). In other cases, including studies in the Western Interior basin (Bralower, 1988), alternate zonations were developed. Numerous non-zonal events, or biohorizons, were used to provide additional biostratigraphic information (e.g., Perch-Nielsen, 1985). In this investigation, we use the zonation of Bralower (1988) in the Cenomanian-Turonian boundary interval. In the overlying section, we apply zonations of Sissingh (1977) and Roth (1978) wherever possible. We suggest a suite of non-zonal markers applicable in future studies of the Western Interior basin.

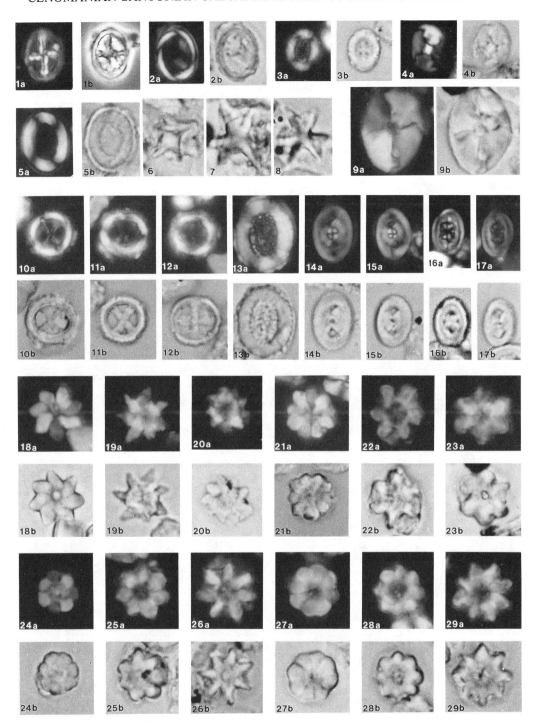

PLATE 1—1) *Vagalapilla dicandidula* n. sp., holotype, RB-920 ft; a. XP, #309-31; b. Ph, #310-31. 2) *Loxolithus* sp. 2, RB-860 ft; a. XP, #311-9; b. Tr, #312-10. 3) *Cretarhabdus multicavus*, lower Cenomanian, Djebel Feguira, Tunisia. a. XP, #311-26; b. Tr, #312-25. *4) Calculites axosuturalis* n. sp., holotype, RB-810 ft; a. XP, #311-15; b. Tr, #312-15. 5) *Loxolithus* sp. 1, RB-920 ft; a. XP, #316- 25; b. Tr, #317-30. 6) *Rhomboaster svabenickiae* n. sp., holotype, RB-820 ft, Tr, #315- 4. 7) *Liliasterites angularis* Stradner and Steinmetz, 1984, RB-840 ft, Tr, #315-13. 8) *Liliasterites angularis* Stradner and Steinmetz, 1984, RB-840 ft, Tr, #315-17. 9) *Calculites axosuturalis* n. sp., RB-820 ft; a. XP, #311-19; b. Tr, #312-19. 10) *Cylindralithus biarcus* Bukry, 1969, RB-660 ft; a. XP, #316-7; b. Tr, #317-12. 11) *Cylindralithus coronatus* Bukry, 1969, bar angle < 90°, RB- 610 ft; a. XP, #316-19; b. Tr, #317-24. 12) *Cylindralithus coronatus* Bukry, 1969, bar angle = 90°, RB-660 ft; a. XP, #316-3; b. Tr, #317-3. 13) *Miravetesina ficula* (Stover, 1966) n. comb., RB-650 ft; a. XP, #311-11; b. Tr, #312-11. 14) *Tranolithus bitraversus* (Stover, 1966) n. comb., RB-820 ft; a. XP, #309-23; b. Tr, #310-23. 15) *Tranolithus bitraversus* (Stover, 1966) n. comb., RB-820 ft; a. XP, #309-25; b. Tr, #310-25. 16) *Chiastozygus spissus* n. sp., holotype, upper Albian, Blieux, SE France; a. XP, #288-16; b. Tr, #289-16. 17) *Chiastozygus spissus* n. sp., RB-890 ft; a. XP, #311-27; b. Tr, #312-27. 18) *Lithastrinus moratus* Stover, 1966, RB-640 ft; a. XP, #313-19; b. Tr, #314-20. 19) *Lithastrinus moratus* Stover, 1966, RB-600 ft; a. XP, #313-23; b. Tr, #314-23. 20) *Radiolithus undosus* (Black, 1973) Varol, 1992, RB-990 ft; a. XP, #313-35; b. Tr, #314-35. 21) Eprolithus floralis (Stradner, 1963) Stover, 1966, RB-660 ft; a. XP, #316-13; b. Tr, #317-14. 22) *Eprolithus rarus* Varol, 1992, RB-860 ft; a. XP, #311- 34; b. Tr, #312-33. 23) *Eprolithus rarus* Varol, 1992, RB-900 ft; a. XP, #311-30; b. Tr, #312-30. 24) *Eprolithus eptapetalus* Varol, 1992, RB-940 ft; a. XP, #313-1; b. Tr, #314-2. 25) *Eprolithus eptapetalus* Varol, 1992, RB-910 ft; a. XP, #313-3; b. Tr, #314- 4. 26) *Eprolithus eptapetalus* Varol, 1992, RB-820 ft; a. XP, #313-13; b. Tr, #314-14. 27) *Eprolithus octopetalus* Varol, 1992, RB- 960 ft; a. XP, #313-27; b. Tr, #314-28. 28) *Eprolithus octopetalus* Varol, 1992, RB-940 ft; a. XP, #313-31; b. Tr, #314-32. 29) *Eprolithus octopetalus* Varol, 1992, RB-920 ft; a. XP, #313-25; b. Tr, #314-26. RB - Bounds. XP- cross-polarized light, Tr - transmitted light, Ph - Phase. All light micrographs are approximately 2500X.

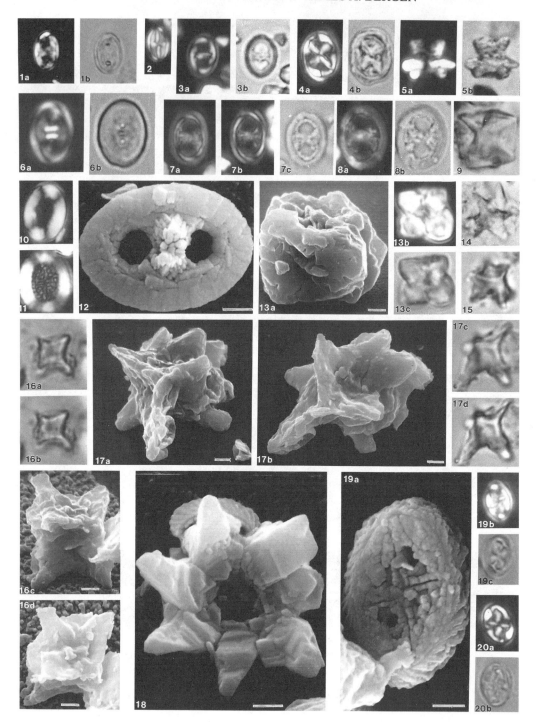

PLATE 2—1) *Percivalia hauxtonensis* Black, 1972, upper Albian, Blieux, SE France; a. XP, #235-7; b. Tr, #235-6. 2) *Acaenolithus galloisii* Black, 1972, upper Albian, Blieux, SE France, XP, #267-23. 3) *Placozygus* sp. aff. *P. spiralis*, Cenomanian, *S. varians* Zone, Culver Cliff, GB; a. XP, #276-24; b. Tr, #279-24. 4) *Helicolithus anceps* (Reinhardt and Gorka, 1967) Noël, 1970, RB-600 ft; a. XP, #323-12; b. Tr, #323-14. 5) *Discorhabdus watkinsii* n. sp., holotype, upper Cenomanian, Hammam Mellegue, Tunisia; a. XP, #276-36; b. Tr, #278-36. 6) *Reinhardtites biperforatus* (Gartner, 1968) Shafik, 1978, lower Campanian, Mississippi; a. XP, #132- 19; b. Tr, #132-22. 7) *Amphizygus brooksii* Bukry, 1969, early morphotype, Turonian, Culver Cliff, GB; a. XP, #309-15; b. XP, #309-16; c. Tr, #310-16. 8) *Amphizygus brooksii* Bukry, 1969, late morphotype, Santonian, Culver Cliff, GB; a. XP, #309-9; b. Tr, #310-10. 9) *Rhomboaster svabenickiae* n. sp., RB-820 ft; Tr; #315-2. 10) *Loxolithus* sp. 1, RB-920 ft; XP, #316-27. 11) *Miravetesina ficula* (Stover, 1966) n. comb., RB-660 ft; XP, #356-7. 12) *Amphizygus brooksii* Bukry, 1969, early morphotype, RB-600 ft; LEM 181, 10KX. 13) *Quadrum gartneri* Prins and Perch-Nielsen, 1977. RB- 600 ft; a. LEM 179B, 6KX, oblique view, 50 deg.; b. XP, LEM #353-35; c. Tr, LEM, #353-36. 14) *Liliasterites angularis* Stradner and Steinmetz, 1984, RB-790 ft; Tr; #315-10. 15-17) *Rhomboaster svabenickiae* n. sp., RB-840 ft; 15) Tr; #354-1. 16) a. Tr, LEM 354-12, 55 deg., 6100X; b. Tr, LEM 354-11, 6100X; c. LEM 187D, 5600X; d. LEM, 187C, 5600X. 17) a. LEM 186C, 50 deg., 5300X; b. LEM 186A, 5300X; c. LEM 354-7; d. Tr, LEM 354-8. 18) *Lithastrinus moratus* Stover, 1966, RB-600 ft; LEM182, 10KX. 19) *Helicolithus compactus* (Bukry, 1969) Varol and Girgis, 1994, RB-910 ft; a. LEM 197, proximal view, 10800X; b. XP, LEM 354-31; c. Tr, LEM 354-33. 20) Helicolithus compactus (Bukry, 1969) Varol and Girgis, 1994, RB-910 ft; a. XP, #354-23; b. Tr, #354-25. RB - Bounds. XP- cross-polarized light, Tr - transmitted light, Ph - Phase. LEM- transferred specimens. All light micrographs are approximately 2500X. Markers on SEMs are 1m.

Escalante Core

The Escalante core was drilled at the base of the Kaipirowits Plateau near the town of Escalante, Utah (Fig. 1). Coring retrieved a thick section of the Tropic Shale and the top of the Dakota Sandstone (Fig. 2). The Tropic Shale is marine in origin and contains nannofossils except between 75 ft (23 m) and 280 ft (85 m). The sequence can also be dated using inoceramids (E. Kauffman, pers. commun., 1994) and planktonic foraminifera (West and Leckie, this volume). The Dakota Sandstone is barren of nannofossils. The Escalante core was sampled between 50 ft (15 m) and 835 ft (255 m) every 8 in (20 cm) on average, and smear slides were prepared of all samples. Sixty-seven samples were selected for detailed biostratigraphy (Table 1). Major nannofossil biohorizons are given in Table 2.

Preservation of nannofossils in the Escalante core is moderate to good. High clay and moderate to low carbonate content inhibited significant overgrowth. Etching is moderate in many samples. Low abundances in certain intervals is probably a result of increased detrital input. The nannofossiliferous part of this expanded section, between 831 ft (253 m) and 282 ft (86 m), extends from the upper Cenomanian *Axopodorhabdus albianus* Zone to the lower Turonian *Eprolithus floralis* Zone (zonation of Bralower, 1988). *Vagalapilla octoradiata*, the first occurrence (FO) of which defines the base of the *A. albianus* Zone, is not present in the lowermost samples; however, this species is rare throughout the section. Absence of *Lucianorhabdus maleformis* in the uppermost samples studied (280 to 300 ft; 61 to 91 m) indicates that this section does not extend into the overlying upper Turonian *L. maleformis* (CC12) Zone of Sissingh (1977). The Cenomanian-Turonian boundary as defined by nannofossil biostratigraphy (last occurrence (LO) of *Microstaurus chiastius*; Bralower, 1988) lies at 816.33 ft (248.8 m), significantly below the boundary defined by inoceramids at 770 ft (334.6 m) (E. Kauffman, pers. commun., 1994).

Portland Core

The Portland core was drilled near Caon City, Colorado (Fig. 1). The sequence recovered includes the lower part of the Niobrara Formation, the Carlile Shale, the Greenhorn Formation, the Graneros Shale, and the upper part of the Dakota Sandstone (Fig. 2). Samples were taken approximately every foot between 559 ft (170 m) and 20 ft (6 m) and included claystone, marlstone and limestone. One hundred and five samples were chosen for detailed biostratigraphy (Table 3). Major nannofossil biohorizons are given in Table 2.

Preservation in the Portland core is poorer than in the other holes. Nannofossils in most samples are considerably etched and overgrown, which limited their abundance and diversity. The interval between 377 ft (103 m) and 258 ft (79 m), which corresponds to the upper Carlile Shale, Blue Hill Shale, Codell Sandstone, and Juana Lopez Calcarenite Members of the Carlile Shale, is barren of nannofossils.

The part of the Portland core sampled (559 to 20 ft; 170 to 6 m) ranges from the upper Cenomanian (probably the *V. octoradiata* Zone (Bralower, 1988) to the Santonian. It is not possible to correlate the top of the section with the zonations of Sissingh (1977) and Roth (1978) due to problems with markers of interest. *C. obscurus* occurs below the base of the underlying

zone (Table 4), and it is not possible to differentiate *L. cayeuxii* from *L. maleformis*. However, the absence of both Eprolithus floralis and Broinsonia parca indicates correlation to the Santonian. The Cenomanian-Turonian boundary is placed at the LO of *Microstaurus chiastius* at 473 ft (144 m), the Turonian-Coniacian boundary is placed between the FOs of *Kamptnerius magnificus* and *Marthasterites furcatus* within the fossil-free interval between 377 ft (103 m) and 258 ft (79 m), and the Coniacian- Santonian boundary is placed just above the LO of *E. floralis* at 170 ft (52 m).

Bounds Core

The Bounds core was drilled in Greeley County, western Kansas (Fig. 1). The sequence cored includes the Niobrara Formation, the Carlile Shale, the Greenhorn Formation, the Graneros Shale, the Dakota Sandstone (Fig. 2) and the Jurassic Morrison Formation. The core was sampled every 5 ft (1.5 m), and 86 samples were selected for detailed biostratigraphic study (Table 5). Major nannofossil biohorizons are given in Table 2.

Preservation in the Bounds core ranges from moderate to good and nannofossils are generally abundant. The interval between 675 ft (206 m) and 780 ft (238 m), which corresponds to the Blue Hill Shale, Codell Sandstone, and Juana Lopez Calcarenite Members of the Carlile Shale, is barren of nannofossils.

As in the Portland core, the top of the section cannot be correlated with the zonations of Sissingh (1977) and Roth (1978) because of problems with the markers of interest, *C. obscurus* and *L. cayeuxii*. The Cenomanian-Turonian boundary is placed at the LO of *Microstaurus chiastius* at 975.55 ft (297.3 m), the Turonian-Coniacian boundary is placed between the FOs of *Kamptnerius magnificus* and *Marthasterites furcatus*, within the fossil-free interval between 675 ft and 780 ft (206 and 238 m), and the Coniacian- Santonian boundary lies just above the LO of *Eprolithus floralis* at 570 ft (174 m).

Big Water Section

We studied an outcrop of Tropic Shale at Big Water, Utah (Fig. 1), on the southern edge of the Kaipirowits Plateau to help resolve biostratigraphic problems at the Cenomanian-Turonian boundary. The section is comprised of mudstone with a few fine siltstone layers that mark the tops of transgressive- regressive cycles (Leithold, 1994). Ten samples were studied as part of this investigation from the basal 10 m of section (Table 6). Nannofossils range from few to common and they are poorly to moderately preserved.

The Big Water section ranges from the uppermost Cenomanian part of the *Axopodorhabdus albianus* Zone to the lowermost Turonian *Eprolithus floralis* Zone (Bralower, 1988), although the order of extinctions is reversed from other Western Interior basin sections. The significance of this result is discussed below. The nannofossil definition of the Cenomanian-Turonian boundary, the LO of *M. chiastius*, lies at 3.2 m. This is below the macrofossil definition of this boundary (Leithold, 1994).

Nannofossil Biohorizons and Zonations

Few studies have addressed the nannofossil biostratigraphy of the Cenomanian-Santonian interval in the Western Interior

TABLE 1.—NANNOFOSSILS IN THE ESCALANTE CORE

SAMPLE (feet)	SAMPLE (meters)	ABUNDANCE	PRESERVATION	Ahmuellerella octoradiata	Assipetra infracretacea	Bidiscus rotatorius	Biscutum constans	Braarudosphaera bigelowii	Braloweria judithae	Broinsonia enormis	Broinsonia signata	Bukrylithus ambiguus	Calculites sp.	Chiastozygus litterarius	Chiastozygus spp.	*Corollithion achylosum*	*Corollithion ellipticum*	Corollithion exiguum	Corollithion signum	Cretarhabdus conicus	Cretarhabdus loriei	Cretarhabdus surrellus	Cribrosphaerella ehrenbergii	*Eiffelithus eximius*	Eiffelithus turriseiffellii	Eprolithus floralis	*Eprolithus octopetalus*	*Eprolithus eptapetalus*	Eprolithus rarus	Flabellites oblongus	Gartnerago segmentatum	Grantarhabdus coronadventis	Helicolithus anceps	Lithastrinus moratus	*Lithraphidites acutum*	Lithraphidites carniolensis	Manivitella pemmatoidea	Markalius circumradiatus	Microrhabdulus decoratus	*Microstaurus chiastius*	Nannoconus sp.	Placozygus spiralis	Prediscosphaera cretacea	Prediscosphaera spinosa	Quadrum gartneri	Radiolithus planus	Rhagodiscus achlyostaurion	Rhagodiscus angustus	*Rhagodiscus asper*	Rhagodiscus splendens	Rotellapillus laffittei	Scapholithus fossilis	Sollasites horticus	Tegumentum stradneri	Tetrapodorhabdus decorus	Tranolithus gabalus	Tranolithus orionatus	Vagalapilla sp.	Vagalapilla stradneri	Watznaueria barnesae	Watznaueria biporta	Watznaueria ovata	Zeugrhabdotus embergeri	Zygodiscus diplogrammus	Zygodiscus elegans	Zygodiscus erectus	Zygodiscus sp.
74.00	22.52	VF	M																																							R																										
282.00	85.84	R	M												R										R																																											R		
290.00	88.27	R	M				R													R					R																				R																	R				R		

(Table continues — remaining sample rows from 296.66 / 90.30 through 831.00 / 252.94 give rare (R), few (F), common (C), or abundant (A) occurrences across the listed species columns.)

Nannofossil abundances: R, rare, F, few, C, common, A, abundant.
Nannofossil preservation: G, good; M, moderate; P, poor.
Species shown in bold are markers.

TABLE 2.–MAJOR NANNOFOSSIL BIOHORIZONS

EVENT	Bounds	Portland	Escalante
LO *E. floralis*	570.00	170.00	--
FO *M. staurophora*	630.00	195.00	--
LO *E. eptapetalus*	780.00	195.00	314.66
FO *M. furcatus*	670.00	258.00	--
FO *K. magnificus*	670.00	258.00	--
FO *L. maleformis*	920.00	258.00	--
LO *C. achylosum*	880.00	393.00	490.03
LO *E. octopetalus*	920.00	--	560.00
FO *E. eptapetalus*	962.00	426.00	680.66
LO *R. asper*	974.25	470.00	816.33
LO *M. chiastius*	975.55	473.17	816.33
FO *E. octopetalus*	978.16	--	770.33
LO *C. loriei*	980.00	477.17	827.00
FO *E. eximius*	--	483.00	810.66
LO *C. kennedyi*	980.00	488.08	--
LO *L. acutum*	980.00	--	830.33
LO *A. albianus*	983.00	488.08	--

Bentonite levels

C	967.0	468.0	771.0
B	976.0	477.0	805.7
A	983.0	483.5	825.0

All levels are given in feet. Dashes (--) indicate that biohorizon was not determined.
LO, last occurrrence; FO, first occurrence

Seaway. Covington (1986; summarized in Watkins et al., 1993) studied the Coniacian-Campanian interval of the type Smoky Hill Chalk in western Kansas. Major nannofossil zones of Sissingh (1977) were delineated, even though some zonal markers were exceptionally rare.

Problems with application of the Sissingh (1977) and Roth (1978) zones in Upper Cretaceous Western Interior basin sections are summarized in Table 4. These include: (1) reversed order of markers, which makes zonal units undefinable; (2) variability in taxonomic concepts between workers; and (3) rare and sporadic occurrences of zonal markers which renders zonal definitions inaccurate.

Results from the Bounds core indicate that the FOs of *Marthasterites furcatus* and *Micula staurophora* are close to the base and in the middle of the Fort Hays Limestone, respectively, while Covington (1986) reported these events from near the top of the Fort Hays Limestone and at the base of the overlying Smoky Hill Chalk, respectively. The Fort Hays Limestone in the Bounds core lies unconformably on the Codell Sandstone, and this might explain the discrepancy in the FO of *M. furcatus*. However, we do not expect the Fort Hays-Smoky Hill transition to be this diachronous within Kansas, nor the range of the common species, *M. staurophora* to be so time- transgressive. Evidence that the FO of *M. staurophora* was determined precisely in the Bounds

core is the similar relationship of this event with the LO of *Miravetesina ficula* in this section and at Culver Cliff, Isle of Wight, England (Bergen, unpublished data). In both sections, the LO of *M. ficula* is immediately below the FO of *M. staurophora*.

In a study of the Cenomanian/Turonian boundary interval (including sections outside of the Western Interior basin), Bralower (1988) resolved problems with existing zonations by erecting new zones. The current investigation includes only sections within the basin, and we are therefore reluctant to define new zones, since they may not be applicable elsewhere. However, the Bounds and Escalante cores, in particular, provide us with the opportunity to determine the ranges of a variety of non-zonal markers in the Turonian-Coniacian interval, which may allow us to establish new zones in the future once these ranges are confirmed elsewhere. For example, the biostratigraphy of species of *Eprolithus* is similar to Varol (1992), but minor differences still need to be resolved.

Potential biohorizons include (Fig. 3, from bottom to top): the LO of *Gartnerago nanum* (Rebecca Bounds (RB)-1025.75 ft, 312.6 m), the LO of *Cretarhabdus loriei* (RB-980 ft, 298.7 m; Escalante (ES)-827 ft, 252.1 m), the FO of *E. octopetalus* (RB-978.16 ft, 298.1 m; ES-770.33 ft, 234.8 m), the LO of *Eprolithus eptapetalus* (RB-780 ft, 237.7 m; ES- 314.66 ft, 95.9 m; Port-

TABLE 3.—NANNOFOSSILS IN THE PORTLAND CORE

SAMPLE (feet)	SAMPLE (meters)	ABUNDANCE	PRESERVATION	Ahmuellerella octoradiata	Assipetra infracretacea	Axopodorhabdus albianus	Bidiscus rotatorius	Biscutum constans	Braarudosphaera regularis	Braloweria judithae	Broinsonia enormis	Broinsonia signata	Calculites obscurus	Chiastozygus litterarius	Chiastozygus spp.	Corollithion achylosum	Corollithion ellipticum	Corollithion exiguum	Corollithion kennedyi	Corollithion signum	Cretarhabdus conicus	Cretarhabdus loriei	Cretarhabdus surirellus	Cribrosphaerella ehrenbergii	Cyclagelosphaera margerelii	Eiffellithus eximius	Eiffellithus turriseiffelii	Eprolithus floralis	Eprolithus rarus	Eprolithus eptapetalus	Eprolithus octopetalus	Flabellites oblongus	Gartnerago segmentatum	Grantarhabdus coronadventis	Helicolithus anceps	Kamptnerius magnificus	Lithastrinus grillii	Lithastrinus moratus	Lithraphidites carniolensis	Lucianorhabdus maleformis	Manivitella pemmatoidea	Markalius circumradiatus	Marthasterites furcatus	Microstaurus chiastius	Micula staurophora	Nannoconus sp.	Placozygus spiralis	Prediscosphaera cretacea	Prediscosphaera spinosa	Quadrum gartneri	Radiolithus planus	Reinhardtites anthophorus	Rhagodiscus achlyostaurion	Rhagodiscus angustus	Rhagodiscus asper	Rhagodiscus splendens	Rotellapillus laffitei	Tetrapodorhabdus decorus	Tranolithus gabalus	Tranolithus orionatus	Vagalapilla sp.	Vagalapilla stradneri	Watznaueria barnesae	Watznaueria ovata	Zeugrhabdotus embergeri	Zygodiscus diplogrammus	Zygodiscus elegans	Zygodiscus sp.
20.00	6.09	A	P	R				R	R			C										R	R	R	R								R					R	R					R	R								R		C					F				
25.00	7.61	C	P	R			R			R	R	R				R				R	R			R					R								R	R					R	R	R								R		R	R	R		F		R	R	R
30.00	9.13	F	P				R		R			R		R						R	R							R	R	R		R				R	R					R					R				R		F				
35.00	10.65	A	P			R	R		R		R									R	R	R			R	R		R					R						R	R	R		R	R	C										
40.00	12.18	F	P					R	R					R						R	R	R			R					R				R	R	R				R	R		F	R		R					
45.00	13.70	F	P	R				R	R	R										R			R					R	R	R					R	R	F				R				
50.00	15.22	C	P			R				R					R	R	R	R			R				R		R	R		R			F							R					R
55.00	16.74	F	P	R		R	R	R	R	R				R	R			R	R				R		R			R	R		F				
60.00	18.26	C	P	R		R	R	R	R			R	R				R		R	R	R	R			R		F				R	
65.00	19.78	C	P	R		R			R	R			R	R	R	R		R		F				C		R	R	
70.00	21.31	C	P	R		R	R	R	R	R		R	R			R		R	R	R	R	R			R		R	R	F		R	
75.00	22.83	C	P	R		R		R			R			R	R	R		R			R		C		R	
80.00	24.35	C	P	R		R	R		R	R	R		R	R	R			R		R	F	R	
85.00	25.87	C	P	R		R	R	R	R		R	R		R	R	R			R		C	
90.00	27.39	A	P	R		R	R	R	R	R		R		R	R	R		R		R		F	
95.00	28.92	F	P	R		R	R	R			R	R			R		C	R	
100.00	30.44	C	P			R	R	R		R	R		R	R	R		R	R		R	R	F	R	
106.00	32.26	C	P			R	R	R	R		R	F		R			R		C	R	R	
110.00	33.48	F	P	R	R	R	R	R	R	R	R		R	R		R	R		R		R	F		R	
116.00	35.31	C	P			R	R	R		R	R		F		R		R	R	C	
120.00	36.53	C	P			R	R	R	R	R	R	R		R		R	F		R	R	R		R	F	
124.00	37.74	C	P			R	R		R		F		R	R	R	A	
130.00	39.57	C	P	R		R	R	R	R	R		R		R	R	R	F		R	R	R	R	R	C		R	
136.00	41.40	C	M	R		R	R	R	R		R		R	R	C		R	
140.00	42.61	C	P	R		R	R	R	R	R	R		R	R		R	R	R	R	R	R	R	R	F		R	R
146.00	44.44	A	M	F	R	R	R	R		R		R	R	R	R	A	
150.00	45.66	A	P	R		R	R	R		R	R	R		F	A	
155.00	47.18	A	P	R		R	R	R		R	R	R		R		R	C	R	
160.00	48.70	A	P	R		R	R		R		R	R	R	R		R	A	R	R	
165.00	50.22	C	P	R		R	R	R		R	R		R		A	R	
170.00	51.74	F	P	R		R	R	R		R	R	R		R		R	C	R	
175.00	53.27	C	P	R		R	R		R	R		R	R	C	R	R	
180.00	54.79	C	P	R		R	R	R		R	R	R	R		R	R	C		
185.00	56.31	A	P	R	R	R	R	R	R		R	R		R	R		R	R	A	R	R	R	
191.00	58.14	A	P	R	R	R	R	R	R	R	R		R		R	R	F		C	R	R	
195.00	59.35	C	P	R	R	R	R	R	R	R		R		R	R	R	R	R		R	C	R	R	
200.00	60.88	C	P	R	R	R	R	R	R	R	R		R		R	R	R	R	R		R	C	R	R	
205.00	62.40	A	P	R	R	R	R	R	R	R	R	R		R		R	R		R	A	R	R	
210.00	63.92	A	P	R	R	R	R	R	R	R	R		R	R	R	R	R		R	A	R	
215.00	65.44	C	P	R	R	R		R		R	R		R	C	
220.00	66.96	A	P	R	R	R		R		R	R	R		R	A	
225.00	68.49	A	P	R	R	R	R	R	R	R		R	R		R	R	R	R	C	R	
230.00	70.01	F	P	C	R		R		
235.00	71.53	A	P	R	R	R	R	R	R		R	R		R	A	R	R	
240.00	73.05	F	P	R	R	R		R		R	F	R	
245.00	74.57	F	P	R	R	R		R		F	
250.00	76.10	C	P	R	R	R	R	R		R	R		R	R	F	R	R	
256.00	77.92	F	P	R	R	R	R		R		R	F	
258.00	78.53	C	M	R	R	R	R	R	R	R	R	R	R	R	R		R	F	R	R	C	R	R	R	R
377.00	114.75	R	M	F	R		R		R		R		R	
380.00	115.66	A	G	R	R	F	R	R	R	R	R	R		R	F	R	R	R	R	C	R	C	R	F	F
383.00	116.58	C	M	R	R	R	R	R	R	R	R	R	R		R	F	R	R	F	R	C	R	R	R
386.00	117.49	A	P	R	R	R	R	R	R	R	R	R	R	R		F	R	R	C	R	C	R	R	R	F
390.00	118.71	A	M	R	R	R	R	R	R	R		F		F	R	R	R	A	C	R	R	F
393.00	119.62	A	M	R	F	R	R	R	R	R	R	R	R		F	R	R	F	C	R	C	R	R
396.00	120.53	C	M	R	R	R	R	R	R	R		F	R	R	F	C	R	R
400.00	121.75	C	G	R	R	F	R	R	R	R	R	C	R	R		R	R	R	C	R	C	R	R	R
403.00	122.67	C	P	R	R	R	R	R	R	R	R		R	R	R	C	R	R	R
406.00	123.58	A	M	R	R	R	R	R		R	F	R	C	R	R	F
410.00	124.80	A	M	R	R	R	R	R	R	R	R		R	F	C	R	R
413.00	125.71	C	M	R	R	R	R	R	R	R	R	R	R		R	R	F	R	C	F	R

Nannofossil abundances: R, rare; F, few; C, common; A, abundant.
Nannofossil preservation: G, good; M, moderate; P, poor.

SAMPLE (feet)	SAMPLE (meters)	ABUNDANCE	PRESERVATION	Ahmuellerella octoradiata	Assipetra infracretacea	Axopodorhabdus albianus	Bidiscus rotatorius	Biscutum constans	Braarudosphaera regularis	Braloweria judithae	Broinsonia enormis	Broinsonia signata	Calculites obscurus	Chiastozygus litterarius	Chiastozygus spp.	Corollithion achylosum	Corollithion ellipticum	Corollithion exiguum	Corollithion kennedyi	Corollithion signum	Cretarhabdus conicus	Cretarhabdus loriei	Cretarhabdus surirellus	Cribrosphaerella ehrenbergii	Cyclagelosphaera margerelii	Eiffellithus eximius	Eiffellithus turriseiffelii	Eprolithus floralis	Eprolithus rarus	Eprolithus eptapetalus	Eprolithus octopetalus	Flabellites oblongus	Gartnerago segmentatum	Grantarhabdus coronadventis	Helicolithus anceps	Kamptnerius magnificus	Lithastrinus grillii	Lithastrinus moratus	Lithraphidites carniolensis	Lucianorhabdus maleformis	Manivitella pemmatoidea	Markalius circumradiatus	Marthasterites furcatus	Microstaurus chiastius	Micula staurophora	Nannoconus sp.	Placozygus spiralis	Prediscosphaera cretacea	Prediscosphaera spinosa	Quadrum gartneri	Radiolithus planus	Reinhardtites anthophorus	Rhagodiscus achlyostaurion	Rhagodiscus angustus	Rhagodiscus asper	Rhagodiscus splendens	Rotellapillus laffittei	Tetrapodorhabdus decorus	Tranolithus gabalus	Tranolithus orionatus	Vagalapilla sp.	Vagalapilla stradneri	Watznaueria barnesae	Watznaueria ovata	Zeugrhabdotus embergeri	Zygodiscus diplogrammus	Zygodiscus elegans	Zygodiscus sp.
416.00	126.62	A	M				R			R	R		R							R		R		R	R	R	R		R		R		R						R		R							R	R				R								F	R	A	R		F	R	R
420.00	127.84	A	M	R			R			R	R											R	R		R	R		F		R																		F								R	R				F	R	A			R	R
423.00	128.75	A	P				R	R		R	R		R			R						R	R	R	R	R		R	R																	R	R									F			C	R			R	R	R
426.00	129.67	A	M				R				R		R			R	R		R	R		R	R					R																R						R		F			C				R	R	R
430.00	130.88	A	M				R		R		R	R	R			R			F	R				R	R														R	R						F			C	R	R	R
433.00	131.80	A	P	R			R		R		R	R			R	R	R	R			R	R					R	R							F	R	A	R	R	R	R
436.00	132.71	A	M				R	R	R	R		R	R			R	R	R	F	R		R	F					R	F			R		R		R	F	C	R	R	R	R
440.00	133.93	A	M	R			R	R		R		R		R	R	F	R	R	R	R	R	R	R	R	R	C	R	R
444.00	135.14	C	P			F		R	R	R	R	R	R	R	R	R	R	R	C	R	R	R	R
447.00	136.06	C	P			F		R	R	R	R	R	R	R	R	R	R	R	C	R	R	R	R
450.00	136.97	F	P	R	C	R	R	R	R	R	R	R	F	R	R	R	R
456.80	139.04	F	M	R	R	R	R	R	R	R	R	R	R	R	R	C	R	R	R
459.00	139.71	A	P	R	R	R	R	R	R	R	R	R	F	R	R
461.00	140.32	C	P	R	R	R R	R	R	R R	R	R	R	R	C	R
463.20	140.99	F	P	R	R	R	R	F	R	R
465.17	141.59	C	P	R	R	R	R	R	R	R	C	R
467.00	142.15	C	P	R	R R	F	R R	R	R	R R	R	C	R R R
467.83	142.40	C	P	R	R R	R	R R	R	R	A	R	R
469.00	142.75	F	P	R	R R	R	R R R	R	R	R R	C	R R R
470.00	143.06	C	P	R	R R	R	R R	R	R	R R	C	R R R
471.00	143.36	C	P	F	R	R	R	R R	R	C	R
472.08	143.69	C	P	R	R	R	R R	R	R R	R	R	R	C	R R
473.17	144.02	C	P	R	R R	R	R	R R	R	R	R R R	R R	F	R R R
475.00	144.58	F	P	R R	R R	R	R	R R	F	R	R
477.17	145.24	C	M	R	F	R R R	R F	R R	R R R	F	R R	R R R R	C	F R F
479.00	145.80	C	P	R	R R R	R	R	R R	R	R	C	R R R
479.83	146.05	R	P	R	R	R	R	C	R
480.92	146.38	C	P	F R	F R	F	R R	R F	R R R R	R	C	R R F
482.08	146.74	A	P	R	R	R	R	R	R R F	R R	R	R F	F R R R R R R F	C	R R F
483.00	147.02	A	P	R	F	R	R	R	F	R F	R	R F	F R R	R R	A	R R F
483.92	147.30	A	P	F	R R	R R	F	F R	R	R R	F	R R R	R	C	R F
484.92	147.60	A	P	F	R R R	R R	R	R R	R	R R	R	A R R R R R R
486.17	147.98	F	P	R	R	R	R	F	R	F
487.08	148.26	F	P	R	R	R	R	R	R	F
488.08	148.56	A	P	R	R	R	R	R	R	R R	R	R	R	R	A	R R R
489.00	148.84	C	P	F	R	R R	R	R R	R	R	R	R R	R	R	C	R R F
491.00	149.45	A	P	R	F	R R	R	R	R F	R	R	R R R	R	R C	R
500.00	152.19	A	M	R	R	R R	R	R R R R	R R	R	R	R R	R R	R	A	R R
510.00	155.23	C	P	R	F	R	R	R R	R	R R	R	C	C	R R R F
520.00	158.28	A	P	R R R	R	R R R	R R	R	R	R F	R	A	R R C R
530.00	161.32	F	P	R R	R R	R R R	R R	R	R	R R R	R	F	R R R
540.00	164.37	A	M	R	F	R	R	R	R R R	R R	R	R	R R	R F R	R	R R	F	C	R F F R
550.00	167.41	A	M	R R F	R R	R R R	R	R	F R	R R	R R R	R R	R R R R	C	R C	R R R F
559.00	170.15	A	M	R F	R	R	R R	R R	R R	R	F	R	R R R R R	F	C	R F R

Nannofosil abundances: R, rare; F, few; C, common; A, abundant.
Nannofossil preservation: G, good; M, moderate; P, poor.

TABLE 4.– PROBLEMS WITH CURRENT CENOMANIAN-SANTONIAN ZONAL SCHEMES

Zone	Definition		Problems with base event
	Base	Top	
Sissingh (1977)			
CC17-*Calculites obscurus*	FO nominate	FO *B. parca*	Lies below base of zone below
CC16-*Lucianorhabdus cayeuxii* preserv.	FO nominate	FO *C. obscurus*	Confusion with *L. maleformis* in poor
CC15-*Reinhardtites anthophorus*	FO nominate	FO *L. cayeuxii*	Taxonomic uncertainties affect range
CC14-*Micula staurophora*	FO nominate	FO *R. anthophorus*	None
CC13-*Marthasterites furcatus*	FO nominate	FO *M. staurophora*	None
CC12-*Lucianorhabdus maleformis*	FO nominate	FO *M. furcatus*	None
CC11-*Tetralithus pyramidus*	FO nominate	FO *L. maleformis*	Rare, sporadic in early part
CC10-*Microrhabdulus decoratus*	FO nominate	FO *T. pyramidus*	Rare, sporadic in early part
Roth (1978)			
NC17-*Tetralithus obscurus-Micula concava*	FO nominate	FO *B. parca*	Lies below base of zone below
NC16-*Broinsonia lacunosa*	FO nominate	FOs *T. obscurus-M. concava*	Lies in Albian
NC15-*Marthasterites furcatus*	FOs nominate-*E. eximius*	FO *B. lacunosa*	FO of *E. eximius* lies near Cenomanian/Turonian boundary
NC14-*Kamptnerius magnificus*	FO nominate-*E. eximius*	FOs *M. furcatus-*	None
NC13-*Micula staurophora*	FOs nominate-*T. pyramidus*	FO *K. magnificus*	Lies above base of zone above
NC12-*Gartnerago obliquum*	FO nominate-LO *L. acutum*	FOs *M. staurophora-T. pyramidus*	Lies below base of zone below
NC11-*Lithraphidites acutum*	FO nominate	LO nominate-FO *G. obliquum*	None

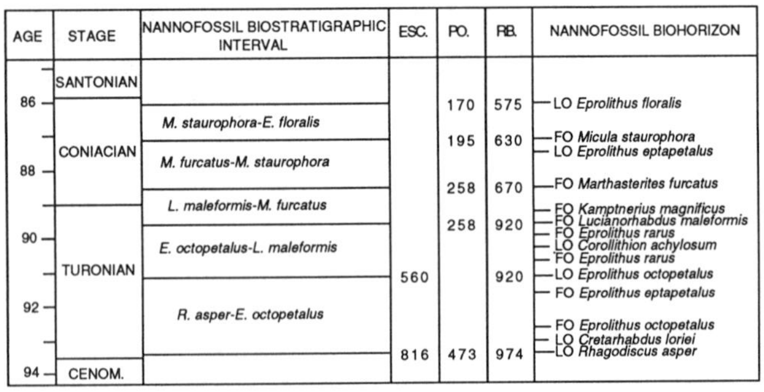

FIG. 3.--Stratigraphic summary of the Turonian-Santonian interval showing informal nannofossil intervals defined here and position of other biohorizons. Depths of horizons are given in feet for the Escalante (ESC.), Portland (PO.), and Bounds (BO.) cores. FO, first occurrence; LO, last occurrence.

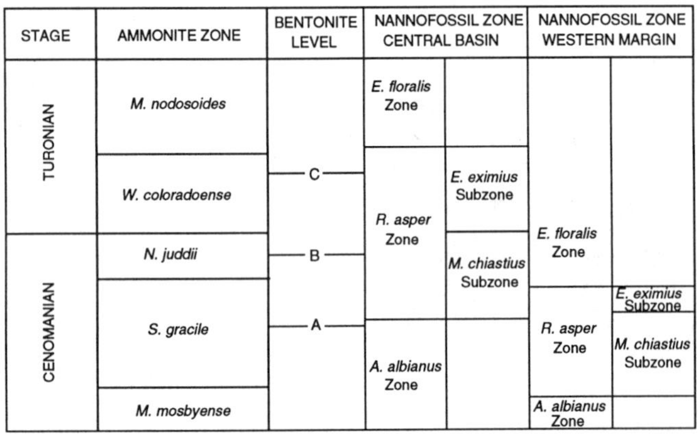

FIG. 4.--Diachroneity of nannofossil zones and subzones of Bralower (1988) in central basin (e.g., Portland, Bounds and Pueblo Rock Canyon section) and western margin locations (e.g., Escalante core and Big Water section) compared to ammonite zones and bentonite levels. Stratigraphic levels of bentonite beds A, B, and C are from Elder (1988).

land (PO)-195 ft, 59.4 m), the LOs of *Percivalia fenestrata* (RB-920 ft, 280.4 m) and *E. octopetalus* (RB-920 ft, 280.4 m; ES-560 ft, 170.7 m), the FO of *C. exiguum* (RB-920 ft, 280.4 m), the FO of *E. rarus* (RB-910 ft, 277.4 m), the LO of *Corollithion achylosum* (RB-880 ft, 268.2 m; ES- 490.03 ft, 149.4 m), the FO of *Calculites axosuturalis* (RB-860 ft, 262.1 m), the LO of *Radiolithus planus* (RB-850 ft, 259.1 m), the LO of *Calculites axosuturalis* (RB-800 ft, 243.8 m), the FO of *Eprolithus eptapetalus* (RB-962 ft, 293.2 m; ES- 680.66 ft, 207.5 m; PO-426 ft, 129.8 m), the LO of *E. rarus* (RB-860 ft, 262.1 m), the FOs of *Lithastrinus moratus* (RB-670 ft, 204.2 m) and *Reinhardtites biperforatus* (RB 660 ft, 201.2 m).

We define informal biostratigraphic intervals for correlation within the Western Interior and name these intervals after the two species which define their bases and tops. We propose no

intervals for the Cenomanian-Turonian boundary interval because an applicable zonation already exists (Bralower, 1988). For the lower Turonian-Santonian, these units include (Fig. 3): *Rhagodiscus asper-Eprolithus octopetalus* interval, *E. octopetalus-Lucianorhabdus maleformis* interval, *L. maleformis-Marthasterites furcatus* interval, *M. furcatus-Micula staurophora* interval (equivalent to M. furcatus (CC13) Zone of Sissingh (1977)), and *M. staurophora-Eprolithus floralis* interval. Problems remain with using several of the biohorizons to correlate between the sections studied because the relative order of events differs (Table 2). This results from poor nannofossil preservation in the Portland core and environmental factors that appear to have affected the ranges of taxa in the Escalante core.

TABLE 5.—NANNOFOSSILS IN THE BOUNDS CORE

SAMPLE (feet)	SAMPLE (meters)	Acaenolithus galloisii	Ahmuellerella octoradiata	Amphizygus brooksii (early form)	Amphizygus brooksii (late form)	Assipetra terebrodentarius	Axopodorhabdus albianus	Axopodorhabdus dietzmanii	Bidiscus rotatorius	Bifidalithus geminicatillus	Biscutum constans	Braarudosphaera bigelowii	Broinsonia dentata	Broinsonia enormis	Broinsonia hanfieldii	Bukrylithus ambiguus	Calculites axosuturalis	Calculites obscurus	Calculites ovalis	Chiastozygus fessus	Chiastozygus garrisonii	Chiastozygus litterarius	Chiastozygus platyrhethus	Chiastozygus propagulis	Chiastozygus spissus	Corollithion achylosum	Corollithion exiguum	Corollithion kennedyi	Corollithion madagascarensis	Corollithion signum	Cretarhabdus conicus	Cretarhabdus loriei	Cretarhabdus multicavus	Cretarhabdus octofenestratus	Cretarhabdus schizobrachiatus	Cretarhabdus surirellus	Cribrosphaerella circula	Cribrosphaerella ehrenbergii	Cyclagelosphaera margerelii	Cylindralithus asymmetricus	Cylindralithus biarcus	Cylindralithus coronatus (=90 degrees)	Cylindralithus coronatus (<90 degrees)	Cylindralithus nudus	Darwinilithus pentarhethum	Discorhabdus watkinsii	Eiffellithus eximius	Eiffellithus turriseiffelii	Ellipsagelosphaera britannica	Eprolithus eptapetalus	Eprolithus floralis	Eprolithus octopetalus	Eprolithus rarus
500.0	152.2		R	R	F			R			S	R	C					R	R	C	S				S					S	R	R				F							C		F		C	F				
510.0	155.2		R	F	F		S	S			S		F					S	R	F	S	R						S				R		R		S			R	R			F		C	F				
520.0	158.3		R	R	F			R				R	F		S			S	F		S	S		S				S	R			F	S	F	S	C			R	F			C		F					s
530.0	161.3		R	F	R			S		S	R	R	F		S			S	R	F	R			S				S	S			R	S	S		R	R	S		R	F	S		C		F				
540.0	164.4		S	R		S		F		F		F	F	R	S		S		C	S	R		S				R	R	S		R	R		R	S	R			R	S	R		F		R				
550.0	167.4		S	R				S			S		F					C							R				R	S				R		S	R			F	F					
560.0	170.5		R	R					S	F	?	S		S	R	C									R				S		R	S	R		R	F				C	F					
570.0	173.5		R	F	R			S	S	S		R	F		S				C		S				R			S	R		S	R		R		S			F	S	F			C		F				S
580.0	176.5			F			S	S	S	S		F	R	S					F	R	S				R	S			R	S		S	S	F					F	S	F	R		C		F				R
590.0	179.6		R					S				S	R						R	S								S		F	R		F	S	R			F	S	R			F		R				R
600.0	182.6		F	C				S		R		F	F	R					C		R							S	R		F		R	S	R	F		C	R	C	S		F		C				R
610.0	185.7		R	F		R		R	R	R		R	F	S					C		R			S	r			S	S	S	R	S			F	S	F	R	F	R	F	R		C		R				F
620.0	188.7		S	R		F		S	S	S	R	S		S	R	S						S	R						S			S	F	S	R	S	S		F	S	R	R		R	R	S	s		F	
630.0	191.8			R				R	S	R	R	R	F		S		R	R	F		S	S		S					S	S				S	R	R	R		R	R			F		F				R
640.0	194.8			S				R	S	R	R	F	F		S					F		F					S				R	S	S		R	R	R	F	R	R	R		F		F				R
650.0	197.8		S	R				S		R	R	F	F	R					F		F								S			F			R		S	R		R			R	F		F				
660.0	200.9		R	F				S	S	R	S	F	F		S			S	S		F	S							S			S	S		F	F		R	R	R	R	R		F	F				
670.0	203.9		S	F				S		S	R	R	R						F	F		R							S		R	R		R	R	R	F	S	S	R			F	R		C				
790.0	240.5		R					R	S										S						S				S										S							S				
800.0	243.5		R					F	F				R													S				R													R		S				
810.0	246.5		R					R	R	F																			S					R	S								S						
820.0	249.6		F					R	F	F			R	S		S			S						R			R	R						F	R							F		F			R	S
830.0	252.6							R	R	R	S	R	S	•		R		R			R	S	R	R				R	S	R	R					F	R	R						R		S	R			
840.0	255.7		R	S				S	R	R	F	F		R	S					R								R			R	R											F		R	R				
850.0	258.7		R	S				R		R	S	F	R		R	S		R	R			S						R	S			R	S		S		R	R					R		R	R				
860.0	261.8		R	S				S		R	S	R	R	S	S	R		R										R	S	S								F	S					F		R	R			S
870.0	264.8		S	R					S	S	F	R	R	S								R						R									R	F					R		F	R	S			S
880.0	267.9		R					S		R	R	F	S			R		F			R	F				S	R			S								R							F	R	R				
890.0	270.9		R	R				S			R	R	R		R		R		R	F					S	R			R	R	S		F		F	R			S			F	R	R		S
900.0	273.9		R	S				R	R	R	R	R	S	S		R		R	F							S	R	S	R	R			R		R	R				F	R	R		R
910.0	277.0		R	R		R		S	R	R	R	R	R				R	F	F		R	R			S	S	R				R					R	R								F	R	R		
920.0	280.0		R					R	S	R	R		R	F		F		R	S				R	R		R	R				F		R							C		R	R	S	
930.0	283.1		R	R				S	R	R	R		R			S	R		R	S						S		R			R	R									F		R		R	
940.0	286.1		R					S	R	R	R		R		F	S		S	R						S						S				F	R	R	S				R	F	F	
950.0	289.2		S					R	S	R	R				R						S				R						F							F		R			F	F	R	
960.0	292.2		R	S				R	R	R	R		R			S		S	S						S	S					S	R	S				F				S	R	S		C	R	F	
962.0	292.8		R	R				R	R	S	R	S	c				S				S	R	S		R	C	R	C								S		F	R	F	R	
964.8	293.7		S						S	S	R	S	R							R	R		S				R	R				F		F					F		S	S	
970.0	295.2		S	S	S		R		R	R				R	R			R		R			S					R	R	F				F				F			
972.6	296.0							R		R	S	S					R				R									F				F					F			
974.3	296.5	?		R		R		R	S	R	S			?			S	R			S	R					R			F	R						F				F	S
976.5	297.2	R		R				R	R				R	F		R	R			S	S		R	C		F						C					
977.1	297.4	S		F				R	R	R	R				R	S	R	R			S	S	F	C				S			C		C	S	
978.2	297.7			S					S					S				S	S	s						S			R	S	F	S	
979.0	298.0	R		R				S	R	R			S			R	R			R	S					R	S	R			C		F	
980.0	298.3	S	R					R	R	R				R				R	R	S		R	S			R	S	R			F			
981.1	298.6	R	R		R			R			R			F				R	R	S		S				C		S	R		C			
983.0	299.2	S	R				S		R	S			R	S		R	S				R	S	R	S				F			F		F	
984.5	299.7	S				R	S			R	S			S	R		R	S	R		S	S	?			F		R			F	F	
987.5	300.6	R		R		R	R		F			R	S		R	S	S	R	S	R	R	S	R	R	R	F	S	F		R	F	R		F	
990.0	301.3	R	R		R	F	R		R	R	R				F		R	R	R				R		S	R	R	F		R	F	
994.2	302.6		R					R		R	R	R		R	F			R	R	R	S	F	R		S	F	S		C		R	
1000.0	304.4	S		R				R	R	R		S		R				S	F		S	S				S		S			F		C	
1005.2	306.0		S		R			R	R	R	S		R	F		S		S	R		S	R	S	S	R		F		R	F		R	S		F		C
1010.0	307.4	R		S	F			R	R	R		S		R	R	S				S	R	S	R			S	R	F		R	R		R		F		C
1016.3	309.3			F				R	R	S	R	R		R		C				R	R	R	S	S	F		C		R	F		F	R		F		C
1020.0	310.5	R						R	R	R	R	R			F	S		R		R	S	R	S		R	F		R	F		R	S		F		C
1025.8	312.2			R	S	S		R	S	R	R	R		R		F		R		R	R	F	R	F		R	F		R	R		F	
1030.0	313.5			R	R			R	S	R	R	R		R	S		R	R	R	F		R	S	R		R			F		C	
1035.0	315.0			S	R			R		R	R	F		R	R					R			R	R		R	R		F		C	
1040.0	316.6	R		S	R	F		R	R	R		R	S		R	R	R	S	R	R	S		R		S			F		C	
1045.0	318.1	S		S				R	R	R	R		S	F	S	R	S	R		S	R				R			F		F	
1050.0	319.6	S		S				R	R			S	S	S					R					F			F	
1055.2	321.2	S		R	R			S	S	R		R	R	S				S	R	R	S	R					F	?	F	
1060.0	322.6			S				R	S	R		S	S		S		R	R			R	R				F		F	
1065.0	324.2			S	S			R	S	R			R	R	S	R	R		F				S			F	S	R	
1070.0	325.7	R		R	R			R		F	R		R	R		S	S				S		S			F	S	F	
1074.8	327.2			S	S			R	F	?			S	R	S			R				R			F	R	S	
1080.0	328.7	S		S				R				S		S	R	S	F			S			R	S	
1084.8	330.2													S										R	S	

Nannofossil abundances: R, rare; F, few; C, common; A, abundant; ?, questionable occurrence. Lower case letters denote reworked specimens.

Preservation: G, good; M, moderate; P, poor.

TABLE 5.—(Continued)

SAMPLE (feet)	SAMPLE (meters)	*Flabellites oblongus*	*Gartnerago nanum*	*Gartnerago obliquum*	*Grantarhabdus coronadventis*	*Helicolithus anceps*	*Helicolithus compactus*	*Helicolithus torunicus*	***Kamptnerius magnificus***	*Kamptnerius punctatus*	*Liliasterites angularis*	*Lithastrinus grillii*	*Lithastrinus moratus*	***Lithraphidites acutum***	*Lithraphidites alatus*	*Lithraphidites carniolensis*	*Loxolithus sp. 1*	*Loxolithus sp. 2*	*Lordia xenota*	*Lucianorhabdus cayeuxii*	*Lucianorhabdus compactus*	***Lucianorhabdus maleformis***	*Manivitella pemmatoidea*	*Markalius circumradiatus*	***Marthasterites furcatus***	*Microrhabdulus decoratus*	*Microrhabdulus helicoideus*	***Microstaurus chiastius***	*Micula cubiformis*	***Micula staurophora***	*Miravetesina ficula*	*Nannoconus elongatus*	*Nannoconus multicadus*	*Octocyclus magnus*	*Orastrum campanensis*	*Percivalia fenestrata*	*Percivalia hauxtonensis*	*Pervilithus varius*	*Placozygus spiralis*	*Placozygus aff. P. spiralis*	*Polypodorhabdus madingleyensis*	*Prediscosphaera columnata*	*Prediscosphaera cretacea*	*Prediscosphaera spinosa*	*Prediscosphaera stoveri*	*Quadrum eneabrachium*	***Quadrum gartneri***	*Quadrum intermedium*	*Quadrum octobrachium*	*Radiolithus orbiculatus*	*Radiolithus planus*	*Radiolithus undosus*	***Reinhardtites anthophorus***
500.0	152.2			F	R	C		S	F	R			R			S	R			R		F	R		R		S		R	F					S								F		R								R
510.0	155.2					C				R			F			S				R		F			S	S	S		S	C					S				R	S			C	S	S								R
520.0	158.3			F	R	C	S	S	R	R			F			R	S					S	R		R	S	R		R	C										S			F		R		S						S
530.0	161.3			C	R	C		S	R	S			F				S			S		F	R	S	R				R	C		S			R					R			C	S	S								S
540.0	164.4			R	S	F	S	S	R	R		R	C			R	S	S		R		F	R			S	R		S	R		S								S			C	S	S		S						S
550.0	167.4			F	S	R			F	R		S	F									S		S	R				R	C										S			R	S								r	
560.0	170.5			F		F	S	S		R		R	F							R		R	R		F		R		R	C										S			C	S									S
570.0	173.5			R		R				S		S	F			S							R		F				R	F		S											F		S	S							?
580.0	176.5	S		F	S	F	S		S	S		R	F			S				S		S	F	S	F	S			R	R									R				C	S		S	R						?
590.0	179.6	S		F	R	S			S			F	R											S	R				S	C													C				S						?
600.0	182.6	R		F	F	C	R		S				C			R	R						R	S	F	R				S		S							R	S	R		C				R	R					?
610.0	185.7	R		R	R	F	R		S				F			R	R			S		S	R	R	R	S				R		S						S	S	R	R		C	S	S	S	R	S	S				?
620.0	188.7	R		R	R	R	S		R				C			R	S	R					R	S	C		S			F										S	R		F	S		S	R	?	R	S			?
630.0	191.8	R		R	R	R	S		R				R			R	R			F		R	R	R	S					S			R							R	S		C	S		R	R	S	R				?
640.0	194.8	R		R	R	R			S				F			S	R			R		S	R	R	F	S					R		R		S			S		S			C	S	S	S	F	S		S			
650.0	197.8	R		F	R	R	S		S				R			S	S						R	R	C						R	S								S			C				F	R		R			
660.0	200.9	R		R	R	F	S		F				F			R	R					S	R	R	A						F												C	S			R	S		S			
670.0	203.9	R		R	R	F	R		R				F			R						S	R	S	R						R									S			C				F			S			
790.0	240.5			F																																				S			R	S									
800.0	243.5			R							s																																F										
810.0	246.5			R																																							R										
820.0	249.6			F												R																								R	R		F	R			S						
830.0	252.6						S																																				F	R			S						
840.0	255.7			F		R	R				R							S																						S			F	R			S						
850.0	258.7			R							R					S		R				R									S									R			F	R							S		
860.0	261.8			F	R		R											S				R																		R	R		F	S							S	R	
870.0	264.8	R		R			R										S					F	S																	S			C				S				?		
880.0	267.9	R		F	S		R										S					R	R																	R			F				S				S		
890.0	270.9	R		R	S	S	R										R					R		R							S									R	R		C	R			S						
900.0	273.9	R		R			F										R					R	R	S																F			F	R							S		
910.0	277.0	R		F		R	F										R					R	R	S																R	R		C	S			S				S		
920.0	280.0	R		R	R	S	R									R	R					F	R													R				R			F	R							R		
930.0	283.1	R		R	S	R	F									S	R						R	R												S					R		F	S			R				R		
940.0	286.1	R		R	S	S	R										R						S																	R			F				S				R		
950.0	289.2	R		R	R		R																R													S				R			F	S			S				R		
960.0	292.2	R		R	R		R									R	R						R													S				R	R		F				S				S		
962.0	292.8	R		R	R	S	R										R						R	R																S		R	C	R			R				F		
964.8	293.7	S			S		R										S					c																			S		R								S		
970.0	295.2	R		R	S	S	F									S	R						R	S												S				R	R		R	S							R		
972.6	296.0	S					R																																			R	F										
974.3	296.5	S		R	S	S	F									S	R						R	S								S										R	F	S							R		
976.5	297.2	R			R		R									S	S						R	R				R															F	S							S		
977.1	297.4	R		S	R	S	R									R	R						R	R				S								S		S		R	S		C				S				R		
978.2	297.7				S		S										S						S																				R								S		
979.0	298.0	R		S	S	S	R									R	S				S		S	R												S							R	S							R		
980.0	298.3	R			R	R	R							S		S	R						R	R				R												R	R		F								R		
981.1	298.6	R			R		F							S		S	R						S	R				F								S				S			C								R		
983.0	299.2	R		S	S	S	R							S		S	R				S		R	R				R								S				S			R								S		
984.5	299.7	S				S	R							S	S	S	S						R					F								S				S			F								R		
987.5	300.6	R		S	R	R	S							R	R	R	S						R	R				R												S	S	R	F								R		
990.0	301.3	R		S		S	R							R		S	S						R	S				F								S				R			C									R	
994.2	302.6	S		R	F	S	R							S	R	R	R						R					R									R			R	S	R	F	R							R		
1000.0	304.4			S			R							R		S							R	R				S												S			F								R		
1005.2	306.0				S	S	R							S	R	S	S						R	S				R									S			S	S	R	F	S							R		
1010.0	307.4	R					S								S	S	S											S												R			F										
1016.3	309.3				R		R								R	R	R						R														S			S	S	?	C	S							R		
1020.0	310.5	F			R																		R																	S			R										
1025.8	312.2		R		R		S									S							R											S		R					R	?	C	S							R		
1030.0	313.5	S			F		R																R	S				R									R			S			C	S									
1035.0	315.0	S			R		R							S			S				S		R													S	R			S	S	S	F	R							F		
1040.0	316.6	S	S		F		S								S	S							R														S			R	R		C	R							R		
1045.0	318.1				F		R								S	S	S	R					R					R								R				S	R	R	F	S							R		
1050.0	319.6		R		R									R									R					S												S	R		F	R							S		
1055.2	321.2		R		R											R							R					R									S			R		S	F	S							R		
1060.0	322.6		R		R		R												?				R																	R			C	R							R		
1065.0	324.2	S	S		S												R		S				S					S												S		S	F	R							R		
1070.0	325.7		S				R										S		R		S							R								S	S			R			F	R							S		
1074.8	327.2		R				S										S		R				R					S				S								R		R	F	S							R		
1080.0	328.7		R		R														R				R																				F	S									
1084.8	330.2		S																				R																				F										

Nannofossil abundances: R, rare; F, few; C, common; A, abundant; ?, questionable occurrence. Lower case letters denote reworked specimens.
Preservation: G, good; M, moderate; P, poor.

SAMPLE (feet)	SAMPLE (meters)	Reinhardtites biperforatus	Reinhardtites scitula	Reinhardtites sisyphus	Repagulum parvidentatum	Rhagodiscus achlyostaurion	Rhagodiscus angustus	*Rhagodiscus asper*	Rhagodiscus reniformis	Rhagodiscus splendens	Rotelapillus crenulatus	Rhomboaster svabenickiae	Scampanella cornuta	Sollasites horticus	Tegumentum stradneri	Tetrapodorhabdus coptensis	Tetrapodorhabdus decorus	Tranolithus bitraversus	Tranolithus exiguus	Tranolithus gabalus	Tranolithus minimus	Tranolithus orionatus	Tubodiscus jurapelagicus	Vagalapilla ara	Vagalapilla dibrachiata	Vagalapilla dicandidula	Vagalapilla elliptica	Vagalapilla gausorhethium	Vagalapilla imbricata	Watznaueria barnesae	Watznaueria biporta	Watznaueria fossacincta	Watznaueria ovata	Watznaueria porta	Watznaueria quadriradiata	Watznaueria virginica	Zeugrhabdotus embergeri	Zeugrhabdotus theta	Zeugrhabdotus trivectis	Zeugrhabdotus sp. A	Zygodiscus bicrescenticus	Zygodiscus diplogrammus	Zygodiscus elegans	Zygodiscus wynnhayii
500.0	152.2	R	?	R		R		S	S		R				R	R	R	S	F					S		R	C	S		R				R					R	R	S		R		
510.0	155.2	F	S	F			S		S		S	S					S	S	F	S		R		S			S	S	C	S		S	R	R				R	S				R		
520.0	158.3	F	S	F			S		S					S				S	F	S		F					S	R	A	S			R	S				S	R	R				R	
530.0	161.3	F	R	F	S		R				S	R	S				S		F	R		F				R		R	A	S		S	S	R				R	S	R				R	
540.0	164.4	F	R	C	S		R		S	S	R		S		R		R	S		S	R	F		S	S		R	S	R	A			R	R		R		R	R	S				R	
550.0	167.4	S	R	F	S		S		S	S	S	R				R	S	S	F	S				S	S	C	S			R				R	R	S				S		
560.0	170.5	F	R	C			S	S		S	R				S	S	S	S	F	S	S	R		R	C			S	S		S	R	S				S	R			
570.0	173.5	R	S	F			S		S	S	R	F			S	S	R	S	R		S		R	S		C			S	S	S		S		S	S	R		
580.0	176.5	R	S	R			S		S		R	S	S	S	R		R	S	R	R		C	S		R	S	R	C			S		R	R	S	F	R	
590.0	179.6	R	S	F					S	R	S				S	R	S		F				S	C	S		S	S	S		
600.0	182.6	R	R	F			S		S	R	R			S	S	R	R	R	R	R	F	?	R	R	A	S		S	R	S		F		S	R	R	S	R
610.0	185.7	S	R	F		R	S		S	S	R			S	R	R	S	F	R	R	F	R	S	S	A		R		R		R	F	R	R
620.0	188.7	R	S	R			S		S		S	S	R	S	R	S	R	C	S		S	R	R		
630.0	191.8	R	S	R	S	S	S	R	S	R	S	F	R	S	C	R	R	C	S	F	S	F	R	S
640.0	194.8	R	R	F	R	R	S	S	R	S	R	R	S	F	S	S	C	S	R	S	F	R	F	R	S
650.0	197.8	S	F	R			S	R	S	S	R	S	F	?	A	R	F	F	S	R	R
660.0	200.9	R	F	R			S	S	S	S	F	R	R	A	R	R	S	R	?	F	R	S
670.0	203.9	S	R	R	S		S	?	S	R	R	S	F	A	S	R	R	S	R	R	S
790.0	240.5			S		S	F	R	R	F	R	S
800.0	243.5	S		S	S	R	F	R	F	R	S
810.0	246.5	S		R	F	F	S	F	R	R
820.0	249.6	S	R	R	R	C	R	F	R	R
830.0	252.6	R	S	R	S	C	S	R	F	R	F	S	R	R
840.0	255.7	R	R R	S	R	F	S	R	S	F	R	S	S	s	R	R	R	R
850.0	258.7	F	F	S	s	S	R	S	F	R	R	S	F	S	R	R	F	R
860.0	261.8	F	R	R	S	S	F	F	S	R	R	S	S	R	F
870.0	264.8	F	S	s	S	R	R	F	R	S	S	R	S	S	F	R
880.0	267.9	F	S	S	R	S	S	S	R	C	S	S	S	R	R	C	R	S	R	R	F	F
890.0	270.9	R	R	R	S	S	S	S	R	F	R	S	R	r	R	R	C	S	R	S	R	R	F
900.0	273.9	F	R	S	R	S	R	R	F	S	S	R	R	C	R	R	R	R	F	F
910.0	277.0	R	R	R	R	S	R	S	R	R	R	S	R	C	S	R	R	R	R	F
920.0	280.0	R	F	R	R	F	R	R	R	F	S	S	R	R	C	S	R	R	R	R	C	R
930.0	283.1	F	S	R	R	R	R	F	R	S	R	S	R	C	R	S	R	R	F
940.0	286.1	F	S	R	S	R	S	R	R	S	R	S	C	S	R	R	R	R	F
950.0	289.2	F	R	R	S	R	R	R	S	S	C	S	S	R	R	R	R	C
960.0	292.2	F	F	R	R	R	R	C	R	S	S	R	F	F	R
962.0	292.8	F	R	R	s	R	S	S	R	S	F	S	C	R	R	F	R	R	F	R
964.8	293.7	R	F	R	S	S	S	F	S	C	S	S	S	R	R	F
970.0	295.2	S	R	S	S	R	S	R	S	R	C	S	R	R	R	F	F
972.6	296.0	F	S	R	C	R	R	S	R	F
974.3	296.5	R	F	R	R	R	R	C	R	R	R	C	S	S	F	F
976.5	297.2	R	S	S	R	R	S	S	S	S	S	R	?	A	S	S	F	S	F
977.1	297.4	R	F	S	R	R	?	S	S	R	R	R	R	A	R	R	C	R	R	F	F
978.2	297.7	R	F	S	S	S	S	C	S	S	R	S	R	R	R
979.0	298.0	R	R	R	S	R	R	S	S	S	S	R	C	F	R	S	F	F
980.0	298.3	R	R	S	S	R	R	R	S	S	F	R	R	C	R	F	R	S	F	F
981.1	298.6	R	C	R	F	S	R	S	S	S	R	R	S	A	S	F	F	R	R
983.0	299.2	R	F	R	R	R	S	S	R	R	S	F	C	R	F	S	R	F
984.5	299.7	R	F	R	R	R	S	S	R	S	R	S	C	R	R	F	C
987.5	300.6	R	F	R	R	R	R	S	S	R	S	R	R	S	R	C	S	R	R	R	F	R	C
990.0	301.3	R	F	R	R	R	R	S	R	R	S	R	S	S	R	C	R	S	R	R	R
994.2	302.6	F	R	R	F	R	R	S	S	S	S	R	F	R	S	C	R	R	R	R	R	F	F
1000.0	304.4	R	R	S	R	R	S	S	R	S	R	S	C	R	R	R	R	R	F	F
1005.2	306.0	F	R	R	R	R	R	S	R	S	F	S	C	R	S	F	R	C
1010.0	307.4	R	S	S	S	R	R	R	F	S	S	R	R	C	F	S	S	R	R	R	C
1016.3	309.3	F	R	S	R	R	S	S	F	R	C	R	A	R	R	S	R	R	C	R	C
1020.0	310.5	R	R	R	C	R	S	F	R	S	A	R	R	F	R	S
1025.8	312.2	F	C	S	F	R	S	R	S	R	R	S	F	R	A	R	S	R	S	C	R	R
1030.0	313.5	R	R	R	R	R	R	R	S	R	F	S	R	A	S	F	F	R	F	R
1035.0	315.0	F	R	R	R	S	S	R	C	S	S	C	R	F	S	R	S	R	R	F	S
1040.0	316.6	R	R	R	R	R	R	S	R	C	S	S	R	C	R	F	S	R	R	S	R	S	R	R
1045.0	318.1	R	R	S	R	R	S	S	F	R	R	S	R	C	R	F	S	S	S	S	F	R	R
1050.0	319.6	F	R	R	R	S	S	R	F	S	C	R	R	S	R	R	R
1055.2	321.2	F	R	R	S	R	S	S	R	S	C	R	F	R	S	R	R	F	R
1060.0	322.6	C	R	R	S	R	R	S	R	F	C	F	S	R	S	R	R	R
1065.0	324.2	F	R	R	R	R	R	R	S	S	R	C	R	R	S	R	R	S
1070.0	325.7	F	R	R	R	R	R	S	S	S	F	C	R	R	S	R	R	R	S
1074.8	327.2	F	R	R	S	R	S	S	R	C	F	S	R	R	F	R
1080.0	328.7	F	S	S	S	R	R	S	C	C	R	R	S	S	R	C	R
1084.8	330.2	S	R	F	C	S	R	F

Nannofossil abundances: R, rare; F, few; C, common; A, abundant; ?, questionable occurrence. Lower case letters denote reworked specimens.
Preservation: G, good; M, moderate; P, poor.

Sample (meters above base)	Abundance	Preservation	Ahmuellerella octoradiata	Axopodorhabdus albianus	Biscutum constans	Braloweria judithae	Broinsonia enormis	Broinsonia signata	Chiastozygus litterarius	Chiastozygus spp.	Corollithion achylosum	Corollithion signum	Cretarhabdus surirellus	Cribrosphaerella ehrenbergii	Eiffellithus turriseiffellii	Eprolithus floralis	Flabellites oblongus	Gartnerago segmentatum	Grantarhabdus coronadventis	Helicolithus anceps	Manivitella pemmatoidea	Microrhabdulus decoratus	Microstaurus chiastius	Prediscosphaera cretacea	Prediscosphaera spinosa	Radiolithus planus	Rhagodiscus achlyostaurion	Rhagodiscus angustus	Tranolithus gabalus	Tranolithus orionatus	Vagalapilla sp.	Vagalapilla stradneri	Watznaueria barnesae	Watznaueria ovata	Zeugrhabdotus embergeri	Zygodiscus diplogrammus	Zygodiscus elegans
10.10	C	M			R				R	R			R	R						R		R			R				R		R			F			R	R
9.00	C	P			R					R			R	R					R	R					R				R		R			F		R	R	R
8.20	C	P	R		R						R	R			R					R					R				R					C		R	R	R
7.00	C	M	R	R	R			R		R			R	R						R					R		R		R	R	R		C	R			R	F
6.00	C	P		R	R					R		R	R		F					R	R				R				R		R		C			R	R	R
5.10	C	M		R				R			R	C			R	R				R					R		R		F	R	C			R			R	R
4.00	C	M	F	R	R			R	R		R		R	R	R	R	R	R		R	R				R	F	R		C				R	R
3.20	C	M			R	R	R		R		R	R		R		R	R	R	R		R	R	R	R	F	R	R	C	R	R	F	F
1.90	F	M			R			R		R		R	R		R	R	R			R				R		R		F	R		F	R
0.90	F	P			R						R	F			R				R				F			F		R	R

Cenomanian-Turonian Boundary Problem

The apparent diachroneity of calcareous nannofossil events and traditional macrofossil zones around the Cenomanian-Turonian boundary was discussed by Bralower (1988). Although markers are often sparse, ammonite and inoceramid zones appear to fit with lithostratigraphic correlations between sections provided by bentonite and limestone marker beds (e.g., Elder, 1987, 1988, 1991) whereas nannofossil zones and events do not. If we assume that the suite of bentonites and limestone markers that exist in this part of the section (e.g., Elder, 1988) are correlated correctly between outcrop sections and the cores (see Dean and Arthur, this volume), then this problem persists at the Cenomanian-Turonian boundary. Patterns of diachroneity emerge once the transect of cores is combined and studied with sites already investigated (Bralower, 1988).

The sequence of events described in the section at Rock Canyon near Pueblo Colorado (Fig. 1) (Watkins, 1985; Bralower, 1988) was determined in the Portland and Bounds cores (Table 2; Fig 4) with almost no difference in order. The biostratigraphic correlations in these central-basin locations appear to be almost consistent with those provided by three major bentonite marker beds A, B, and C (Elder, 1988). A minor disparity exists in the Portland core. Here, the LOs of *Axopodorhabdus albianus* and *Corollithion kennedyi* occur almost 5 ft below bentonite level A, whereas at Rock Canyon and in the Bounds core these events occur at or slightly above, bentonite A. This disparity can be explained by poor preservation in the Portland core.

Significant correlation problems exist in marginal, western-basin locations, the Escalante core and the Big Water section. In the Escalante core, marker species are as abundant as those in central-basin locations (Table 1); in the Big Water section markers are exceptionally rare and show sporadic occurrences (Table 6). In both sections the nannofossil biohorizons of interest occur substantially lower with respect to marker bentonites and inoceramid zonal boundaries than in the central-basin localities (Fig. 4). This is similar to other western margin sections, the Blue Point and Lohali Point Sections in Black Mesa, Arizona (Bralower, 1988).

Two possible explanations exist for this discrepancy. First, the bentonites may have been correlated incorrectly and the nannofossil events are broadly synchronous. Even though no unequivocal data (trace, rare earth elements, mineralogy) have been published finger-printing the bentonites, their correlation is constrained by macrofossil zonal boundaries, the distribution of marker limestones in the sections of interest (e.g., Elder, 1987,

1988), and the position of the ä¹³C excursion (e.g., Pratt and Threlkeld, 1985). If the nannofossil events are viewed as synchronous, then all of these other markers must be time-transgressive.

The second possible explanation for the lack of correlation of nannofossil biozones among marginal and basin-center sections is that an environmental factor caused the early extinction of several nannoplankton taxa in the latest Cenomanian, a few hundred thousand years before they became extinct in the main part of the seaway. This explanation is more likely given that nannofossil diversity decreases towards the margin of the basin and that planktonic foraminiferal faunas also change dramatically (e.g., West and Leckie, this volume).

Three possible environmental explanations exist. The first is a decrease in salinity below a critical level in the western margin of the basin, which is a feasible possibility given the proximity of the western margin sequences to large river/delta systems (e.g., Pratt, 1984; Leithold, 1994). The interval of interest is dominantly transgressive; with distance to shoreline increasing, one would expect the proportion of run-off to decrease and salinity to rise. However, this effect could be outweighed by an interval of increased precipitation. The ecological affinities of all but one of the species of interest are unknown. That species, *Rhagodiscus asper*, is associated with warm surface waters (e.g., Mutterlose, 1989). The abundance of *R. asper* in the brackish Early Cretaceous North Sea Basin (e.g., Williams and Bralower, 1995) indicates that it was probably tolerant of low salinity conditions.

A second environmental factor that might explain the lack of correlation of nannofossil biozones is a change in the source of surface water along the western margin. Surface waters in the Western Interior Seaway came from the Arctic, the Gulf of Mexico, and run-off from mountainous areas to the west (e.g., Leckie, 1985; Watkins, 1986; Hay et al., 1993; Fisher et al., 1994; Leckie et al., this volume). Influx of a new surface-water mass with different properties could have led to premature extinction of nannoplankton species. However, most of the taxa of interest were widely distributed in the seaway prior to their extinctions (e.g., Watkins, 1986; Bralower 1988), and thus this cause seems less likely than the others.

The third environmental condition which might have affected nannoplankton in these marginal, western locations is fertility. As fertility increased in the latest Cenomanian (e.g., Arthur et al., 1987), marginal environments may have been affected more than the open ocean, which decreased the diversity of nannoplankton and planktonic foraminifera. The species that became extinct, already rare in this marginal location, might have been among those that disappeared first. These extinctions occurred later in the central basin locations when fertilities reached critical levels. None of the species that disappeared is known to be tolerant of high fertility.

CONCLUSIONS

A study of the calcareous nannofossil biostratigraphy of three Cenomanian-Santonian sequences was carried out as a part of the Western Interior drilling transect. We find that existing zonations cannot be used to subdivide much of the section. We propose several new biohorizons which have potential for correlation between the cores and for use in future zonation. Environmental factors are thought to have led to reduced nannofossil diversity and premature extinction of nannofossil markers in the Cenomanian-Turonian boundary interval along the western margin of the seaway.

ACKNOWLEDGMENTS

We are grateful to Michael Arthur and Walter Dean for inviting us to participate in the WIKS drilling transect. We thank the staff of the USGS Core Research Center for their efficient sampling. We are grateful to Mark Leckie, Erle Kauffman, and Lonnie Leithold for providing biostratigraphic and lithostratigraphic information, Joel Bond, Celeste Burns, and Jonathan Grubbs for smear slide preparation, and Walt Dean, Jim Pospichal and Dave Watkins for helpful reviews.

APPENDIX

TAXONOMY

In the following, informal remarks are made concerning the identification of numerous species that have been described in detail elsewhere. Five new species are described and four new combinations are proposed. Slides containing the holotypes of the new species are stored in the permanent calcareous nannofossil reference collections in Applied Paleontologic Technology (Amoco, Houston, Texas). Reference numbers indicated for the holotypes pertain to their storage in the Amoco collections.

Acaenolithus galloisii Black, 1972 (Plate 2, fig. 2)

1972 *Acaenolithus galloisii* n. sp. Black (partim), pp. 57-58, Pl. 21, Fig. 12 (non Pl. 21, Figs. 8, 9, 13, 14).

Discussion: The holotype illustrated by Black (1972) has a different rim construction than other illustrated specimens and appears atypical of the genus *Acaenolithus*. Specimens attributed to this species exhibit an extinction pattern more akin to murolith species placed in the genus *Gartnerago*. In cross-polarized light, the rim exhibits a tricyclic extinction pattern in which: (1) the outer cycle is narrow and faintly birefringent; (2) the middle cycle is narrow and displays a bright, first-order white birefringence; and (3) the inner cycle is broad and faintly birefringent. The small central area of this species is diamond-shaped and nearly filled by a simple, axial cross. The length of this species is between 4-5 m. It is distinguished from *Gartnerago praeobliquum* Jakubowski (1986) by its small central area and the absence of footed terminations on the axial cross.
Stratigraphic Range: upper Albian to lower Turonian.

Ahmuellerella octoradiata (Gorka, 1957) Reinhardt, 1964

Amphizygus brooksii Bukry, 1969 (Plate 2, Figs. 7, 8, 12)

Discussion: In cross-polarized light, early morphotypes of this species (Cenomanian to Coniacian) are characterized by their more prominent, spiraled inner rim cycles and relatively narrow outer rim cycles (Plate, 2, Fig. 7). The birefringent inner rim cycle is greatly reduced at the expense of the faintly birefringent outer rim cycle in the late morphotype, which ranges from late Coniacian to early Maastrichtian (Plate 2, Fig. 8). The species is further distinguished by the two central yoke cycles described by Bukry (1969).
Stratigraphic Range: upper Cenomanian to lower Maastrichtian.

Assipetra infracretacea (Thierstein, 1972) Roth, 1973
Assipetra terebrodentarius (Applegate et al., 1987) Rutledge and Bergen, 1994
Axopodorhabdus albianus (Black, 1967) Wind and Wise, 1977
Axopodorhabdus dietzmannii (Reinhardt, 1965) Wind and Wise, 1977
Bidiscus rotatorius Bukry, 1969 Bifidalithus geminicatillus Varol, 1991
Biscutum constans (Gorka, 1957) Black, 1959
Braarudosphaera africana Stradner, 1961
Braarudosphaera bigelowii (Gran and Braarud, 1935) Deflandre, 1947
Braarudosphaera regularis Black, 1973
Braloweria judithae (Black, 1972) Crux, 1991
Broinsonia dentata Bukry, 1969
Broinsonia enormis (Shumenko, 1968) Manivit, 1971
Broinsonia handfieldii Bukry, 1969
Broinsonia signata (Nöel, 1969) Nöel, 1970 Bukrylithus ambiguus Black, 1971

Calculites axosuturalis Bergen, n. sp. (Plate 1, Figs. 4, 9)

Diagnosis: An elliptical, medium- to large-sized species of *Calculites* exhibiting a faint, gray birefringence and possessing irregular sutures nearly aligned with the major ellipse axes. A central stem may be present.

Description (Light Microscope): Specimens are divided into four quadrants by irregular sutures that are nearly aligned with the major ellipse axes. Specimens exhibit a faint gray birefringence in plan view. A solid, circular stem-base may be present at the center of the holococcolith.

Discussion: This species is distinguished from similar holococcoliths by its faint birefringence. It is distinguished from *Calculites anfractus* Jakubowski (1986) by the orientation of its sutures and central stem-base.

Known Stratigraphic Range: Turonian section of the Bounds core.

Holotype: Plate 1, Fig. 4.

Type level: Turonian, 810 ft.

Type locality: Bounds core, Kansas.

Calculites obscurus (Deflandre, 1959) Prins and Sissingh, 1977
Calculites ovalis (Stradner, 1963) Prins and Sissingh, 1977
Calculites sp.
Chiastozygus fessus (Stover, 1966) Shafik, 1979
Chiastozygus garrisonii Bukry, 1969
Chiastozygus litterarius (Gorka, 1957) Manivit, 1971
Chiastozygus platyrhethus Hill, 1976
Chiastozygus propagulis Bukry, 1969

Chiastozygus spissus Bergen, n. sp. (Plate 1, Figs. 16, 17)

Diagnosis: A medium-sized, normally elliptical murolith with a broad rim and small central area nearly filled by a diagonal central structure. Faintly birefringent.

Description (Light Microscope): The broad rim exhibits a faint gray birefringence and is divided into two cycles. The small central area is nearly filled by a diagonal central structure whose birefringence is similar to the rim. The bars of the diagonal central structure are broad and divided by longitudinal sutures.

Etymology: Latin: spissus = close, thick, dense.

Discussion: The uniform, faint birefringence of both the broad rim and central structure (both are divided by extinction lines) distinguish this species from other known species of *Chiastozygus*. *Zygodiscus bicrescenticus* (Stover, 1966) has a similar construction, but its central structure spans the transverse axis.

Known Stratigraphic Range: upper Albian to upper Turonian.

Holotype: Plate 1, Fig. 16.

Type level: upper Albian, 74.45 m.

Type locality: Blieux, southeastern France.

Chiastozygus sp.
Corollithion achylosum (Stover, 1966) Thierstein, 1971
Corollithion ellipticum Bukry, 1969
Corollithion exiguum Stradner, 1961
Corollithion kennedyi Crux, 1981
Corollithion madagascarensis Perch-Nielsen, 1973
Corollithion signum Stradner, 1963
Cretarhabdus conicus Bramlette and Martini, 1964 *Cretarhabdus loriei* Gartner, 1968

Cretarhabdus multicavus Bukry, 1969 (Plate 1, Fig. 3)

Discussion: This species has a central plate with a reinforced longitudinal suture. Two longitudinal rows of perforations are present, although only four central perforations are observed in early morphotypes (e.g., Plate 1, Fig. 3). In contrast, Maastrichtian specimens are much larger and possess two distinct rows of longitudinal perforations.

Stratigraphic Range: Albian-Maastrichtian.

Cretarhabdus octofenestratus Bralower, 1989
Cretarhabdus schizobrachiatus Gartner, 1969
Cretarhabdus surirellus (Deflandre, 1954) Reinhardt, 1970
Cribrosphaerella circula (Risatti, 1973) Verbeek, 1976
Cribrosphaerella ehrenbergii (Arkhangelsky, 1912) Deflandre, 1952
Cyclagelosphaera margerelii Noël, 1965

Cylindralithus asymmetricus Bukry, 1969

Discussion: This species is distinguished from other *Cylindralithus* by its narrow cross-bars, which form an H-shape.

Stratigraphic Range: lower Turonian-lower Campanian.

Cylindralithus biarcus Bukry, 1969 (Plate 1, Fig. 10)

Discussion: This species is distinguished from other species of *Cylindralithus* by its broad arcuate central structure which forms a low angle.

Stratigraphic Range: Coniacian-middle Maastrichtian.

Cylindralithus coronatus Bukry, 1969 (Plate 1, Figs. 11, 12)

Discussion: This species is distinguished from other species of *Cylindralithus* by its central structure which forms a cross. Bukry (1969) accounted for variations in bar angle between 67 and 90o for the species. Forms with bar angles of 90o do not range above the Coniacian, while those with bar angles less than 90o range into the early Campanian. *C. coronatus* is distinguished from *Corollithion achylosum* Stover, 1966 by its higher distal shield.

Cylindralithus nudus Bukry, 1969
Darwinilithus pentarhethum Watkins, 1984
Diazomatolithus lehmannii (Nol, 1965)

Discorhabdus watkinsii Bergen, n. sp. (Plate 2, Fig. 5)

Diagnosis: A species of *Discorhabdus* with a very low, broad, and hollow distal projection. The proximal shield is reduced and its width similar to that of the distal projection. Description (Light Microscope): Circular, placolith constructed of two shields of non-imbricate elements with radial sutures. The width of the distal shield is approximately twice that of the proximal shield. The large central area is roughly equivalent to the distal shield width. A short, broad, hollow distal projection occupies the central area and flares distally.

Etymology: named in honor of Dr. David Watkins.

Discussion: True *Discorhabdus* species (i.e., radiate placoliths having a distal projection) are more typical of the Jurassic and earliest Cretaceous.

Known Stratigraphic Range: This extremely rare species is restricted to Cenomanian section in the Bounds core, southeastern France (Vergons) and Tunisia (Hammam Mellegue).

Holotype: Plate 2, Fig. 5.

Type level: upper Cenomanian.

Type locality: Hammam Mellegue, Tunisia.

Eiffellithus eximius (Stover, 1966) Perch-Nielsen, 1968
Eiffellithus turriseiffelii (Deflandre, 1954) Reinhardt, 1965
Ellipsagelosphaera britannica (Stradner, 1963) Perch-Nielsen, 1968

Genus *Eprolithus* Stover, 1966

Discussion: *Eprolithus* and *Radiolithus* are Polycyclolithaceae with non-imbricate wall elements. *Eprolithus* is retained herein for the Aptian to Santonian portion of a lineage including the four species discussed in this paper (see below). This lineage involves the successive reduction from 9 rays (*E. floralis*) to 6 rays (*E. rarus*), but also records the progressive change in dominance from brick-like elements to petal-like elements. Varol (1992) used the difference in brick-like and petal-like elements to separate *Eprolithus* and *Radiolithus*. However, *Radiolithus* is retained for species having a low wall, whose birefringence is equal to that of the central diaphragm.

Eprolithus eptapetalus Varol, 1992 (Plate 1, Figs. 24-26)

Discussion: Polycyclolithaceae with a central diaphragm and seven non-imbricate wall elements. Specimens have both brick-like and petal-like elements (see Plate 1). Previous authors (e.g., Perch-Nielsen, 1985) identified this species as *Lithastrinus moratus*; however *L. moratus* clearly has imbricate, curved wall elements (see Stover, 1966; Varol, 1992).

Stratigraphic Range: Turonian. The highest occurrence of this species is truncated by a facies change in the Western Interior cores (Codell Sandstone). However, its highest occurrence is uppermost Turonian in southern England (Culver Cliff) and Tunisia.

Eprolithus floralis (Stradner, 1963) Stover, 1966 (Plate 1, Fig. 21)

Discussion: The taxonomic concept of this species used herein is conservative and would include Polycyclolithaceae with a central diaphragm and nine, fairly broad and birefringent (1st order yellow-orange), non-imbricate wall elements. *Radiolithus planus* and *Radiolithus undosus* also possess nine non-imbricate wall elements, but both species have low (i.e., less-birefringent) walls. *Radiolithus undosus* is further distinguished by its individual elements that are both pointed and rounded (Varol, 1992).

Stratigraphic Range: *Eprolithus floralis* has been recovered from the uppermost lower Aptian in the stratotype section at La Bedoule (Bergen, 1994). Its extinction is within the upper Coniacian.

Eprolithus octopetalus Varol, 1992 (Plate 1, Figs. 27-29)

Discussion: Polycyclolithaceae with a central diaphragm and eight non-imbricate wall elements. Specimens have both brick-like and petal-like elements (see Plate 1).
Stratigraphic Range: uppermost Cenomanian to lower Turonian.

Eprolithus rarus Varol, 1992 (Plate 1, Figs. 22, 23)

Discussion: Polycyclolithaceae with a central diaphragm and six non-imbricate wall elements. All specimens observed had petal-like elements. *Lithastrinus grillii* Stradner, 1962 has imbricate, curved wall elements and a thick central diaphragm.
Stratigraphic Range: This species is restricted to the Turonian section in the Western Interior cores. Its lowest occurrence is higher than the three other *Eprolithus* species having a higher number (7-9) of wall elements.

Flabellites oblongus (Bukry, 1969) Crux, 1982
Gartnerago nanum Thierstein, 1974
Gartnerago obliquum (Stradner, 1963) Thierstein, 1974
Gartnerago segmentatum Thierstein, 1973
Grantarhabdus coronadventis Reinhardt, 1966

Helicolithus anceps (Gorka ex Reinhardt and Gorka, 1967) Noël, 1970 (Plate 2, Fig. 4)

Discussion: The taxonomic concept of this species used by Varol and Girgis (1994) is followed herein, which includes all medium-sized (>6µm) *Helicolithus* having a diagonal cross.
Stratigraphic Range: Albian to Campanian.

Helicolithus compactus (Bukry, 1969) Varol and Girgis, 1994 (Plate 2, Figs. 19, 20)

Discussion: Medium-sized (>6µm) *Helicolithus* having an axial cross nearly aligned with the major ellipse axes. *Helicolithus anceps* is another medium-sized *Helicolithus*, but has a diagonal cross and a younger extinction.
Stratigraphic Range: lower Albian to lower Santonian.

Helicolithus turonicus Varol and Girgis, 1994
Kamptnerius magnificus Deflandre, 1959
Kamptnerius punctatus Forchheimer, 1972

Liliasterites angularis Stradner and Steinmetz, 1984 (Plate 1, Figs. 7, 8; Plate 2, Fig. 14)

Discussion: Stradner and Steinmetz (1984) noted similarities between *Liliasterites* and early Eocene species *Tribrachiatus contortus*, but also noted *Liliasterites* as a possible ancestor to *Marthasterites* (Coniacian-early Campanian). The lowest occurrence of *Rhomboaster svabenickae* n. sp. is below that of *Liliasterites angularis* in the Bounds core, suggesting that *Liliasterites* may have descended from this cubic form (see discussion of *R. svabenickae*).
Stratigraphic Range: Turonian.

Lithastrinus grillii Stradner, 1962

Lithastrinus moratus Stover, 1966 (Plate 1, Figs. 18, 19; Plate 2, Fig. 18)

Discussion: Varol (1992) noted that *Lithastrinus septenarius* Forchheimer, 1972 is a taxonomic junior synonym of *Lithastrinus moratus*. *Lithastrinus moratus* is distinguished from other Polycyclolithaceae by its seven elements, a thick central diaphragm, and curved, imbricate wall elements. Incomplete specimens are most often observed, either as detached cycles of wall elements (Plate 1, Fig. 18) or central diaphragms.
Stratigraphic Range: uppermost Turonian to lower Santonian.

Lithraphidites acutum Manivit et al., 1977
Lithraphidites carniolensis Deflandre, 1963

Loxolithus sp. 1 (Plate 1, Fig. 5; Plate 2, Fig. 10)

Discussion: Large elliptical muroliths with broad rims exhibiting a bright first order white birefringence, radial extinction lines and an open central area. Many specimens have an irregular inner rim margin, which suggests that a delicate central plate may have been present. Extremely well-preserved assemblages recovered from the upper Gault Clay and lower Greensand at Folkstone (uppermost Albian-basal Cenomanian) contain small to medium-sized muroliths exhibiting a rim birefringence pattern identical to the large *Loxolithus* sp. 1. These smaller specimens all have thin (non-birefringent) central plates and may be conspecific with *Percivalia*

hintonesis, which was described from the upper Gault and Cambridge Greensand by Black (1972).
Stratigraphic Range: upper lower Cenomanian to uppermost Maastrichtian.

Loxolithus sp. 2 (Plate 1, Fig. 2)

Discussion: Medium- to large-sized elliptical muroliths having a bicyclic rim extinction pattern (inner rim cycle is bright and spiraled) and an open central area. Much smaller specimens having delicate (faintly birefringent) central plates and a rim extinction pattern identical to *Loxolithus* sp. 2 were recovered from the upper Gault Clay and lower Greensand. However, these smaller specimens could not be referred to any described species.
Stratigraphic Range: Turonian to Maastrichtian

Lordia xenota (Stover, 1966) Varol and Girgis, 1994
Lucianorhabdus cayeuxii Deflandre, 1959
Lucianorhabdus compactus (Verbeek, 1976) Prins and Sissingh, 1977
Lucianorhabdus maleformis Reinhardt, 1966
Manivitella pemmatoidea (Manivit, 1965) Thierstein, 1971
Markalius circumradiatus (Stover, 1966) Perch- Nielsen, 1968
Marthasterites furcatus (Deflandre in Deflandre and Fert, 1954) Deflandre, 1959
Microrhabdulus decoratus Deflandre, 1959
Microrhabdulus heliocoideus Deflandre, 1959
Microstaurus chiastius (Worsley, 1971) Bralower in Bralower, Monechi and Thierstein, 1989 *Micula cubiformis* Forchheimer, 1972
Micula staurophora (Gardet, 1955) Stradner, 1963

Miravetesina ficula (Stover, 1966) Bergen, n. comb. (Plate 1, Fig. 13; Plate 2, Fig. 11)

1966 *Coccolithites ficula* n. sp. Stover, p. 138, Pl. 5, Figs. 5,6; Pl. 9, Fig. 11

Discussion: *Miravetesina favula* Grün, 1975 is the type species and was described from the lower Berriasian. This genus was distinguished from other Speetonaciae by its granular central area elements, but it also possesses an axial central structure and a rim more typical of the genera *Cruciellipsis* and *Polypodorhabdus*. *Miravetesina ficula* (Stover, 1966) n. comb. was described from the Turonian section in the Paris Basin and shares the same morphologic characteristics, although specimens recovered from the Western Interior are typically larger than the holotype illustrated by Stover (1966).
Stratigraphic Range: Turonian-lower Coniacian section in the Bounds core; its highest occurrence at 640 ft in this core is very distinct.

Nannoconus elongatus Brönnimann, 1955
Nannoconus multicadus Deflandre and Deflandre, 1959
Nannoconus sp.
Octocyclus magnus Black, 1972
Orastrum campanensis (Cepek, 1970) Wind and Wise, 1977
Percivalia fenestrata (Worsley, 1971) Wise, 1983

Percivalia hauxtonensis Black, 1972 (Plate 2, Fig. 1)

Discussion: Medium-sized, elliptical muroliths having a bright central plate and two longitudinal perforations are tentatively assigned to *Percivalia hauxtonensis*, although they appear to lack the distal stem-base present in the original illustrations of that species by Black (1972; Pl. 31, Figs. 10-14).
Stratigraphic Range: upper Albian to lower Turonian.

Pervilithus varius Crux, 1981
Placozygus spiralis Bramlette and Martini, 1964

Placozygus sp. aff. *P. spiralis* (Bramlette and Martini, 1964) Hoffman, 1970 (Plate 2, Fig. 3)

Discussion: Early morphotypes of *Placozygus spiralis* are differentiated by their: (1) larger central areas; (2) oblong transverse bars; and (3) distinct, bicyclic rim extinction patterns. The extinction lines of their inner rim cycles also appear less tightly spiraled than the Maastrichtian holotype illustrated by Bramlette and Martini (1964).
Stratigraphic Range: The lowest occurrence of this morphotype is upper Albian. Hill (1976; Pl. 12, Figs. 28-37) illustrated specimens from the upper Albian Pawpaw Shale and lower Cenomanian Grayson Marl of the western Gulf Coast.

Polypodorhabdus madinglyensis Black, 1971
Prediscosphaera columnata (Stover, 1966) Perch- Nielsen, 1984
Prediscosphaera cretacea (Arkhangelsky, 1912) Gartner, 1968

Prediscosphaera spinosa (Bramlette and Martini, 1964) Gartner, 1968
Prediscosphaera stoveri (Perch-Nielsen, 1968) Shafik and Stradner, 1971

Quadrum eneabrachium Varol, 1992

Quadrum gartneri Prins and Perch-Nielsen, 1977 (Plate 2, Fig. 13)

Stratigraphic Range: The first occurrence of this species has been used as a basal Turonian marker (e.g., Sissingh, 1977), but its stratigraphic position varies significantly (e.g., Watkins, 1985; Bralower, 1988; this study).

Quadrum intermedium Varol, 1992
Quadrum octobrachium Varol, 1992
Radiolithus orbiculatus (Forchheimer, 1972) Varol, 1992
Radiolithus planus Stover, 1966

Radiolithus undosus (Black, 1973) Varol, 1992 (Plate 1, Fig. 20)

Discussion: The taxonomic concept of this species used by Varol (1992) is maintained herein. This species has nine wall elements. It was identified by its individual elements that show variation from rounded to pointed terminations in the proximo-distal direction. The extremely rare specimens observed during the course of this study are related to *Lithastrinus*, as the pointed terminations are also curved.
Stratigraphic Range: Varol (1992) indicated a Middle Albian to upper Cenomanian occurrence for the species. Rare specimens were observed in a single Cenomanian sample (990 ft) from the Bounds core.

Reinhardtites anthophorus (Deflandre, 1959) Perch- Nielsen, 1968

Reinhardtites biperforatus (Gartner, 1968) Shafik, 1978 (Plate 2, Fig. 6)

Discussion: This large, distinct murolith is distinguished by its extremely broad rim, two central longitudinal perforations, and two brightly birefringent transverse bars.
Stratigraphic Range: basal Coniacian to lowermost Campanian.

Reinhardtites scitula Bergen, 1994

Reinhardtites sisyphus (Gartner, 1968) Bergen, n. comb.
Discussion: This species was reassigned due to its similarity to other species of *Reinhardtites* (e.g., *R. anthophorus*).

Repagulum parvidentatum (Deflandre and Fert, 1954) Forchheimer, 1972
Rhagodiscus achlyostaurion (Hill, 1976) Doeven, 1983
Rhagodiscus angustus (Stradner, 1963) Reinhardt, 1971
Rhagodiscus asper (Stradner, 1963) Reinhardt, 1967
Rhagodiscus reniformis Perch-Nielsen, 1973
Rhagodiscus splendens (Deflandre, 1953) Verbeek, 1977

Rhomboaster svabenickiae Bergen, n. sp. (Plate 1, Fig. 6; Plate 2, Figs. 9, 15-17)

Diagnosis: A Late Cretaceous species of *Rhomboaster*.
Description: Medium-sized, nannolith having a cubic outline. Each of the six faces of the cube are concave (the edges are raised), and the corners of the cube may extend far beyond the basic cubic outline in a radial direction. Recovered specimens appear to be composed of a single fused element as they are either entirely non-birefringent or exhibit a first order yellow-orange birefringence perpendicular to the various cube faces.
Etymology: named in honor of Dr. Lilian Svabenicka, Geological Survey, Prague.
Discussion: This species, which appears identical to the late Paleocene genus *Rhomboaster*, is believed to be the ancestral form of a Late Cretaceous lineage that evolved into *Marthasterites* (*Liliasterites* is the intermediate morphotype). If so, this would provide evidence of iterative evolution, as the succession of morphotype appears identical to that observed in the *Rhomboaster-Tribrachiatus* lineage of the late Paleocene to early Eocene.
Differentiation: This species is distinguished from other cubic Late Cretaceous nannolith genera, such as *Quadrum* and *Micula*, by its uniform birefringence or non-birefringence (depending on orientation) and apparent fused nature (one element) of the nannolith in cross-polarized light. *Quadrum* is brightly birefringent and subdivided into eight primary elements stacked in two layers. Cubic *Micula* species (e.g. *M. staurophora* and *M. concava*) display the same variations in outline and are virtually indistinguishable with the electron microscope, but these taxa are brightly birefringent and display distinct extinction patterns in cross-polarized light.
Known Stratigraphic Range: Turonian-Coniacian of the Bounds core.
Holotype: Plate 1, Fig. 6.
Type level: Turonian, 820 ft.
Type locality: Bounds core, Kansas.

Rotelapillus crenulatus (Stover, 1966) Perch-Nielsen, 1984
Rotelapillus laffittei Noël, 1956.
Scampanella cornuta Forchheimer and Stradner, 1973
Scapholithus fossilis Deflandre in Deflandre and Fert, 1954
Sollasites horticus (Stradner et al., 1966) Cepek and Hay, 1969
Tegumentum stradneri Thierstein, 1972
Tetrapodorhabdus coptensis Black, 1971
Tetrapodorhabdus decorus (Deflandre, 1954) Wind and Wise, 1976

Tranolithus bitraversus (Stover, 1966) Bergen, n. comb. (Plate 1, Figs. 14, 15)

1966 *Parhabdolithus? bitraversus* n. sp. Stover, p. 145, Pl. 6, Figs. 20-22, Pl. 9, Fig. 19.

Discussion: This species of Tranolithus is distinguished by its distal projection, which is solid and circular in transverse section. The stem is supported by four central elements, which are blocky in specimens recovered from the Western Interior cores.
Stratigraphic Range: upper Albian to Campanian.

Tranolithus exiguus Stover, 1966
Tranolithus gabalus Stover, 1966
Tranolithus minimus (Bukry, 1969) Perch-Nielsen, 1984
Tranolithus orionatus (Reinhardt, 1966) Perch- Nielsen, 1968
Tranolithus phacelosus Stover, 1966
Tranolithus sp. aff. *T. phacelosus* (form with stem)
Tubodiscus jurapelagicus (Worsley, 1971) Roth, 1973

Vagalapilla ara (Gartner, 1968) Bergen, n. comb.

Discussion: This species exhibits a bicyclic rim extinction pattern; the inner cycle is bright and the outer cycle is faint. The slender axial cross also exhibits a bright 1st order white birefringence and its "footed" terminations are diagnostic of the species in the light microscope. This species has often been identified as *Vagalapilla imbricata*, but that species is restricted to the late Coniacian to early Campanian. *Vagalapilla imbricata* is identified by its broad axial cross composed of numerous longitudinally arranged lathes and its large stem.
Stratigraphic range: upper Coniacian to mid-Maastrichtian.

Vagalapilla dibrachiata (Gartner, 1968)

Vagalapilla dicandidula Bergen, n. sp.
(Plate 1, Fig. 1)

Diagnosis: A medium-sized, elliptical murolith with a faintly birefringent rim and a brightly birefringent axial central structure having its bars divided by longitudinal extinction lines.
Description (Light Microscope): The narrow rim exhibits a first order gray birefringence. The large central area is spanned by an axial cross-structure. The four arms of the axial cross exhibit a bright 1st order birefringence and are divided by longitudinal extinction lines. Opposing bars may be slightly offset. A solid, distal projection is present and is circular in transverse outline.
Etymology: Latin: (di-) = separate; candidula = shining white (dim.).
Discussion: This species is distinguished from other muroliths with axial central structures by its faintly birefringent rim and brightly birefringent bars that are divided by longitudinal extinction lines. *Chiastozygus platyrhethus* Hill, 1976 displays similar optical characteristics, but has a diagonal cross-structure.
Known stratigraphic range: Turonian section of the Bounds core.
Holotype: Plate 1, Fig. 1.
Type level: Turonian, 920 ft.
Type locality: Bounds core, Kansas.

Vagalapilla elliptica (Gartner, 1968) Bukry, 1969
Vagalapilla gausorhethium Hill, 1975
Vagalapilla imbricata (Gartner, 1968) Bukry, 1969
Vagalapilla stradneri (Rood et al., 1971) Thierstein, 1973
Vagalapilla sp.
Watznaueria barnesae (Black, 1959) Perch-Nielsen, 1968
Watznaueria biporta Bukry, 1969
Watznaueria communis Reinhardt, 1964
Watznaueria fossacincta (Black, 1971) Bown and Cooper, 1989
Watznaueria ovata Bukry, 1969
Watznaueria porta Bukry, 1969
Watznaueria quadriradiata Bukry, 1969
Watznaueria virginica Bukry, 1969
Zeugrhabdotus embergeri (Noël, 1959) Perch-Nielsen, 1984

Zeugrhabdotus theta (Black, 1959) Black, 1973
Zeugrhabdotus trivectis Bergen, 1994
Zeugrhabdotus sp. A (see Bergen, 1994)
Zygodiscus bicrescenticus (Stover, 1966) Wind and
Wise, 1977
Zygodiscus diplogrammus (Deflandre and Fert, 1954)
Gartner, 1968.
Zygodiscus elegans (Gartner, 1968) Bukry, 1969
Zygodiscus wynnhayii Risatti, 1973
Zygodiscus sp.

REFERENCES

ARTHUR, M. A., SCHLANGER, S. O., AND JENKYNS, H. C., 1987, The Cenomanian-Turonian Oceanic Anoxic Event, II. Palaeoceanographic controls on organic-matter production and preservation, in Brooks, J., and Fleet, A., eds., Marine Petroleum Source Rocks: London, Special Publication Geological Society of London, v. 24, p. 401-420.

BERGEN, J. A., 1994, Berriasian to early Aptian calcareous nannofossils from the Vocontian Trough (SE France) and Deep Sea Drilling Site 534: New nannofossil taxa and a summary of low-latitude biostratigraphic events: Journal of Nannoplankton Research, v. 16, p. 59-69.

BLACK, M., 1972, British Lower Cretaceous coccoliths. I Gault Clay: London, Monographs Palaeontographical Society of London, v. 126, p. 1-48.

BRALOWER, T. J., 1988, Calcareous nannofossil biostratigraphy and assemblages of the Cenomanian-Turonian boundary interval: implications for the origin and timing of oceanic anoxia: Paleoceanography, v. 3, p. 75-316.

BRALOWER, T. J., AND SIESSER, W. G., 1992, Cretaceous calcareous nannofossil stratigraphy of sediments recovered on ODP Leg 122, Exmouth Plateau, N.W. Australia: College Station, Proceedings Ocean Drilling Program Scientific Results, v. 122, p. 529-556.

BRAMLETTE, M. N., AND MARTINI, E., 1964, The great change in calcareous nannoplankton fossils between the Maestrichtian and Danian: Micropaleontology, v. 10, p. 291-322.

BUKRY, D., 1969, Upper Cretaceous coccoliths from Texas and Europe: Lawrence, University of Kansas Paleontological Contributions, v. 51, 1-79.

CEPEK, P. AND HAY, W. W., 1969, Calcareous nannoplankton and biostratigraphic subdivision of the Upper Cretaceous: Transactions Gulf Coast Association of Geological Societies, v. 19, p. 323-336.

COBBAN, W. A., 1951, Scaphitoid cephalopods of the Colorado Group: Washington, D.C., U.S. Geological Survey Professional Paper 239, p. 1-39.

COBBAN, W. A., AND SCOTT, G. R., 1972, Stratigraphy and ammonite fauna of the Graneros Shale and Greenhorn Limestone near Pueblo, Colorado: Washington, D.C., U.S. Geological Survey Professional Paper 645, 1-108.

COVINGTON, J. M., 1986, Upper Cretaceous nannofossils from the Niobrara Formation of Kansas-Biostratigraphy and cell paleomorphology: Unpublished M.S. Thesis, Florida State University, 88 p.

DAWSON, G. M., 1874, Foraminifera, coccoliths and rhabdoliths from Cretaceous of Manitoba: Canadian Naturalist, p. 252-257.

ELDER, W. P., 1985, Biotic patterns across the Cenomanian-Turonian extinction boundary near Pueblo Colorado, in Pratt, L. M., Kauffman, E. G., and Zelt, F. B., eds., Fine-grained Deposits and Biofacies of the Cretaceous Western Interior Seaway: Evidence of Cyclic Sedimentary Processes, Field Trip Guidebook, 4: Tulsa, Society of Economic Paleontologists and Mineralogists, p. 157-169.

ELDER, W. P., 1987, The paleoecology of the Cenomanian-Turonian (Cretaceous) stage boundary extinctions at Black Mesa, Arizona: Palaios, v. 2, p. 24-40.

ELDER, W. P., 1988, Geometry of Upper Cretaceous bentonite beds: implications about volcanic source areas and paleowind patterns, western interior, United States: Geology, v. 16, p. 835-838.

ELDER, W. P., 1991, Molluscan paleoecology and sedimentation patterns of the Cenomanian-Turonian extinction interval in the southern Colorado Plateau region: Boulder, Geological Society of America Special Paper 260, p. 113-137.

FISHER, C. G., HAY, W. W., AND EICHER, D. L., 1994, Oceanic front in the Greenhorn Sea (late middle through late Cenomanian): Paleoceanography, v. 6, p. 879-892.

GARTNER, S., Jr., 1968, Coccoliths and related calcareous nannofossils from Upper Cretaceous deposits of Texas and Arkansas: Lawrence, University of Kansas Paleontological Contributions, v. 48, 56 p.

HANCOCK, J. M., AND KAUFFMAN, E. G., 1979, The great transgressions of the Late Cretaceous: Journal of the Geological Society of London, v. 136, p. 175-186.

HAY, W. W., EICHER, D. L., AND DINER, R., 1993, Physical oceanography and water masses in the Cretaceous Western Interior Seaway, in Caldwell, W. G. E., and Kauffman, E. G., eds., Evolution of the Western Interior Basin: St. Johns, Geological Association of Canada Special Paper 39, p. 297-318.

HILL, M. E., 1976, Lower Cretaceous calcareous nannofossils from Texas and Oklahoma: Palaeontographica B., v. 156, p. 103-179.

JAKUBOWSKI, M., 1986, New calcareous nannofossil taxa from the Lower Cretaceous of the North Sea: International Nannoplankton Association Newsletter, v. 8, p. 36-42.

JIANG, M.-J., 1989, Biostratigraphy and geochronology of the Eagle Ford Shale, Austin Chalk, and lower Taylor Marl in Texas based on calcareous nannofossils: Unpublished Ph.D. Dissertation, Texas A&M University, 496 p.

KAUFFMAN, E. G., 1977, Geological and biological overview: Western Interior Cretaceous Basin: Mountain Geologist, v. 14, p. 75-99.

KAUFFMAN, E. G., AND CALDWELL, W. G. E., 1993, The Western Interior Basin in space and time, in Caldwell, W. G. E., and Kauffman, E. G., eds., Evolution of the Western Interior Basin: St. Johns, Geological Association of Canada Special Paper 39, p. 1-30.

LECKIE, R. M., 1985, Foraminifera of the Cenomanian-Turonian boundary interval, Greenhorn Formation, Rock Canyon anticline, Pueblo, Colorado, in Pratt, L. M., Kauffman, E. G., and Zelt, F. B., eds., Fine-grained Deposits and Biofacies of the Cretaceous Western Interior Seaway: Evidence of Cyclic Sedimentary Processes, Field Trip Guidebook, 4: Tulsa, Society of Economic Paleontologists and Mineralogists, p. 139-150.

LEITHOLD, E. L., 1994, Stratigraphical architecture at the muddy margin of the Cretaceous Western Interior Seaway, southern Utah: Sedimentology, v. 41, p. 521-542.

MONECHI, S., PRATT THIERSTEIN, H. R., 1985, Late Cretaceous-Eocene nannofossil and magnetostratigraphic correlations near Gubbio, Italy: Marine Micropaleontology, v. 9, p. 419-440.

MUTTERLOSE, J., 1989, Temperature controlled migration of calcareous nannofloras in the northwest European Aptian, in Crux, J.A., and van Heck, S.E., eds., Nannofossils and Their Applications: Chichester, Ellis Horwood, p. 122.

PERCH-NIELSEN, K., 1985, Mesozoic calcareous nannofossils, in Bolli, H. M., Saunders, J. B., and Perch-Nielsen, K., eds., Plankton Stratigraphy: New York, Cambridge University Press, p. 329-426.

PRATT, L. M., 1984, Influence of paleoenvironmental factors on preservation of organic matter in Middle Cretaceous Greenhorn Formation, Pueblo, Colorado: American Association of Petroleum Geologists Bulletin, v. 68, p. 1146-1159.

PRATT, L. M., AND THREKELD, C. N., 1985, Stratigraphic significance of $^{13}C/^{12}C$ ratios in mid-Cretaceous strata of the Western Interior Basin, in Stott, D. F. and Glass, D. J., eds., Mesozoic of middle North America: Canadian Society of Petroleum Geology Memoir 9, p. 305-312.

ROTH, P. H., 1978, Cretaceous nannoplankton biostratigraphy and oceanography of the northwestern Atlantic Ocean: Washington, D.C., Initial Reports of the Deep Sea Drilling Project, v. 44, p.731-759.

SISSINGH, W., 1977, Biostratigraphy of Cretaceous calcareous nannoplankton: Geologie Mijnbouw, v. 56, p. 37-65.

STOVER, L. E., 1966, Cretaceous coccoliths and associated nannofossils from France and the Netherlands: Micropaleontology, v. 12, p. 133-167.

STRADNER, H., AND STEINMETZ, J., 1984, Cretaceous calcareous nannofossils from the Angola Basin, Deep Sea Drilling Project Site 530: Washington, D.C., Initial Reports of the Deep Sea Drilling Project, v. 75, p. 565-649.

VAROL, O., 1992, Taxonomic revision of the Polycyclolithaceae and its contribution to Cretaceous biostratigraphy: Newsletters in Stratigraphy, v. 27, p. 93-127.

VAROL, O., AND GIRGIS, M. H., 1994, New taxa and taxonomy of Jurassic and Cretaceous calcareous nannofossils: Neues Jahrbuch fr Geologie und Palaeontologie Abhandlungen, v. 192, p. 221-253.

WATKINS, D. K., 1985, Biostratigraphy and paleoecology of calcareous nannofossils in the Greenhorn marine cycle, in Pratt, L. M., Kauffman, E. G., and Zelt, F. B., eds., Fine-grained Deposits and Biofacies of the Cretaceous Western Interior Seaway: Evidence of Cyclic Sedimentary Processes, Field Trip Guidebook, 4: Tulsa, Society of Economic Paleontologists and Mineralogists, p. 151-156.

WATKINS, D. K., 1986, Calcareous nannofossil paleoceanography of the Cretaceous Greenhorn Sea: Geological Society of America Bulletin, v. 97, p. 1239-1249.

WATKINS, D. K., 1989, Nannoplankton productivity fluctuations and rhythmically-bedded pelagic carbonates of the Greenhorn Limestone (Upper Cretaceous): Palaeogeography, Palaeoclimatology, Palaeoecology, v. 74, p. 75-86.

WATKINS, D. K., BRALOWER, T. J., COVINGTON, J. M.,AND FISHER, C. G., 1993, Biostratigraphy and paleoecology of the Upper Cretaceous calcareous nannofossils in the Western Interior Basin, North America, in Caldwell, W. G. E. and Kauffman, E. G., eds., Evolution of the Western Interior Basin: St. Johns, Geological Association of Canada Special Paper 39, p. 521-538.

WILLIAMS, J. R., AND BRALOWER, T. J., 1995, Nannofossil assemblages, fine fraction stable isotopes, and the paleoceanography of the Valanginian-Barremian (Early Cretaceous) North Sea Basin: Paleoceanography, v. 10, p. 815-839.

FORAMINIFERAL PALEOECOLOGY AND PALEOCEANOGRAPHY OF THE GREENHORN CYCLE ALONG THE SOUTHWESTERN MARGIN OF THE WESTERN INTERIOR SEA

OONA L.O. WEST,[1,2] R. MARK LECKIE,[2] AND MAXINE SCHMIDT[2]

[1]*Graduate Program in Organismic and Evolutionary Biology, University of Massachusetts, Amherst, Massachusetts 01003*

[2]*Department of Geosciences, Box 35820, University of Massachusetts, Amherst, Massachusetts 01003-5820.*

ABSTRACT: Foraminifera in shales and mudrocks of the Greenhorn Cycle (late Cenomanian-middle Turonian age) in the Cretaceous Western Interior Basin were strongly influenced by sea level change. This long-term record of third-order sea level rise and fall is superposed by fourth-order relative sea level cycles as delimited by carbonate and sedimentological data. The study interval includes the Cenomanian-Turonian boundary and the early Turonian record of the highest stand of sea level in the western interior. We document stratigraphic variations in foraminiferal assemblages and their response to changing sea level for one drill core through the Tropic Shale (Escalante, Utah) and two outcrop sections of the Mancos Shale (Lohali Point, Arizona; Mesa Verde, Colorado) from the Colorado Plateau. The three sections record deposition along the southwestern margin of the Greenhorn Sea and provide a temporal and spatial framework for interpretations of paleoecology and paleoceanography.

Earlier studies demonstrate that fluctuations in planktic foraminifera and calcareous and agglutinated benthic foraminifera track the transgression and regression of the Greenhorn Cycle. Results of assemblage analyses presented here show that benthic taxon dominance also correlates to fourth-order sea level changes, and to the type of systems tract. Assemblages of calcareous benthic foraminifera are dominated by two species, *Gavelinella dakotensis* and *Neobulimina albertensis*. *Neobulimina*, an infaunal taxon, dominated during the late transgression and highstand of the Greenhorn Sea (early Turonian) when warm, normal salinity, oxygen-poor Tethyan waters advanced northwards into the seaway. In contrast, the epifaunal/shallow infaunal taxon *Gavelinella* proliferated briefly during times of water mass renewal and when deposition of organic matter increased at the transition between fourth-order cycles. Peaks in abundance of other calcareous benthic species delimit transgressive pulses prior to the spread of oxygen-poor Tethyan water masses. These broad-based correlations may result from an intricate relationship among changing water masses, flux of terrestrial and marine organic matter, sedimentation rates, and benthic oxygenation.

Regression of the Greenhorn Sea resulted in a greater restriction of oceanic circulation and in the withdrawal of Tethyan waters that were replaced by cooler, lower salinity water masses of Boreal affinity. An abrupt change to dominance by agglutinated benthic foraminifera and loss of nearly all planktic foraminifera marks this paleoceanographic event. Enhanced biological productivity accompanied regression in south-central Utah. Depauperate benthic foraminiferal assemblages reflect the stress of low-oxygen conditions despite an abundance of food. Enhanced salinity stratification during later stages of regression may have reduced ventilation on the seafloor and led to dysoxic bottom waters.

Sea level change helped produce distinctive assemblages of benthic foraminifera that can be used to delimit successive systems tracts. Foraminiferal assemblages also provide insight into their evolutionary responses to rapidly changing paleoenvironments. Our results indicate no evolution in the foraminiferal biota of the study sections, which we think points to evolutionary stasis.

INTRODUCTION

The Late Cretaceous was a time of dynamic change in the marine environment with much of it recorded in shale and limestone sequences of the Western Interior Basin of North America. Sediments accumulated in an extensive epicontinental seaway and provide a thick stratigraphic record for the interpretation of Late Mesozoic paleoenvironments. Many workers have studied the depositional and tectonic history of these sequences. Considerable attention has also focused on reconstructing the seaway's climatic and oceanographic history, as revealed by sedimentology, geochemistry, and biotic composition.

Planktic and benthic foraminifera furnish useful paleoecologic data for interpreting Cretaceous paleoenvironments in the Western Interior Sea. This includes information on salinity, productivity, stratification of ancient water masses, and benthic oxygenation (e.g., Eicher and Worstell, 1970; Frush and Eicher, 1975; Eicher and Diner, 1985; Leckie et al., 1991; Caldwell et al., 1993; Fisher et al., 1994). Foraminifera provide critical information for testing paleoceanographic and sequence stratigraphic models. This contribution describes the paleoecology of foraminiferal assemblages in three upper Cenomanian-middle Turonian sections (Greenhorn Cycle) from the southwestern side of the "Greenhorn Sea". We also show the usefulness of foraminifera for understanding and interpreting relationships between sea level change, productivity, and benthic ventilation in ancient marine environments.

Geologic Setting

The Western Interior Sea covered most of the west-central North American craton during the Late Cretaceous. This was a time of warm greenhouse (CO_2-rich) climate, high global sea level, and widespread burial and preservation of organic matter. Six third-order tectonoeustatic cycles of marine transgression and regression are recorded in Cretaceous strata of the Western Interior (Kauffman, 1977, 1984, 1985; following the terminology of Vail et al., 1977). The best developed and most extensive of these cycles is the early Cenomanian-middle Turonian Greenhorn Cycle.

The early Turonian was the time of the highest sea level of the first-order Mesozoic-Cenozoic, tectonoeustatic cycle (Hancock and Kauffman, 1979; Haq et al. 1987). The seaway reached its maximum extent during early Turonian time. Tectonic and tectonoeustatic controls were tightly coupled with sedimentation patterns within the western interior (Kauffman, 1977, 1985). Warm southern water masses invaded the Western Interior Sea during transgressive episodes (Kauffman, 1984; Kauffman and Caldwell, 1993). With transgression came the mixing and/or juxtaposition of very different water masses, that is, cool, northern Boreal waters, and warm, southern Tethyan waters. The interaction of these water masses was probably also influenced by north-south and east-west differences in evaporation, precipitation, and runoff, as well as seasonality and the movement of storms across the seaway (Parrish et al., 1984; Glancy et al., 1986, 1993;

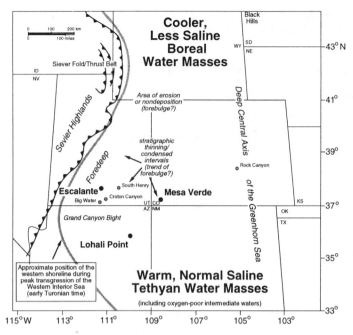

FIG. 1—Map of the four corners area of Utah, Colorado, New Mexico, and Arizona showing the location of the three sections studied here: the Escalante (ES) core was drilled on the northern edge of the Kaiparowits Plateau near Escalante, Utah; the Lohali Point (LP) section is located on the eastern side of Black Mesa in northeastern Arizona; and the Mesa Verde (MV) section is located near the entrance to Mesa Verde National Park, between Cortez and Mancos, Colorado. Also shown is the approximate position of the western shoreline of the Greenhorn Sea at the time of peak transgression in the early Turonian (after Molenaar, 1983).

Ericksen and Slingerland, 1990; Hay et al., 1993; Jewell, 1993; Parrish, 1993; Slingerland et al., 1996). Water mass boundaries and circulation patterns fluctuated with changes in relative sea level, climate, and changes in the physiographic size and shape of the seaway. Many rivers provided fresh water, especially along the tectonically active western margin of the seaway. These contrasts in water mass were best developed close to peak transgressive episodes. Rhythmic bedding of limestone/chalk and marlstone/calcareous shale occurred during these times (e.g., Pratt, 1984; Barron et al., 1985; Arthur et al., 1985; Eicher and Diner, 1989; Sageman et al., 1997).

Previous Foraminiferal Studies

Cenomanian-Turonian rocks of the Western Interior of the United States and Canada contain an excellent, well documented foraminiferal record (e.g., Tappan, 1940; Cushman, 1946; Young, 1951; Jones, 1953; Fox, 1954; Frizzell, 1954; Eicher, 1965, 1966, 1967, 1969; Stelck and Wall,1954, 1955; Wall, 1960, 1967; Lamb, 1968; Eicher and Worstell, 1970; Hazenbush, 1973; Lessard, 1973; Frush and Eicher, 1975; North and Caldwell, 1975; Caldwell et al., 1978, 1993; McNeil and Caldwell, 1981; Eicher and Diner, 1985; Leckie, 1985; Leckie et al., 1991 and this volume; Olesen, 1991; Fisher et al., 1994; Schroeder-Adams et al., 1996). In the United States, many of these studies focused on the central and eastern portions of the seaway (e.g., Eicher and Worstell, 1970) and there are less published data for the western margin. In addition, although the taxonomy and to a lesser extent the spatial distributions of Cenomanian-Turonian foraminifera were studied (e.g., Caldwell et al., 1978, 1993; Eicher and

Diner, 1985), the paleobiogeography and paleoecology of these biotas were not evaluated in the context of recent advances in our understanding of foraminiferal biology and ecology.

We present new paleoecologic interpretations based on upper Cenomanian-middle Turonian (Greenhorn Cycle) foraminiferal assemblages from the Colorado Plateau. This, and a complementary study by Leckie et al. (this volume), represent the first comprehensive paleoecologic analyses of foraminiferal assemblages from this southwestern region. Our data are interpreted with the aid of recent conceptual models of microhabitat ecology based on living benthic foraminifera (e.g., Corliss, 1985; Corliss and Chen, 1988; Corliss and Emerson, 1990; Jorissen et al., 1995), and then used to deduce paleoenvironment in relation to water mass interactions, affinities, and sea level change within the Greenhorn Sea.

Objectives

Our primary objective was to determine the response of foraminifera in an atypical marine environment stressed by sea level change in a dynamic depositional system. Questions that we sought to answer were: (1) Which foraminiferal taxa dominate Greenhorn assemblages and why?, (2) What kinds of paleoenvironmental conditions might exclude other taxa?, (3) Can some of the paleoecologic constraints acting on foraminiferal communities be established? In particular, are certain benthic taxa indicative of benthic oxygen levels?, (4) Can changes in foraminiferal assemblage structure be used to recognize and constrain third-order and fourth-order cycles of relative sea-level change?, (5) Can foraminifera be used to provide ground-truth data supporting climatic and paleoceanographic models for the Western Interior Sea? (6) And in a broad sense, what are the evolutionary responses of individual taxa and communities to rapid changes in paleoenvironmental conditions induced by changes in sea level?

MATERIALS AND METHODS

Foraminiferal assemblages were studied from three localities representing a transect across the southwestern side of the Greenhorn Sea: a cored section of Tropic Shale near Escalante, Utah (ES) on the Kaiparowits Plateau and two outcrop sections of Mancos Shale, one at Lohali Point, Arizona (LP) in the eastern Black Mesa Basin and the other near Mesa Verde National Park, Colorado (MV) on the Four Corners Platform (Fig. 1). These three stratigraphic sections allow a regional comparison of foraminiferal assemblage composition. This type of spatial comparison provides a comprehensive approach for analyzing biotic trends within the seaway and permits reliable interpretations of paleoecology and paleoenvironment.

We use foraminiferal assemblages to infer paleoenvironment. One-hundred and ninety-three samples from the Mancos and Tropic Shales were analyzed for both planktic and benthic foraminifera. The interval contained within these samples is approximately 3.5 million years. Foraminiferal assemblage data collected for each sample included: (1) planktic to benthic ratio (% planktics), (2) planktic morphotype analysis (that is % biserial, triserial, trochospiral, planispiral, and keeled morphologies), and (3) abundance of major benthic species (*Neobulimina albertensis* and *Gavelinella dakotensis*) or groups of taxa (agglutinated taxa and other calcareous benthics). Leckie (1985), West et al. (1990), and Leckie et al. (1991 and this volume) demonstrate the utility

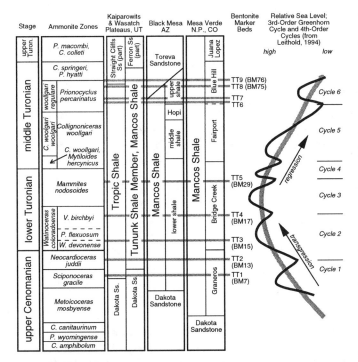

FIG. 2—Time-stratigraphic relationships and proposed correlation of rock units of the Greenhorn Cycle on the southwestern side of the Greenhorn Sea. Ammonite biostratigraphy is after Kirkland (1991) and Kauffman et al. (1993). Elder (1985) and Kirkland (1991) included *Vascoceras birchbyi* and *Pseudospidoceras flexuosum* as subzones of the *Watinoceras* Zone. Stratigraphy: southern Utah, Leithold (1994); Black Mesa, Kirkland (1991); Mesa Verde, Kirkland et al. (1995), Leckie et al. (1997). Bentonite marker beds: TT (Tropic-Tununk) from Leithold (1993, 1994) and BM (Black Mesa) from Kirkland (1991) (Table 1). Kirkland (1991) notes the concurrence of *P. percarinatus* and *C. woollgari regulare* in the Hopi Sandy Member of the Mancos Shale in Black Mesa and places the boundary between the *C. woollgari woollgari* and *C. woollgari regulare* Subzones within the Hopi Sandy interval. We correlate the Hope Sandy Member with the regressive phase of Cycle 5 in southern Utah, which Leithold (1994) interpreted as a forced regression.

of this type of data for paleoenvironmental analysis of Cenomanian-Turonian age rocks from the Western Interior. Foraminiferal assemblage data are based on counts of the >63mm size fraction of approximately 200-300 specimens where practical. Many samples from the middle Turonian parts of the sections (Cycles 5 and 6) contain depauperate benthic foraminiferal assemblages. The numbers of foraminifera picked from these latter samples are based on one or two trays of a sample split (see Appendices 1-3). Rare species and other biogenic and mineral components were also examined in all of the samples. Additional details on sample processing, picking, and counting methods can be found in Leckie et al. (1991).

STRATIGRAPHY

Escalante Core (ES)

A 695 ft.-thick-section (212 m) of Tropic Shale was cored near the town of Escalante in south-central Utah (Fig. 1). The Tropic Shale consists primarily of calcareous shale and mudstone representing muddy prodeltaic depositional environments of the western margin of the Greenhorn Sea (Leithold, 1993, 1994; Leithold and Dean, this volume). It conformably overlies nonmarine and marginal marine facies of the Dakota Sandstone

near the Kaiparowits Plateau (Leithold, 1994). Its upper contact grades with the overlying shallow marine and deltaic deposits of the Tibbet Canyon Member of the Straight Cliffs Formation (Fig. 2). The contact with the Dakota Sandstone is placed at about 835 ft. (255 m) in the Escalante core; the contact with the Straight Cliffs Formation is tentatively placed at about 140 ft. (43 m) based on the first meter-thick sandstone in the Tropic/Straight Cliffs transition (Leithold, pers. commun., 1996). Samples representing the transgression of the Greenhorn Sea and the early phases of its regression were examined in this study.

Leithold (1994) recognized six fourth-order depositional sequences superposed on the third-order transgressive-regressive Greenhorn Cycle (Fig. 2). The depositional sequences consist of fifth-order parasequences, of which at least 37 were delimited by Leithold (1994) in the Tropic Shale and correlative Tununk Shale Member of the Mancos Shale in southern Utah. Cycles 1-3 record the transgression of the Greenhorn Sea. The highest percentages of carbonate occur in Cycle 3. This interval in the lower Turonian *Mammites nodosoides* Zone may represent peak transgression (Leithold, 1994). Cycle 4 records the highstand and early phase of regression. During Cycle 5, the shoreline was displaced rapidly eastward during a major regressive phase. A rise of relative sea level (renewed subsidence?) forced the shoreline to rapidly retreat westward and then prograde during Cycle 6. According to Leithold (1993, 1994) this latter flooding event was associated with anoxic bottom waters. Leithold and Dean (this volume) note the similarity between the stratigraphy of the Escalante core and that of Leithold's (1993, 1994) Big Water section, which is located on the southern rim of the Kaiparowits Plateau (Fig. 1). At Big Water there is a disconformity in the upper part of Cycle 5 (topset beds are truncated) due to subaerial or shallow subaqueous erosion as the shoreline was displaced rapidly eastward (Leithold, 1994). However, there is no evidence of this disconformity in the Escalante core (Leithold and Dean, this volume).

Lohali Point Section (LP)

LP is located on the eastern side of the Black Mesa in northeastern Arizona. The Mancos Shale at LP is 203 m thick. The shale is underlain disconformably by the upper Cenomanian Dakota Sandstone and overlain conformably by the middle Turonian Toreva Sandstone (Fig. 2) (Kirkland, 1991). Kirkland (1991) distinguished three informal members of the Mancos Shale and formally proposed the Hopi Sandy Member for the numerous sandstone beds between the middle and upper shale members. The lower shale member is the most carbonate-rich interval of the Mancos at LP. This unit records the transgression of the Greenhorn Sea and its maximum extent during early Turonian time (Elder, 1991; Kirkland, 1991; Leckie et al., 1991), and it correlates to Utah Cycles 1-4 of Leithold (1994). The middle shale member correlates to the transgressive and early regressive, lower part of Cycle 5 in the Tropic Shale and in the Tununk Member of the Mancos Shale in the southern Utah sections. We correlate the Hopi Sandy Member to the regressive upper part of Cycle 5. The upper shale member reflects an abrupt relative rise of sea level-perhaps subsidence-and its subsequent fall, which correlates to Cycle 6 in Utah (Kirkland, 1991; Leckie et al., 1991).

Mesa Verde Section (MV)

The MV section is near the northern edge of Mesa Verde National Park in the southwestern corner of Colorado. This sec-

TABLE 1.—BENTONITE BEDS[1]

Utah Bentonite Marker Beds	Escalante Core	Black Mesa Basin, AZ	Black Mesa Basin, AZ	Lohali Point Section	Mesa Verde Nat'l. Park, CO	Mesa Verde Section	Rock Canyon Pueblo, CO
Leithold, 1994	Kauffman and Leithold, unpubl.	Kirkland, 1991	(C/T interval only) Elder, 1987	Kirkland, 1991	Leckie et al., 1997	Leckie et al., 1997	Kauffman, Pratt et al., 1985
TT9	273' (83.2 m)	BM76		164.4 m			PF34?
TT8	309' (94.2 m)	BM75		154.1 m			PF31/33?
TT7	329' (100.3 m)	Unit 183?		120.1 m	Unit 171?	71.0 m	PF30/31?
TT6	339' (103.4 m)	Unit 182?		119.9 m	Unit 169?	70.0 m	PF24/26?
TT5	657' (200.3 m)	BM29		35.2 m	Unit 72	27.8 m	PBC32
TT4	749' (228.5 m)	BM17	D	25.0 m	Unit 53	25.2 m	PBC20
TT3	771' (235.1 m)	BM15	C ("boundary")	16.4 m	Unit 49	24.4 m	PBC17
TT2	804.5' (245.3 m)	BM13	B ("Neocard")	13.2 m	Unit 44	23.2 m	PBC11
TT1	824.5' (251.4 m)	BM7	A4 ("Skip")	7.2 m	Unit 36	20.7 m	PBC5
	830' (250 m) depth in core (feet/m)		A1	meters above Dakota Sandstone		meters above Dakota Sandstone	

[1]Tentative correlation of bentonite beds TT1–TT9 of Leithold (1993, 1994) to bentonites of Black Mesa and the Lohali Point section at Black Mesa (Elder, 1987; Kirkland, 1991), and to the Mesa Verde section (Leckie et al., 1997).

tion is the principal reference section of the Mancos Shale in its type area (Kirkland et al., 1995; Leckie et al., 1997). The Greenhorn Cycle is 141 m thick at MV and is represented by the Graneros, Bridge Creek, Fairport, and Blue Hill Members of the Mancos Shale (Fig. 2). Increasing carbonate content and an overall fining upwards sequence in the Graneros Shale Member indicate transgression of the Greenhorn Sea. The interbedded calcareous shale, limestone, and calcarenite of the Bridge Creek Limestone Member record peak transgression and high stand. Carbonate content drops off sharply in the middle and upper parts of the Fairport Shale Member (Leithold, pers. commun., 1997). We correlate this drop in carbonate in the Fairport to the Hopi Sandy Member of the Mancos Shale at LP and the major regressive episode (Cycle 5) along the southwestern side of the seaway. The Blue Hill Shale Member records an episode of relative sea level rise and subsequent fall with regression of the Greenhorn Sea during the middle Turonian, and it correlates to the upper shale unit at LP and to Leithold's Cycle 6 in south-central Utah.

Marker Beds and Correlation

Macrofossil assemblages and numerous bentonite beds (e.g., TT1–TT5) provide reliable correlation through the transgressive phase of the Greenhorn Cycle, particularly in the upper Cenomanian *Sciponoceras gracile* to lower Turonian *Mammites nodosoides* zones (e.g., Cobban and Scott, 1972; Kauffman, 1977; Elder, 1985, 1987, 1991; Elder and Kirkland, 1985; Kirkland, 1991; Leithold, 1993, 1994; Leckie et al., 1997) (Fig. 2). This interval correlates to Cycles 1–4 of the Tropic Shale in southern Utah (Leithold, 1993, 1994), the lower shale unit of the Mancos Shale in Black Mesa (Kirkland, 1991), and to the upper Graneros and Bridge Creek Members of the Mancos Shale at Mesa Verde (Kirkland et al., 1995; Leckie et al., 1997).

Regression of the Greenhorn Sea began in early middle Turonian time, which correlates to the *Collignoniceras woollgari* zone. Cycle 5, in the southern Utah sections, records a forced regression. According to Leithold (1994), the shoreline migrated eastward at least 150 km during this fourth-order cycle and perhaps as much as 300 km based on the distribution of the correla-

tive Coon Spring Sandstone Bed in eastern Utah and westernmost Colorado (Molenaar and Cobban, 1991). We propose that the Hopi Sandy Member of the Mancos Shale in northeastern Arizona (Black Mesa Basin) represents the same regressive event. The Black Mesa sections (Kirkland, 1991) are located south and southeast of the southern Utah sections (Leithold, 1993, 1994) and at similar or slightly more distal locations relative to the shoreline. Therefore, regressive deposits could be roughly isochronous between the two areas. The correlation of the Hopi Sandy Member with the regressive part of Leithold's Cycle 5 is supported by (1) the relatively coarse ammonite biostratigraphy available in the regressive phase of the Greenhorn Cycle, and (2) by the foraminiferal population data presented here.

If this lithostratigraphic correlation is correct, then marker bentonite beds TT6–TT9 in southern Utah likely correlate to bentonite beds Unit 182 and Unit 183 at LP and marker bentonite beds BM75 and BM76 in Black Mesa (Table 1). The first occurrence of *Prionocyclus hyatti* is between bentonites TT8 and TT9 in southern Utah (Leithold, 1994) and between bentonites BM75 and BM76 in Black Mesa (Kirkland, 1991), supporting the proposed correlation. However, the correlation of these marker beds to MV is more problematic for several reasons: (1) the Mesa Verde section lies further offshore and lithofacies changes were perhaps diachronous relative to southern Utah and Black Mesa, (2) macrofossil zones are of broader duration, diversity is lower, and the paucity of adult ammonite specimens in this part of the Greenhorn Cycle renders biostratigraphy less robust than in the transgressive upper Cenomanian–lower Turonian interval, and (3) sandy regressive deposits representing distal equivalents to the Hopi Sandy Member or Coon Spring Sandstone Bed are not present in the MV section (Leckie et al., 1997).

The Blue Hill Shale Member of the Mancos Shale at MV is correlated to the upper shale member at Black Mesa based primarily on the distinctive noncalcareous shale, paucity of macrofossils, and sharp drop in bentonite/limonite frequency. Bentonite correlations between LP and MV in Hopi Sandy/Blue Hill equivalents must be considered tentative due to the paucity of macrofossils. Despite the uncertainties in exact bentonite correlations between ES/LP and MV, the foraminiferal population data presented below help to delimit the distal equivalents of Cycle 5 and the Hopi Sandy Member.

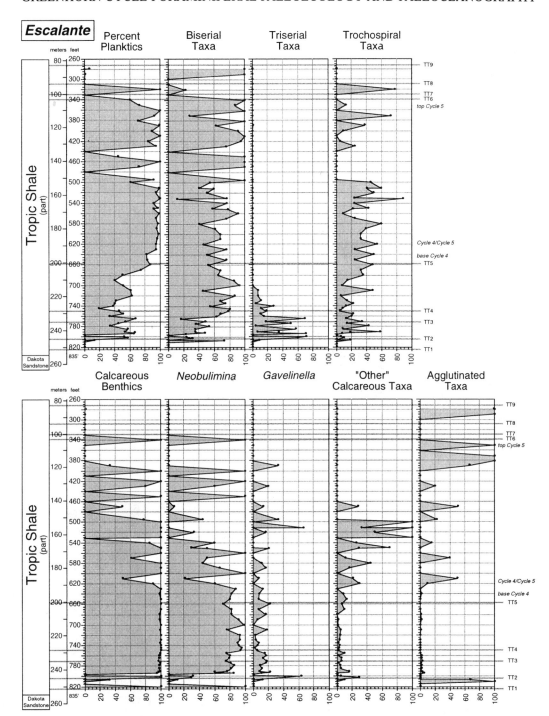

FIG. 3— Foraminiferal data for much of the Greenhorn Cycle in the Escalante core. Thickness corresponds to depth in the core (in feet). Top panel: planktic foraminiferal data include the proportion of planktics to total foraminifera (percent planktics), biserial taxa (*Heterohelix* spp.), triserial (*Guembelitria cenomana*), and trochospiral taxa (species of *Hedbergella* and *Whiteinella*). Bottom panel: benthic foraminiferal data include the proportion of calcareous benthics to total benthic foraminifera (percent calcareous benthics), *Neobulimina albertensis* (a common infaunal calcareous benthic), *Gavelinella dakotensis* (a common epifaunal/shallow infaunal calcareous benthic), other calcareous benthic taxa (e.g., *Cassidella tegulata*, *Buliminella fabilis*), and the proportion of agglutinated benthics to total benthic foraminifers. Bentonite marker beds and the intervals that correlate to Cycles 4 and 5 are shown (after Leithold, 1994; Leithold and Dean, this volume; Leithold, pers. comm., 1996).

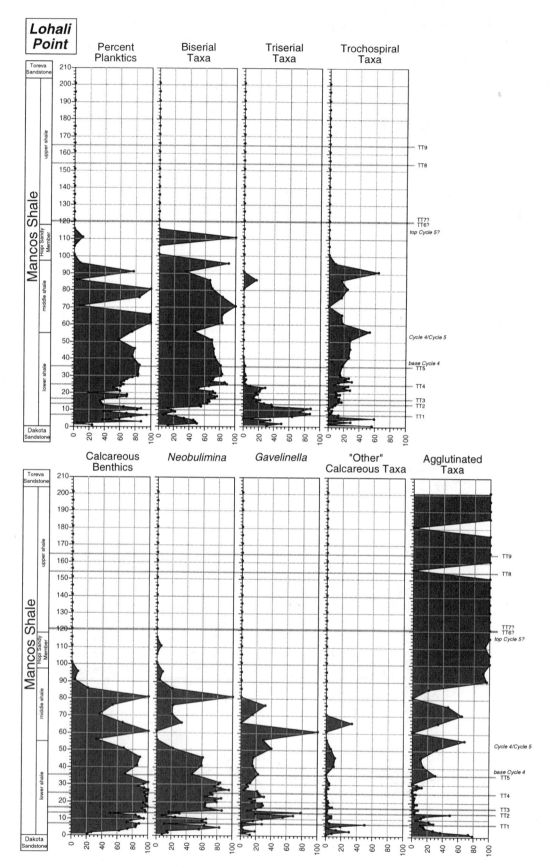

FIG. 4—Foraminiferal data for the Greenhorn Cycle at the Lohali Point section. Thickness corresponds to meters above the Dakota Sandstone. Top panel: planktic foramin-iferal data; bottom panel: benthic foraminiferal data (for details refer to Figure 3). Proposed correlation to Cycles 4 and 5 of southern Utah (Leithold, 1994).

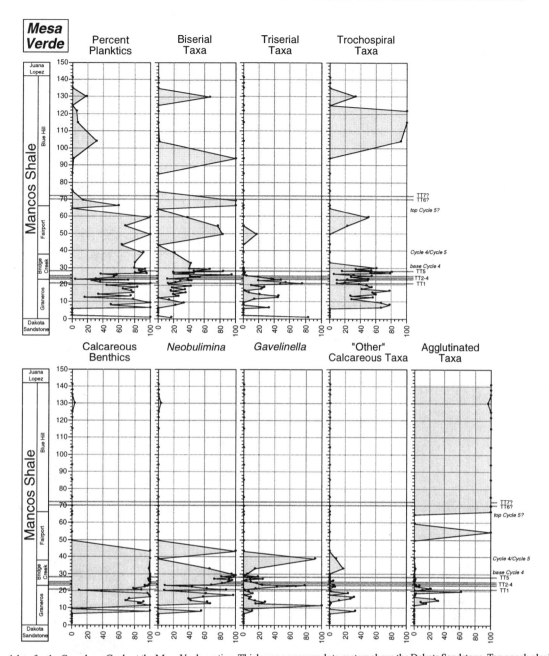

FIG. 5—Foraminiferal data for the Greenhorn Cycle at the Mesa Verde section. Thickness corresponds to meters above the Dakota Sandstone. Top panel: planktic foraminiferal data; bottom panel: benthic foraminiferal data (for details refer to Figure 3). Proposed correlation to Cycles 4 and 5 of southern Utah (Leithold, 1994).

RESULTS

Planktic Foraminifera

Major biotic trends in the planktic foraminiferal assemblages for all three sections analyzed are expressed as percentage data (Figs. 3, 4, 5; see Appendices for raw data tables). The first appearance of planktic foraminifera is diachronous from east to west, occurring first at MV, then at LP, and finally at ES (Fig. 6). At all three sites the foraminifera appear suddenly and abundantly within the sections, accounting for 60-90% of the total assemblage. The proportions of planktics to benthics fluctuate during transgression but stabilize during highstand (Cycle 4). Fluctuations during the transgression (Cycles 1-3) may reflect dynamic environmental shifts associated with the numerous fifth-

order parasequences superposed on the fourth-order cyclicity (Leithold, 1994; O.L.O. West, in prep.). Relative abundance of planktics increases offshore from ES to LP to MV. In the upper Cenomanian there is a decrease in planktic foraminifera at each locality followed by recovery in the lower Turonian. They then decline in abundance after TT6, but disappear sooner at LP than at the other sites.

Analysis of percent planktic morphotypes reveals that the first occurrence of biserial, triserial, and trochospiral morphotypes is also diachronous. Biserial taxa (*Heterohelix*) are most abundant in the lower Turonian (above TT4; Cycle 3) at all three sites and account for the majority of the increase seen in total planktics through the lower Turonian. In contrast, triserial taxa (*Guembelitria*) appear first at MV and LP near TT1 in high abundances, followed by a sharp decline in abundance below TT2. At ES, they occur later. Populations at this site fluctuate more than

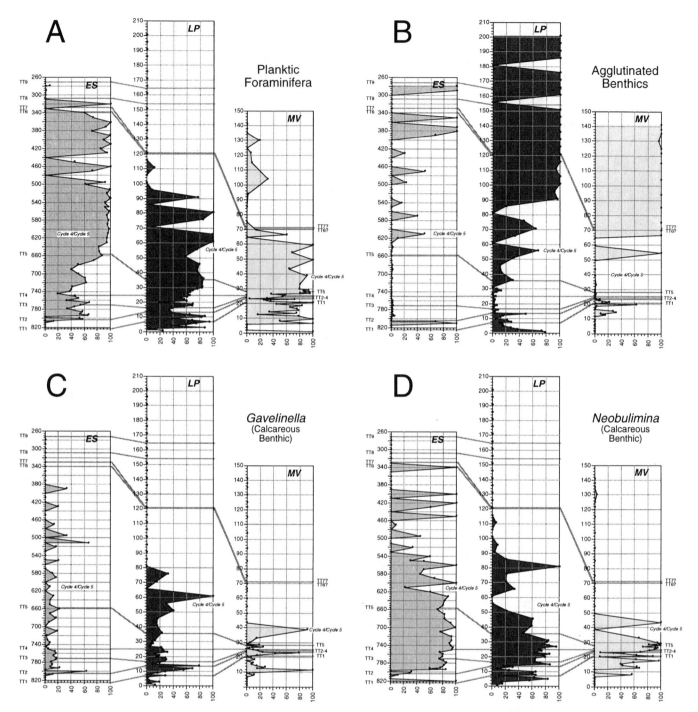

Fig. 6—Correlation of the Escalante core with the Lohali Point and Mesa Verde outcrop sections showing the stratigraphic distribution of planktic foraminifera, agglutinated benthic foraminifera, and the calcareous benthic species *Gavelinella dakotensis* and *Neobulimina albertensis*. Note the following: (A) the increase in relative abundance of planktic foraminifera between TT4 and TT5 in all three sections corresponding to the approach of peak transgression in Cycle 3, (B) agglutinated benthics are very subordinate to calcareous benthics until the onset of regression of the Greenhorn Sea; in particular, the regressive phase of Cycle 5 (below TT6), represented by the Hopi Sandy Member of the Mancos Shale at Lohali Point, marks the shift to agglutinated dominance in benthic foraminiferal assemblages, (C) the two correlative acmes of *Gavelinella*: one near TT2 corresponds with the transition from Cycle 1 to Cycle 2, and the other above TT5 corresponds with the transition between Cycle 4 and Cycle 5 (each acme precedes a shift to *Neobulimina* dominance), and (D) the dominance of *Neobulimina* throughout Cycles 2, 3, and 4 in all three sections at a time when the seaway was at or near its maximum extent; a short-lived recovery also occurs in the transgressive part of Cycle 5 at all three sections (between TT5 and TT6).

at the other two sites and persist longer. Trochospiral taxa (species of *Hedbergella* and *Whiteinella*) exhibit high variability at all three sites. The high stratigraphic occurrence of relatively abundant trochospiral morphotypes at MV is an artifact due to very small population numbers.

All three sites are dominated by biserial and triserial taxa,

but there is a brief period of trochospiral dominance in the upper Cenomanian (below TT1) at LP and MV. In the uppermost Cenomanian (*Neocardioceras* Zone) there is a decrease in triserial taxa and a coincident increase in the abundance of biserial taxa. This turnover is attributed to the invasion of southern water masses, and perhaps heightened marine productivity, as water mass mixing proceeded across the southwestern side of the sea-way (Leckie et al., this volume). No keeled taxa were recorded for ES and LP and only two specimens were found at MV in the upper Cenomanian.

Benthic Foraminifera

Major biotic trends in the benthic foraminiferal assemblages for all three sections are expressed as percentage data (Figs. 3, 4, 5; see Appendices for raw data tables). Within the upper Cenomanian and basal Turonian (TT1-TT4), benthic assemblages show similar stratigraphic trends at ES, LP, and MV. There is a sharp increase in relative abundance of *Gavelinella dakotensis* in the upper Cenomanian (near TT2), followed by a rapid shift to dominance by *Neobulimina albertensis* (between TT2 and TT3) near the Cenomanian-Turonian boundary (Fig. 6). The *Neobulimina* abundance is sustained until shortly after TT5, and the dramatic change in taxon dominance is approximately iso-chronous at all three western sites. This ecologic shift from a *Gavelinella* acme to *Neobulimina* dominance, is also documented at Rock Canyon, Colorado, in the central part of the seaway (Leckie et al., 1991, this volume) and is the basis for our posi-tioning of the Cenomanian-Turonian boundary at approximately 242 m (795 ft) in the ES core.

Benthic foraminiferal numbers drop off markedly after Cycle 3 and again after Cycle 4 in the middle Turonian (Fig. 7). Be-tween TT5 and TT6 at MV and LP, there is a second short-lived increase in *Gavelinella* (Figure 6) that is roughly isochronous and associated with a concurrent increase in other calcareous benthics at ES and LP. An abrupt decline in overall dominance of this taxon abruptly follows and later abundances are more variable. At ES, no increase in *Gavelinella* is observed, perhaps due to the depauperate nature of benthic foraminifera in the samples. However, planktic foraminifera are abundant in the same samples, which indicates that sediment dilution and dissolution cannot account for the paucity of benthics. Other nearshore sec-tions are being studied to test this idea. *Neobulimina* dominates benthic assemblages through Cycles 1-4 and makes an attenu-ated reappearance at all three sites within the lower part of Cycle 5 (Fig. 6).

Other calcareous benthic foraminifera are present in very low abundance throughout all three sections. Diverse calcareous taxa characteristic of the benthonic zone of Eicher and Worstell (1970) are notably absent. Calcareous benthic assemblages are dominated at different times by one of only two calcareous taxa, *Gavelinella* or *Neobulimina*.

Agglutinated taxa suddenly appear in abundance at all three localities following the reappearance of *Neobulimina* in Cycle 5, which marks a shift from calcareous-dominated to agglutinated-dominated benthic assemblages. Agglutinated taxa are highly variable and less abundant at ES than at LP while at MV, some samples lack benthic foraminifera below TT6. The interval above TT6 correlates to Cycle 6 and regression of the Greenhorn Sea. Benthic foraminiferal assemblages are dominated almost exclu-sively by agglutinated taxa through this interval.

DISCUSSION

Overview

Foraminiferal paleoecology is widely applicable for recon-structing paleoenvironments. It is of special value in paleoceanographic studies for inferring water mass affinities and for tracking changes in sea level within the Western Interior Sea. In epicontinental seas, oceanographic variables fluctuated sig-nificantly through time but may also have been amplified by en-hanced seasonality, climatic cyclicity, and mixing of different water masses. Because foraminifera are sensitive to these oceano-graphic factors, attendant changes in foraminiferal assemblages can be used to track water mass evolution both temporally and spatially. For example, different water masses, defined qualita-tively in terms of salinity, temperature, origin, and oxygenation can be deciphered from planktic and benthic foraminiferal as-semblages and paleobiogeography (see Leckie et al., this vol-ume).

A major aim of this study was to apply foraminiferal paleoecologic analysis to questions about water masses in the southwestern region of the Western Interior Sea during the depo-sition of the Greenhorn Cyclothem. This period is represented by a major transgressive-regressive episode that records the ad-vance and retreat of the Greenhorn Sea (Fig. 2). Some conten-tion surrounds the paleoenvironmental conditions that existed during deposition of the Greenhorn Cycle. Most workers agree that bottom waters became oxygen-deficient at times during trans-gression and regression but the cause, severity, and distribution of oxygen deficiency are uncertain.

Some authors (e.g., Pratt, 1984; Arthur et al., 1985; Barron et al., 1985; Pratt et al., 1993; Savrda and Bottjer, 1993) propose that a low-salinity surface layer resulting from fluvial discharge from the western highlands and driven by cyclic changes in cli-mate, led to density stratification and reduced benthic ventila-tion in the Greenhorn Sea. Others favor this freshwater cap sce-nario in combination with the incursion of an oxygen minimum zone during eustatic sea level rise that was concomitant with a global oceanic anoxic event (e.g., Leckie et al., 1991; Sageman et al., 1997). Benthic oxygen stress can also result from high biological productivity in surface waters. Along the western margin, particularly during the regressive episode of Cycle 5, high productivity created oxygen-poor benthic conditions due to the high flux of marine organic matter (Leithold, 1993, 1994; Leithold and Dean, this volume). We tried to test these paleoenvironmental interpretations using foraminiferal data. We assessed the conditions in the water column and at the sediment surface by analyzing foraminiferal assemblages within a mod-ern ecologic and sequence stratigraphic framework.

Previous studies of ancient communities of both macrobiota and microbiota in the Western Interior, found that the communi-ties have high species dominance and low species diversity. Our data indicate that this also holds true for the foraminiferal com-munities along the southwestern side of the Greenhorn Sea. This may indicate that the salinity of the seaway was lower than nor-mal or that low oxygen conditions were widespread in the basin at times. For example, salinity stratification, combined with en-hanced productivity due to elevated fluvial input and/or incur-sion of warm, oxygen-poor waters into the seaway with trans-gression may have reduced benthic ventilation, resulting in dysoxia or anoxia of bottom waters and underlying sediments.

At all three sections, planktic and benthic foraminiferal di-versity are low and there is an inverse correlation between fora-miniferal abundances and diversity. This may be because high total abundances result from an increase in one or a few species. This especially seems to be the case for the benthic assemblages. Planktic foraminiferal assemblage structure clearly reflects the transgression and regression of the third-order Greenhorn Cycle.

Of more significance, perhaps, patterns of benthic foraminiferal succession within the sections correlate broadly to fourth-order cycles of relative sea level (cycles 1-6 of Leithold, 1994) that are superposed on the Greenhorn Cycle.

Planktic Foraminiferal Paleoecology

Analysis of assemblage structure of planktic foraminifera provides key data for understanding ancient oceans. For example, Leckie (1987) showed that there are marked differences in the composition of low-latitude, open-ocean planktic assemblages and epicontinental sea assemblages. Planktic foraminifera reveal the nature of the upper water column. When used in conjunction with information on bottom waters garnered from benthic foraminiferal data, they provide a powerful tool for paleoceanographic reconstruction (Leckie et al., 1991, also this volume).

Our results indicate that the percentage of planktic foraminifera fluctuates between samples in the upper Cenomanian (Fig. 6). Diversity is low (4-5 species), with the epicontinental sea taxa *Heterohelix* and *Guembelitria* comprising more than 70% of specimens. These surface-dwelling genera indicate shallow marine environments (Leckie, 1987; Leckie et al., 1991). The diachronous appearance of planktic taxa records the advance of warm, normal marine southern waters as the Greenhorn Sea transgressed rapidly across the region. For example, *Guembelitria* appears to be absent from the nearshore ES site until just after bentonite bed TT2 while the percentage of this morphotype at LP and MV approaches 80% between TT1 and TT2. After TT4, during the early Turonian, this morphotype virtually disappears from assemblages at all three sites. The planispiral genus *Globigerinelloides* is restricted to the upper Cenomanian *Sciponoceras* Zone. Species of *Globigerinelloides* are characteristic of more open marine pelagic environments and indicate the establishment of normal marine conditions coincident with peak transgression. The fluctuation of the planktic to benthic ratio during transgression suggests that conditions were stressful in the upper water column at this time. We suggest that a combination of lowered salinity and high rates of sedimentation from increased fluvial input caused fluctuations in planktic populations. The effects of these factors would be especially pronounced in nearshore environments; indeed, at the Escalante site, planktic to benthic proportions are lower during transgression than at the more offshore locales.

Based on the percentage of morphotype groups, we interpret planktic assemblages as characteristic of particular oceanographic environments. Higher ratios of trochospiral and keeled morphotypes indicate a more open marine environment, while assemblages dominated by biserial *Heterohelix* likely indicate stressful conditions (Leckie, 1985, 1987; Leckie et al. 1991, and this volume). The increased abundance of this morphotype suggests less saline conditions in the upper water column and/or oxygen-poor waters associated with salinity stratification. An alternative explanation is that *Heterohelix* was associated with the incursion of Tethyan water masses into the seaway and perhaps, in particular, with low-oxygen conditions including stratification and enhanced productivity (Leckie et al., this volume). The rapid rise to dominance of this genus during the Cenomanian-Turonian Oceanic Anoxic Event (Cycle 2) and its continued dominance through peak transgression and highstand (Cycles 3 and 4) supports this hypothesis.

Through the lower Turonian, there is an increase up section in planktic foraminiferal abundance and diversity (Leckie et al., 1991; Olesen, 1991). A similar trend is seen in sections from the more distal, central part of the seaway (Eicher and Worstell, 1970; Eicher and Diner, 1985). Sustained maximum abundances at all three sites occur within Cycle 3 and Cycle 4 of Leithold (1994). During this interval of peak transgression and highstand, planktic foraminifera comprise 60-100% of foraminiferal assemblages and there is a general increase in trochospiral morphotypes (*Whiteinella* and *Hedbergella*) representative of the open marine, shallow-water biota of Leckie (1987).

Abundant fecal pellets and a marked decrease in bioturbation intensity in Cycle 5, during the early middle Turonian, particularly at ES (Leithold and Dean, this volume), point to enhanced productivity and oxygen depletion in the benthos associated with regression of the Greenhorn Sea. Elevated proportions of the biserial planktic foraminifera through this interval further support the association of *Heterohelix* with unstable pelagic conditions, which include reduced salinity, high productivity, and oxygen reduction in the water column. Boreal waters pushed southward with the withdrawal of warm, normal marine Tethyan waters during the late regressive phase of the Greenhorn Cycle. Associated with this major change in water masses (Fig. 6), planktic foraminifera were very rare in the southwestern seaway during Cycle 6.

Benthic Foraminiferal Paleoecology

Benthic foraminifera are sensitive indicators of change in the marine environment. Benthic foraminiferal microhabitats in modern taxa are primarily determined by a delicate interplay between the amount of organic matter and oxygen present in the sediment and at the sediment-water interface. Epifaunal foraminifera (those that live on top of the sediment) and infaunal forms (those living within the sediment) also have distinctive morphologies that are related to their life habit (e.g., Corliss, 1985; Corliss and Chen, 1988; Corliss and Emerson, 1990; Corliss and Fois, 1991; Jorissen et al., 1995). By analogy, fossil benthic foraminifera can be used to indicate the amount of organic matter and/or degree of oxygenation within sediments and at the sediment-water interface in ancient marine systems.

We find the conceptual model of Jorissen et al. (1995) to be especially useful for interpreting benthic foraminiferal paleoecology in our sections. This model explains microhabitats of benthic foraminifera in terms of trophic conditions and oxygen concentrations. The model states that foraminiferal microhabitat preferences are a function of the negative interplay between oxygen and food availability that results from differences in the downward organic flux. Jorissen et al. (1995) posit that, under oligotrophic conditions, microhabitat depth is controlled by the availability of food particles, but in eutrophic settings, it is limited by a critical oxygen level. A similar interpretation was also proposed by Corliss and Emerson (1990) to explain differences in microhabitats in the northwest Atlantic.

In a transect across the Adriatic Sea, Jorissen et al. (1995) found that shallow microhabitats are oxygen-controlled on the shelf and upper slope areas, while in the deeper parts of the basin, the availability of food and not oxygen, determines the depth at which foraminifera live. This model offers a logical approach to paleoenvironmental reconstruction based on benthic foraminifera, and we think that it provides a useful framework within which to interpret benthic foraminiferal paleoecology of the West-

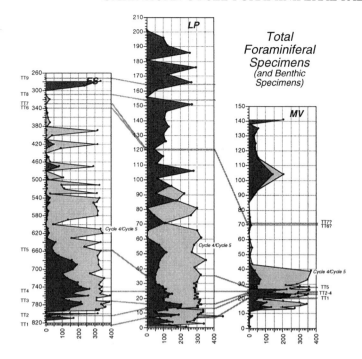

FIG. 7—Number of foraminiferal specimens picked and counted in the Escalante core and in the Lohali Point and Mesa Verde outcrop sections. Light stipple = total foraminifera (benthics and planktics), dark stipple = benthic foraminifera. Because assemblages are based on counts of sample splits (less than one picking tray for samples containing abundant foraminifera and no more than two trays for samples containing fewer specimens), these plots represent a proxy for foraminiferal abundance in the >63mm fraction.

ern Interior.

Our results indicate that along the southwestern margin of the Western Interior Sea, a low species diversity and high species dominance foraminiferal biota prevailed during deposition of the Greenhorn Cycle. Two calcareous species dominate the benthic foraminiferal assemblages at all three study sections: *Gavelinella dakotensis* and *Neobulimina albertensis*. Two distinct benthic habitat characteristics, infaunal or epifaunal, are recognized by biologists. Leckie et al. (1991, and this volume) interpret *Gavelinella dakotensis* as an epifaunal species and *Neobulimina albertensis* as an infaunal species. However, as discussed by Jorissen et al. (1995), it is almost impossible to determine whether smaller foraminifera are strictly epifaunal. Buzas et al. (1993) argued that when dealing with soft sediments, a real substrate does not exist and that all benthic foraminifera are infaunal. Also, few species of foraminifera live exclusively as deep infauna. Indeed, Linke and Lutze (1993) argued that we should not classify foraminiferal microhabitats as static, but instead think of them as the expression of a dynamic adaptation for acquiring food. In line with this more realistic approach to microhabitat determination, we interpret *Gavelinella* as an epifaunal/shallow infaunal taxon and *Neobulimina* as an infaunal taxon.

Within our study sections, we see clear-cut and coherent alternations between short-lived *Gavelinella*-dominated assemblages and those dominated by *Neobulimina*. We suggest that this alternation in taxon dominance indicates an ecologic shift that was caused by changes in the dynamic relationship between microhabitat depth and organic flux, and that it was intimately related to fluctuations of sea level, flux of organic matter, and water mass characteristics of the seaway. Changes in relative sea level not only influenced water mass distribution and circula-

tion, but also affected marine productivity by delivering nutrients and terrestrial organic matter to the seaway.

The rapid transgression of the Greenhorn Sea during the late Cenomanian opened new benthic niches in the core of the seaway, as relatively well oxygenated, normal marine waters invaded from the south. These niches were filled by diverse assemblages of calcareous benthics ("benthonic zone" of Eicher and Worstell, 1970) that immigrated to the Greenhorn Sea with the incursion of warm, Tethyan water masses. The benthonic zone is well developed in offshore sites such as Rock Canyon, Colorado (Fig. 1; Eicher and Worstell, 1970; Frush and Eicher, 1975; Eicher and Diner, 1985; Leckie, 1985). In contrast, in our nearshore sections along the southwestern margin of the seaway, benthic diversity is very low and the diagnostic species of the benthonic zone are not present. This very poor expression of the benthonic zone is perhaps due to the diachronous nature of the transgression. Water mass characteristics such as temperature and salinity or other factors such as water clarity, water depth, and organic character or content of the sediments, may account for the absence of the benthonic zone on the western margin during the time of the *Sciponoceras* zone.

An increase in abundance of the epifaunal-shallow infaunal species *Gavelinella dakotensis*, which dominates assemblages just prior to the Cenomanian-Turonian boundary (Fig. 6), is present at all locales of the southwestern margin and the central part of the seaway (Leckie et al., this volume). We call this the *Gavelinella* acme. We interpret *Gavelinella* as an opportunistic pioneer species capable of taking advantage of the combination of both new niches opened by the incursion of the transgressing seaway and sporadic input of terrestrial and marine detrital organic matter, which coincide with fourth-order, transgressive-regressive pulses. The *Gavelinella* acme developed at the end of Cycle 1 and/or start of Cycle 2. A putative second acme smaller in magnitude developed at the transition from Cycle 4 to Cycle 5 (Fig. 6). There are problems in interpreting data for this latter transition due to the paucity of benthic foraminifera (Figure 7).

According to Gooday (1993), sudden inputs of phytodetritus from plankton blooms promote opportunistic epifaunal taxa because they thrive on high-quality food and reproduce rapidly. This results in a biota with strong epifaunal dominance. Pulses of phytodetritus would be common along the western margin of the seaway, as numerous rivers drained into it. Fluvial input delivered both terrestrial organic matter and nutrients, which stimulated plankton blooms. In shallow, marginal marine environments, this provided a plentiful source of food for opportunistic epifaunal-shallow infaunal species, which could outcompete other species and rapidly prosper. As long as bottom waters remained oxygenated these taxa dominated benthic assemblages when food was abundant. However, a reduction in these food pulses kept more opportunist taxa (like *Gavelinella dakotensis*) in check allowing other species to flourish. We think that this reduction in food supply coupled with well-oxygenated bottom waters, accounts for the high diversity of the benthonic zone from the central and eastern part of the seaway.

We propose that the *Gavelinella* acme in our sections resulted from rapid multiplication of this opportunistic, epifaunal-shallow infaunal taxon in response to sudden inputs of food. In all sections we see a *Gavelinella* acme in the uppermost Cenomanian and a smaller one at LP and MV in the lower middle Turonian. Both acmes coincide with the end of regressive pulses or the start of transgressive pulses of Leithold's (1994) fourth-order cycles (Fig. 8). Increased riverine influx associated with fourth-order regressions provided pulses of organic matter from plankton blooms and high sedimentation rates, which may have precluded infaunal microhabitats. Because of its epifaunal-shallow infaunal microhabitat, *Gavelinella* could take advantage of fresh, easily metabolized food fragments concentrated at the sediment surface. We suggest that benthic bottom waters were mesotrophic and oxygenated at these times. Under these conditions benthic assemblages were food-dominated; that is the availability of organic matter, and not oxygen, determined the microhabitat depth (Corliss and Emerson, 1990; Jorissen et al., 1995). However, if rates of productivity and sedimentation are too high, oxygen depletion at the seafloor may greatly limit benthic foraminiferal proliferation, despite the abundance of food. This may account for the absence of a second *Gavelinella* acme at ES.

Across the Cenomanian-Turonian boundary the dominance of *Gavelinella* in the benthic assemblages at all sections declines due to an abrupt increase in the infaunal species *Neobulimina albertensis* (Figs. 6, 8). As mentioned previously, we think that this alternation in taxon dominance indicates an ecologic shift. It occurs at all three sites and indicates paleoceanographic change associated with the Cenomanian-Turonian boundary interval (Leckie et al., this volume). We interpret the shift from *Gavelinella*-dominated assemblages to those dominated by *Neobulimina* as indicative of a change in which microhabitat is controlled by a critical oxygen level instead of food availability.

Neobulimina is very abundant and dominates the foraminiferal assemblages of Cycles 2-4 of Leithold (1994) (Figs. 3, 4, 5). Dominance of this taxon also correlates to the approach of peak transgression of the Greenhorn Sea in the early Turonian (Figure 8). Based on planktic foraminiferal assemblage structure and carbonate data, peak transgression occurred within Cycle 3 (Leckie et al., 1991; Leithold, 1994; Leithold and Dean, this volume). During this time the southern portion of the seaway was dominated by a warm, perhaps oxygen-poor Tethyan intermediate water mass (Leckie et al., 1991, this volume). Primary productivity may also have increased based on the abundance of *Heterohelix* in the planktic assemblages (Figs. 3-5).

According to the model of Jorissen et al. (1995), shallow microhabitats are oxygen-controlled in eutrophic settings on shelf and upper slope areas, which leads to a shallowing of microhabitat depth and a decrease in the size of the infaunal niche. Progressing from an oligotrophic to a mesotrophic environment, the depth of foraminiferal microhabitats becomes deeper and the size of the infaunal niche gets larger as food supplies increase. This will obviously result in an increase in the percentage of infaunal species in an assemblage. Jorissen et al. (1995) argue that in oxygen-controlled environments, the increasing oxygen-stress preferentially eliminates the less resistant epifaunal-shallow infaunal taxa. The net result, is a continued rise in the percentage of infaunal species until they approach 100% in highly dysoxic environments such as sapropels (e.g., Verhallen, 1990). Under such oxygen-poor conditions, infaunal taxa will most likely live at the surface of the sediment rather than within it (Jorissen et al., 1995). Some benthic foraminifera may be facultative anaerobes (Bernhard, 1989, 1993 1996; Sen Gupta and Machain-Castillo, 1993), and it is possible that this may have been the case for *Neobulimina* (West et al., 1993). This ability facilitates survival when bottom waters are anoxic. The possibility that this taxon also harbored bacterial chemosymbionts should not be ruled out; chemoautotrophy is a viable trophic strategy in anaerobic, sulfide-rich sediments (West, 1993).

Other calcareous taxa immigrated to the Western Interior Sea from the south. However, the dominance of *Gavelinella* and *Neobulimina* and the notable paucity of other calcareous benthics in the sections under study indicates that this region was at least partly influenced by a cooler northern water mass and/or coastal water masses, that were periodically hyposaline and dysoxic and thus inhibited the development of a diverse benthic assemblage. Pulses of organic matter from riverine input and increased productivity, combined with the development of dysoxic bottom waters, favor opportunistic taxa (*Gavelinella*) and those able to survive in low oxygen environments (*Neobulimina*). This gives rise to a succession of *Gavelinella* followed by *Neobulimina* dominated assemblages, which result from the delicate interplay between the relative importance of food versus oxygen levels (Fig. 8). We predict that a similar biotic succession exists in the overlying upper Turonian-Santonian Niobrara Cycle.

There is a sharp decrease in the abundance of benthic foraminifera after Cycle 3, that is after the time of peak transgression. A further decrease in benthic foraminiferal abundance occurs in the upper part of Cycle 4 at ES and near the base of Cycle 5 at LP and MV (Fig. 7). A similar stratigraphic pattern is observed in correlative rocks from the central part of the seaway, in the Fairport Member of the Carlile Shale (Eicher, 1966; Eicher and Worstell, 1970; Eicher and Diner, 1985). These depauperate benthic assemblages are associated with abundant fecal pellets in the shale and relatively abundant planktic foraminifera through the interval of Cycle 4 and the lower part of Cycle 5 (Fig. 6). The presence of abundant planktics rules out the possibility that the benthic assemblages were diluted or dissolved.

We think that low benthic abundances were caused by the combined effects of changing benthic water mass - perhaps the retreat of Tethyan water masses and increasing influence of Boreal waters? - and renewed oxygen depletion in bottom waters associated with high productivity and high sedimentation rates along the western margin. These changes were concomitant with the major Coon Spring-Hopi regression of Cycle 5 (Leithold, 1994) (Fig. 8). Increased productivity is indicated by the abundant planktic foraminifera and fecal pellets and an increase in the flux of marine organic carbon documented at ES (Leithold and Dean, this volume). With an increase in productivity one would expect to see an increase in benthic foraminifera. This begs the question of why no such increase is seen. We suggest that high productivity in surface waters, coupled with higher sedimentation rates, resulted in rapid depletion of benthic oxygen. A marked decrease in bioturbation intensity upward through the shale of Cycle 5 at ES further supports an interpretation of benthic oxygen depletion (Leithold and Dean, this volume). Benthic foraminifera are unable to proliferate under anoxic conditions even though food was abundant.

Following the minor increase in *Gavelinella* seen at the Cycle 4/Cycle 5 transition at LP and MV, the relative abundance of *Neobulimina* near the base of Cycle 5 increases at all three sites (Fig. 6). This trend is associated with increased carbonate values in the lower part of Cycle 5 at all three sites (Leithold and Dean, this volume; Leithold, pers. commun., 1997; McCormic, pers. commun., 1997) and in correlative rocks of the basal Fairport Member of the Carlile Shale in the central part of the seaway (Glenister and Kauffman, 1985) (Fig. 2). We interpret this as indicative of the final influence of Tethyan water masses during the Greenhorn Cycle.

Within Cycle 5 strata (below TT6) there is a clear shift from calcareous-dominated to agglutinate-dominated benthic assemblages with more northern affinities (Fig. 6). At LP, the shift in

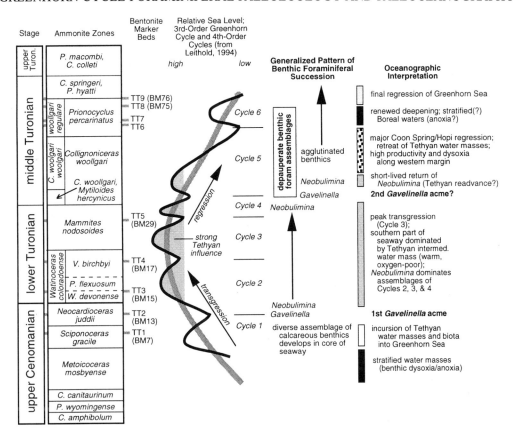

FIG. 8— General trends in benthic foraminiferal populations relative to the fourth-order transgressive-regressive cycles of Leithold (1994). Calcareous benthics are common on the southwestern side of the Greenhorn Sea during transgression and highstand (late Cenomanian-early Turonian), when the seaway was strongly influenced by a southern water mass. Agglutinated benthics characterize the seaway during regression (middle Turonian) which reflects the displacement of Tethyan waters by cooler, less saline water masses from the north. The calcareous benthics display a pattern of succession through the fourth-order cycles. However, these cycles of succession are masked by *Neobulimina* dominance when the seaway is large and dominated by an oxygen-poor Tethyan water mass (Cycles 2, 3, and 4). A major regression during Cycle 5 (Coon Spring/Hopi) brings the strong Tethyan influence to an end and Boreal water masses dominate the seaway for the remainder of the Greenhorn Cycle.

foraminiferal assemblages corresponds with the middle shale-Hopi Sandy Member contact of the Mancos Shale. This dramatic shift in benthic foraminiferal communities is very similar to the trends documented in the central part of the seaway (Eicher, 1966; Eicher and Worstell, 1970; Eicher and Diner, 1985). We interpret this shift as representing the continued displacement of Tethyan waters by Boreal water masses across the southwestern side of the seaway, coincident with an early phase of regression of the Greenhorn Sea (Fig. 8).

Cycle 6, which correlates to the upper shale member of the Mancos Shale at LP and to the Blue Hill Member of the Mancos at MV, records the final regression of the Greenhorn Sea that led to restricted circulation of less saline and, perhaps, stratified Boreal waters. These conditions are indicated by very low carbonate in the shale and its equivalents from the central part of the seaway (Blue Hill Member of the Carlile Shale; Glenister and Kauffman, 1985; Leithold, 1994; Leithold and Dean, this volume), depauperate assemblages of benthic molluscs and ammonites (Glenister and Kauffman, 1985; Kirkland, 1991; Leckie et al., 1997), very rare and sporadic planktic foraminifera, and benthic foraminiferal assemblages dominated exclusively by agglutinated taxa.

Sequence Stratigraphy

Benthic foraminiferal taxon dominance can be used to de-limit fourth-order sequence stratigraphic systems tracks within the third-order Greenhorn Cycle. We find that diagnostic benthic taxa characterize the systems tracks recognized by Leithold (1994) (Fig. 9). During Cycle 1 of the transgressive phase of the Greenhorn Cycle, the transgressive systems tract (TST) is marked by a mixed calcareous and agglutinated assemblage, which is equivalent to the benthonic zone (Eicher and Worstell, 1970) further out in the basin. The thin, highstand systems tract (HST) and the shelf margin system tract (SMST), are characterized by *Neobulimina* and *Gavelinella* respectively (see also Leckie et al., this volume). Within Cycle 2, the TST, HST, and SMST are all dominated by *Neobulimina*; which indicates rising sea level and the incursion of an oxygen minimum zone of Tethyan affinity (Leckie et al., this volume). Cycles 3 and 4 are associated with peak transgression and highstand of the Greenhorn Sea. Continued persistence of a warm, oxygen-poor Tethyan water mass is indicated by the high abundances of *Neobulimina*. A similar pattern of biotic succession is repeated in the regressive deposits prior to the final retreat of Tethyan waters, where again *Gavelinella* is diagnostic of the SMST of Cycle 4 and *Neobulimina* is characteristic of the TST of Cycle 5. This pattern is what one might predict for a transgressive-regressive succession, but further testing is needed. For example, the *Gavelinella* acmes at the Cycle 1/2 and Cycle 4/5 transitions may prove to characterize a broader spectrum of systems tracks including late HST, SMST, and early TST. Agglutinated benthics

Third-Order Greenhorn Cycle

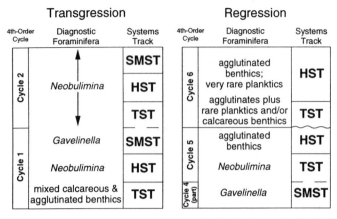

FIG. 9—Correlation of foraminiferal assemblages to the fourth-order cycles of Leithold (1994) and their preliminary sequence stratigraphic interpretation. TST = transgressive systems track, HST = highstand systems track, SMST = shelf margin systems track (Van Wagoner et al., 1988). *Gavelinella* is associated with the transition from Cycle 1 to Cycle 2 and possibly with the transition from Cycle 4 to Cycle 5. We tentatively correlate these occurrences with the SMST. Influx of planktic and/or calcareous benthic foraminifera is associated with the TST of Cycles 1 and 6, while the advance of warm, oxygen-poor waters during deposition of Cycle 2, 3, and 4 accounts for the dominance of *Neobulimina* in both transgressive and regressive phases of these cycles (that is TST, HST, and SMST). Regression of the Greenhorn Sea and the retreat of Tethyan water masses accounts for the shift to agglutinated benthic foraminiferal assemblages in the upper part of Cycle 5 (HST) and in Cycle 6.

characterize the HST and TST at the top of Cycle 5 and within Cycle 6. This is the expression of the influence of Boreal waters and the removal of Tethyan water masses from the region.

Evolutionary Response to Changing Sea Level

The Western Interior Sea represents a dynamic environment. Such a system with rapid changes in environmental conditions might be expected to elicit a correspondingly rapid evolutionary response in the biota and induce speciation within the dominant lineages. However, our data show that the foraminiferal assemblages remained stable. Benthic foraminifera show a pattern of no change in terms of species diversity; the same species persist throughout the sections. Although the assemblages track environmental change in terms of species dominance and composition there appears to be no evolution within the individual species lineages or change in the composition of the foraminiferal assemblages across the study area. Although environment seems to determine which assemblage dominated at any point in time throughout the sections, the same distinctive species and foraminiferal communities persisted, which indicates a pattern of evolutionary stasis.

A perusal of recent paleobiological literature reveals a surge of renewed interest in this well-recognized pattern of long periods of stasis and evolutionary stability disrupted by episodes of geologically rapid evolutionary change (Ivany and Schopf, 1996). Recent research reveals that stasis and episodes of rapid evolutionary turnover may occur concurrently in entire communities (Brett, 1995). Stasis is now seen as a pattern of linked stability and linked change where the principle unit of selection is the community/ecosystem rather than individual species. This pattern emerging from the fossil record is reframed in terms of community and group selection and is called coordinated stasis (Brett

and Baird, 1995).

All studies to date documenting coordinated stasis in the fossil record were done on higher metazoa. A major unanswered question is, does stasis manifest differently in the fossil record of microorganisms than in the fossil record of animals? For example, studies conducted of coordinated stasis in animals reveal consistent change in taxonomic composition in recurring fossil assemblages and no change in species composition during periods of stasis (Brett and Baird, 1995; Leiberman et al., 1995). Also, morphological stasis is thought to be a response to widely fluctuating physical environments on geological timescales (Sheldon, 1996). Is this the case for the microbiota? Study of the relationship between the dynamic environmental and depositional conditions manifest in Upper Cretaceous strata of the Western Interior Sea, and the evolutionary response of the foraminifera of this large epicontinental sea, provide an ideal opportunity to test for stasis in the microbiota (O.L.O. West, in prep.).

CONCLUSIONS

1. Foraminiferal assemblages and swings in species dominance are closely linked to third- and fourth-order transgressive and regressive episodes in the Cenomanian-Turonian Greenhorn Cycle. Cyclic changes in benthic species dominance (i.e., *Gavelinella* versus *Neobulimina*) are recorded in sequences of the southwestern margin of the Western Interior Sea. Diagnostic foraminiferal taxa can be used to delimit several of the six fourth-order, transgressive-regressive cycles and systems tracks recognized by Leithold (1994). A *Gavelinella* acme at all study sites is associated with the transition from regressive to transgressive pulses of fourth-order cycles. The epifaunal-shallow infaunal species *Gavelinella dakotensis* indicates more normal marine waters, higher benthic oxygen levels, and pulses of organic matter into the system. Assemblages dominated by *Neobulimina albertensis* are associated with the late transgressive and highstand phases of the third-order Greenhorn Cycle of the early Turonian, when a warm, oxygen-poor intermediate water mass of Tethyan affinity spread across much of the southern seaway. Under these presumed dysoxic conditions, a critical level of oxygen was more important than food; low-oxygen tolerant infaunal taxa could proliferate and exclude epifaunal-shallow infaunal taxa that require higher oxygen levels. *Neobulimina* may also have been a facultative anaerobe and thus preadapted to survive in low-oxygen conditions.

2. We see a pattern of regional biotic change that signals paleoenvironmental change across the Cenomanian-Turonian boundary. We think it results from the expansion of an oxygen minimum zone and increased primary productivity at the Cenomanian-Turonian boundary. We place the boundary at approximately 242 m (795 ft) in the Escalante core on the basis of an ecologic shift from epifaunal-shallow infaunal-dominated foraminiferal assemblages (*Gavelinella*) to infaunal-dominated assemblages (*Neobulimina*).

3. Regression in Cycle 5 (early middle Turonian) increased productivity along the western margin. This lead to the development of dysoxic bottom waters, which resulted in depauperate benthic foraminiferal assemblages despite an abundance of food.

4. The transition from calcareous-dominated to agglutinate-dominated assemblages within Cycle 5 indicates the displacement of Tethyan water masses that invaded the Western Interior Basin during eustatic sea level rise by cooler, less saline northern waters of Boreal affinity. This displacement is associated with

increased sedimentation rates, freshwater runoff from the western highlands, and perhaps the development of salinity stratification and dysoxic bottom waters further out in the basin.

5. Variations in foraminiferal assemblage composition result from changes in salinity, oxygenation, and perhaps temperature, that followed from changing sea level and coincident variability in freshwater runoff and productivity. The Greenhorn Sea was influenced by Tethyan water masses during transgression that eventually were displaced by Boreal waters. Foraminiferal data from the southwestern margin support the paleoceanographic model of Slingerland et al. (1996), which posits that estuarine circulation with a strong counterclockwise gyre occupied the entire north-south extent of the seaway.

6. Our initial results on the evolutionary response of foraminifera to changing sea level indicate that the Western Interior provides an outstanding model system for testing and refining evolutionary concepts. In particular, foraminiferal communities could help us test how coordinated stasis, which has mainly been studied in macrobiotic communities, may manifest in the microbiota.

ACKNOWLEDGMENTS

We thank Michael Arthur and Walter Dean for allowing us to be a part of the WIK drilling transect and the staff of the USGS Core Laboratory in Denver for sampling the Escalante core. Conscientious reviews by Walt Dean, Pamela Hallock Muller, Isabella Premoli-Silva, and Neil Tibert improved the final version of this manuscript. We also thank Lynn Margulis for providing microscopy facilities to O.L.O. West, and David Snoeyenbos, Paul Morris, Emily CoBabe, and Lonnie Leithold for stimulating discussion that helped refine the ideas presented herein. This paper forms part of the senior author's doctoral dissertation and was supported by the National Science Foundation (Earth Sciences Division) to Leckie and Yuretich, and Leckie and Leithold, and by the Department of Energy to Arthur and Dean. Partial support for this research was also provided by a Geological Society of America Student Research Grant and the Elinor Fierman Award, bestowed by the Department of Geosciences, University of Massachusetts to O.L.O. West.

APPENDIX 1.—ESCALANTE DATA

Sample Number	279.08	280.06	290.10	299.81	310.00	320.04	330.00	340.00	350.21	361.09	371.06	381.09	389.50	400.08	410.04	420.02	429.83	440.10	450.05	460.06	470.00	479.86	494.97	500.10	511.10	511.28	520.10
Raw Data																											
Planktic Morphotypes																											
Biserial	22	2	2	0	0	5	0	3	7	13	7	5	196	30	5	253	65	0	8	26	210	0	98	6	132	56	54
Triserial	0	0	0	0	0	0	0	0	0	0	0	0	3	3	0	1	0	0	0	0	0	0	0	0	0	0	0
Trochospiral	0	0	0	0	0	17	0	0	1	0	18	0	117	3	0	13	21	0	0	0	0	0	0	5	191	38	52
Planispiral	0	0	0	0	0	0	0	0	0	0	0	0	0	0	0	0	0	0	0	0	0	0	0	0	0	0	0
Keeled	0	0	0	0	0	0	0	0	0	0	0	0	0	0	0	0	0	0	0	0	0	0	0	0	0	0	0
Benthic Morphotypes																											
Calcareous Other	0	0	0	0	0	0	0	0	0	0	0	0	0	0	0	0	0	0	0	0	23	0	0	7	2	1	3
Neobulimina	2	0	0	0	0	0	0	2	0	0	0	0	0	4	0	50	3	0	10	0	6	0	4	0	0	0	2
Gavelinella	1	0	0	0	0	0	0	0	0	0	0	0	1	0	0	0	1	0	0	0	11	0	3	0	4	0	1
Total Calcareous Specimens	3	0	0	0	0	0	0	2	0	0	0	0	1	4	0	50	4	0	10	0	40	0	7	7	6	1	6
Total Agglutinated Specimens	311	306	236	0	0	0	0	0	3	0	2	2	2	0	0	0	1	0	0	0	42	0	2	0	0	0	0
Total Number of Benthics	314	306	236	0	0	0	0	2	3	0	2	2	3	4	0	50	5	0	10	0	82	0	9	7	6	1	6
Total Number of Planktics	22	2	2	0	0	22	0	3	8	13	25	5	313	33	5	267	86	0	8	26	210	0	98	11	323	94	106
Total Number of Specimens	336	308	238	0	0	22	0	5	11	13	27	7	316	37	5	317	91	0	18	26	292	0	107	18	329	95	112
% Data																											
Planktic Morphotypes																											
Biserial	100.0	100.0	100.0	0.0	0.0	22.7	0.0	100.0	87.5	100.0	28.0	100.0	62.6	90.9	100.0	94.8	75.6	0.0	100.0	100.0	100.0	0.0	100.0	54.5	40.9	59.6	50.9
Triserial	0.0	0.0	0.0	0.0	0.0	0.0	0.0	0.0	0.0	0.0	0.0	0.0	1.0	0.0	0.0	0.4	0.0	0.0	0.0	0.0	0.0	0.0	0.0	0.0	0.0	0.0	0.0
Trochospiral	0.0	0.0	0.0	0.0	0.0	77.3	0.0	0.0	12.5	0.0	72.0	0.0	37.4	9.1	0.0	4.9	24.4	0.0	0.0	0.0	0.0	0.0	0.0	45.5	59.1	40.4	49.1
Planispiral	0.0	0.0	0.0	0.0	0.0	0.0	0.0	0.0	0.0	0.0	0.0	0.0	0.0	0.0	0.0	0.0	0.0	0.0	0.0	0.0	0.0	0.0	0.0	0.0	0.0	0.0	0.0
Keeled	0.0	0.0	0.0	0.0	0.0	0.0	0.0	0.0	0.0	0.0	0.0	0.0	0.0	0.0	0.0	0.0	0.0	0.0	0.0	0.0	0.0	0.0	0.0	0.0	0.0	0.0	0.0
% Planktics	6.5	0.6	0.8	0.0	0.0	100.0	0.0	60.0	72.7	100.0	92.6	71.4	99.1	89.2	100.0	84.2	94.5	0.0	44.4	100.0	71.9	0.0	91.6	61.1	98.2	98.9	94.6
Benthic Morphotypes																											
Calcareous Other	0.0	0.0	0.0	0.0	0.0	0.0	0.0	0.0	0.0	0.0	0.0	0.0	0.0	0.0	0.0	0.0	0.0	0.0	0.0	0.0	28.0	0.0	0.0	100.0	33.3	100.0	50.0
Neobulimina	0.6	0.0	0.0	0.0	0.0	0.0	0.0	100.0	0.0	0.0	0.0	0.0	0.0	100.0	0.0	100.0	60.0	0.0	100.0	0.0	7.3	0.0	44.4	0.0	0.0	0.0	33.3
Gavelinella	0.3	0.0	0.0	0.0	0.0	0.0	0.0	0.0	0.0	0.0	0.0	0.0	33.3	0.0	0.0	0.0	20.0	0.0	0.0	0.0	13.4	0.0	33.3	0.0	66.7	0.0	16.7
Total Calcareous Specimens	1.0	0.0	0.0	0.0	0.0	0.0	0.0	100.0	0.0	0.0	0.0	0.0	33.3	100.0	0.0	100.0	80.0	0.0	100.0	0.0	48.8	0.0	77.8	100.0	100.0	100.0	100.0
Total Agglutinated Specimens	99.0	100.0	100.0	0.0	0.0	0.0	0.0	0.0	100.0	0.0	100.0	100.0	66.7	0.0	0.0	0.0	20.0	0.0	0.0	0.0	51.2	0.0	22.2	0.0	0.0	0.0	0.0

APPENDIX 1.—ESCALANTE DATA (continued)

	765.00	760.00	755.00	750.00	745.00	740.00	735.00	730.00	720.00	710.00	700.00	690.00	680.00	670.00	660.00	650.00	640.00	630.00	620.10	610.00	600.06	589.90	580.10	570.10	560.10	551.50	549.81	540.04	531.10	530.10
	34	80	114	121	49	70	97	94	194	74	161	111	105	159	152	146	138	112	150	218	29	78	124	228	281	208	175	230	3	236
	144	19	18	28	7	36	5	16	16	12	0	0	0	0	0	0	0	0	0	0	0	0	0	0	0	0	0	0	0	
	27	25	37	7	5	16	29	19	13	79	13	18	58	76	138	48	135	36	177	102	14	50	183	73	26	58	128	72	22	74
	2	0	0	0	0	0	0	0	0	0	0	0	0	0	0	0	0	0	0	0	0	0	0	0	0	0	0	0	0	0
		0.2?	0	0	0	0	0	0	0	0	0	0	0	0	0	0	0	0	0	0	0	0	0	0	0	0	0	0	0	0
	6	5	15	3	7	7	4	7	6	5	2	3	0	8	3	5	5	5	6	3	0	1	4	2	0	7	7	7	0	2
	79	208	122	176	254	196	175	184	123	79	163	175	131	66	32	32	50	7	12	3	1	4	4	9	4	12	3	16	0	0
	13	43	24	9	5	12	14	6	3	19	1	11	27	6	10	3	5	1	0	1	0	1	1	0	0	5	0	0	0	0
	98	256	161	188	266	215	193	197	132	103	166	189	158	80	45	40	60	8	18	7	0	6	9	11	4	24	10	23	0	2
	2	5	1	0	0	0	1	0	0	1	0	3	2	0	0	0	1	0	2	7	0	0	0	7	0	0	0	4	0	0
	100	261	162	188	266	215	194	197	132	104	166	192	160	80	45	40	61	8	20	14	1	6	9	18	4	24	10	27	0	2
	209	129	169	156	61	129	131	139	224	165	174	129	163	235	290	194	273	148	327	320	43	128	307	301	307	266	303	302	25	310
	309	390	331	344	327	344	325	336	356	269	340	321	323	315	335	234	334	156	347	334	44	134	316	319	311	290	313	329	25	312
	16.3	62.0	67.5	77.6	80.3	54.3	74.0	67.6	86.6	44.8	92.5	86.0	64.4	67.7	52.4	75.3	50.5	75.7	45.9	68.1	67.4	60.9	40.4	75.7	91.5	78.2	57.8	76.2	12.0	76.1
	68.9	14.7	10.7	17.9	11.5	27.9	3.8	11.5	7.6	7.3	0.0	0.0	0.0	0.0	0.0	0.0	0.0	0.0	0.0	0.0	0.0	0.0	0.0	0.0	0.0	0.0	0.0	0.0	0.0	0.0
	12.9	19.4	21.9	4.5	8.2	12.4	22.1	13.7	5.8	47.9	7.5	14.0	35.6	32.3	47.6	24.7	49.5	24.3	54.1	31.9	32.6	39.1	59.6	24.3	8.5	21.8	42.2	23.8	88.0	23.9
	0.0	0.0	0.0	0.0	0.0	0.0	0.0	0.0	0.0	0.0	0.0	0.0	0.0	0.0	0.0	0.0	0.0	0.0	0.0	0.0	0.0	0.0	0.0	0.0	0.0	0.0	0.0	0.0	0.0	0.0
	67.6	33.1	51.1	43.3	18.7	37.5	40.3	41.4	62.9	61.3	51.2	40.2	50.5	74.6	86.6	82.9	81.7	94.9	94.2	95.8	97.7	95.5	97.2	94.4	98.7	91.7	96.8	91.8	100.0	99.4
	6.0	1.9	9.3	1.6	2.6	3.3	2.1	3.6	4.5	4.8	1.2	1.6	0.0	10.0	6.7	12.5	8.2	0.0	30.0	21.4	0.0	16.7	44.4	11.1	0.0	29.2	70.0	25.9	0.0	100.0
	79.0	79.7	75.3	93.6	95.5	91.2	90.2	93.4	93.2	76.0	98.2	91.1	81.9	82.5	71.1	80.0	82.0	87.5	60.0	21.4	100.0	66.7	44.4	50.0	100.0	50.0	30.0	59.3	0.0	0.0
	13.0	16.5	14.8	4.8	1.9	5.6	7.2	3.0	2.3	18.3	0.6	5.7	16.9	7.5	22.2	7.5	8.2	12.5	0.0	7.1	0.0	16.7	11.1	0.0	0.0	20.8	0.0	0.0	0.0	0.0
	98.0	98.1	99.4	100.0	100.0	100.0	99.5	100.0	100.0	99.0	100.0	98.4	98.8	100.0	100.0	100.0	98.4	100.0	90.0	50.0	100.0	100.0	100.0	61.1	100.0	100.0	100.0	85.2	0.0	100.0
	2.0	1.9	0.6	0.0	0.0	0.0	0.5	0.0	0.0	1.0	0.0	1.6	1.3	0.0	0.0	0.0	1.6	0.0	10.0	50.0	0.0	0.0	0.0	38.9	0.0	0.0	0.0	14.8	0.9	14.8

	770.00	774.00	779.00	785.00	790.00	792.00	794.00	796.00	798.00	800.00	802.00	806.00	807.00	810.00	815.00	820.00
	81	63	61	73	57	103	16	0	0	39	10	0	8	0	0	0
	29	87	5	119	11	75	50	0	0	119	19	0	1	0	0	0
	56	24	49	17	95	40	5	0	0	9	3	0	2	0	0	0
	1	0	0	0	0	0	0	0	0	0	0	0	0	0	0	0
	0	0	0	0	0	0	0	0	0	0	0	0	0	0	0	0
	8	4	10	8	23	17	0	0	0	7	7	0	7	0	0	0
	106	173	190	127	108	64	32	0	0	47	7	0	8	0	0	1
	25	36	19	18	11	24	4	0	0	95	10	0	9	0	0	0
	139	213	219	153	142	105	36	0	0	149	24	0	24	0	0	1
	4	2	4	3	1	3	2	0	0	0	0	0	50	170	0	0
	143	215	223	156	143	108	38	0	0	149	24	0	74	170	0	0
	167	174	115	209	163	218	71	0	0	167	32	0	11	0	0	0
	310	389	338	365	306	326	109	0	0	316	56	0	85	170	0	0
	48.5	36.2	53.0	34.9	35.0	47.2	22.5	0.0	0.0	23.4	31.3	0.0	72.7	0.0	0.0	0.0
	17.4	50.0	4.3	56.9	6.7	34.4	70.4	0.0	0.0	71.3	59.4	0.0	9.1	0.0	0.0	0.0
	33.5	13.8	42.6	8.1	58.3	18.3	7.0	0.0	0.0	5.4	9.4	0.0	18.2	0.0	0.0	0.0
	0.6	0.0	0.0	0.0	0.0	0.0	0.0	0.0	0.0	0.0	0.0	0.0	0.0	0.0	0.0	0.0
	53.9	44.7	34.0	57.3	53.3	66.9	65.1	0.0	0.0	52.8	57.1	0.0	12.9	0.0	0.0	0.0
	5.6	1.9	4.5	5.1	16.1	15.7	0.0	0.0	0.0	4.7	29.2	0.0	9.5	0.0	0.0	0.0
	74.1	80.5	85.2	81.4	75.5	59.3	84.2	0.0	0.0	31.5	29.2	0.0	10.8	0.0	0.0	100.0
	17.5	16.7	8.5	11.5	7.7	22.2	10.5	0.0	0.0	63.8	41.7	0.0	12.2	0.0	0.0	0.0
	97.2	99.1	98.2	98.1	99.3	97.2	94.7	0.0	0.0	100.0	100.0	0.0	32.4	0.0	0.0	0.0
	2.8	0.9	1.8	1.9	0.7	2.8	5.3	0.0	0.0	0.0	0.0	0.0	67.6	100.0	0.0	100.0

APPENDIX 2.—LOHALI POINT DATA

Sample number	LP1	LP2	LP3	LP4	LP5	LP6	LP7	LP8	LP9	LP10	LP11	LP12	LP13	LP14	LP15	LP16
Raw data																
Planktic Morphotypes																
Biserial	7.0	3.0	142.0	41.0	55.0	50.0	17.0	33.0	25.0	46.0	34.0	120.0	36.0	68.0	67.0	79.0
Triserial	17.0	3.0	67.0	32.0	0.0	136.0	230.0	271.0	86.0	208.0	257.0	82.0	25.0	29.0	17.0	27.0
Trochospiral	31.0	0.0	84.0	0.0	98.0	27.0	38.0	6.0	5.0	13.0	1.0	13.0	6.0	10.0	17.0	16.0
Planispiral	0.0	0.0	3.0	17.0	12.0	0.0	0.0	0.0	0.0	0.0	0.0	0.0	0.0	0.0	0.0	0.0
Keeled	0.0	0.0	0.0	0.0	0.0	0.0	0.0	0.0	0.0	0.0	0.0	0.0	0.0	0.0	0.0	0.0
Benthic Morphotypes																
Calcareous other	0.0	10.0	13.0	20.0	0.0	14.0	9.0	7.0	3.0	1.0	3.0	9.0	1.0	2.0	4.0	3.0
Neobulimina	15.0	21.0	6.0	86.0	97.0	47.0	2.0	95.0	114.0	38.0	11.0	10.0	35.0	34.0	172.0	145.0
Gavelinella	23.0	4.0	8.0	5.0	1.0	2.0	5.0	2.0	31.0	16.0	24.0	82.0	22.0	155.0	17.0	39.0
Total Calcareous specimens	38.0	35.0	27.0	111.0	98.0	63.0	16.0	104.0	148.0	55.0	38.0	101.0	58.0	191.0	193.0	187.0
Total Agglutinated specimens	143.0	96.0	16.0	41.0	19.0	11.0	2.0	40.0	42.0	3.0	5.0	19.0	56.0	6.0	7.0	16.0
Total number of Benthics	181.0	131.0	43.0	152.0	117.0	74.0	18.0	144.0	190.0	58.0	43.0	120.0	114.0	197.0	200.0	203.0
Total number of Planktics	55.0	6.0	296.0	90.0	165.0	213.0	285.0	310.0	116.0	267.0	292.0	215.0	67.0	107.0	101.0	122.0
Total number of Specimens	236.0	137.0	339.0	242.0	282.0	287.0	303.0	454.0	306.0	325.0	335.0	335.0	181.0	304.0	301.0	325.0
% Data																
Planktic Morphotypes	23.3	4.4	87.3	37.2	58.5	74.2	94.1	68.3	37.9	82.2	87.2	64.2	37.0	35.2	33.6	37.5
Biserial	12.7	50.0	48.0	45.6	33.3	23.5	6.0	10.6	21.6	17.2	11.6	55.8	53.7	63.6	66.3	64.8
Triserial	30.9	50.0	22.6	35.6	0.0	63.8	80.7	87.4	74.1	77.9	88.0	38.1	37.3	27.1	16.8	22.1
Trochospiral	56.4	0.0	28.4	0.0	59.4	12.7	13.3	1.9	4.3	4.9	0.3	6.0	9.0	9.3	16.8	13.1
planispiral	0.0	0.0	1.0	18.9	7.3	0.0	0.0	0.0	0.0	0.0	0.0	0.0	0.0	0.0	0.0	0.0
Keeled	0.0	0.0	0.0	0.0	0.0	0.0	0.0	0.0	0.0	0.0	0.0	0.0	0.0	0.0	0.0	0.0
% Planktics	23.3	4.4	87.3	37.2	58.5	74.2	94.1	68.3	37.9	82.2	87.2	64.2	37.0	35.2	33.6	37.5
Benthic Morphotypes	76.7	95.6	12.7	62.8	41.5	25.8	5.9	31.7	62.1	17.8	12.8	35.8	63.0	64.8	66.4	62.5
Calcareous other	0.0	7.6	30.2	13.2	0.0	18.9	50.0	4.9	1.6	1.7	7.0	7.5	0.9	1.0	2.0	1.5
Neobulimina	8.3	16.0	14.0	56.6	82.9	63.5	11.1	66.0	60.0	65.5	25.6	8.3	30.7	17.3	86.0	71.4
Gavelinella	12.7	3.1	18.6	3.3	0.9	2.7	27.8	1.4	16.3	27.6	55.8	68.3	19.3	78.7	8.5	19.2
Total Calcareous specimens	21.0	26.7	62.8	73.0	83.8	85.1	88.9	72.2	77.9	94.8	88.4	84.2	50.9	97.0	96.5	92.1
Total Agglutinated specimens	79.0	73.3	37.2	27.0	16.2	14.9	11.1	27.8	22.1	5.2	11.6	15.8	49.1	3.0	3.5	7.9

Sample number	LP17	LP18	LP19	LP20	LP21	LP22	LP23	LP24	LP25	LP26	LP27	LP28	LP29	LP30	LP31
Raw data															
Planktic Morphotypes															
Biserial	127.0	163.0	146.0	141.0	127.0	80.0	81.0	114.0	165.0	153.0	124.0	142.0	159.0	191.0	209.0
Triserial	22.0	31.0	41.0	30.0	37.0	35.0	44.0	16.0	5.0	0.0	3.0	7.0	1.0	6.0	4.0
Trochospiral	24.0	20.0	33.0	20.0	15.0	44.0	30.0	44.0	14.0	24.0	56.0	42.0	59.0	36.0	38.0
Planispiral	0.0	0.0	0.0	0.0	0.0	0.0	0.0	0.0	0.0	0.0	0.0	7.0	2.0	0.0	0.0
Keeled	0.0	0.0	0.0	0.0	0.0	0.0	0.0	0.0	0.0	0.0	0.0	0.0	0.0	0.0	0.0
Benthic Morphotypes															
Calcareous other	11.0	3.0	1.0	3.0	3.0	0.0	2.0	0.0	1.0	0.0	0.0	3.0	1.0	2.0	2.0
Neobulimina	87.0	65.0	64.0	74.0	95.0	109.0	96.0	84.0	73.0	87.0	93.0	78.0	55.0	40.0	41.0
Gavelinella	37.0	30.0	28.0	14.0	23.0	15.0	47.0	28.0	24.0	24.0	4.0	6.0	7.0	4.0	6.0
Total Calcareous specimens	135.0	98.0	93.0	91.0	121.0	124.0	145.0	112.0	98.0	111.0	97.0	87.0	63.0	46.0	49.0
Total Agglutinated specimens	1.0	3.0	6.0	2.0	2.0	3.0	5.0	1.0	5.0	0.0	0.0	5.0	9.0	3.0	0.0
Total number of Benthics	136.0	101.0	99.0	93.0	123.0	127.0	150.0	113.0	103.0	111.0	97.0	92.0	72.0	49.0	49.0
Total number of Planktics	173.0	214.0	220.0	191.0	179.0	159.0	155.0	174.0	184.0	177.0	183.0	201.0	221.0	233.0	251.0
Total number of Specimens	309.0	315.0	319.0	284.0	302.0	286.0	305.0	287.0	287.0	288.0	280.0	293.0	293.0	282.0	300.0
% Data															
Planktic Morphotypes	56.0	67.9	69.0	67.3	59.3	55.6	50.8	60.6	64.1	61.5	65.4	68.6	75.4	82.6	83.7
Biserial	73.4	76.2	66.4	73.8	70.9	50.3	52.3	65.5	89.7	86.4	67.8	70.6	71.9	82.0	83.3
Triserial	12.7	14.5	18.6	15.7	20.7	22.0	28.4	9.2	2.7	0.0	1.6	3.5	0.5	2.6	1.6
Trochospiral	13.9	9.3	15.0	10.5	8.4	27.7	19.4	25.3	7.6	13.6	30.6	20.9	26.7	15.5	15.1
planispiral	0.0	0.0	0.0	0.0	0.0	0.0	0.0	0.0	0.0	0.0	0.0	3.5	0.9	0.0	0.0
Keeled	0.0	0.0	0.0	0.0	0.0	0.0	0.0	0.0	0.0	0.0	0.0	0.0	0.0	0.0	0.0
% Planktics	56.0	67.9	69.0	67.3	59.3	55.6	50.8	60.6	64.1	61.5	65.4	68.6	75.4	82.6	83.7
Benthic Morphotypes	44.0	32.1	31.0	32.7	40.7	44.4	49.2	39.4	35.9	38.5	34.6	31.4	24.6	17.4	16.3
Calcareous other	8.1	3.0	1.0	3.2	2.4	0.0	1.3	0.0	1.0	0.0	0.0	3.3	1.4	4.1	4.1
Neobulimina	64.0	64.4	64.6	79.6	77.2	85.8	64.0	74.3	70.9	78.4	95.9	84.8	76.4	81.6	83.7
Gavelinella	27.2	29.7	28.3	15.1	18.7	11.8	31.3	24.8	23.3	21.6	4.1	6.5	9.7	8.2	12.2
Total Calcareous specimens	99.3	97.0	93.9	97.8	98.4	97.6	96.7	99.1	95.1	100.0	100.0	94.6	87.5	93.9	100.0
Total Agglutinated specimens	0.7	3.0	6.1	2.2	1.6	2.4	3.3	0.9	4.9	0.0	0.0	5.4	12.5	6.1	0.0

LP36	LP41	LP46	LP51	LP56	LP61	LP66	LP71	LP76	LP81	LP86	LP91	LP96	LP101	LP106	LP111	LP116	LP121	LP126	LP131	LP136	LP141	LP146	LP151	LP156	LP161	LP166	LP171	LP176	LP181	LP186	LP191	LP196	LP201
219.0	158.0	198.0	106.0	97.0	268.0	246.0	8.0	185.0	207.0	4.0	61.0	10.0	0.0	0.0	6.0	0.0	0.0	0.0	0.0	0.0	0.0	0.0	0.0	0.0	0.0	0.0	0.0	0.0	0.0	0.0	0.0	0.0	0.0
4.0	1.0	0.0	0.0	0.0	0.0	0.0	0.0	0.0	0.0	1.0	0.0	0.0	0.0	0.0	0.0	0.0	0.0	0.0	0.0	0.0	0.0	0.0	0.0	0.0	0.0	0.0	0.0	0.0	0.0	0.0	0.0	0.0	0.0
48.0	56.0	77.0	43.0	112.0	57.0	50.0	0.0	37.0	71.0	1.0	109.0	1.0	0.0	0.0	0.0	0.0	0.0	0.0	0.0	0.0	0.0	0.0	0.0	0.0	0.0	0.0	0.0	0.0	0.0	0.0	0.0	0.0	0.0
1.0	0.0	0.0	4.0	1.0	0.0	0.0	0.0	0.0	0.0	0.0	0.0	0.0	0.0	0.0	0.0	0.0	0.0	0.0	0.0	0.0	0.0	0.0	0.0	0.0	0.0	0.0	0.0	0.0	0.0	0.0	0.0	0.0	0.0
0.0	0.0	0.0	0.0	0.0	0.0	0.0	0.0	0.0	0.0	0.0	0.0	0.0	0.0	0.0	0.0	0.0	0.0	0.0	0.0	0.0		0.0	0.0	0.0	0.0	0.0	0.0	0.0	0.0	0.0	0.0	0.0	0.0
0.0	7.0	9.0	7.0	1.0	0.0	1.0	0.0	0.0	0.0	0.0	0.0	0.0	0.0	0.0	0.0	0.0	0.0	0.0	0.0	0.0	0.0	0.0	0.0	0.0	0.0	0.0	0.0	0.0	0.0	0.0	0.0	0.0	0.0
23.0	41.0	45.0	25.0	1.0	0.0	1.0	22.0	9.0	2.0	37.0	2.0	11.0	0.0	0.0	3.0	0.0	0.0	0.0	0.0	0.0	0.0	0.0	0.0	0.0	0.0	0.0	0.0	0.0	0.0	0.0	0.0	0.0	0.0
11.0	11.0	12.0	46.0	21.0	1.0	0.0	16.0	13.0	0.0	0.0	0.0	0.0	0.0		0.0	0.0	0.0	0.0	0.0	0.0	0.0	0.0	0.0	0.0	0.0	0.0	0.0	0.0	0.0	0.0	0.0	0.0	0.0
34.0	59.0	66.0	78.0	23.0	1.0	2.0	38.0	22.0	2.0	37.0	2.0	11.0	0.0	0.0	0.0	0.0	0.0	0.0	0.0	0.0	0.0	0.0	0.0	0.0	0.0	0.0	0.0	0.0	0.0	0.0	0.0	0.0	0.0
15.0	10.0	8.0	37.0	48.0	0.0	1.0	67.0	19.0	0.0	137.0	51.0	123.0	62.0	278.0	50.0	97.0	42.0	113.0	101.0	131.0	105.0	114.0	269.0	0.0	55.0	266.0	128.0	294.0	0.0	256.0	125.0	101.0	1.0
49.0	69.0	74.0	115.0	71.0	1.0	3.0	105.0	41.0	2.0	174.0	53.0	134.0	62.0	278.0	53.0	97.0	42.0	113.0	101.0	131.0	105.0	114.0	269.0	0.0	55.0	266.0	128.0	294.0	0.0	256.0	125.0	101.0	1.0
271.0	215.0	277.0	153.0	210.0	323.0	296.0	8.0	222.0	298.0	6.0	170.0	11.0	0.0	0.0	6.0	0.0	0.0	0.0	0.0	0.0	0.0	0.0	0.0	0.0	0.0	0.0	0.0	0.0	0.0	0.0	0.0	0.0	0.0
320.0	284.0	351.0	268.0	281.0	324.0	299.0	113.0	263.0	300.0	180.0	223.0	145.0	62.0	278.0	59.0	97.0	42.0	113.0	101.0	131.0	105.0	114.0	269.0	0.0	55.0	266.0	128.0	294.0	0.0	256.0	125.0	101.0	1.0
84.7	75.7	78.9	57.1	74.7	99.7	99.0	7.1	84.4	99.3	3.3	76.2	7.6	0.0	0.0	10.2	0.0	0.0	0.0	0.0	0.0	0.0	0.0	0.0	0.0	0.0	0.0	0.0	0.0	0.0	0.0	0.0	0.0	0.0
80.8	73.5	71.5	69.3	46.2	83.0	83.1	100.0	83.3	69.5	66.7	35.9	90.9	0.0	0.0	100.0	0.0	0.0	0.0	0.0	0.0	0.0	0.0	0.0	0.0	0.0	0.0	0.0	0.0	0.0	0.0	0.0	0.0	0.0
1.5	0.5	0.0	0.0	0.0	0.0	0.0	0.0	0.0	0.0	16.7	0.0	0.0	0.0	0.0	0.0	0.0	0.0	0.0	0.0	0.0	0.0	0.0	0.0	0.0	0.0	0.0	0.0	0.0	0.0	0.0	0.0	0.0	0.0
17.7	26.0	27.8	28.1	53.3	17.6	16.9	0.0	16.7	23.8	16.7	64.1	9.1	0.0	0.0	0.0	0.0	0.0	0.0	0.0	0.0	0.0	0.0	0.0	0.0	0.0	0.0	0.0	0.0	0.0	0.0	0.0	0.0	0.0
0.4	0.0	0.0	2.6	0.5	0.0	0.0	0.0	0.0	0.0	0.0	0.0	0.0	0.0	0.0	0.0	0.0	0.0	0.0	0.0	0.0	0.0	0.0	0.0	0.0	0.0	0.0	0.0	0.0	0.0	0.0	0.0	0.0	0.0
0.0	0.0	0.0	0.0	0.0	0.0	0.0	0.0	0.0	0.0	0.0	0.0	0.0	0.0	0.0	0.0	0.0	0.0	0.0	0.0	0.0	0.0	0.0	0.0	0.0	0.0	0.0	0.0	0.0	0.0	0.0	0.0	0.0	0.0
84.7	75.7	78.9	57.1	74.7	99.7	99.0	7.1	84.4	99.3	3.3	76.2	7.6	0.0	0.0	10.2	0.0	0.0	0.0	0.0	0.0	0.0	0.0	0.0	0.0	0.0	0.0	0.0	0.0	0.0	0.0	0.0	0.0	0.0
15.3	24.3	21.1	42.9	25.3	0.3	1.0	92.9	15.6	0.7	96.7	23.8	92.4	100.0	100.0	89.8	100.0	100.0	100.0	100.0	100.0	100.0	100.0	100.0	0.0	100.0	100.0	100.0	100.0	0.0	100.0	100.0	100.0	100.0
0.0	10.1	12.2	6.1	1.4	0.0	33.3	0.0	0.0	0.0	0.0	0.0	0.0	0.0	0.0	0.0	0.0	0.0	0.0	0.0	0.0	0.0	0.0	0.0	0.0	0.0	0.0	0.0	0.0	0.0	0.0	0.0	0.0	0.0
46.9	59.4	60.8	21.7	1.4	0.0	33.3	21.0	22.0	100.0	21.3	3.8	8.2	0.0	0.0	5.7	0.0	0.0	0.0	0.0	0.0	0.0	0.0	0.0	0.0	0.0	0.0	0.0	0.0	0.0	0.0	0.0	0.0	0.0
22.4	15.9	16.2	40.0	29.6	100.0	0.0	15.2	31.7	0.0	0.0	0.0	0.0	0.0	0.0	0.0	0.0	0.0	0.0	0.0	0.0	0.0	0.0	0.0	0.0	0.0	0.0	0.0	0.0	0.0	0.0	0.0	0.0	0.0
69.4	85.5	89.2	67.8	32.4	100.0	66.7	36.2	53.7	100.0	21.3	3.8	8.2	0.0	0.0	0.0	0.0	0.0	0.0	0.0	0.0	0.0	0.0	0.0	0.0	0.0	0.0	0.0	0.0	0.0	0.0	0.0	0.0	0.0
30.6	14.5	10.8	32.2	67.6	0.0	33.3	63.8	46.3	0.0	78.7	96.2	91.8	100.0	100.0	94.3	100.0	100.0	100.0	100.0	100.0	100.0	100.0	100.0	0.0	100.0	100.0	100.0	100.0	0.0	100.0	100.0	100.0	100.0

APPENDIX 3.—MESA VERDE DATA

Sample Number	27.6	27.1	26.7	26.2	25.8	25.2	24.7	23.6	23.3	22.7	22.4	21.5	20.8	20.5	19.5	18.5	17.5	16.5	15.5	14.5	13.5	12.5	11.4	9.5	8.2	6.5	5.8	4.7	3.6	2	1
Raw Data																															
Planktic Morphotypes																															
Biserial	53.0	90.0	48.0	123.0	142.0	78.0	84.0	10.0	5.0	102.0	25.0	24.0	25.0	19.0	16.0	30.0	67.0	42.0	85.0	13.0	53.0	0.0	1.0	1.0	2.0	0.0	0.0	0.0	0.0	0.0	1.0
Triserial	0.0	3.0	0.0	0.0	0.0	7.0	30.0	32.0	5.0	115.0	17.0	58.0	156.0	45.0	4.0	45.0	45.0	4.0	17.0	14.0	93.0	4.0	1.0	0.0	0.0	1.0	0.0	0.0	0.0	0.0	5.0
Trochospiral	221.0	104.0	191.0	7.0	64.0	101.0	43.0	44.0	3.0	22.0	43.0	21.0	19.0	52.0	18.0	91.0	80.0	203.0	132.0	40.0	55.0	5.0	2.0	2.0	7.0	2.0	0.0	0.0	0.0	0.0	0.0
Planispiral	0.0	0.0	0.0	0.0	0.0	1.0	0.0	2.0	0.0	0.0	3.0	3.0	6.0	1.0	0.0	1.0	2.0	9.0	5.0	0.0	3.0	0.0	3.0	0.0	0.0	0.0	0.0	0.0	0.0	0.0	0.0
Keeled	0.0	0.0	0.0	0.0	0.0	0.0	0.0	0.0	0.0	0.0	0.0	0.0	0.0	0.0	0.0	0.0	0.0	2.0	0.0	0.0	0.0	0.0	0.0	0.0	0.0	0.0	0.0	0.0	0.0	0.0	0.0
Benthic Morphotypes																															
Calcareous other	0.0	0.0	0.0	1.0	7.0	0.0	0.0	12.0	16.0	1.0	18.0	1.0	0.0	2.0	1.0	8.0	0.0	26.0	18.0	16.0	0.0	1.0	0.0	0.0	3.0	0.0	0.0	0.0	0.0	0.0	0.0
Neobulimina	16.0	35.0	11.0	213.0	141.0	142.0	95.0	141.0	23.0	32.0	106.0	148.0	78.0	35.0	14.0	21.0	102.0	48.0	26.0	55.0	44.0	31.0	0.0	0.0	5.0	0.0	0.0	0.0	0.0	0.0	0.0
Gavelinella	1.0	12.0	1.0	9.0	3.0	15.0	20.0	50.0	218.0	75.0	104.0	37.0	1.0	6.0	3.0	4.0	2.0	8.0	5.0	19.0	19.0	7.0	2.0	2.0	1.0	0.0	0.0	0.0	0.0	0.0	0.0
Total Calcareous Specimens	17.0	47.0	12.0	223.0	151.0	157.0	115.0	203.0	257.0	108.0	228.0	186.0	79.0	43.0	18.0	33.0	104.0	82.0	49.0	90.0	63.0	39.0	2.0	2.0	9.0	0.0	0.0	0.0	0.0	0.0	0.0
Total Agglutinated Specimens	0.0	1.0	0.0	0.0	2.0	0.0	0.0	3.0	19.0	3.0	16.0	51.0	10.0	6.0	29.0	1.0	1.0	0.0	18.0	41.0	6.0	7.0	0.0	0.0	0.0	0.0	0.0	0.0	0.0	0.0	0.0
Total number of Benthics	17.0	48.0	12.0	223.0	153.0	157.0	115.0	206.0	276.0	111.0	244.0	237.0	89.0	47.0	47.0	34.0	105.0	82.0	67.0	131.0	69.0	46.0	2.0	0.0	9.0	0.0	0.0	0.0	0.0	0.0	0.0
Total number of Planktics	274.0	197.0	242.0	130.0	208.0	187.0	157.0	88.0	13.0	239.0	88.0	106.0	206.0	138.0	38.0	167.0	194.0	260.0	239.0	67.0	204.0	9.0	7.0	3.0	9.0	3.0	0.0	0.0	0.0	0.0	6.0
Total number of specimens	291.0	245.0	254.0	353.0	361.0	344.0	294.0	289.0	350.0	332.0	343.0	295.0	187.0	187.0	85.0	201.0	299.0	342.0	306.0	198.0	273.0	55.0	9.0	3.0	18.0	3.0	0.0	0.0	0.0	0.0	6.0
%Data																															
Planktic Morphotypes																															
Biserial	19.3	45.7	19.8	94.6	68.3	41.7	53.5	11.4	38.5	42.7	28.4	22.6	12.1	13.8	42.1	18.0	34.5	16.2	35.6	19.4	26.0	0.0	14.3	33.3	22.2	0.0	0.0	0.0	0.0	0.0	16.7
Triserial	0.0	1.5	1.2	0.0	1.0	3.7	19.1	36.4	38.5	48.1	19.3	54.7	75.7	47.8	10.5	26.9	23.2	1.5	7.1	20.9	45.6	44.4	14.3	0.0	0.0	33.3	0.0	0.0	0.0	0.0	83.3
Trochospiral	80.7	52.8	78.9	5.4	30.8	54.0	27.4	50.0	23.1	9.2	48.9	19.8	9.2	37.7	47.4	54.5	41.2	78.1	55.2	59.7	27.0	55.6	28.6	66.7	77.8	66.7	0.0	0.0	0.0	0.0	0.0
Planispiral	0.0	0.0	0.0	0.0	0.0	0.5	0.0	2.3	0.0	0.0	3.4	2.8	2.9	0.7	0.0	0.6	1.0	3.5	2.1	0.0	0.0	0.0	0.0	0.0	0.0	0.0	0.0	0.0	0.0	0.0	0.0
Keeled	0.0	0.0	0.0	0.0	0.0	0.0	0.0	0.0	0.0	0.0	0.0	0.0	0.0	0.0	0.0	0.0	0.0	0.8	0.0	0.0	0.0	0.0	0.0	0.0	0.0	0.0	0.0	0.0	0.0	0.0	0.0
% Planktics	94.2	80.4	95.3	36.8	57.6	54.4	53.4	30.4	3.7	72.0	25.7	35.9	100.0	73.8	44.7	83.1	64.9	76.0	78.1	33.8	74.7	16.4	77.8	100.0	50.0	100.0	0.0	0.0	0.0	0.0	100.0
Benthic Morphotypes																															
Calcareous other	0.0	0.0	0.4	0.4	4.6	0.0	0.0	5.8	5.8	0.9	7.4	0.4	0.0	0.4	2.1	23.5	0.0	31.7	26.9	12.2	0.0	2.2	0.0	0.0	33.3	0.0	0.0	0.0	0.0	0.0	0.0
Neobulimina	94.1	72.9	91.7	95.5	92.2	90.4	82.6	68.4	8.3	28.8	43.4	62.4	87.6	7.8	29.8	61.8	97.1	58.5	38.8	42.0	63.8	67.4	0.0	100.0	55.6	33.3	0.0	0.0	0.0	0.0	0.0
Gavelinella	5.9	25.0	8.3	4.0	2.0	9.6	17.4	24.3	79.0	67.6	42.6	15.6	1.3	1.3	6.4	11.8	1.9	9.8	7.5	14.5	27.5	15.2	100.0	0.0	11.1	0.0	0.0	0.0	0.0	0.0	0.0
Total Calcareous Specimens	100.0	97.9	100.0	100.0	98.7	100.0	100.0	98.5	93.1	97.3	93.4	78.5	88.8	9.6	38.3	97.1	99.0	100.0	73.1	68.7	91.3	84.8	100.0	100.0	100.0	0.0	0.0	0.0	0.0	0.0	0.0
Total Agglutinated Specimens	0.0	2.1	0.0	0.0	1.3	0.0	0.0	1.5	6.9	2.7	6.6	21.5	11.2	1.3	61.7	2.9	1.0	0.0	26.9	31.3	8.7	15.2	0.0	0.0	0.0	0.0	0.0	0.0	0.0	0.0	0.0

Sample Number	28.1	28.4	29.2	29.8	33	38.8	43.4	49.7	54.5	59.6	64.6	66.6	69.6	74.8	85	94	104	115	121.6	125	130	135	138	141
Raw Data																								
Planktic Morphotypes																								
Biserial	263.0	119.0	133.0	90.0	118.0	70.0	0.0	5.0	10.0	3.0	0.0	3.0	2.0	0.0	0.0	1.0	1.0	0.0	0.0	0.0	4.0	0.0	0.0	0.0
Triserial	0.0	1.0	0.0	0.0	0.0	0.0	0.0	1.0	0.0	0.0	0.0	0.0	0.0	0.0	0.0	0.0	0.0	0.0	0.0	0.0	0.0	0.0	0.0	0.0
Trochospiral	50.0	139.0	68.0	137.0	158.0	257.0	12.0	0.0	3.0	4.0	0.0	0.0	0.0	0.0	0.0	0.0	60.0	4.0	2.0	0.0	2.0	0.0	0.0	0.0
Planispiral	2.0	0.0	0.0	0.0	4.0	7.0	0.0	0.0	0.0	1.0	0.0	0.0	0.0	0.0	0.0	0.0	4.0	0.0	0.0	0.0	0.0	0.0	0.0	0.0
Keeled	0.0	0.0	0.0	0.0	0.0	0.0	0.0	0.0	0.0	0.0	0.0	0.0	0.0	0.0	0.0	0.0	0.0	0.0	0.0	0.0	0.0	0.0	0.0	0.0
Benthic Morphotypes																								
Calcareous other	0.0	0.0	0.0	0.0	12.0	4.0	0.0	0.0	0.0	0.0	0.0	0.0	0.0	0.0	0.0	0.0	0.0	0.0	0.0	0.0	1.0	0.0	0.0	0.0
Neobulimina	45.0	36.0	15.0	55.0	45.0	0.0	0.0	0.0	6.0	0.0	0.0	2.0	0.0	0.0	2.0	0.0	0.0	0.0	0.0	0.0	27.0	3.0	0.0	0.0
Gavelinella	1.0	8.0	3.0	3.0	10.0	43.0	7.0	0.0	13.0	0.0	0.0	0.0	0.0	0.0	0.0	0.0	0.0	0.0	0.0	26.0	6.0	0.0	0.0	0.0
Total Calcareous Specimens	46.0	44.0	15.0	58.0	67.0	47.0	7.0	0.0	19.0	0.0	0.0	2.0	0.0	0.0	2.0	0.0	0.0	0.0	0.0	2.0	27.0	3.7	0.0	0.0
Total Agglutinated Specimens	0.0	0.0	0.0	0.0	1.0	0.0	0.0	0.0	0.0	0.0	0.0	0.0	0.0	0.0	0.0	0.0	0.0	0.0	0.0	0.0	0.0	0.0	0.0	0.0
Total number of Benthics	46.0	44.0	15.0	58.0	68.0	47.0	7.0	0.0	19.0	0.0	0.0	2.0	0.0	0.0	2.0	0.0	0.0	0.0	0.0	40.0	27.0	55.0	0.0	0.0
Total number of Planktics	313.0	259.0	201.0	227.0	280.0	337.0	12.0	6.0	13.0	8.0	3.0	3.0	2.0	0.0	0.0	1.0	65.0	4.0	2.0	40.0	6.0	55.0	9.0	204.0
Total number of specimens	359.0	303.0	216.0	285.0	348.0	384.0	19.0	6.0	32.0	8.0	0.0	5.0	14.0	0.0	2.0	0.0	206.0	55.0	37.0	40.0	33.0	55.0	9.0	204.0
%Data																								
Planktic Morphotypes																								
Biserial	84.0	45.9	66.2	39.6	42.1	20.8	0.0	83.3	76.9	37.5	0.0	100.0	100.0	0.0	0.0	100.0	1.5	0.0	0.0	0.0	66.7	0.0	0.0	0.0
Triserial	0.4	0.4	0.0	0.0	0.0	0.0	0.0	16.7	0.0	0.0	0.0	0.0	0.0	0.0	0.0	0.0	0.0	0.0	0.0	0.0	0.0	0.0	0.0	0.0
Trochospiral	53.7	53.7	33.8	60.4	56.4	76.3	100.0	0.0	23.1	50.0	100.0	0.0	0.0	0.0	0.0	0.0	92.3	100.0	100.0	0.0	33.3	0.0	0.0	204.0
Planispiral	0.6	0.0	0.0	0.0	1.4	2.1	0.0	0.0	0.0	12.5	0.0	0.0	0.0	0.0	0.0	0.0	6.2	0.0	0.0	0.0	0.0	3.7	0.0	0.0
Keeled	0.0	0.0	0.0	0.0	0.0	0.0	0.0	0.0	0.0	0.0	0.0	0.0	0.0	0.0	0.0	0.0	0.0	0.0	0.0	0.0	0.0	0.0	0.0	0.0
% Planktics	87.2	85.5	93.1	79.6	80.5	87.8	63.2	100.0	40.6	100.0	100.0	60.0	14.3	0.0	0.0	31.6	7.3	5.4	18.2	96.3	100.0	100.0	100.0	100.0
Benthic Morphotypes																								
Calcareous other	0.0	0.0	0.0	0.0	17.6	8.5	0.0	0.0	0.0	0.0	0.0	0.0	0.0	0.0	0.0	0.0	0.0	0.0	0.0	0.0	3.7	0.0	0.0	0.0
Neobulimina	97.8	81.8	100.0	94.8	66.2	0.0	0.0	0.0	31.6	0.0	0.0	100.0	0.0	0.0	100.0	0.0	0.0	0.0	0.0	0.0	96.3	0.0	0.0	0.0
Gavelinella	2.2	18.2	0.0	5.2	14.7	91.5	100.0	0.0	68.4	0.0	0.0	0.0	0.0	0.0	0.0	0.0	0.0	0.0	0.0	100.0	0.0	3.7	0.0	0.0
Total Calcareous Specimens	100.0	100.0	100.0	100.0	98.5	100.0	100.0	0.0	100.0	0.0	0.0	100.0	0.0	0.0	100.0	0.0	100.0	100.0	100.0	100.0	100.0	96.3	100.0	100.0
Total Agglutinated Specimens	0.0	0.0	0.0	0.0	1.5	0.0	0.0	0.0	0.0	0.0	0.0	0.0	0.0	0.0	0.0	0.0	0.0	0.0	0.0	0.0	0.0	3.7	0.0	0.0

REFERENCES

ARTHUR, M. A., DEAN, W. E., POLLASTRO, R. M., CLAYPOOL, G. E., and SCHOLLE, P. A., 1985, Comparative geochemical and mineralogical studies of two cyclic transgressive pelagic limestone units, Cretaceous western interior basin, U.S., in Pratt, L.M., Kauffman, E.G., and Zelt, F.B., eds., Fine-grained Deposits and Biofacies of the Cretaceous Western Interior Seaway: Evidence of Cyclic Sedimentary Processes: Field Trip Guidebook No. 4 Tulsa, Society of Economic Paleontologists and Mineralogists, p. 16-27.

BARRON, E. J., ARTHUR, M. A., and KAUFFMAN, E. G., 1985, Cretaceous rhythmic bedding sequences: A plausible link between orbital variations and climate: Earth and Planetary Science Letters, v. 72, p. 327-340.

BERNHARD, J. M., 1989, The distribution of benthic foraminifera with respect to oxygen concentration and organic levels in shallow-water Antarctic sediments: Limnology and Oceanography, v. 34, p. 1131-1141.

BERNHARD, J. M., 1993, Experimental and field evidence of Antarctic foraminiferal tolerance to anoxia and hydrogen sulfide: Marine Micropaleontology, v. 20, p. 203-213.

BERNHARD, J.M., 1996, Microaerophilic and facultative anaerobic benthic foraminifera: A review of experimental and ultrastructural evidence: Revue de Paleobiologie, v. 15, p. 261-275.

BRETT, C. E., 1995, Stasis: life in the balance: Geotimes, v. 40, p. 18-20.

BRETT, C. E, and BAIRD, G. C., 1995, Coordinated stasis and evolutionary ecology of the Silurian to Middle Devonian faunas of the Appalachian Basin, in Erwin, D. H., and Anstey, R. L., eds., New Approaches to Speciation in the Fossil Record: New York, Columbia University Press, p. 285-315.

BUZAS, M. A., CULVER, S. J., and JORISSEN, F. J., 1993, A statistical evaluation of the microhabitats of living (stained) infaunal benthic foraminifera: Marine Micropaleontology, v. 20, p. 311-320.

CALDWELL, W. G. E., NORTH, B. R., STELCK, C. R., and WALL, J. H., 1978, A foraminiferal zonal scheme for the Cretaceous System in the interior plains of Canada, in Stelck, C. R., and Chatterton, B. D. E., eds., Western and Arctic Canadian Biostratigraphy: St John's, Geological Association of Canada Special Paper 18, p. 495-575.

CALDWELL, W. G. E., DINER, R., EICHER, E. L., FOWLER, S. P., NORTH, B. R., STELK, C. R., and VON HOLDT W. L., 1993, Foraminiferal biostratigraphy of Cretaceous marine cyclothems, in Caldwell, W. G. E., and Kauffman, E. G., eds., Evolution of the Western Interior Basin: St John's Geological Association of Canada Special Paper 39, p. 477-520.

COBBAN, W. A., and SCOTT, G. R., 1972, Stratigraphy and Ammonite fauna of the Graneros Shale and Greenhorn Limestone Near Pueblo, Colorado: Washington, D.C.,U.S. Geological Survey Professional Paper 645, 108 p.

CORLISS, B.H., 1985., Microhabitats of benthic foraminifera within deep-sea sediments: Nature, v. 314, p. 435-438.

CORLISS, B. H., and CHEN, C., 1988, Morphotype patterns of Norwegian Sea deep-sea benthic foraminifera and ecological implications: Geology, v. 16, p. 716-719.

CORLISS, B. H., and EMERSON, S., 1990, Distribution of Rose Bengal stained deep-sea benthic foraminifera from the Nova Scotian continental margin and Gulf of Maine: Deep-Sea Research, v. 37, p. 381-400.

CORLISS, B. H., and FOIS, E., 1991, Morphotype analysis of deep-sea foraminifera from the northwest Gulf of Mexico: Palaios, v. 6, p. 589-605.

CUSHMAN, J. A., 1946, Upper Cretaceous foraminifera of the Gulf Coastal region of the United States and adjacent areas: Washington, D.C., U.S. Geological Survey Professional Paper 206, 241 p.

EICHER, D. L. 1965, Foraminifera and biostratigraphy of the Graneros Shale: Journal of Paleontology, v. 39, p. 875-909.

EICHER, D. L. 1966, Foraminifera from the Cretaceous Carlile Shale of Colorado: Cushman Foundation for Foraminiferal Research Contributions, v. 17, p. 16-31.

EICHER, D. L. 1967, Foraminifera from Belle Fourche Shale and equivalents, Wyoming and Montana: Journal of Paleontology, v. 41, p. 167-188.

EICHER, D. L. 1969, Cenomanian and Turonian planktonic foraminifera from the western interior of the United States, in Bronnimann, P., and Renz, H. H., eds., Proceedings of the first international conference on planktonic microfossils: Leiden, E. J. Brill, v. 2, p. 163-174.

EICHER, D. L., and DINER, R., 1985, Foraminifera as indicators of water mass in the Cretaceous Greenhorn Sea, western interior, in Pratt, L. M., Kauffman, E. G., and Zelt, F. B., eds., Fine-grained Deposits and Bofacies of the Cretaceous Western Interior Seaway: Evidence of Cyclic Sedimentary Processes Field Trip Guidebook No. 4: Tulsa, Society of Economic Paleontologists and Mineralogists, p. 60-71.

EICHER, D. L., and DINER, R., 1989, Origin of the Cretaceous Bridge Creek cycles in the western interior, United States: Palaeogeography, Palaeoclimatology, Palaeoecology, v. 74, p. 127-146.

EICHER, D. L., and WORSTELL, P., 1970, Cenomanian and Turonian foraminifera from the Great Plains, United States: Micropaleontology, v. 16, p. 269-324.

ELDER, W. P., 1985, Biotic patterns across the Cenomanian-Turonian boundary near Pueblo, Colorado, in Pratt, L. M., Kauffman, E. G., and Zelt, F. B., eds., Fine-grained Deposits and Biofacies of the Cretaceous Western Interior Seaway: Evidence of Cyclic Sedimentary Processes Field Trip Guidebook No. 4: Tulsa, Society of Economic Paleontologists and Mineralogists, p. 157-169.

ELDER, W. P., 1987, The paleoecology of the Cenomanian-Turonian (Cretaceous) stage boundary extinctions at Black Mesa, Arizona: Palaios, v. 2, p. 24-40.

ELDER, W. P., 1991, Molluscan paleoecology and sedimentation patterns of the Cenomanian-Turonian extinction interval in the southern Colorado Plateau region, in Nations, J. D., and Eaton, J. G., eds., Stratigraphy, depositional environments, and sedimentary tectonics of the western margin, Cretaceous Western Interior Seaway: Boulder,Geological Society of America Special Paper 260, p. 113-137.

ELDER, W. P., and KIRKLAND, J. I., 1985, Stratigraphy and depositional environments of the Bridge Creek Limestone Member of the Greenhorn Formation at Rock Canyon Anticline near Pueblo, Colorado, in Pratt, L. M., Kauffman, E. G., and Zelt, F. B., eds., Fine-grained Deposits and Biofacies of the Cretaceous Western Interior Seaway: Evidence of Cyclic Sedimentary Processes Field Trip Guidebook No. 4: Tulsa, Society of Economic Paleontologists and Mineralogists, p. 122-134.

ERICKSEN, M. C., and SLINGERLAND, R., 1990, Numerical simulations of tidal and wind-driven circulation in the Cretaceous interior seaway of North America: Geological Society of America Bulletin, v. 102, p. 1499-1516.

FISHER, C. G., HAY, W. W., and EICHER, D. L., 1994, Oceanic front in the Greenhorn Sea (late middle through late Cenomanian): Paleoceanography, v. 9, p. 879-892.

FOX, S. K., JR., 1954, Cretaceous Foraminifera from the Greenhorn, Carlile, and Cody Formations, South Dakota, Wyoming: Washington, D.C., U.S. Geological Survey Professional Paper, No. 254-E, p. 97-124.

FRIZZELL, D. L., 1954, Handbook of Cretaceous Foraminifera of Texas: Austin, University of Texas, Bureau of Economic Geology, Report of Investigations No. 22, 232 p.

FRUSH, M. P., and EICHER, D. L., 1975, Cenomanian and Turonian foraminifera and paleoenvironments in the Big Bend region of Texas and Mexico, in Caldwell, W. G. E., ed., The Cretaceous System in the Western Interior of North America: St John's, Geological Association of Canada Special Paper 13, p. 277-301.

GLANCY, T. J., JR., BARRON, E. J., and ARTHUR, M. A., 1986, An initial study of the sensitivity of modeled Cretaceous climate to cyclical insolation forcing: Paleoceanography, v. 1, p. 523-537.

GLENISTER, L. M., and KAUFFMAN, E.G., 1985, High resolution stratigraphy and depositional history of Greenhorn regressive hemicyclothem, Rock Canyon Anticline, Pueblo, Colorado, in Pratt, L. M., Kauffman, E. G., and Zelt, F. B., eds., Fine-grained Deposits and Biofacies of the Cretaceous Western Interior Seaway: Evidence of Cyclic Sedimentary Processes Field Trip Guidebook No. 4: Tulsa, Society of Economic Paleontologists and Mineralogists, p. 170-183.

GOODAY, A. J., 1993, Deep-sea benthic foraminiferal species which exploit phytodetritus: Characteristic features and controls on distribution: Marine Micropaleontology, v. 22, p. 187-205.

HANCOCK, J. M., and KAUFFMAN, E. G., 1979, The great transgressions of the Late Cretaceous: Journal of the Geological Society, v. 136, p. 175-186.

HAQ, B. U., HARDENBOL, J., and VAIL, P. R., 1987, The chronology of fluctuating sea level since the Triassic: Science, v. 235, p. 1156-1167.

HAY, W. W., EICHER, D. L., and DINER, R., 1993, Physical oceanography and water masses in the Cretaceous Western Interior Seaway, in Caldwell, W. G. E., and Kauffman, E. G., eds., Evolution of the Western Interior Basin: St John's, Geological Association of Canada Special Paper 39, p. 297-318.

HAZENBUSH, G. C., 1973, Stratigraphy and depositional environments of the Mancos Shale (Cretaceous), Black Mesa, Arizona, in Fassett, J. E., ed., Cretaceous and Tertiary Rocks of the southern Colorado Plateau: Durango, Four Corners Geological Society Memoir, p. 57-71.

IVANY, L. C., and SCHOPF, K. M., EDS., 1996, New perspectives on faunal stability in the fossil record: Palaeogeography, Palaeoclimatology, Palaeoecology, Special Issue, v. 127, 361p.

JEWELL, P. W., 1993, Water-column stability, residence times, and anoxia in the Cretaceous North American seaway: Geology, v. 21, p. 579-582.

JONES, D. J., ED., 1953, Microfossils of the Upper Cretaceous of Northeastern Utah and Southwestern Wyoming: Salt Lake, Utah Geological and Mineralogical Survey Bulletin 47 (Contributions to Micropaleontology No. 1), 158p.

JORISSEN, F. J., DE STIGTER, H. C., and WIDMARK, J. G. V., 1995, A conceptual model explaining benthic foraminiferal microhabitats: Marine Micropaleontology, 26, p. 3-15.

KAUFFMAN, E. G., 1977, Geological and biological overview: Western interior Cretaceous basin: The Mountain Geologist, v. 14, p. 75-100.

KAUFFMAN, E. G., 1984, Paleobiogeography and evolutionary response dynamic in the Cretaceous Western Interior Seaway of North America, in Westermann, G.

E. G., ed., Jurassic-Cretaceous Biochronology and Paleogeography of North America: St. John's, Geological Association of Canada Special Paper 27, p. 273-306.

KAUFFMAN, E. G., 1985, Cretaceous evolution of the western interior basin of the United States, in Pratt, L. M., Kauffman, E. G., and Zelt, F. B., eds., Fine-grained Deposits and Biofacies of the Cretaceous Western Interior Seaway: Evidence of Cyclic Sedimentary Processes Field Trip Guidebook No. 4: Tulsa, Society of Economic Paleontologists and Mineralogists, p. iv-xiii.

KAUFFMAN, E. G. and CALDWELL, W. G. E., 1993, The western interior basin in space and time, in Caldwell, W. G. E., and Kauffman, E. G., eds., Evolution of the Western Interior Basin: St. John's, Geological Association of Canada Special Paper 39, p. 1-30.

KIRKLAND, J. I., 1991, Lithostratigraphic and biostratigraphic framework for the Mancos Shale (late Cenomanian to middle Turonian) at Black Mesa, northeastern Arizona, in Nations, J. D., and Eaton, J. G., eds., Stratigraphy, Depositional Environments, and Sedimentary Tectonics of the Western Margin, Cretaceous Western Interior Seaway: Boulder, Geological Society of America Special Paper 260, p. 85-111.

KIRKLAND, J. I., LECKIE, R. M., and ELDER, W. P., 1995, A new principal reference section for the Mancos Shale (Late Cretaceous) at Mesa Verde National Park, in Santucci, V. L., and McClelland, L., eds., National Park Service Paleontological Research: Washington, D.C., National Park Service Technical Report NPS/NRPO/NRTR-95/16, p. 77-81.

LAMB, G. M., 1968, Stratigraphy of the lower Mancos Shale in the San Juan Basin: Geological Society of America Bulletin, v. 79, p. 827-854.

LECKIE, R. M., 1985, Foraminifera of the Cenomanian-Turonian boundary interval, Greenhorn Formation, Rock Canyon Anticline, Pueblo, Colorado, in Pratt, L. M., Kauffman, E. G., and Zelt, F. B., eds., Fine-grained Deposits and Biofacies of the Cretaceous Western Interior Seaway: Evidence of Cyclic Sedimentary Processes Field Trip Guidebook No. 4: Tulsa, Society of Economic Paleontologists and Mineralogists, p. 139-149.

LECKIE, R. M., 1987, Paleoecology of mid-Cretaceous planktonic foraminifera: A comparison of open ocean and epicontinental sea assemblages: Micropaleontology, v. 33, p. 164-176.

LECKIE, R. M., KIRKLAND, J. I., and ELDER, W. P., 1997, Stratigraphic framework and correlation of a principal reference section of the Mancos Shale (Upper Cretaceous), Mesa Verde, Colorado, in Anderson, O. J., Kues, B. S., and Lucas, S. G., eds., Mesozoic Geology and Paleontology of the Four-Corners Region, 48th Guidebook: Alburqueque, New Mexico Geological Society, p. 163-216.

LECKIE, R. M., SCHMIDT, M. G., FINKELSTEIN, D., and YURETICH, R., 1991, Paleoceanographic and paleoclimatic interpretations of the Mancos Shale (Upper Cretaceous), Black Mesa Basin, Arizona, in Nations, J. D., and Eaton, J. G., eds., Stratigraphy, Depositional Environments, and Sedimentary Tectonics of the Western Margin, Cretaceous Western Interior Seaway: Boulder, Geological Society of America Special Paper 260, p. 139-152.

LEITHOLD, E. L., 1993, Preservation of laminated shale in ancient clinoforms; comparison to modern subaqueous deltas: Geology, v. 21, p. 359-362.

LEITHOLD, E. L., 1994, Stratigraphical architecture at the muddy margin of the Cretaceous Western Interior Seaway, southern Utah: Sedimentology, v. 41, p. 521-542.

LESSARD, R. H., 1973, Micropaleontology and paleoecology of the Tununk Member of the Mancos Shale: Utah Geological and Mineral Survey, Special Studies 45, 28 p.

LIEBERMAN, B. S., BRETT, C. E., and ELDREDGE, N., 1995, A study of stasis and change in two species lineages from the middle Devonian of New York state: Paleobiology, v. 21, p. 15-27.

LINKE, P., and LUTZE, G. F., 1993, Microhabitat preferences of benthic foraminifera: a static concept or a dynamic adaptation to optimize food acquisition? Marine Micropaleontology, v. 20, p. 215-234.

MCNEIL, D. H., and CALDWELL, W. G. E., 1981, Cretaceous Rocks and Their Foraminifera in the Manitoba Escarpment: St. John's Geological Association of Canada Special Paper 21, 438 p.

MOLENAAR, C. M., 1983, Major depositional cycles and regional correlations of Upper Cretaceous rocks, southern Colorado Plateau and adjacent areas, in Reynolds, M. W., and Dolly, E. B., eds., Mesozoic Paleogeography of West-central United States: Denver, Colorado, Rocky Mountain Section, Society of Economic Mineralogists and Paleontologists, p. 201-224.

MOLENAAR, C. M., and COBBAN, W. A., 1991, Middle Cretaceous Stratigraphy on the South and East Sides of the Uinta Basin, northeastern Utah and northwestern Colorado: Washington, D.C., U.S. Geological Survey Bulletin 1787-P, 34 p.

NORTH, B. R., and CALDWELL, W. G. E., 1975, Foraminiferal faunas in the Cretaceous System of Saskatchewan, in Caldwell, W. G. E., ed., The Cretaceous System in the Western Interior of North America: St. John's, Geological Association of Canada Special Paper 13, p. 303-331.

OLESEN, J., 1991, Foraminiferal biostratigraphy and paleoecology of the Mancos Shale (Upper Cretaceous), southwestern Black Mesa Basin, Arizona, in Nations, J. D., and Eaton, J. G., eds., Stratigraphy, Depositional Environments, and Sedimentary Tectonics of the Western Margin, Cretaceous Western Interior Sea-

way: Boulder, Geological Society of America Special Paper 260, p. 153-166.

PARRISH, J. T., 1993, Mesozoic climates of the Colorado Plateau, in Morales, M., ed., Aspects of Mesozoic Geology and Paleontology of the Colorado Plateau: Flagstaff, Museum of Northern Arizona Bulletin 59, p. 1-11.

PARRISH, J. T., GAYNOR, G. C., and SWIFT, D. J. P., 1984, Circulation in the Cretaceous Western Interior Seaway of North America, a review, in Stott, D. F., and D. J., eds., The Mesozoic of Middle North America: St. John's,Canadian Society of Petroleum Geologists Memoir 9, p. 221-231.

PRATT, L. M., 1984, Influence of paleoenvironmental factors on preservation of organic matter in middle Cretaceous Greenhorn Formation, Pueblo, Colorado: American Association of Petroleum Geologists Bulletin, v. 68, p. 1146-1159.

PRATT, L. M., ARTHUR, M. A., DEAN, W. E., and SCHOLLE, P. A., 1993, Paleo-oceanographic cycles and events during the Late Cretaceous in the Western Interior Seaway of North America, in Caldwell, W. G. E., and Kauffman, E. G., eds., Evolution of the Western Interior Basin: St. John's, Geological Association of Canada Special Paper 39, p. 333-353.

SAGEMAN, B. B., RICH, J., ARTHUR, M. A., BIRCHFIELD, G. E., and DEAN, W. E., 1997, Evidence for Milankovitch periodicities in Cenomanian-Turonian lithologic and geochemical cycles, western interior U.S.A.: Journal of Sedimentary Research, v. 67, p. 286-302.

SAVRDA, C. E., and BOTTJER, D. J., 1993, Trace fossil assemblages in fine-grained strata of the Cretaceous western interior, in Caldwell, W. G. E., and Kauffman, E. G., eds., Evolution of the Western Interior Basin: St. John's, Geological Association of Canada Special Paper 39, p. 621-639.

SCHROEDER-ADAMS, C. J., LECKIE, D. A., BLOCH, J., CRAIG, J., MCINTYRE, D. J., and ADAMS, P. J., 1996, Paleoenvironmental changes in the Cretaceous (Albian to Turonian) Colorado Group of western Canada: Microfossil, sedimentological and geochemical evidence: Cretaceous Research, v. 17, p. 311-365.

SEN GUPTA, B. K. and MACHAIN-CASTILLO, M. L., 1993, Benthic foraminifera in oxygen-poor habitats: Marine Micropaleontology, v. 20, p. 183-201.

SHELDON, P. R., 1996, Plus ca change-a model for stasis and evolution in different environments, in Ivany, L. C., and Schopf, K. M., eds., New perspectives on faunal stability in the fossil record: Palaeogeography, Palaeoclimatology, Palaeoecology, v. 127, p. 209-227.

SLINGERLAND, R., KUMP, L. R., ARTHUR, M. A., FAWCETT, P. J., SAGEMAN, B. B., and BARRON, E. J, 1996, Estuarine circulation in the Turonian Western Interior seaway: Geological Society of America Bulletin, v. 108, p. 941-952.

STELCK, C. R., and WALL, J. H., 1954, Kaskapau Foraminifera From Peace River Area of Western Canada: Edmonton, Alberta Research Council Report No. 68, p. 2-38.

STELCK, C. R., and WALL, J. H., 1955, Foraminifera of the Cenomanian Dunveganoceras Zone from the Peace River area of western Canada: Edmonton, Alberta Research Council Report No. 70, p. 1-81.

TAPPAN, H., 1940, Foraminifera from the Grayson Formation of northern Texas: Journal of Paleontology, v. 14, p. 93-126.

VAIL, P. R., MITCHUM, R. M., JR., and THOMPSON, S., III, 1977, Global cycles of relative changes of sea level, in Payton, C. E., ed., Seismic Stratigraphic Applications to Hydrocarbon Exploration: Tulsa, American Association of Petroleum Geologists Memoir 26, p. 83-98.

VAN WAGONER, J. C., POSAMENTIER, H. W., MITCHUM, R. M., VAIL, P. R., SARG, J. F., LOUTIT, T. S., and HARDENBOL, J., 1988, An overview of the fundamentals of sequence stratigraphy and key definitions, in Wilgus, C. K., Hastings, B. S., Kendall, C. G. St. C., Posamentier, H. W., Ross, C. A., and Van Wagoner, J. C., eds., Sea-Level Changes: An Integrated Approach: Tulsa, Society of Economic Paleontologists and Mineralogists Special Publication 42, p. 39-45.

VERHALLEN, P. J. J. M., 1990, Late Pliocene to early Pleistocene Mediterranean mud-dwelling foraminifera; influence of a changing environment on community structure and evolution: Utrecht Micropaleontological Bulletin, v. 40, 219 pp.

WALL, J. H., 1960, Upper Cretaceous Foraminifera From the Smoky River Area, Alberta: Edmonton, Research Council of Alberta Bulletin 6, 43 p.

WALL, J.H., 1967, Cretaceous Foraminifera of the Rocky Mountain Foothills, Alberta: Edmonton, Research Council of Alberta Bulletin 20, 185 p.

WEST, O. L. O., LECKIE, R. M., and SCHMIDT, M. G., 1990, Cenomanian-Turonian foraminiferal changes in a transect across the Cretaceous Western Interior seaway (abs.): The Geological Society of America Abstracts with Programs, v. 22, p. A234.

WEST, O. L. O., 1993, Do foraminifera in sulfide-rich marine environments harbor symbiotic chemoautotrophic bacteria? (abs.): Oneonta, Society of Protozoology, Fourth East Coast Conference on Protozoa, Program with Abstracts, p.9.

WEST, O. L. O., LECKIE, R. M., BERK, W., YURETICH R., and COBABE, E, 1993, Benthic foraminiferal morphology and paleoenvironmental change in the Cretaceous U.S. Western Interior (abs.): The Geological Society of America, Abstracts with programs, v. 25, p. A430.

YOUNG, K., 1951, Foraminifera and stratigraphy of the Frontier Formation (Upper Cretaceous), southern Montana: Journal of Paleontology, v. 25, p. 35-68.

PALEOCEANOGRAPHY OF THE SOUTHWESTERN WESTERN INTERIOR SEA DURING THE TIME OF THE CENOMANIAN-TURONIAN BOUNDARY (LATE CRETACEOUS)

R. MARK LECKIE[1], RICHARD F. YURETICH[1], OONA L. O. WEST[1,2], DAVID FINKELSTEIN[3], AND MAXINE SCHMIDT[1]

[1]*Department of Geosciences, Box 35820, University of Massachusetts, Amherst, MA 01003-5820*
[2]*Graduate Program in Organismic and Evolutionary Biology, University of Massachusetts, Amherst, MA 01003*
[3]*Department of Geology, University of Illinois, Urbana, IL 61801*

ABSTRACT: The Cenomanian-Turonian boundary interval (93-94 Ma) was a time of rapid oceanographic change in the U.S. Western Interior Sea ("Greenhorn Sea"). Previous studies documented changes in $\delta^{18}O$ in carbonates and shifts in macrofossil (molluscan) populations indicating the incursion of a subsaline surface-water mass into the region and dysoxic to anoxic benthic conditions across wide areas of the seaway. These changes were accompanied by an expanding oxygen minimum zone in concert with global-scale burial of organic matter, which was driven in part by elevated rates of marine productivity. The oceanography of the southern seaway was undoubtedly complex with signals of a global anoxic event overprinted by regional influences of water mass stratification, mixing, productivity, changes in relative sea level, and biotic turnover.

Our studies of planktic and benthic foraminiferal assemblages and clay-mineral distribution in calcareous mudrocks and dark marlstones from the southwestern side of the Greenhorn Sea provide further constraints on the paleoceanography of this region. Specifically, there is a major change in planktic foraminiferal population structure (*"Heterohelix* shift") coupled with an influx of kaolinite and illite at both a neritic site (Lohali Point, Arizona) and a distal basin site (Rock Canyon, Colorado). These characteristics are not observed at a proximal basin site (Mesa Verde, Colorado) in between the other locales. The distinctive clay mineral assemblages suggest two disparate sources. One of these sources most likely was from the Sevier orogenic belt along the western side of the seaway, and another was either the southwestern corner of the seaway or the southern part of the stable craton. The distribution of clay mineral and planktic foraminiferal assemblages provide information on circulation of the upper water column.

The benthic foraminiferal assemblages of Lohali Point and Mesa Verde are very similar and have northern affinities, suggesting the influence of cool bottom waters along the western side of the seaway. We suggest that a submerged tectonic forebulge or bathymetric high near Mesa Verde caused "edge-effect" mixing and upwelling of cool water masses originating from the north. To the west, a foredeep shelf in northeastern Arizona and south-central Utah provided a conduit for northward-flowing, warmer surface water masses over the southward-flowing, cooler waters. These southern waters were bifurcated by the Mesa Verde high. The resultant oceanographic front, or mixing zone, caused the contrast in ecological and sedimentological patterns at the sites. With rising sea level came the incursion of oxygen-poor Tethyan intermediate waters into the Greenhorn Sea during latest Cenomanian-early Turonian time and the development of widespread, low diversity benthic foraminiferal assemblages dominated by *Neobulimina*.

INTRODUCTION

The Cretaceous Western Interior Sea had a long and dynamic oceanographic history. Its marine waters were strongly influenced by northern water masses for much of this history. The paucity of many normal marine invertebrates, such as echinoderms, bryozoans, corals, and rudist bivalves, suggests that the salinity of the sea was lower than normal (Kauffman, 1975, 1977, 1984). The Western Interior Sea received runoff from numerous rivers that drained highlands on the western side of the seaway and lowlands of the stable craton on the eastern side. During episodes of high eustatic sea level and transgression, subtropical normal marine water masses invaded the seaway from the Tethys Sea to the south. The warm water masses penetrated as far north as the prairie provinces of southern Canada (Kauffman, 1984; Caldwell et al., 1993; Kauffman and Caldwell, 1993). With these transgressions came a more normal marine biota, mixing and juxtaposition of different water masses, and changes in regional climate and sedimentation patterns. One such episode occurred during the late Cenomanian and early Turonian when the seaway, also called the "Greenhorn Sea", expanded to nearly 2000 km wide during the highest eustatic sea level rise of the Mesozoic Era (Hancock and Kauffman, 1979; Haq et al., 1988; Hay et al., 1993).

The axis, or deepest part of the seaway was probably along the corridor from northeastern New Mexico to the Black Hills region of northeastern Wyoming and southwestern South Dakota (e.g., Eicher, 1969b; Kauffman, 1977; Sageman and Arthur, 1994) (Fig. 1). This is where the salinity-sensitive planktic foraminifera first invaded the hyposaline interior sea when sea level rose during the late Cenomanian and where they persisted the

longest with the subsequent fall of eustatic sea level during the middle Turonian (Eicher, 1969a; Eicher and Worstell, 1970; Eicher and Diner, 1985). It is also the corridor through which benthic foraminifera with southern affinities invaded the seaway.

The rapid incursion of subtropical water masses into the seaway during late Cenomanian time abruptly improved benthic conditions and brought an invasion of warm water molluscs, foraminifera, and calcareous nannofossils (Cobban and Reeside, 1952; Reeside, 1957; Eicher and Worstell, 1970; Kauffman, 1977; McNeil and Caldwell, 1981; Eicher and Diner, 1985; Elder, 1985, 1991; Elder and Kirkland, 1985; Bralower, 1988; Caldwell et al., 1993; Hay et al., 1993; Kauffman and Caldwell, 1993; Watkins et al., 1993; Fisher et al., 1994). This event also coincided with the onset of a global perturbation in the carbon system, referred to as an Oceanic Anoxic Event (OAE-2), which was related in part to a brief interval of elevated marine productivity and burial of organic carbon (Arthur et al., 1987, 1988, 1990; Schlanger et al., 1987). The carbon event, measured in terms of $\delta^{13}C$ of carbonates and organic matter, was probably intimately linked with tectonics, eustatic sea level rise, oceanic circulation, and mode of deep water mass formation.

The global OAE-2 had a profound effect on some of the marine plankton, nekton, and benthon of the Greenhorn Sea. For example, Elder (1991) delineated multiple extinction steps in western interior sections through the uppermost Cenomanian and lower Turonian (Fig. 2), and Eicher (1969a) and Leckie (1985) documented a major turnover in planktic foraminifera through the Cenomanian-Turonian boundary interval. Therefore, the Western Interior Sea responded to major global changes that were modified by regional climatic, oceanographic, and deposi-

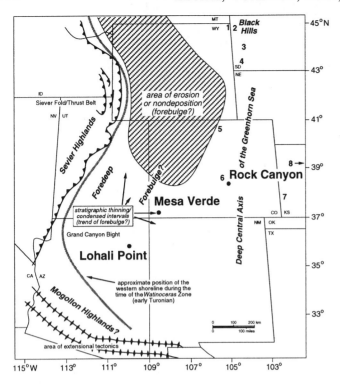

FIG. 1— Base map showing the southwestern portion of the Cenomanian-Turonian Western Interior Sea ("Greenhorn Sea"). Map is centered over Wyoming, Utah, Colorado, Arizona, and New Mexico. Locations of the three primary sites discussed in the text are: Lohali Point (LP), Black Mesa Basin, northeastern Arizona, a neritic site; Mesa Verde (MV), northern San Juan Basin, southwestern Colorado, a proximal basin site; Rock Canyon (RC), near Pueblo, Colorado, a distal basin site. Numbered localities show sections of Eicher and Worstell (1970) Section 8 is located in central Kansas. Map information from McGookey et al. (1972), Molenaar (1983), Cobban and Hook (1984), Kauffman (1984, 1985), Hattin (1985, 1987), Merewether and Cobban (1986), Eaton et al. (1990), Eaton and Nations (1991), Kirkland (1991),

tional signals. A major focus of this paper is to examine the record of regional oceanographic change along the southwestern side of the Greenhorn Sea, as documented by clay mineralogy and foraminiferal paleoecology and paleobiogeography. Together with earlier studies of stable isotopes and molluscan paleoecology and paleobiogeography, we evaluate the impact of Cretaceous global change on this large epicontinental seaway.

LOCATION AND METHODS

Measured Sections

Three sections through the Cenomanian-Turonian stage boundary were examined, that represent a transect across the southwestern side of the Cenomanian-Turonian Greenhorn Sea (Fig. 1). According to Elder's (1991) lithofacies regions, Lohali Point (LP) is in the clay-dominated western region, Mesa Verde (MV) is in the transitional lithofacies region, and Rock Canyon (RC) is in the carbonate-dominated central axis of the Greenhorn Sea. There is a progressive change in the siliciclastic and carbonate content of the strata from the western margin to the basin center (Hattin, 1985, 1986b) which reflects proximity to the tectonically active Sevier orogenic belt on the west. In general, sediment accumulation rates were two to four times greater in the western, clay-dominated lithofacies than in the transitional and carbonate-dominated lithofacies (Elder, 1985; Elder and

Kirkland, 1985). The carbonate-rich sections are characterized by interbedded, light-colored limestones and dark marlstones or calcareous shales of the Bridge Creek Member through the Cenomanian-Turonian boundary interval (Mancos Shale at Mesa Verde; Greenhorn Formation at Rock Canyon). Our samples come from the muddy and marly intervals only; no limestones were investigated.

Lohali Point (LP), located on the eastern side of the Black Mesa Basin in northeastern Arizona, represents a neritic depositional environment. This area was part of a north-south trending "foredeep" that developed seaward of the Sevier orogenic belt. The foredeep contained the thickest accumulation of sediments shed off the adjacent highlands, but this region probably did not represent the greatest water depths (Kauffman, 1977, 1984; Kirkland, 1991). LP is near the distal end of a broad, shallow, seaward-sloping shelf that covered northern Arizona and south-central Utah ("Grand Canyon Bight") during latest Cenomanian-middle Turonian time (Eaton and Nations, 1991; Kirkland, 1991; Leckie et al., 1991; Leithold, 1993). The lithostratigraphy and ammonite biostratigraphy of the lower shale member of the Mancos Shale are from Elder (1987) and Kirkland (1991).

The Mesa Verde (MV) section represents a proximal basin setting located at the northern end of Mesa Verde National Park in southwestern Colorado. There is evidence for stratigraphic thinning and condensed intervals in the upper Graneros and lower Bridge Creek Members of the Mancos Shale (uppermost Cenomanian-lower Turonian), similar to the Red Wash section south of MV in the northwestern corner of New Mexico (Elder, 1985, 1991; Leckie et al., 1997). The four-corners area may have been along the trend of a bathymetric high, or forebulge, at this time. The stratigraphic framework for this study comes from Kirkland et al. (1995) and Leckie et al. (1997).

The Rock Canyon section (RC) includes the uppermost Hartland Shale and lower Bridge Creek Limestone Members of the Greenhorn Formation exposed west of Pueblo, Colorado. This represents a distal basin depositional setting from within the deepest part of the seaway (Eicher, 1969a, b; Cobban and Scott, 1972; Hattin, 1971, 1985, 1986b; Kauffman, 1977, 1984; Kauffman, Pratt, et al., 1985). The molluscan biostratigraphy and lithostratigraphy used in this study are from Elder (1985), Elder and Kirkland (1985), and Sageman (1985).

Foraminiferal Assemblage Analyses

We use planktic and benthic foraminiferal sediment assemblages to infer environmental changes in the upper water column and near the seafloor. The analyses focus on the following: (1) proportion of planktics to benthics in the total foraminiferal assemblage (percent planktics), (2) major groups of planktic genera or morphotypes (biserial *Heterohelix*, triserial *Guembelitria*, trochospiral *Hedbergella* and *Whiteinella*, planispiral *Globigerinelloides*, and the keeled genera *Rotalipora*, *Praeglobotruncana*, and *Dicarinella*), and (3) major groups or species of benthics (*Neobulimina albertensis*, *Gavelinella dakotensis*, other calcareous benthic taxa, and agglutinated taxa). Previous studies documented the utility of these taxa or morphogroups in Cenomanian-Turonian rocks from the western interior (e.g., Eicher and Diner, 1985; Leckie, 1985; Leckie et al., 1991; Fisher et al., 1994; Schroeder-Adams et al., 1996). The assemblages are based on counts of approximately 300 specimens from the >63μm size fraction. Many of the samples from the Rock Canyon section were picked a second time to increase

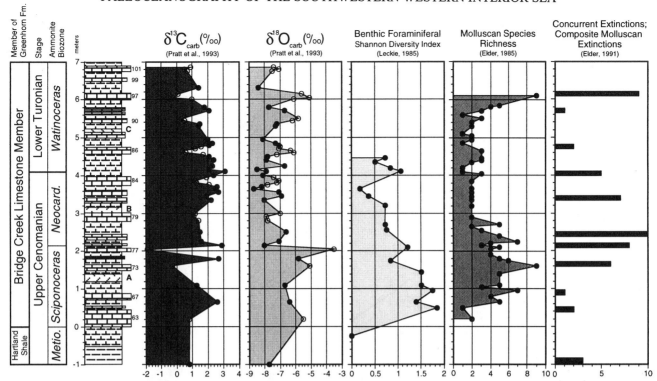

FIG. 2—Details of the Cenomanian-Turonian boundary interval at the Rock Canyon section (near Pueblo, Colorado). Ammonite biostratigraphy, lithostratigraphy, and bentonite beds (A-C) are from Elder (1985) and Elder and Kirkland (1985); limestone bed numbers are from Cobban and Scott (1972). Isotope data are from Pratt et al. (1993); open circles are limestones and closed circles are dark marlstones or calcareous shales. Benthic foraminiferal diversity data [Shannon-Wiener diversity index, H(S)] are from Leckie (1985). Molluscan extinction steps are from Elder (1991).

the benthic assemblages to at least 300 specimens (Leckie, 1985). All samples were also examined for rare species and other biogenic and mineral components. For details of sample processing, picking, and counting, refer to Leckie et al. (1991).

Clay Mineralogy

Our analyses focused on the 1-2μm size fraction, a bit of a departure from most investigations of clay-mineral abundance that examine the entire <2 or <1μm size. Because of the dominance of mixed-layer illite/smectite in the broad fractions, we examined the coarsest subset to enhance the relative contributions of other clays, which often have larger grains (Gibbs, 1977). The grain sizes were separated by centrifuging for the appropriate length of time. For details of sample preparation, see Leckie et al. (1991). Oriented mounts were prepared by smearing a clay paste on glass slides (Gibbs, 1971). X-ray diffraction of the samples was accomplished with a Siemens diffractometer updated with a Databox digital data processing system. Samples were scanned untreated and after solvation with ethylene glycol. Selected samples were also heated to 550°C or Mg-saturated and solvated with glycerol to further help identify the clay-mineral phases (Brindley and Brown, 1980; Hardy and Tucker, 1988).

Obtaining quantitative mineral abundance from X-ray diffractograms has always been a vexing problem. Although the techniques for sample preparation and analysis are being refined so that X-ray intensities can be converted more meaningfully into mineral percentages, the techniques are not always amenable to processing large numbers of samples (Moore and Reynolds, 1989). Simple measurements of relative peak height (or area) are adequate for semi-quantitative comparisons of simi-

lar samples (Yuretich, 1979; Leckie et al., 1991). This is especially true when relative changes are the critical component, as opposed to absolute amounts of a particular mineral in a sample. For these reasons, our clay studies focus on the most easily recognized and measurable X-ray diffraction peaks in the clays. Relative abundance of clay minerals was determined by ratios of diffractogram peak height of the 001 diffraction line of each clay (Griffin, 1971).

BACKGROUND

Tectonics

The Cretaceous, the mid- to Late Cretaceous in particular, was characterized by heightened plate tectonic activity, including elevated rates of seafloor spreading, convergent margin volcanism, and intraplate volcanism (e.g., Larson, 1991a, b; Arthur et al., 1985, 1990). More than 700 volcanic ashes (bentonites) are described from upper Albian to middle Maastrichtian strata of the U.S. western interior (Kauffman, 1985). The Sevier Orogeny refers to tectonic deformation during Cretaceous time that occurred along a narrow but continuous fold and thrust belt, extending from southern Nevada to the Canadian Rockies (Armstrong, 1968; McGookey et al., 1972). The tectonism was associated with large-scale magmatic activity and terrain accretion along the convergent western margin of the North American cordillera (e.g., Schwartz and DeCelles, 1988). Magmatism reached its greatest extent at about 90-95 Ma, roughly coincident with the peak of Cretaceous sea level (Armstrong and Ward, 1993). The eastward-vergent tectonic loading was responsible for creating a large, asymmetric foreland basin in the Canadian and U.S. western interior (e.g., Price, 1973; Cross and Pilger, 1978; Beaumont, 1981; Jordan, 1981; Wiltschko and Dorr, 1983; Schedl and Wiltschko, 1984). The axis of maximum sediment accumulation in the western interior foreland basin was immediately adjacent to the thrust belt, also called the foredeep (e.g., Lawton, 1985; Villien and Kligfield, 1986).

Within the foreland basin there is abundant sedimentologic, lithostratigraphic, and biostratigraphic evidence of common subtle intraforeland uplifts. Many of these structural highs were intermittent and short-lived features on the seafloor of the interior seaway. For example, Merewether and Cobban (1986) documented numerous elongate arches and mild swells of middle Cenomanian to middle Coniacian age, many of which are progressively younger to the east. Heller et al. (1993) suggest that such topographic features are due to modest changes in intraplate stress levels and are to be expected in a heterogeneous lithosphere with a long, but punctuated history of changing plate motion stresses. In addition, differential movement along preexisting planes of weakness may have accentuated the topographic relief of some of these features. There is also evidence that some of the Cretaceous intraforeland uplifts reflect reactivated basement uplifts, some of which are precursors to later Laramide deformation (e.g., Ryer and Lovekin, 1986; Hattin, 1987; Schwartz and DeCelles, 1988; Eaton et al., 1990; Heller et al., 1993).

Numerous authors have speculated about the existence of a forebulge adjacent to the axis of maximum subsidence and sediment accumulation (e.g., Lorenz, 1982; Kauffman, 1984, 1985; Merewether and Cobban, 1986; Eaton et al., 1990; Leithold, 1994; Christie-Blick and Driscoll, 1995). One possible candidate for a forebulge or topographic high during middle Cenomanian-early Turonian time is the arch or series of arches that developed across western Wyoming, northeastern Utah, and western Colorado (Merewether and Cobban, 1986; Ryer and Lovekin, 1986; Molenaar and Cobban, 1991; Franczyk et al., 1992). This area is delineated by a widespread lacuna (Elder and Kirkland, 1993 a, b; Sageman and Arthur, 1994). The broad area of uplift apparently migrated across central Wyoming, and central and northeastern Colorado during middle and late Turonian time (Merewether and Cobban, 1986).

Further to the south there is also evidence of stratigraphic thinning and erosion on presumed topographic highs during late Cenomanian-early Turonian time in eastern Utah along the present trend of the San Rafael Swell and Monument Uplift (Eaton et al., 1990; Elder, 1991; Leithold, 1994), in southwestern Colorado (Lamb, 1968; Elder, 1991; Leckie et al., 1997), and in easternmost Arizona along the present trend of the Defiance Uplift (Elder, 1991). Elder (1991) commented on the importance of this general north-south structural trend in controlling Cenomanian-Turonian lithofacies distribution on the Colorado Plateau, with the thicker, higher sedimentation-rate, clay-dominated sequences confined to the west (foredeep). Toward the basin center to the east, stratigraphic thinning in uppermost Cenomanian and basal Turonian strata of the Bridge Creek Member (Greenhorn Formation) delineate another probable bathymetric high trending northwest-southeast across south-central Colorado and northeast New Mexico (Hattin, 1987).

Eustasy and Relative Sea Level

The transgression and regression of the latest Albian-middle Turonian Greenhorn Sea represents a third-order, largely eustatic, rise and fall of sea level (Kauffman, 1977, 1984, 1985; Kauffman

and Caldwell, 1993). Superimposed on this dominantly global sea level record are smaller scale changes in relative sea level that are controlled by a combination of factors including eustasy, episodic uplift and basin subsidence, and climatically driven changes in sediment supply (Kauffman, 1985; Leithold, 1993, 1994; Ryer, 1993; Elder et al., 1994). Relative sea level rise, as recorded in upper Cenomanian and lower Turonian sediments from the southwestern side of the seaway, is characterized by fourth- and fifth-order marine flooding surfaces and back-stepping, progradational sequences that comprise the transgressive systems tract of the third-order Greenhorn cycle (Elder et al., 1994; Gardner and Cross, 1994; Leithold, 1994; Christie-Blick and Driscoll, 1995; West et al., this volume). In offshore mudrock facies, the flooding surfaces and parasequence boundaries are recognized by thin intervals of sediment condensation (winnowing, calcisiltite/calcarenite horizons, shell lags, concretion horizons), a decrease in grain size, elevated levels of carbonate, and/or influx of warm water taxa (Elder, 1991; Leithold, 1993, 1994; Elder et al., 1994). Elder (1991) notes that the ammonite biozone boundaries of the Cenomanian-Turonian boundary interval are associated with transgressive pulses and he suggests that the faunal turnover observed at these boundaries may be closely linked to relative sea level events in the seaway in terms of changing sedimentation rates, benthic oxygenation, substrate firmness, and water temperature. Similarly, Li and Habib (1996) attributed changes observed in dinoflagellate diversity, assemblage composition, and organic facies to changes in sea level during late Cenomanian and earliest Turonian time.

Climate

The striking development of interbedded, light-colored limestones and dark marlstones or calcareous shales in the Bridge Creek Limestone Member of the Greenhorn Formation has long been attributed to climatic cyclicity (Gilbert, 1895; Kauffman, 1977; Fischer, 1980; Pratt, 1984; Barron et al., 1985; Fischer et al., 1985; Eicher and Diner, 1989). Individual limestone beds can be traced over 1000 km, which attests to the remarkable uniformity across the vast central part of the seaway (Hattin, 1971, 1985, 1986b). Periodicities in the Bridge Creek Member are believed to roughly correspond with Milankovitch climate forcing dominated by the 41 ka obliquity cycle (Kauffman, 1977; Fischer, 1980; Fischer et al., 1985).

Rhythmic bedding of the Bridge Creek Member is best developed in the deep, distal, central part of the seaway late in transgression, following incursion of warm water masses. This represents the time and place of heightened contrasts between water masses and west-east patterns of sedimentation. Enhanced limestone-shale cyclicity was due, in part, to decreased sedimentation rates with rising sea level and increased calcareous plankton productivity with the incursion of warm normal marine waters into the seaway with transgression (e.g., Elder, 1985, 1991; Hattin, 1986b). The marlstones and calcareous shales are darker in color due to higher relative concentrations of clay minerals and organic matter (Pratt, 1984). These lithologies are either laminated or only weakly bioturbated. The limestones are composed mostly of biogenic carbonate from calcareous nannofossils and planktic foraminifera (Hattin, 1971; 1986a). These latter beds are highly bioturbated and contain less organic matter and terrigenous material (Pratt, 1984; Barron et al., 1985).

According to the dilution model of Pratt (1984), the shalier or muddier lithologies of the Bridge Creek Member represent times of greater rainfall and runoff. Accordingly, episodes of high riverine discharge to the seaway result in the development of a sediment-laden, reduced-salinity surface layer, heightened terrigenous influx, poor benthic ventilation due to salinity stratification, and weak deep water circulation (Pratt, 1984, 1985; Barron et al., 1985; Hattin, 1985, 1986b; Bottjer et al., 1986; Pratt et al., 1993). The limestones, on the other hand, represent times of drier climate, reduced terrigenous influx, near-normal marine salinities from surface to deep, and improved circulation and benthic oxygenation. The darker marlier and shalier lithologies are commonly thicker than the adjacent limestone beds in the central part of the seaway (e.g., Hattin, 1985, 1987), which suggests higher rates of siliciclastic sedimentation during periods of higher precipitation and runoff. In addition, increasing burrow size, density, and depth of burrow penetration are correlated with decreasing organic carbon and increasing carbonate content (Savrda and Bottjer, 1993). This evidence suggests that climate-driven cycles of clastic dilution and benthic ventilation are the principal causes of rhythmic bedding in the Greenhorn Formation (Hattin, 1971; Pratt, 1984; Barron et al., 1985; Savrda and Bottjer, 1993; Pratt et al., 1993).

Clay minerals often indicate the weathering conditions in the source area from which they were derived and, in the absence of subsequent diagenetic alteration, the relative abundance of different kinds of clays can be used to indicate paleoclimate. Chamley (1989) and Singer (1980, 1984) summarized the relationships gathered from numerous studies of modern sediments and sedimentary rock sequences. In cold climates, physical weathering predominates and the fine-grained components of soils are generally fragments of parent minerals. Illite and chlorite are frequently the most abundant clay minerals, since the lithologies in many high-latitude areas are dominated by igneous and metamorphic bedrock. In more temperate climates, chlorite decomposes into vermiculite, and illite will hydrolyze to form smectitic clays. Under these climatic conditions, the abundance of smectite generally increases as the amount of annual rainfall increases. Under warm, tropical conditions, chemical breakdown of parent materials produces aluminous chemical residua with larger proportions of kaolinite and gibbsite. Again, the more aluminous end members result from the highest rates of leaching, which is generally a function of rainfall amount. These generalizations can be modified owing to the source lithology and drainage characteristics, but they are effective as initial guidelines for interpreting paleoclimate. For example, Pratt (1984) found that discrete illite is much more abundant in the dark calcareous shales of the Hartland Shale Member of the Greenhorn Formation than in the limestone beds of the Bridge Creek Limestone Member. She attributed the stratigraphic changes in illite distribution to greater river discharge during deposition of the Hartland Shale. Leckie et al. (1991) documented a large increase in the proportion of kaolinite in the Lohali Point section during early Turonian time that coincides with peak transgression of the Greenhorn Sea. They suggested that regional climate became warmer and wetter as sea level rose, facilitating the formation of kaolinite adjacent to the seaway. Sethi and Leithold (1994) found cyclic changes in the relative abundance of clay minerals in the lower Turonian Mancos Shale and Tropic Shale in Utah. They noted an increase of mixed-layer illite-smectite in the more carbonate-rich beds that presumably formed during drier climatic periods.

In a core and in outcrop sections of the Greenhorn Formation near Pueblo, Colorado, a heavy trend in whole-rock $\delta^{13}C_{carb}$ values closely tracks the signal for organic matter ($\delta^{13}C_{org}$)

through the Cenomanian-Turonian boundary interval of the lower Bridge Creek Member (Pratt et al., 1993) (Fig. 2). The limestone beds record the most negative $\delta^{13}C_{carb}$ values in the cyclic lithofacies of the Bridge Creek. Barron et al. (1985) proposed that the lighter $\delta^{13}C_{carb}$ values of the limestones reflect times of greater mixing and carbonate productivity due to availability of isotopically light and nutrient-rich deep waters. Eicher and Diner (1989) also support an interpretation of increased productivity during limestone deposition based on the presence of abundant calcispheres, which indicate high productivity, in the limestone beds; the concentration of clay-size quartz in many of the limestones, which are believed to be derived from the dissolution of radiolarian skeletons, and contrasts in $\delta^{13}C_{carb}$ and $\delta^{18}O_{carb}$ between the limestones and adjacent marlstones.

Eicher and Diner (1985, 1989) propose that the rhythmic bedding of the Bridge Creek Member is the result of productivity cycles rather than dilution cycles, which were driven by periodic changes in water mass production and vertical mixing. These authors emphasize that salinity-sensitive planktic foraminifera characterize the marlstones as well as the limestones. They reject the hypothesis that the marlstones represent times of increased runoff with the attendant development of a sediment-laden brackish cap across the seaway. In support of this argument, molluscan trophic and species diversity - including ammonites, epifaunal and infaunal bivalves, and gastropods - are greatest along the western clay-dominated corridor of the seaway (northeastern Arizona and southern Utah) where the brackish cap would be best developed, compared with the transitional and carbonate-dominated central sectors (Elder, 1990, 1991; Kirkland, 1991).

These observations suggest that if precipitation and runoff from the Sevier orogenic belt or any other area increased during the wetter part of the climate cycle, when the marlstone or calcareous shale were being deposited, then only minor salinity changes in the upper water column are feasible given the stenohaline nature of most marine organisms. Pratt et al. (1993) suggested that the amplitude of $\delta^{18}O_{carb}$ change (2-3$^o/_{oo}$) across the limestone-shale couplets translates into salinity variability of 10-20% (i.e., 3.4-7.0$^o/_{oo}$). Fischer et al. (1985) concluded that the rhythmic bedding of the Bridge Creek Member was due to a combination of dilution and carbonate productivity cycles. We concur with this interpretation and suggest that the darker marlier and shalier parts of the bedding couplets - those lithofacies examined in this study - accumulated under minor reductions in near-surface water salinities (<7$^o/_{oo}$). Small salinity gradients between the surface and bottom in continental shelf or semi-enclosed sea settings are sufficient to stratify a relatively stable water column and therefore influence benthic ventilation (Mann and Lazier, 1991; Jewell, 1993).

Oceanography

A global positive excursion in whole-rock carbonate $\delta^{13}C$ values is well-developed in uppermost Cenomanian-basal Turonian marine strata of the U.S. western interior (Pratt and Threlkeld, 1984; Barron et al., 1985; Pratt, 1985; Arthur et al., 1987, 1988, 1990; Schlanger et al., 1987; Pratt et al., 1993). This excursion lasted about 0.5 m.y. and is attributed to enhanced marine productivity and the burial and preservation of organic matter with rising global sea level. These conditions correspond to times of increased low-latitude deep water production on flooded continental margins and heightened intraplate and convergent margin volcanism. Sediments of the North Atlantic also record dramatic changes in deep water ventilation and carbonate preservation related to the opening of the deep water gateway between the North and South Atlantic at this time (Tucholke and Vogt, 1979; Leckie, 1989). Heightened productivity expanded oxygen minimum zones (OMZ) in the mid- to upper water column along continental margins and under areas of oceanic divergence. Oxygen-poor water masses may have invaded epicontinental seas with rising sea level, enhancing preservation of organic matter (Arthur et al., 1987; Schlanger et al., 1987; Slingerland et al., 1996). Frush and Eicher (1975) proposed that the incursion of an OMZ into the Greenhorn Sea may have limited benthic foraminifera through much of the upper Cenomanian-lower Turonian strata of the deep central axis of the seaway. An alternate hypothesis is that the mixing of polar waters and subtropical waters in the Western Interior Sea produced its own oxygen-depleted, intermediate water mass through cabbeling (Hay et al., 1993; Fisher et al., 1994).

Parrish et al. (1984) suggested that oceanographic circulation in the southern part of the Western Interior Sea may have alternated seasonally between north-flowing and south-flowing, whereas Jewell (1993) proposed that a wind-driven subtropical gyre dominated flow south of latitude 50°N; the latter implies northward flow along the western side of the seaway and southward flow along the eastern side. In contrast, the distribution of planktic foraminifera in the Cenomanian-Turonian Greenhorn Sea suggests that warm, normal marine waters moved northward along the central and eastern sides of the seaway during transgression and cooler, slightly less saline waters moved southward along the western side (Eicher, 1969a; Eicher and Diner, 1985; Fisher et al., 1994). Numerical simulations of wind-driven circulation in the seaway suggest that oceanic circulation was dominated largely by storms, particularly winter storms, and that storm-driven shelf currents on the western side of the seaway were predominantly to the south (Ericksen and Slingerland, 1990).

Circulation experiments using a global climate model (GENESIS) and a coastal ocean circulation model (CIRC) indicate that circulation in the early Turonian Greenhorn Sea was driven largely by freshwater runoff (Arthur et al., 1996; Slingerland et al., 1996). Runoff from the eastern side of the seaway flowed northward as a coastal jet and runoff from the western side flowed southward. The seaway exported freshwater and drew in both Boreal and Tethyan waters, creating a strong counterclockwise gyre (Slingerland et al., 1996). These model results are in good agreement with planktic and benthic foraminiferal distribution patterns (Eicher and Diner, 1985; Caldwell et al., 1993; Fisher et al., 1994; West et al., this volume).

PLANKTIC FORAMINIFERAL PALEOECOLOGY AND
PALEOBIOGEOGRAPHY IN THE WESTERN INTERIOR SEA

Sediment assemblages of fossil planktic foraminifera provide useful information about the nature of the ancient uppermost water column, such as its stratification and productivity. Most species of modern planktic foraminifera are adapted to relatively narrow ranges of temperature and salinity (Be, 1977); this relationship that holds true for most marine organisms and probably for Cretaceous planktic foraminifera as well. The richness of species, genera, or morphotypes is greatest in well preserved sediment assemblages deposited under normal salinity, generally low-nutrient waters of the low- to mid-latitudes where seasonal or year-round temperature gradients in the upper water

column provide a variety of trophic and density-specific niches (Lipps, 1979; Leckie, 1989; Hallock et al., 1991) (Fig. 3). Sediment assemblages that accumulate under ecotones - areas of the ocean where two surface water masses meet and co-mingle - may actually have slightly higher simple diversity due to mixing of biocenoeses (Cifelli and Benier, 1976; Hallock et al., 1991). Stratification in the upper water column, which is caused by seasonal changes in the strength or position of the thermocline, is known to be fundamentally important in maintaining marine plankton communities. It affects, for example, nutrient availability and recycling, productivity, seasonal succession, reproduction, and predation (Mann and Lazier, 1991).

The relative abundance of planktic foraminifera to total foraminifera in depositional settings above the lysocline and carbonate compensation depth are very useful in delineating water depth or surface water productivity. Planktic foraminifera are rare in nearshore environments but increase in abundance relative to benthic taxa across the continental shelf (e.g., Phleger, 1951; Bandy, 1956; Murray, 1976; Gibson, 1989). In the modern ocean, planktics typically constitute 90-99% of outer neritic to mid-bathyal foraminiferal sediment assemblages (Fig. 3). However, elevated surface water productivity can significantly reduce the relative abundance of planktics. The enhanced flux of organic matter from the surface waters to the seafloor stimulates benthic productivity, and while the flux of planktic foraminiferal shells too may be higher, the greatest increase in relative abundances is in the benthic foraminifera and other benthic organisms, for example, ostracods, echinoderms, and sponges (Leckie, 1987; Berger and Diester-Haass, 1988; Herguera and Berger, 1991).

During the mid-Cretaceous, biserial (*Heterohelix*) and triserial (*Guembelitria*) planktic morphotypes were much more common in the shallower, more proximal waters of epicontinental seas than in the more distal, open ocean settings that were their habitat in the later Cretaceous and Cenozoic (Leckie, 1985; 1987; Nederbragt, 1990). For example, Tappan (1940) and Mancini (1982) noted the common occurrence of *Guembelitria* and *Heterohelix* in the warm, neritic environment of the lower Cenomanian Grayson Formation of north-central Texas. *Heterohelix* was widely distributed and was one of the most abundant planktic foraminiferal groups in the seaway during Cenomanian-Turonian time (e.g., Eicher, 1969a; Caldwell et al., 1978; McNeil and Caldwell, 1981), and was reported as far north as the Arctic Slope of Alaska (Tappan, 1962). On the other hand, *Guembelitria* was more restricted to southern and presumably warmer waters, and its greatest relative abundances are reported from neritic localities (e.g., Hazenbush, 1973; Lessard, 1973; Mancini, 1982; Leckie et al., 1991; Olesen, 1991). On the western side of the seaway, *Guembelitria* has been reported from as far north as southeastern Utah, where it is common to abundant at some levels (Lessard, 1973), and in the central part of the seaway, *Guembelitria* has been found in calcareous shales as far north as the Black Hills (Fisher, pers. commun., 1995). Forms referable to *Guembelitria* sp. from the prairie provinces of Canada (e.g., Stelck and Wall, 1954, 1955; Wall, 1960, 1967; Caldwell et al., 1978) may prove to be a benthic buliminid.

Leckie (1987) called *Heterohelix* and *Guembelitria*, among others, the "Epicontinental Sea Fauna" on the basis of their characteristic mid-Cretaceous distribution pattern. Perhaps these genera, particularly the earliest species, had a benthic stage to their life-cycle thereby placing water depth constraints on their distributions. Alternatively, perhaps these genera were broadly tolerant (eurytopic) of variable or unstable neritic waters with ex-

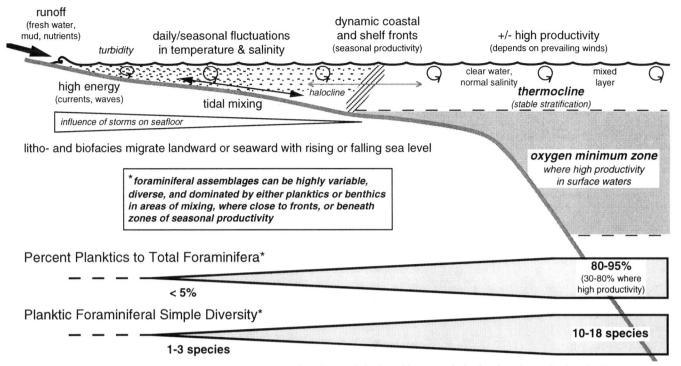

FIG. 3—Generalized oceanographic conditions and foraminiferal assemblage characteristics in neritic to upper bathyal settings along a dominantly siliciclastic margin. Offshore trends in the proportion of planktic to total foraminifera and simple planktic diversity apply to the Cenomanian-Turonian (see text for details).

tremes of seasonal productivity, salinity, and/or temperature (Fig. 3).

The distribution of Cretaceous and Cenozoic biserial and triserial planktic foraminifera has been enigmatic compared with most trochospirally-coiled species. For example, *Heterohelix* proliferated with the expansion of epicontinental seas during the late Cenomanian and early Turonian, and during the time of OAE-2 (e.g., Eicher, 1969a; Sliter, 1972; Jarvis et al., 1988; Nederbragt, 1990). Its Cenozoic biserial look-alikes, *Chiloguembelina* and *Streptichilus* may have been associated with low oxygen waters in the lower part of the seasonal thermocline (Boersma and Premoli-Silva, 1989; Resig, 1993). *Gallitellia*, the present-day homeomorph of the triserial *Guembelitria*, is very sparse in the modern ocean but occurs in fairly significant abundances in the Red Sea and in upwelling areas of the northern Indian Ocean (Kroon and Nederbragt, 1990).

Trochospiral morphotypes, particularly species of *Hedbergella* and *Whiteinella*, were the "weeds" of the mid-Cretaceous oceans, found in great abundances in both epicontinental sea and open ocean settings. These taxa are grouped as the major components of the "Shallow Water Fauna" because, like the more common and widespread species of the oceans today, they probably lived in the sun-lit near-surface waters of the mixed layer and/or upper reaches of the thermocline, where trophic resources are concentrated (Leckie, 1987, 1989). Along with the biserial forms, the trochospiral morphotypes were among the most abundant and widespread planktics in the Greenhorn Sea, especially *Hedbergella delrioensis*, *H. planispira*, and *H. loetterlei* (Eicher, 1969a; Eicher and Worstell, 1970; Caldwell et al., 1978; McNeil and Caldwell, 1981; Schroeder-Adams et al., 1996). *Hedbergella loetterlei* is also reported from the north slope of Alaska (Tappan, 1962). *Globigerinelloides*, a planispiral morphotype, occurs widely, although they are rarely common and tend to disappear before *Hedbergella* or *Heterohelix* in the

shoreward direction. Other mid-Cretaceous genera such as *Clavihedbergella* and *Schackoina* have distributions and relative abundances similar to *Globigerinelloides*. Species of all these genera ranged as far north as Alberta, Saskatchewan, and Manitoba at the time of peak transgression of the Greenhorn Sea during the early Turonian (e.g., Caldwell et al., 1978; McNeil and Caldwell, 1981). These last three taxa are also included in the "Shallow Water Fauna" based on their paleobiogeography, but they were probably more stenotopic than some species of *Hedbergella* and *Whiteinella*.

Cenomanian-Turonian keeled morphotypes, species of *Rotalipora*, *Praeglobotruncana*, *Dicarinella*, and *Marginotruncana*, are typically diagnostic components of high-diversity assemblages and indicate warm, normal marine, stratified water masses (Leckie, 1987, 1989). These taxa are collectively grouped as the "Deep Water Fauna", not necessarily because they were the deepest-dwelling as originally proposed by Leckie (1987), but because they were more common in deeper, more distal 'blue water" settings, and because they were probably the most stenotopic planktic foraminifera of this age. As a group, keeled species probably occupied a diverse range of habitats from the mixed layer to various depths along the thermocline, or perhaps even deeper (e.g., Douglas and Savin, 1978; Hart, 1980; Caron and Homewood, 1982; Jarvis et al., 1988; Corfield et al., 1990). In the Greenhorn Sea, this group of taxa had the most limited biogeographic range of all planktic foraminifera (e.g., Eicher, 1969a). Keeled taxa are not found north of the Black Hills (Fisher, pers. commun., 1996). Their distribution was restricted primarily to the eastern half of the seaway, centered on the axial part of the basin in eastern Colorado (Eicher, 1969a; Eicher and Diner, 1985, 1989; Fisher et al., 1994). Eicher (1969a) reported *Rotalipora* from west-central Colorado and Hazenbush (1973) reported both *Rotalipora* and *Praeglobotruncana* from northeastern Arizona, but these and other keeled taxa are only represented by two specimens in the Cenomanian-Turonian boundary interval at our MV section in the southwestern corner of Colorado. No keeled taxa were reported from the Red Wash section in the northwestern corner of New Mexico (Lamb, 1968).

BENTHIC FORAMINIFERAL PALEOECOLOGY AND
PALEOBIOGEOGRAPHY IN THE WESTERN INTERIOR SEA

Benthic foraminifera have long been used to infer paleoenvironmental conditions, including water depth, salinity, and oxygenation (e.g., Phleger, 1951; Bandy, 1956; Douglas and Woodruff, 1981; Bernhard, 1986; Culver, 1988; Murray, 1991). More recently, modern benthic foraminifera have been found to occupy different microhabitats - for example, epifaunal, shallow infaunal, and deep infaunal - and there is a significant relationship between microhabitats and foraminiferal test shape, mode of coiling, and distribution of pores (Corliss, 1985; Corliss and Chen, 1988; Corliss and Emerson, 1990; Corliss and Fois, 1991). Koutsoukos and Hart (1990), building on these and earlier studies, including Chamney (1976), proposed a model of Cretaceous benthic foraminiferal morphogroup distribution, paleo communities, and trophic strategies.

The distribution of modern benthic foraminiferal morphotypes, and hence microhabitats, depends on numerous factors including food supply and oxygen content below the sediment/water interface (Corliss and Emerson, 1990; Jorissen et al., 1995) (Fig. 4). For example, Gooday (1993) and Thomas and Gooday (1996) have suggested that some epifaunal taxa are trophic opportunists that respond quickly to the input of food, such as seasonal pulses of phytodetritus from the surface waters associated with the spring bloom. In addition, there is growing evidence that some infaunal foraminiferal taxa may have a competitive advantage in dysoxic or anoxic environments by requiring very little oxygen (microaerophiles) or being capable of surviving without oxygen for an extended interval of time (facultative anaerobes), or by harboring symbiotic chemoautotrophic bacteria (West, 1993; Bernhard, 1996).

The Cretaceous of the western interior contains a rich benthic foraminiferal record. Cenomanian-Turonian foraminifera of the United States and Canada have been well documented (Tappan, 1940; Cushman, 1946; Young, 1951; Jones, 1953; Fox, 1954; Frizzell, 1954; Stelck and Wall, 1954, 1955; Wall, 1960, 1967; Eicher, 1965, 1966, 1967; Lamb, 1968; Eicher and Worstell, 1970; Hazenbush, 1973; Lessard, 1973; Frush and Eicher, 1975; North and Caldwell, 1975; McNeil and Caldwell, 1981; Eicher and Diner, 1985; Leckie, 1985; Bloch et al., 1993; Caldwell et al., 1993; Fisher et al., 1994; and Schroeder-Adams et al., 1996). Aspects of their spatial distribution have been summarized by Caldwell et al. (1978, 1993) and Eicher and Diner (1985). Benthic foraminiferal faunas from the Canadian prairie provinces and from sections in Texas differ significantly from each other, not only in terms of species present, but also in gross population structure and diversity. For example, northern Cenomanian-Turonian assemblages are typically dominated by agglutinated taxa while coeval southern assemblages are dominated by calcareous taxa, both numerically and in species richness. Therefore, the paleobiogeography of calcareous and agglutinated benthic taxa offer an important, yet largely underutilized, proxy for water mass affinities in the Cretaceous Western Interior Sea.

Eicher and Worstell (1970) documented the taxonomy and distribution of foraminifera from eight Cenomanian-Turonian

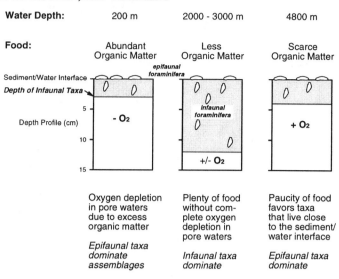

FIG. 4—Relationship between epifaunal and infaunal dominance in benthic foraminiferal assemblages and the amount of organic matter supplied to the seafloor (summarized from data presented by Corliss and Emerson, 1990).

localities of the Great Plains (Table 1; Fig. 1); four northern sections from around the Black Hills of northeastern Wyoming and southwestern South Dakota (localities 1-4) and four southern sections from eastern Colorado and western Kansas (localities 5-8). All sections lie near the axis of the seaway and are composed of calcareous shales and/or marlstones in the Cenomanian-Turonian boundary interval; interbedded limestones occur in all sections but the northernmost in Bull Creek Wyoming. This section is very close to the western edge of a pronounced regional facies change from calcareous to noncalcareous shales (Eicher and Worstell, 1970; Fisher et al., 1994).

Eicher and Worstell (1970) recognized three major assemblage zones in these strata: (1) the upper Cenomanian "lower planktonic zone", in which planktics dominate to the near-exclusion of benthics, (2) the upper Cenomanian "benthonic zone", which is characterized by abundant planktic foraminifera and diverse populations of benthic foraminifera dominated by calcareous taxa, and (3) the uppermost Cenomanian-middle Turonian "upper planktonic zone", which is characterized by high dominance, very low diversity planktic assemblages, and often sparse benthic assemblages. The lower boundary of the benthonic zone is sharp at the southern sections, but it is gradual in the northern sections. The paucity of benthic foraminifera in strata from below and above the "benthonic zone" was attributed to oxygen-poor conditions on the seafloor (Eicher and Worstell, 1970; Frush and Eicher, 1975; Eicher and Diner, 1985).

The benthonic zone represents an important oceanographic event within the Greenhorn Sea. The biggest change in benthonic zone assemblages occurs between the southern Black Hills and east-central Colorado (Fig. 1). Most of the agglutinated species documented from the benthonic zone have northern affinities based on their distributions across the northern Great Plains of the United States and Alberta (Fox, 1954; Stelck and Wall, 1955; Eicher, 1967); most notable of these are *Ammobaculoides mosbyensis*, *Trochammina wetteri*, *Reophax recta*, and *Haplophragmoides topagorukensis* (Table 1). *Spiroplectinata vulpes* occurs as far south as Rock Canyon but has greater relative abundances in the northern sections, which also suggests northern affinities. *Textularia rioensis* and *Gaudryina* cf. *G. quadrans*, on the other hand, are absent from the northern Black Hills sections but are characteristic, albeit very rare, of the benthonic zone of eastern Colorado and western Kansas (Eicher and Worstell, 1970). Southern affinities for *T. rioensis* are strongly supported by its common occurrence in the Grayson Formation of north-central Texas (e.g., Tappan, 1940; Mancini, 1982).

Numerous calcareous benthic taxa are widely distributed across the Great Plains (Eicher and Worstell, 1970), including: *Quinqueloculina moremani, Citharina kochii, Dentalina basiplanata, D. communis, Lenticulina gaultina, Ramulina aculeata, Buliminella fabilis, Neobulimina albertensis, Tappanina laciniosa, Valvulineria loetterlei, Pleurostomella nitida, Fursenkoina croneisi, Cassidella tegulata, Gavelinella dakotensis, G. plummerae, Lingulogavelinella asterigerinoides, L. modesta, L. newtoni,* and *Orithostella viriola* (Table 1). Of these, *Tappanina laciniosa* and *Lingulogavelinella newtoni* are more abundant in the southern sections, and *Valvulineria loetterlei* and *Pleurostomella nitida* are more common in the Kansas sections (Table 1). We propose that these last four taxa have southern, Gulf Coast affinities and are particularly diagnostic of "healthy" benthic conditions: abundant food, ample supply of dissolved oxygen, and warm, clear, and well-circulated waters.

Buliminella fabilis and *Gavelinella plummerae* are most abundant in eastern Colorado and westernmost Kansas. We suggest that these two species are diagnostic of the deepest part of the seaway and are either intermediate water mass or bathymetric indicators (Table 1). To further support the interpretation of Gulf Coast affinities for all of these taxa, we also note that *Valvulineria loetterlei, Gavelinella plummerae,* and *Lingulogavelinella asterigerinoides* are common to abundant constituents of the lower Cenomanian Grayson Formation of north-central Texas (Tappan, 1940; Mancini, 1982). In addition, these three taxa plus *Buliminella fabilis, Lingulogavelinella modesta, Lenticulina gaultina,* and *Neobulimina albertensis* are among nine species of calcareous benthics associated with a brief incursion of warm southern waters into southern Colorado (Thatcher Limestone Member of Graneros Shale) during early Cenomanian time (Eicher, 1965; Eicher and Diner, 1985).

RESULTS

Planktic Foraminifera

The relative abundance of planktic foraminifera (percent planktics to total foraminifera; Fig. 5) increases in the offshore direction from LP (neritic) to MV (proximal basin) to RC (distal basin). The section at RC is dominated by planktics throughout the study interval with an average of 89.5%, a typical value for outer neritic-upper bathyal settings. The sections at LP and MV show a marked decline in planktic foraminifera in the uppermost Cenomanian *Neocardioceras* zone. In all three sections there is a small to moderate drop in planktics for a brief interval within the upper Cenomanian *Sciponoceras* zone.

The section at LP is dominated by biserial (*Heterohelix*) and triserial (*Guembelitria*) taxa, except in the upper Cenomanian *Metoicoceras* zone, where trochospiral taxa (mostly *Hedbergella*) dominate and there is an influx in planispiral forms (*Globigerinelloides*) (Fig. 6). Although we did not find any speci-

mens of keeled species in our samples, Hazenbush (1973) reported *Rotalipora greenhornensis* and *Praeglobotruncana stephani* in two samples from the upper Cenomanian of LP. RC is dominated by trochospiral and biserial species. Significant numbers of planispiral forms are present in the *Sciponoceras* zone. Triserial forms are very rare, but seven keeled species, including *Rotalipora cushmani* and *R. greenhornensis*, are present in the upper Cenomanian of the RC section (Leckie, 1985).

The sections at both LP and RC show an abrupt increase in biserial forms at the base of the uppermost Cenomanian *Neocardioceras* zone. This dramatic change in planktic foraminiferal assemblages ("*Heterohelix* shift") roughly coincides with a major $\delta^{18}O$ depletion event (Pratt, 1985). At Blue Point, a more proximal section in the Black Mesa basin, Olesen (1991) also documented a marked increase in *Heterohelix* in the uppermost Cenomanian. The planktic foraminiferal assemblages at MV are transitional in overall composition between LP and RC, particularly in their relative abundance of the triserial *Guembelitria*. Importantly, however, population structure through the MV section is highly variable from sample to sample, the *Heterohelix* shift is weakly developed, and only two specimens of keeled taxa were found in the upper Cenomanian strata (West et al., 1990).

Benthic Foraminifera

Benthic foraminiferal assemblages at LP and MV are very similar. There is a significant proportion of agglutinated taxa with northern affinities (Fig. 7). Both sections also display nearly identical stratigraphic trends in benthic assemblage changes. By contrast, the upper Cenomanian *Sciponoceras* zone at RC is dominated by a diverse assemblage of calcareous benthics (the "benthonic zone" of Eicher and Worstell (1970)), many of which have southern affinities. A majority of these calcareous species are not present at either LP or MV. In addition, the RC section only contains trace specimens of two agglutinated species, and these have southern affinities.

All three sections show a marked increase in the relative abundance of *Gavelinella dakotensis*, a presumed epifaunal taxon, in the uppermost Cenomanian *Neocardioceras* zone. This is followed by a rapid shift to dominance by *Neobulimina albertensis*, a presumed infaunal taxon, within or at the top of the *Neocardioceras* zone (Cenomanian-Turonian boundary). Note that this abrupt change in benthic assemblages appears to be diachronous from basin center to western margin, occurring first at RC within the *Neocardioceras* zone and last at LP at the Cenomanian-Turonian stage boundary.

In summary, the planktic assemblages at LP are similar to those at RC, despite being some 450 km apart and representing neritic and deep water environments, respectively, but the LP benthics are more similar to those at MV. Additional environmental constraints, such as clay mineralogy, are needed to interpret this pattern.

Clay Mineralogy

The clay mineralogy of the sections at LP and RC are very similar and display identical stratigraphic trends through the Cenomanian-Turonian boundary interval (Fig. 8). In unadjusted diffractograms, mixed-layer illite/smectite is the prominent clay at both sites, with lesser amounts of discrete illite and kaolinite. By contrast, illite more obviously contributes to the MV clay assemblages, but kaolinite and illite/smectite are major components

TABLE 1—BENTHIC FORAMINIFERA OF THE BENTHONIC ZONE[1]

Localities	1	2	3	4	6	RC	7	8	LP	MV
	Bull Creek	Belle Fourche	Black Gap	Hot Springs	Rock Canyon	Rock Canyon	Hartland-BC	Bunker Hill	Lohali Point	Mesa Verde
	NE Wyoming	W S.Dakota	W S.Dakota	SW S.Dakota	S Central Colo.	S Central Colo.	SW Kansas	N Central Kan.	NE Arizona	SW Colorado
Agglutinated Species										
Saccammina alexanderi			0.1						0.2	
Reophax recta	2.3								0.7	
Haplophragmoides topagorukensis	0.0	0.1	0.0						0.6	1.4
Haplophragmium sp.				1.0	0.1					
Ammobaculites junceus	0.0			0.2						
Ammobaculoides mosbyensis	11.6								0.2	
Coscinophragma? codyensis	0.2	0.2	0.1	1.5					1.3	
Textularia rioensis			0.0		0.2	0.1	0.6			
Trochammina rainwateri	0.3		0.0						5.6	0.9
Trochammina wetteri	2.5	0.1	1.1						8.6	3.7
Gaudryina cf. G. quadrans					0.2		0.6	0.1		
Verneuilina sp.	0.0									
Spiroplectinata vulpes	1.7		1.2	0.3	0.1					
Marssonella conica	0.0		0.0							
Ammobaculites impexus									0.4	0.4
Ammobaculites fragmentarius									0.2	
Trochamminoides apricarius									1.9	2.1
Haplophragmoides spp.									0.6	
Calcareous Species										
Quinqueloculina moremani	0.3		0.2		0.1	0.1	0.4		2.2	3.4
Massilina planoconvexa			0.0							
Citharina complanata			0.0		0.1		0.4	0.1		
Citharina kochii	0.3	0.5	0.2	0.7	0.2	0.3	0.6	0.4	1.3	0.1
Citharina petila	0.1		0.1		0.1					
Dentalina basiplanata	0.1			0.3	0.1				0.7	0.1
Dentalina communis	0.2	0.5	0.2	0.2	0.2	0.1	0.2			
Dentalina intrasegma	0.0		0.1		0.2		0.4	0.1		
Frondicularia extensa			0.0				0.2			
Frondicularia imbricata	0.1	0.1	0.0	0.2	0.1					
Lagena apiculata	0.1									
Lagena striatifera	0.0		0.0							
Lagena sulcata	0.1									
Lenticulina gaultina	0.4	0.5	0.1	0.2	0.3	0.3	0.8	0.5	2.1	
Marginulina siliquina	0.0									
Marginulinopsis amplaspira	0.0	0.2				0.1				
Nodosaria bighornensis	0.3		0.2	0.2			0.2		0.9	
Planularia dissona	0.0									
Pseudonodosaria sp.	0.1		0.1				0.2			
Saracenaria reesidei	0.1							0.1		
Vaginulina debilis	0.1		0.0				0.2			
Vaginulina cretacea	0.1									0.1
Lingulina nodosaria	0.1		0.0		0.1		0.4			
Globulina lacrima	0.1		0.0		0.2			0.1		
Pyrulina cylindroides	0.2		0.0			0.1	0.2			
Bullopora laevis	0.1						0.2			
Ramulina aculeata	0.1		0.1		0.1	0.1		0.1		
Ramulina globulifera	0.1		0.0				0.2			
Washitella sp.	0.0									
Buliminella fabilis	14.8	7.7	11.4	12.2	25.7	30.3	20.8	9.6		0.3
Neobulimina albertensis	28.7	50.3	48.0	36.7	18.4	14.3	7.8	4.5	57.5	57.4
Tappanina laciniosa	6.9	6.0	15.3	13.6	19.3	20.3	31.9	56.5		
Valvulineria loetterlei	5.9	0.3	5.0	2.2	6.8	4.6	11.1	11.5		
Spirillina minima	0.1			0.2						
Pleurostomella nitida	0.8		1.2	0.2	0.4	1.8	2.1	3.8		
Fursenkoina croneisi	0.0		0.6	0.2	0.4	0.2	0.2	0.1		
Cassidella tegulata	0.3	0.2	0.6	1.0	0.4	0.2	0.6	0.1		
Gavelinella dakotensis	10.5	32.8	5.1	21.6	6.1	3.4	0.2	0.2	15.0	22.1
Gavelinella plummerae	2.4		0.2	1.2	12.0	13.5	7.3	1.1		
Lingulogavelinella asterigerinoides	1.9	0.2	0.3	0.9	1.5	5.5	1.0	2.8		
Lingulogavelinella modesta	1.0		7.0	3.6		0.7	0.6	0.5		
Lingulogavelinella newtoni	0.2	0.1	0.9	1.5	6.5	4.0	10.7	7.2		
Orithostella viriola	0.7		0.0	0.2	0.1		0.2			
Conorboides minutissima	0.3		0.2			0.2		0.4		
Hoeglundina charlottae	3.1					0.1				0.1
Number of Samples in Analysis	15	6	8	4	4	7	3	3	6	6
Total Number of Specimens	4129	860	2333	588	1139	1989	523	819	534	700

[1] Average benthic foraminiferal percentage data from the upper Cenomanian "benthonic zone" of Eicher and Worstell (1970) from across the U.S. western interior to illustrate paleobiogeographic affinities and water depth. Data are from Eicher and Worstell (1970; localities 1-8), Leckie (1985) for the RC section, and this study. Refer to Fig. 1 for section locations.

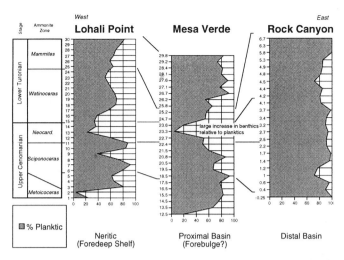

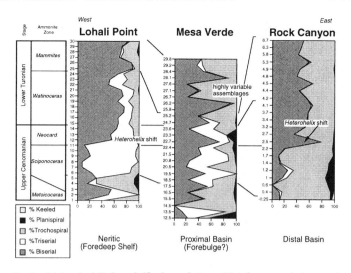

FIG. 5—Proportion of planktic taxa to total foraminifera through the Cenomanian-Turonian boundary interval of Lohali Point (LP), Mesa Verde (MV), and Rock Canyon (RC). Sample numbers for the LP and MV sections of the Mancos Shale correspond to meters above the Dakota Sandstone. Sample numbers for the RC section of the Greenhorn Formation correspond to meters above the base of the Bridge Creek Limestone Member (-0.25 m is from the uppermost Hartland Shale Member). Note the major drop in planktics and increase in benthics in the *Neocardioceras* zone (uppermost Cenomanian) of the LP and MV sections.

FIG. 6—Major planktic foraminiferal morphotypes. Note large sample-to-sample variability at the MV section compared with LP section to the west and RC section to the east. Also note the abrupt change in planktic populations (*Heterohelix* shift) at the base of the *Neocardioceras* zone at both LP and RC. Refer to Fig. 5 caption for additional explanation.

as well. Of particular interest is the chlorite in the section at MV; this clay mineral is not found at the other sites. Chlorite is found throughout the section with a relatively stable peak intensity. Although these intensity values do not represent absolute mineral abundance, the trends depicted are reproduced when the data are adjusted according to the mineral intensity factors (MIF) in the manner proposed by Reynolds (1989).

The stratigraphic fluctuations in relative clay-mineral abundance are most pronounced in the LP section, with the abundance of mixed-layer illite/smectite increasing from the base of the section up through the middle of the *Sciponoceras* zone. This trend reflects westward shoreline migration with transgression of the Greenhorn Sea. A similar, albeit more subdued expression of the same trend is visible in the RC data. At both sites, the *Sciponoceras-Neocardioceras* zone transition is marked by a doublet peak of increased kaolinite and illite. The upper peak, near the base of the *Neocardioceras* zone, is more intense, particularly in the deeper-water, distal RC section. Kaolinite and illite decline at the Cenomanian-Turonian boundary, then increase again in the lower Turonian. This latter effect is more pronounced at the more proximal LP section. The clay-mineral abundance pattern at MV does not show the same degree of synchroneity, suggesting that major depositional processes at this site differed from those at the other two sites.

DISCUSSION

Foraminiferal Assemblages

The high, relatively stable abundances of planktics through the Cenomanian-Turonian boundary interval at the RC section, coupled with its high planktic species richness, especially in the *Sciponoceras* zone, indicate a fully marine, deep water, stratified environment (Fig. 5). The section at RC contains distal epicontinental sea assemblages of planktic foraminifera, with an abundance and diversity of trochospiral, biserial, keeled, and planispiral taxa, and by paucity of triserial forms. The central core of the seaway was essentially an arm of the Tethys Sea (Eicher and Diner, 1989). The southwestern side of the Greenhorn Sea contains warm water neritic assemblages based in part on high abundances of triserial and biserial taxa and paucity of keeled forms at LP (Fig. 6). The section at LP is also characterized by lower relative abundances of planktic foraminifera and thereby indicative of mid- to outer-neritic depths. Despite being located some 450 km apart and containing deep and shallow water assemblages of planktics, respectively, both RC and LP show a synchronous and marked change in population structure at the base of the uppermost Cenomanian *Neocardioceras* zone (Figs. 6, 9). This suggests that the *Heterohelix* shift represents a widespread oceanographic event in the southwestern Greenhorn Sea during latest Cenomanian time and it supports the hypothesis that the seaway at the RC and LP localities had the same surface water mass at this time (Leckie et al., 1991).

Although the section at MV contains abundant planktics and has compositional trends that are, in general, transitional between LP and RC, it also contains a number of important distinguishing features. First, the planktic assemblages vary greatly from sample to sample. Perhaps this is due to winnowing and current activity on a bathymetric high. If winnowing of foraminifera tests was responsible, then why don't the benthic populations or the ratio of planktics to benthics (% planktics) display as much variability, and why do the stratigraphic trends in benthics and in percent planktics parallel those at LP so well? Second, the *Heterohelix* shift is poorly developed at MV and very few warm water keeled species were recovered from the study interval (West et al., 1990). Again, perhaps this is a consequence of shallow water depths. If shallow depths were the cause, then why are the assemblages at MV, in general, transitional between LP and RC in both percent planktics and in overall population structure (Figs. 5, 6)? Water mass mixing or water column instability in a relatively shallow environment may have inhibited stable stratification or reduced surface water salinities (beyond tolerance of steno-

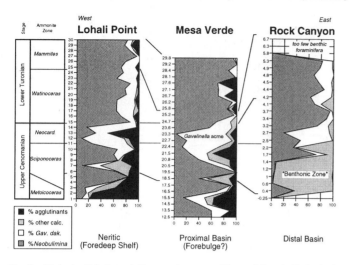

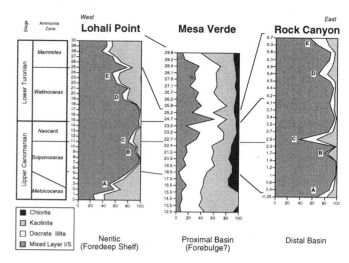

FIG. 7—Major benthic foraminifer morphogroups. Note striking similarity in the composition and stratigraphic trends of benthic foraminifers in the LP and MV sections, particularly in the upper Cenomanian. All three sites display an acme of *Gavelinella dakotensis* in the uppermost Cenomanian followed by an abrupt but diachronous change to *Neobulimina albertensis* dominance through the Cenomanian-Turonian boundary interval. Refer to Fig. 5 caption for additional explanation.

FIG. 8—Clay mineralogy. Letters A-E denote points of correlation between the sections at LP and RC. Note how very similar the composition and stratigraphic trends are between the LP and RC sections, while MV differs significantly from the other two. Refer to Fig. 5 caption for additional explanation.

haline taxa) thereby causing environmental exclusion of some taxa. Mixing of different water masses along a non-stationary oceanic front, or edge effect mixing and upwelling near MV, are two hypotheses that may explain both, the variable planktic assemblages and the weakly developed *Heterohelix* shift.

The benthos provide an entirely different paleoenvironmental view of change through the Cenomanian-Turonian boundary interval than the planktic record does. Based on the population structure of benthic foraminiferal sediment assemblages and their stratigraphic changes, it appears that LP and MV record very similar benthic environmental histories (Fig. 7). (1) Both sections display a marked increase in the proportion of benthics relative to planktics in the uppermost Cenomanian (Fig. 5), (2) both sections contain a distinct component of agglutinated taxa with northern affinities, which are completely lacking in the section at RC (Table 1), (3) both sections show nearly synchronous peak-to-peak changes in population structure through the upper Cenomanian, and (4) both sections lack the highly diverse calcareous assemblages of the benthonic zone (Eicher and Worstell, 1970) that are so well-developed and widespread in the deeper central and carbonate-dominated eastern parts of the seaway. Even the most diagnostic species of the benthonic zone (e.g., *Buliminella fabilis*, *Tappanina laciniosa*, *Valvulineria loetterlei*, *Gavelinella plummerae*, *Lingulogavelinella asterigerinoides*, *L. modesta*, and *L. newtoni*) are few or absent all-together at LP and MV (Table 1). We propose that the sections at LP and MV were situated at similar mid- to outer-neritic water depths and that both sites were exposed to a cool, northern water mass near the seafloor.

Clay Mineralogy

The presence of detectable chlorite in the MV section raises questions about the level of post-depositional thermal alteration of the clay minerals in that section (Fig. 8). Chlorite is commonly found as a product of burial diagenesis of sedimentary sequences. It usually appears during the conversion of mixed-layered illite-smectite to illite (Hower et al., 1976). However, several lines of evidence suggest a detrital origin for this chlorite. First, although ordered illite-smectite occurs in the lower 70 m of the 700 m Mancos Shale section at Mesa Verde (that is, in the upper Cenomanian-lower Turonian rocks that we studied) the bulk of the MV section is dominated by randomly interstratified illite-smectite, and chlorite persists through the entire section. This indicates that significant thermal diagenesis had not occurred in the MV study interval; illite and chlorite are not more abundant in the lower 70 m of the Mancos Shale, although there are stratigraphic changes in the relative abundances of these two clay species through the section (Hayden Scott, 1992; Finkelstein, 1991). Secondly, scanning electron microscope (SEM) examinations of whole-rock specimens failed to reveal evidence of the large (>10 µm) euhedral chlorite grains that suggest in-situ formation. Clay mineralogists still disagree as to whether chlorite is an authigenic mineral at all; the chlorite appears to be detrital even in the drill cores from Gulf Coast wells (Weaver, 1989). Lastly, Rock-Eval data run on shale samples from all three sections show a T_{max} below the oil-generating window (Table 2). Translated into down-hole temperatures, this indicates a burial temperature of <75°C. Accordingly, based upon available evidence, we interpret the chlorite at the MV section as predominantly of detrital origin.

Kaolinite can also have a post-depositional origin; when pyrite and other sulfide minerals weather, they produce acidic solutions that can hydrolyze clay minerals into Al-rich varieties (Pollastro, 1985). Such processes are common in black shale facies. Again, we found little evidence for this process in the clay minerals of the samples examined. SEM examination of the mudstone did not reveal the distinctive books of kaolinite that characterize post-depositional growth of this mineral under acidic conditions. In addition, pyrite is only a minor constituent of the fine-grained sediments and black shales of the sort deposited in highly anoxic environments do not occur in the study interval. The MV section, as discussed below, has fewer indices of anoxia than either the LP or the RC section, so the weathering of sulfide minerals should not have a pervasive effect on the sedimentary mineralogy.

TABLE 2 - ROCK-EVAL DATA[2]

TOC and Rock-Eval Analyses from RC, MV, and LP								
Sample	Tmax	Reliable?	S1	S2	S3	TOC	HI	OI
BC-19	456	no	0.02	0.07	0.25	0.04	175	625
BC-16	439	no	0.05	0.06	0.57	0.06	100	950
BC-11	**433**	yes	0.07	0.8	0.88	0.44	182	200
BC-9	386	no	0.03	0.07	0.43	0.1	70	430
BC-5	417	no	0.07	0.25	0.73	0.14	179	521
BC-1	430	no	0.05	0.28	1.01	0.26	108	388
MV-29.8	**434**	yes	0.38	3.57	1.51	1.89	189	80
MV-25.2	439	no	0.01	0.04	0.41	0.08	50	513
MV-24.7	483	no	0.01	0.05	0.43	0.04	125	1075
MV-23.6	437	no	0.02	0.13	0.53	0.28	46	189
MV-22.7	503	no	0.02	0.12	0.36	0.21	57	171
MV-20.5	408	no	0.01	0.03	0.25	0.15	20	167
MV-18.5	501	no	0.01	0.07	0.26	0.26	27	100
LP-22	**430**	yes	0.02	2.47	0.76	1.11	223	68
LP-17	466	no	0.01	0.1	0.71	0.2	50	355
LP-14	**432**	yes	0.02	3.2	2.86	2.16	148	132
LP-12	**435**	yes	0.04	1.71	3.57	1.44	119	248
LP-9	436	no	0.01	0.42	1.75	0.75	56	233
LP-7	426	no	0.04	0.13	1.78	0.38	34	468

[2]Rock-Eval data from selected samples from the LP, MV, and RC sections. Tmax = temperature of maximum rate of evolution of S2 hydrocarbons (Tmax values are not necessarily reliable with S2<0.50); S1 = free, thermally extractable hydrocarbons; S2 = hydrocarbons from the cracking of kerogen and high molecular weight free hydrocarbons which did not vaporize in the S1 peak; S3 = organic carbon dioxide; TOC = total organic carbon (weight percent of carbon); HI = hydrogen index; and OI = oxygen index. Note that the samples with "reliable" values all yield similar Tmax values (other samples yield unreliable Tmax values due to low S2 values).

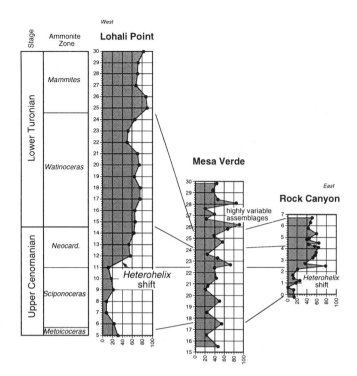

FIG. 9—Percent of the planktic foraminiferal genus *Heterohelix* at all three sites. Stratigraphic position of samples from the Mancos Shale at the LP and MV sections corresponds to meters above the Dakota Sandstone. At the RC section the sample position corresponds to meters above the base of the Bridge Creek Limestone Member of the Greenhorn Formation (lowest sample is from the uppermost Hartland Shale Member). Note the *Heterohelix* shift at the base of the *Neocardioceras* zone of sections at LP and RC and the highly variable populations at MV.

The foregoing considerations compel us to conclude that the clay minerals in the Cenomanian-Turonian interval of the three sections are largely detrital, and reflect conditions in the source areas and at the depositional sites in the seaway. Therefore, the stratigraphic changes in relative abundances record ancient fluctuations in the physical and chemical environment of the Western Interior Sea. The increasing abundance of illite-smectite beginning the base of the LP section is consistent with transgression of the sea during late Cenomanian time. Normally, illite-smectite consists of smaller particles and is deposited farther from shore than either kaolinite or discrete illite (Gibbs, 1977; Chamley, 1989). A decrease in kaolinite at LP is concomitant with transgression. A decrease in kaolinite also occurs at MV and RC, but in a more subdued fashion because those sections were located farther offshore ("A" in Figs. 8 and 10).

The "spikes" of kaolinite found in the *Sciponoceras/ Neocardioceras* biozone transition, particularly in the sections at RC and LP, are a curious phenomenon ("B" and "C" in Figs. 8 and 10). One interpretation is that influx of kaolinite resulted from a short regressive event within this transgressive interval. We do not favor this interpretation because the spikes are more pronounced in the deepest, most distal environment. However, kaolinite production is enhanced during warm, wet climatic episodes and such conditions could also cause the kaolinite increase.

But why the greater effect at the most distal site? One possibility is that the sediment source was not from the Sevier orogenic belt west and northwest of the study sites. Instead, perhaps the kaolinite influx was derived from the south or southeast. The Sevier belt certainly supplied detritus, but the rapid rate of uplift and the short transport distances to the seaway may have delivered less weathered material to the depositional sites. In contrast, a warm, humid climate in less tectonically active southerly

parts of the continent may have produced a greater abundance of kaolinite in the soils.

Increasing kaolinite abundance in the lower Turonian part of the LP section also corroborates this idea (Leckie et al., 1991). There, the supply of kaolinite increased despite a widening of the sea and deepening of the water column through the time of peak transgression in the early Turonian. The kaolinite presumably came from a deeply weathered source. Tethyan waters could have transported these clays into southern parts of the seaway. The section at MV shows some of these influences, but the very different clay assemblages suggest an alternate source area.

Comparisons Among the Sections

The LP and MV sections both show a dramatic decline in the proportion of planktics to total foraminifera in the uppermost Cenomanian *Neocardioceras* zone (Fig. 5). Such a trend may record a decline in planktic abundances, perhaps due to reduced salinities in the upper water column or to dissolution of the thinner-walled planktics. Conversely, the signal could reflect greater abundances of benthics due to improved conditions at the seafloor or an influx of food.

The last scenario is favored for the following reasons: (1) the clay mineralogy is very different in the sections at MV and LP, but if both sections had experienced the same low salinity event, then a similar spike in detrital clay minerals would be expected (Fig. 8); (2) the striking similarities in benthic assemblages at LP and MV (Fig. 7) suggest that both sections experi-

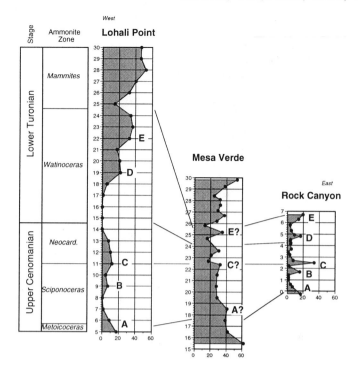

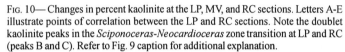

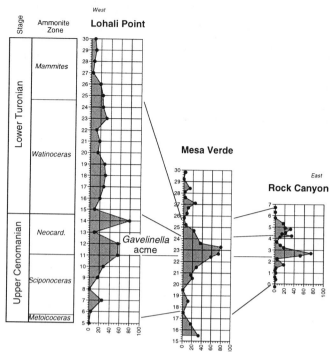

FIG. 10—Changes in percent kaolinite at the LP, MV, and RC sections. Letters A-E illustrate points of correlation between the LP and RC sections. Note the doublet kaolinite peaks in the *Sciponoceras-Neocardioceras* zone transition at LP and RC (peaks B and C). Refer to Fig. 9 caption for additional explanation.

FIG. 11—Changes in percent *Gavelinella dakotensis* through the Cenomanian-Turonian boundary interval at the LP, MV, and RC sections. Note the acme of this taxon in the *Neocardioceras* zone of all three sites. Refer to Fig. 9 caption for additional explanation.

enced similar environmental conditions at the seafloor; (3) there is no evidence for differential dissolution of planktic tests; and (4) a relatively short-lived acme of *Gavelinella dakotensis*, a presumed epifaunal taxon, at all three localities (Fig. 11) may record an opportunistic response to a greater flux of organic matter to the seafloor from either terrestrial or marine sources (e.g., Gooday, 1993; Thomas and Gooday, 1996; West et al., this volume).

In further support of this hypothesis, Elder (1987, 1991) documented a dramatic increase in the detrital feeding gastropod *Drepanochilus* through the middle and upper parts of the *Neocardioceras* zone in sections from northeastern Arizona and south-central Utah. The highest organic carbon contents also occur in the *Neocardioceras* zone, which suggests that the availability of abundant organic matter may have stimulated growth of benthic foraminiferal populations.

At all three localities, the acme of *Gavelinella dakotensis* in the *Neocardioceras* zone is followed by a diachronous but abrupt change in structure of benthic foraminiferal populations (Figs. 7, 12). Assemblages characterized by high dominance of *Neobulimina albertensis*, an infaunal taxon, and very low diversity first appear in the lower part of the *Neocardioceras* zone at RC, in the middle part of that zone at MV, and at the top of the zone, which marks the Cenomanian-Turonian boundary, at LP. This appears to be a widespread event that cuts across paleodepths and facies. Elder (1990, 1991) also notes a diachronous change in molluscan faunas from the west-central part of the seaway toward the western margin during *Neocardioceras* time. Diverse to moderately diverse mixed infaunal-epifaunal biofacies of the *Sciponoceras* zone are replaced by depauperate, mixed infaunal-epifaunal, inoceramid-*Pycnodonte*, or other inoceramid-dominated biofacies. By early *Watinoceras* time (earliest Turonian), a more widespread and ubiquitous mix of inoceramid-

Entolium, inoceramid-*Pseudoperna*, inoceramid-*Phelopteria*, or moderately diverse epifaunal biofacies lacking infaunal components characterized the southwestern and central parts of the seaway. The last stronghold for mixed infaunal and epifaunal molluscan communities through the Cenomanian-Turonian boundary interval was the clay-dominated western facies (Elder, 1990, 1991).

Why are the benthic foraminiferal assemblages from the LP section so similar to those at MV, while the clay mineralogy and stratigraphic trends in the planktic assemblages at LP are more similar to those at RC? If the water column were stratified by water masses from different sources, then this transect of sites across the southwestern side of the Greenhorn Sea records the character of those water masses where they met, interacted, and changed with rising sea level.

Upper Water Column

The *Heterohelix* shift coincides with a sharp depletion in $\delta^{18}O_{carb}$ recorded in the lower Bridge Creek Member of the Greenhorn Formation near Pueblo, Colorado (Pratt, 1985) (Fig. 2). Whole-rock oxygen isotope values drop from a high of about -3.5‰ in a limestone bed near the top of the *Sciponoceras* zone to about -8.0 ‰ in a calcareous mudstone at the base of the *Neocardioceras* zone (Pratt, 1985; Pratt et al., 1993). Light values (<-7 ‰) characterize all lithologies through the uppermost Cenomanian. The $\delta^{18}O_{carb}$ values actually fall to nearly -9 ‰ in marly shales near the Cenomanian-Turonian boundary. Limestones through the basal Turonian show gradually heavier values (-7 to -5 ‰), whereas the dark marlstones and marly shales remain markedly depleted but gradually become heavier (-8 to -6.5 ‰) (Pratt et al., 1993).

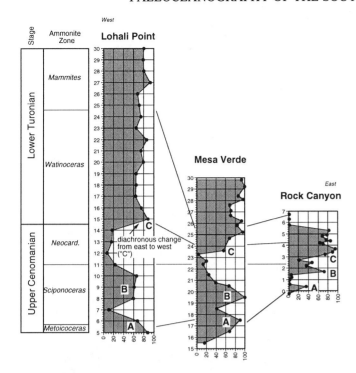

FIG. 12—Changes in percent *Neobulimina* (mostly *N. albertensis*) through the Cenomanian-Turonian boundary interval at the LP, MV, and RC sections. Note the abrupt but diachronous change to *Neobulimina* dominance through the uppermost Cenomanian *Neocardioceras* zone. Refer to Fig. 9 caption for additional explanation.

The abruptness and magnitude of the negative $\delta^{18}O_{carb}$ shift at the base of the *Neocardioceras* zone suggests that this signal records a major influx of fresh water to the seaway from riverine sources during latest Cenomanian time (Pratt, 1984, 1985; Barron et al., 1985; Pratt et al., 1993). This interpretation is supported by the coincident increase in discrete illite and kaolinite in a core from near Pueblo, Colorado (Pratt, 1984) and in our LP and RC sections (Fig. 8). In addition, an increase in apparent sedimentation rates in the *Neocardioceras* and lower *Watinoceras* zones in the central part of the seaway also supports Pratt's model, which points to increased runoff and fluvial input as the cause of the uppermost Cenomanian $\delta^{18}O$ depletion interval (Elder, 1985).

If we accept the hypothesis that LP and RC were influenced by the same southern surface water mass based on planktic foraminiferal assemblages and the same detrital source based on clay mineralogy, then the fresh water influx responsible for the negative excursion in $\delta^{18}O$ and detrital clay spike in the uppermost Cenomanian must have originated along the southeastern or southwestern part of the seaway and not in the Sevier orogenic belt of Utah-Nevada-Idaho-Wyoming. There is no doubt that the Sevier orogenic belt was an important primary source of riverine and clastic sediment input to the sea. But at times, particularly during transgressive pulses, clays across the southern part of the seaway may have come from additional sources. This interpretation is supported, in part, by the greater increase in illite + kaolinite in the basal *Neocardioceras* zone at RC as compared with the more proximal LP section (Fig. 8). Additional evidence comes from the distribution of macrofossils. For example, Elder (1985) suggested that the paucity of ammonites - including limestone -

in the *Neocardioceras* zone at RC is perhaps related to the exclusion of ammonite larvae due to a stable subsaline surface layer on the sea. The few ammonites that occur at RC indicate warm southern affinities (Elder, 1985). However, ammonites are more common in this zone along the western margin (Elder, 1987, 1991; Kirkland, 1991). These observations raise further questions about the Sevier orogenic belt as the source of freshwater and detrital clays at LP.

Simulations of Cenomanian-Turonian climate were run to test if orbitally-induced insolation forcing, caused cyclic changes in precipitation intensity, as well as to test for the existence of high precipitation belts in or adjacent to the Western Interior Sea, which is required by the dilution model (e.g., Barron et al., 1985; Glancy et al., 1986, 1993). These models consistently indicate heavy winter precipitation along the northern continental margin of the Tethys Sea. This area includes the southern part of the Western Interior Sea and adjacent land masses. Glancy et al. (1986) suggest that the cyclonic rotation of winter storm tracks may have moved warm, moist air masses up the eastern slopes of the Sevier highlands, bringing greater precipitation and runoff under "maximum" forcing. Parrish (1993) also suggests that if the Sevier orogenic belt was high enough, it may have resulted in a low-pressure cell capable of drawing in moisture from the south, in spite of the possibility that the highlands may have also provided a rain shadow to zonal air flow.

The climate models do suggest that orbitally-driven changes in precipitation intensity may have influenced sedimentation patterns within the Western Interior Sea. In addition, the models also suggest that highlands of northern Mexico and/or southern Arizona, as well as the southern stable cratonic lowlands of the United States (e.g., Sohl et al., 1991) may have periodically experienced heavy winter rainfall because of its proximity to the northern margin of Tethys.

The preponderance of illite and the significant amount of chlorite at MV suggest that the clay mineral assemblages originated from a very different source area, one that was cooler and more temperate. We envision a circulation pattern in the Western Interior Sea that brought detritus from the Sevier orogenic belt and more northerly sources (Frontier Delta?) to sites in the vicinity of MV, while southerly sources supplied at least some of the muds deposited near LP and RC during the time of the Cenomanian-Turonian boundary. We propose that a broad offshore bathymetric high (forebulge?) separating the foredeep shelf from the deep central axis of the seaway bifurcated warm water currents (Fig. 13). Additional evidence of a bifurcating southern tongue comes from an independent analysis of the trace-element geochemistry of uppermost Cenomanian strata (Orth et al., 1993) (Fig. 14). Their elemental abundance data point toward a sediment source to the south and east.

Also noteworthy is an increase in kaolinite and illite in the upper part of the *Sciponoceras* zone at LP and RC that is associated with a brief but dramatic change in benthic foraminiferal population structure at RC (1.7 m) followed by brief recovery at the top of the zone (2.2 m). This pattern is a precursor to a second, larger detrital clay spike, *Heterohelix* shift, and negative oxygen isotope shift in the basal *Neocardioceras* zone (Figs. 2, 15). A similar pattern of biotic perturbation and subsequent recovery of molluscan communities occurred in the section at RC just prior to a faunal turnover and immigration event at the *Sciponoceras/Neocardioceras* zonal boundary (Elder, 1991) (Fig. 2). This upper *Sciponoceras* zone molluscan perturbation event is common at many localities (Elder, 1985). However, the clay

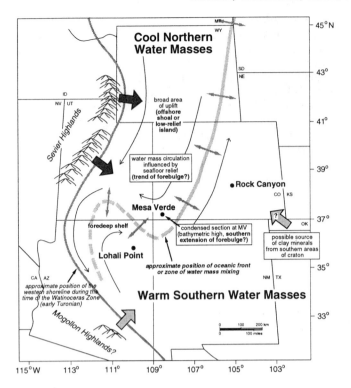

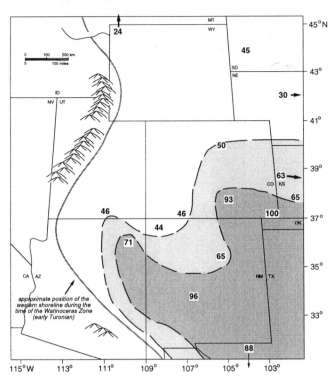

FIG. 13—Oceanographic interpretation of the Greenhorn Sea during latest Cenomanian-earliest Turonian time showing possible sources of clay minerals and a hypothetical circulation pattern that accounts for the temporal and spatial changes in clay mineralogy, and planktic and benthic foraminiferal paleobiogeography. Note the position of the Mesa Verde section along a presumed north-south trending bathymetric high, or forebulge. We suggest that this forebulge influenced the circulation of upper water masses on the western side of the seaway (thin arrows).

FIG. 14—Trace element concentration (Sc + Cr + Co + Ir) in upper Cenomanian rocks from across the Western Interior Sea (data from Orth et al., 1993). Values represent averages of the four elements through the *Sciponoceras* and lower *Neocardioceras* zones. These values appear to be bifurcated in the area around Mesa Verde.

spike and faunal events in the upper *Sciponoceras* zone are not associated with significant $\delta^{18}O$ depletion or with any change in planktic foraminiferal population structure at either LP or RC.

There appears to be a compelling, albeit complex, causal relationship between the *Heterohelix* shift, the $\delta^{18}O$ depletion event, and the clay-mineral changes. At the RC section in the center of the seaway, the *Heterohelix* shift coincides with a major influx of illite and kaolinite (2.5 m), but to the southwest, at LP, the change in planktic foraminiferal populations (12 m) occurs shortly after the small but similar clay spike (11 m) (compare Figs. 9, 10). Did *Heterohelix* respond to increased riverine input and development of a subsaline cap, as suggested by the $\delta^{18}O$ signal and the influx in detrital clays, or did it respond to other environmental variables? Leckie et al. (1991) suggested that the *Heterohelix* shift was related to the expansion or incursion of an oxygen minimum zone (OMZ) into the seaway rather than to slightly lower salinity conditions in the near-surface waters. Their rationale for preferring the OMZ hypothesis was based on the observation that *Heterohelix* continues to dominate the planktic foraminiferal assemblages at both LP and RC *above* the Cenomanian-Turonian boundary despite oxygen isotope data from the seaway that indicates a return to more normal marine conditions *at* the boundary (Pratt and Threlkeld, 1984; Pratt, 1985). In addition, following the detrital clay spike in the basal *Neocardioceras* zone, both kaolinite and illite drop off sharply and remain depressed across the Cenomanian-Turonian boundary. If *Heterohelix* responded mostly to reduced salinities, then

its abundance, too, might be expected to drop off in concert with the detrital clays (Fig. 15).

The development of a widespread OMZ near the time of the Cenomanian-Turonian boundary was proposed to account for the unique biotic, sedimentologic, and geochemical signatures preserved in numerous localities around the world (e.g., Berger and von Rad, 1972; Frush and Eicher, 1975; Schlanger and Jenkyns, 1976; Hilbrecht and Hoefs, 1986; Arthur et al., 1987; Schlanger et al., 1987; Jarvis et al., 1988; Thurow et al., 1992; Arthur and Sageman, 1994; Kaiho and Hasegawa, 1994). In fact, the development or incursion of an OMZ into the southern Western Interior Sea may have been amplified by haline stratification caused by the development of a widespread, subsaline cap that originated from the southern part of the seaway during the latest Cenomanian. This scenario would have further decreased benthic ventilation, particularly in the deep central axis. The freshwater input at this time may be related to a fourth- or fifth-order sea level events as suggested by molluscan and dinoflagellate data (Elder, 1991; Li and Habib, 1996).

We propose that the influence and environmental impact of an OMZ expansion or incursion was diachronous from the central axis of the seaway toward the western margin during the latest Cenomanian. This hypothesis is supported by observed diachroneity in fossil assemblages and extinction events: (1) the *Heterohelix* shift relative to the clay influx noted above; (2) extinction of *Rotalipora* in the Western Interior Sea (Leckie, 1985); and (3) the observed diachroneity of biotic changes in both benthic foraminiferal and molluscan assemblages during *Neocardioceras* time (Elder, 1990, 1991).

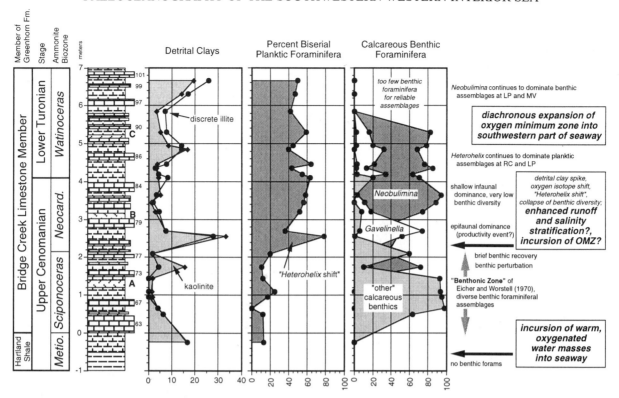

FIG. 15—Percentages of detrital clays, biserial planktic foraminifera, and calcareous benthic foraminifera through the Cenomanian-Turonian boundary interval at the Rock Canyon section, and interpretations of the changes observed in the clay mineralogy and in the foraminiferal assemblages.

Break-up of the subsaline cap, weakening of haline stratification, and a slight reduction in the intensity of the OMZ at the base of the *Watinoceras* zone may be due in part to a major transgressive pulse noted by Elder (1991). Such a hypothesis is supported by an abrupt increase in carbonate content at sections in the western clay-dominated facies and by a dramatic influx of *Mytiloides*, which is a turbidity-sensitive, epifaunal bivalve, at the base of the Turonian (Elder, 1985, 1987, 1991). However, we propose that the impact of an OMZ was still widely felt across southern parts of the seaway through the time of peak transgression in the late early Turonian (*Mammites* zone) based on: (1) the persistence of *Heterohelix* dominance through this interval (Leckie et al., 1991; West et al., this volume); (2) the low diversity assemblages of benthic foraminifera dominated by *Neobulimina* (Eicher and Worstell, 1970; Frush and Eicher, 1975; Eicher and Diner, 1985; Leckie, 1985; West et al., this volume); and (3) the widespread, depauperate communities dominated by epifaunal bivalves, which suggest low benthic oxygen conditions that limited bioturbation and benthic turbidity (Kauffman, 1977, 1984; Elder, 1985, 1987, 1990, 1991; Elder and Kirkland, 1985; Kirkland, 1991).

Lower Water Column

Are the similarities in upper Cenomanian benthic foraminiferal assemblages of LP and MV due to similar water depths, depositional facies, water masses bathing the seafloor in this part of the seaway, or a combination of these variables? And, why do all three sections show strikingly similar patterns of change in the uppermost Cenomanian and basal Turonian (Figs. 11, 12)? It is likely that the benthic biota records a complex history in-

volving not only different water masses and mixing of water masses, including the possible incursion of an oxygen minimum zone as discussed above, but also changes in relative sea level, surface water productivity, and paleobathymetry. For example, transgressive pulses at the bases of the *Sciponoceras*, *Neocardioceras*, and *Watinoceras* zones changed substrate firmness and composition, benthic oxygen levels, and clastic sedimentation rates, all of which influenced the nature of the benthic molluscan communities (Elder, 1991).

During late Cenomanian time (*Sciponoceras* zone), the central axis of the seaway was dominated by diverse, calcareous benthic foraminiferal taxa with strong southern affinities (Eicher and Worstell, 1970; Frush and Eicher, 1975; Eicher and Diner, 1985; Leckie, 1985) (Fig. 7, Table 1). These assemblages included a mix of probable infaunal and epifaunal taxa. Although the sections at LP and MV lack this rich benthic microbiota, species richness is greatest in both sections in the *Sciponoceras* zone, and both include infaunal and epifaunal taxa. In addition, the LP and MV sections contain northern agglutinated taxa as well as sporadic abundances of miliolid benthic foraminifera (*Quinqueloculina moremani*), which may indicate the influence of warm, normal salinity waters. In the western, clay-dominated sections of northeastern Arizona and southern Utah, diverse benthic macrofossils occur in the *Sciponoceras* zone, including stenotopic bivalves, gastropods, and echinoderms, many of which are restricted to this zone (Elder, 1991; Kirkland, 1991). The composition of the molluscan and benthic foraminiferal assemblages indicate mixed biotas of southern and northern affinities during *Sciponoceras* time, but also normal salinity and well-oxygenated conditions at the seafloor. It seems possible that the lower part of the water column in this relatively shallow region of the Colorado Plateau, including the foredeep shelf (LP) and offshore bathymetric high (MV), consisted of mixed northern and southern water masses (Fig. 16A). Based on the rich marine biota, we suggest that there is little compelling evidence for the existence of a thick, well-developed brackish cap along this part of the western clay-dominated lithofacies belt during *Sciponoceras* time.

The benthos at all three localities shows significant and parallel changes through the uppermost Cenomanian *Neocardioceras* zone. The pattern of change from diverse and heterogeneous molluscan and foraminiferal communities of the *Sciponoceras* zone to high dominance, low diversity, ubiquitous communities across the Cenomanian-Turonian boundary suggests that changes occurred in the lower part of the water column related, in part, to benthic oxygenation and trophic resources. This hypothesis is supported by benthic foraminiferal evidence. For example, near the *Sciponoceras-Neocardioceras* zonal boundary the relative abundance of *Gavelinella dakotensis* abruptly increases (Fig. 11). The dominance of a single epifaunal taxon may reflect deterioration of conditions within the sediments -for example, anoxia? - thereby excluding infaunal taxa.

Alternatively, our preferred interpretation is that the "*Gavelinella* acme" records an increase in the delivery of organic matter to the seafloor which stimulated benthic productivity (e.g., Leckie, 1987; Berger and Diester-Haass, 1988; Herguera and Berger, 1991; Corliss and Emerson, 1990; Gooday, 1993; Jorissen et al., 1995; Thomas and Gooday, 1996; West et al., this volume) (Figs. 7, 11). The negative $\delta^{18}O_{carb}$ excursion and kaolinite + illite influx at the base of the *Neocardioceras* zone are attributed to a major influx of freshwater into the seaway. Such an event presumably would also be accompanied by an influx of dissolved nutrients and terrestrial organic matter. We documented a major increase in the proportion of benthic foraminifera to total foraminifera (% planktics) in the *Neocardioceras* zone in the sections at LP and MV (Fig. 5), which could be attributed to such a burst in benthic productivity rather than to increased dissolution of planktics. The increased supply of organic matter to the benthos could have been supplied by an influx of terrestrial material or as a rain of marine organic matter from increased surface water productivity (Fig. 16B). The dramatic increase in the abundances of the detritus-feeding gastropod *Drepanochilus* in the northeastern Arizona and southern Utah sections during *Neocardioceras* time was attributed to an influx of terrestrial organic matter (Elder, 1991).

Benthic productivity in the central axis of the seaway may have been suppressed by very low oxygen conditions near the seafloor and within the sediments, despite a possible concurrent influx of organic matter during the time of the *Neocardioceras* zone. This hypothesis is supported by foraminiferal and molluscan data at the RC section (Eicher and Diner, 1985; Elder, 1985; Elder and Kirkland, 1985; Leckie, 1985). Oxygen-poor intermediate waters from the south may have gradually invaded the western margin as sea level rose.

The incursion or expansion of an oxygen minimum zone may have been most intense during *Neocardioceras* time due to the development of a subsaline surface water mass as evidenced by the detrital clay spike. An influx of riverine waters from a southerly source caused enhanced stratification of the water column (Fig. 16B). The OMZ waters mixed with northern waters along the western margin and may have contributed to diachronous changes observed in the benthic assemblages of foraminifera and mollusks during the time of the *Neocardioceras*

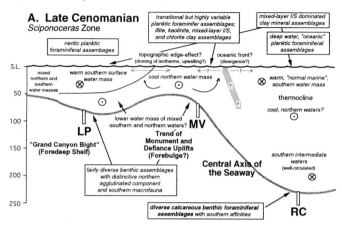

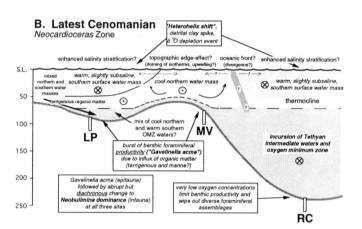

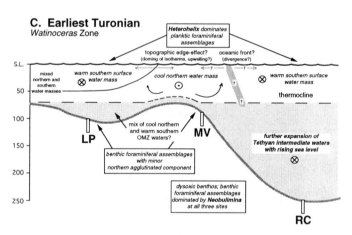

FIG. 16—Model of oceanographic conditions across the southwestern side of the Greenhorn Sea during the time of the Cenomanian-Turonian boundary; (A) late Cenomanian time (*Sciponoceras* zone), (B) latest Cenomanian time (*Neocardioceras* zone), and (C) earliest Turonian time (*Watinoceras* zone). Water depth is estimated in meters. Circles containing an "X" represent southern water masses flowing northward; circles containing a dot represent northern water masses flowing southward. Note rising sea level from panel A to C.

and *Watinoceras* zones (Fig. 16C). Weakening of a widespread OMZ during the early Turonian *Watinoceras* zone is suggested by the slight improvement in benthic conditions across a sea dominated by low diversity, depauperate communities (Eicher and Worstell, 1970; Elder, 1985, 1987, 1990, 1991; Elder and Kirkland, 1985; Eicher and Diner, 1985; Leckie, 1985; Leckie et al., 1991; Kirkland, 1991). A return to more normal marine surface waters is indicated by heavier whole-rock $\delta^{18}O_{carb}$ values during earliest Turonian time (Pratt, 1985; Pratt et al., 1993); one important consequence may have been a weakened salinity stratification of the water column.

Neobulimina, a presumed infaunal, calcareous benthic species, dominates the benthic assemblages (>60%) at all three sites in the *Watinoceras* zone (Figs. 7, 12). The diachronous change from *Gavelinella* dominance (epifaunal) to *Neobulimina* dominance across the Cenomanian-Turonian boundary may indicate subtle changes in the balance between organic matter flux to the seafloor, and dissolved oxygen at the sediment-water interface and in sediment pore waters. Some benthic foraminifera may have evolved symbiotic associations with chemoautotrophic bacteria (West, 1993), while others may be microaerophiles or facultative anaerobes (Bernhard, 1996), which potentially could confer an advantage for survival in dysoxic or anoxic environments. We suggest that *Neobulimina* was tolerant of the low oxygen conditions associated with the warm Tethyan water masses that filled the southern part of the seaway as sea level rose (West et al., this volume). The fact that macrofossil communities continue to be dominated by epifaunal bivalves to the near-exclusion of infaunal taxa in the *Watinoceras* zone suggests that the anoxic-dysoxic boundary in the sediments remained fairly close to the surface.

CONCLUSIONS

1. The late transgressive phase of the Cenomanian-Turonian Greenhorn Sea was characterized by the meeting and mixing of different water masses from northern and southern sources. Warm surface and intermediate waters moved rapidly northwards into the seaway during the late Cenomanian (*Sciponoceras* zone). These southern waters dominated the water column of the deep central axis of the seaway. Cooler, probably slightly less saline waters moved southwards along the western margin. An oceanographic front or mixing zone developed between the northward-flowing and southward-flowing water masses in the deeper, more stable central part of the seaway during the time of the Cenomanian-Turonian boundary. Benthic foraminiferal evidence suggests that warm southern waters overrode the cooler northern waters near Lohali Point in northeastern Arizona.

2. Bathymetric highs on the seafloor may have significantly influenced the circulation and mixing of water masses in the seaway. Seafloor relief on the present-day Colorado Plateau permitted the incursion of warm southern water masses along the foredeep shelf of northeastern Arizona and south-central Utah. Cooler northern waters dominated the broad, north-south bathymetric high (forebulge) to the east due to topographic mixing or surface water divergence and upwelling.

3. Clay mineralogy suggests detrital sources of mud in addition to the Sevier orogenic belt along the western side of the seaway. Significant riverine input from the southwest—possibly

the Mogollon highlands or northern Mexico—or from the southeast—southern U.S. craton—may have occurred during periods of dark marlstone or calcareous shale deposition across the southern Greenhorn Sea (Fig. 13).

4. Changes in planktic and benthic foraminifera communities from the southwestern side of the Greenhorn Sea through the Cenomanian-Turonian boundary interval, particularly through the uppermost Cenomanian *Neocardioceras* zone, reflect a combination of oceanographic variables and continued rising sea level (Fig. 15). Influx of freshwater to the seaway during the latest Cenomanian may have originated from a southern or southeastern source, as indicated by clay mineral data, but was also accompanied by the incursion of oxygen-poor intermediate waters from Tethys.

This latter interpretation is supported by foraminiferal and molluscan data. Influx of terrestrial organic matter and enhanced primary productivity associated with water-mass mixing and the influx of detrital sediments and nutrients at this time resulted in a burst of benthic biomass on the shallower southwestern side of the seaway. A widespread but short-lived abundance maximum of epifaunal *Gavelinella dakotensis* is interpreted as a response to increased organic matter flux to the seafloor. Enhanced salinity stratification in the deeper central part of the seaway may have amplified the affects of an oxygen minimum zone thereby severely limiting benthic productivity there. The influence of the OMZ gradually infiltrated the lower part of the water column on the western side of the seaway based on the diachronous changes in benthic foraminiferal and molluscan assemblages through the *Neocardioceras* zone (Fig. 16). *Neobulimina albertensis*, an infaunal taxon that was probably tolerant of low-oxygen conditions, dominates the benthic foraminiferal assemblages at all three localities in the basal Turonian *Watinoceras* zone when warm oxygen-poor intermediate waters invaded the southern seaway with rising sea level.

ACKNOWLEDGMENTS

RML would like to thank his mentors of Western Interior Cretaceous research, Professors Don Eicher and Erle Kauffman. In addition, RML warmly acknowledges the many individuals who generously shared their ideas and pre-prints over the years, including, Mike Arthur, Liz Balcells-Baldwin, Eric Barron, Tim Bralower, Walt Dean, Richard Diner, Will Elder, Cindy Fisher, Don Hattin, Bill Hay, Jim Kirkland, Lonnie Leithold, "K" Molenaar, Lisa Pratt, Brad Sageman, Bill Sliter, and Chuck Savrda. The paper has greatly benefited from the thoughtful reviews of Walt Dean, Will Elder, Cindy Fisher, Don Hattin, Dale Leckie, and Isabella Premoli-Silva. We would also like to acknowledge the editorial assistance of Andrea Ash and M. Catherine White at SEPM. This research was supported by the National Science Foundation (Earth Sciences). RML also acknowledges the donors to the American Chemical Society-Petroleum Research Fund for partial research support. A special thanks to Jim Kirkland for collecting the Lohali Point samples and assistance in the field at Mesa Verde.

APPENDIX A. Raw data from Lohali Point (LP). Foraminiferal data represent numbers of specimens. Clay mineral data are expressed as percentage of total clays.

Sample	Meters Above Dakota Sandst.	Total Planktics	Total Benthics	Total Foraminifers	Biserial Heterohelix	Triserial Guembelitria	Trochospiral planktics	Planispiral planktics	Keeled planktics	Neobulimina	Gavelinella	Other Calcareous	Agglutinated	% Mixed-Layer Illite-Smectite	% Discrete Illite	% Kaolinite	% Chlorite
LP30	30 +/-	233	49	282	191	6	36	0	0	40	4	2	3	43.9	9.4	46.7	0.0
LP29	29 +/-	221	72	293	159	1	59	2	0	55	7	1	9	44.9	8.8	46.3	0.0
LP28	28 +/-	201	92	293	142	7	42	7	0	78	6	3	5	36.6	11.2	52.2	0.0
LP27	27 +/-	183	97	280	124	3	56	0	0	93	4	0	0	54.1	5.6	40.3	0.0
LP26	26 +/-	177	111	288	153	0	24	0	0	87	24	0	0	57.7	10.4	31.9	0.0
LP25	25 +/-	184	103	287	165	5	14	0	0	73	24	1	5	77.5	7.9	14.6	0.0
LP24	24 +/-	174	113	287	114	16	44	0	0	84	28	0	1	52.7	13.2	34.1	0.0
LP23	23 +/-	155	150	305	81	44	30	0	0	96	47	2	5	57.5	5.9	36.5	0.0
LP22	22 +/-	159	127	286	80	35	44	0	0	109	15	0	3	60.5	6.8	32.7	0.0
LP21	21 +/-	179	123	302	127	37	15	0	0	95	23	3	2	78.5	3.9	17.6	0.0
LP20	20 +/-	191	93	284	141	30	20	0	0	74	14	3	2	69.9	9.4	20.7	0.0
LP19	19 +/-	220	99	319	146	41	33	0	0	64	28	1	6	72.0	6.1	21.9	0.0
LP18	18 +/-	214	101	315	163	31	20	0	0	65	30	3	3	92.2	1.9	5.9	0.0
LP17	17 +/-	173	136	309	127	22	24	0	0	87	37	11	1	97.0	2.0	1.0	0.0
LP16	16 +/-	122	203	325	79	27	16	0	0	145	39	3	16	98.3	1.7	0.0	0.0
LP15	15 +/-	101	200	301	67	17	17	0	0	172	17	4	7	97.0	3.0	0.0	0.0
LP14	14 +/-	107	197	304	68	29	10	0	0	34	155	2	6	97.8	2.2	0.0	0.0
LP13	13 +/-	67	114	181	36	25	6	0	0	35	22	1	56	89.4	2.9	7.7	0.0
LP12	12 +/-	215	120	335	120	82	13	0	0	10	82	9	19	84.2	5.8	10.0	0.0
LP11	11 +/-	292	43	335	34	257	1	0	0	11	24	3	5	80.0	8.0	12.0	0.0
LP10	10 +/-	267	58	325	46	208	13	0	0	38	16	1	3	93.5	2.2	4.3	0.0
LP9	9 +/-	116	190	306	25	86	5	0	0	114	31	3	42	90.3	2.9	6.8	0.0
LP8	8 +/-	310	144	454	33	271	6	0	0	95	2	7	40	98.4	1.1	0.5	0.0
LP7	7 +/-	285	18	303	17	230	38	0	0	2	5	9	11	96.3	2.4	1.3	0.0
LP6	6 +/-	213	74	287	50	136	27	0	0	47	2	14	11	87.8	3.5	8.7	0.0
LP5	5 +/-	165	117	282	55	0	98	12	0	97	1	0	19	77.9	5.2	16.9	0.0
LP4	4 +/-	90	152	242	41	32	0	17	0	86	5	20	41	65.8	7.9	26.3	0.0
LP3	3 +/-	296	43	339	142	67	84	3	0	6	8	13	16	50.0	6.4	43.6	0.0
LP2	2 +/-	6	131	137	3	3	0	0	0	21	4	10	96	58.0	10.5	31.5	0.0
LP1	1 +/-	55	181	236	7	17	31	0	0	15	23	0	143	24.8	16.8	58.4	0.0

APPENDIX B. Raw data from Mesa Verde (MV). Foraminiferal data represent numbers of specimens. Clay mineral data are expressed as percentage of total clays.

Sample	Meters Above Dakota Sandst.	Total Planktics	Total Benthics	Total Foraminifers	Biserial Heterohelix	Triserial Guembelitria	Trochospiral planktics	Planispiral planktics	Keeled planktics	Neobulimina	Gavelinella	Other Calcareous	Agglutinated	% Mixed-Layer Illite-Smectite	% Discrete Illite	% Kaolinite	% Chlorite
MV29.8	29.8	227	56	285	90	0	137	0	0	55	3	0	0	9.1	29.1	53.0	8.8
MV29.2	29.2	201	15	216	133	0	68	0	0	15	0	0	0	12.5	40.6	38.0	8.9
MV28.4	28.4	259	44	303	119	1	139	0	0	36	8	0	0	35.6	33.9	24.1	6.4
MV28.1	28.1	313	46	359	263	0	50	0	0	45	1	0	0	28.6	34.3	31.6	5.5
MV27.6	27.6	274	17	291	53	0	221	0	0	16	1	0	0	30.0	33.3	32.0	4.7
MV27.1	27.1	197	48	245	90	3	104	0	0	35	12	0	1	21.4	42.9	28.9	6.8
MV26.7	26.7	242	12	254	48	3	191	0	0	11	1	0	1	35.2	24.1	36.8	3.9
MV26.2	26.2	130	223	353	123	0	7	0	0	213	9	1	0	30.4	37.0	27.5	5.1
MV25.8	25.8	208	153	361	142	2	64	0	0	141	3	7	2	40.0	40.0	12.8	7.2
MV25.2	25.2	187	157	344	78	7	101	1	0	142	15	0	0	25.0	33.3	34.4	7.3
MV24.7	24.7	157	115	272	84	30	43	0	0	95	20	0	0	60.0	20.0	15.6	4.4
MV23.6	23.6	88	206	294	10	32	44	2	0	141	50	12	3	15.8	50.0	29.8	4.4
MV23.2	23.2	13	276	289	5	5	3	0	0	23	218	16	19	38.1	38.1	20.4	3.4
MV22.7	22.7	239	111	350	102	115	22	0	0	32	75	1	3	32.5	42.5	17.4	7.6
MV22.4	22.4	88	244	332	25	17	43	3	0	106	104	18	16	15.6	40.6	32.0	11.8
MV21.5	21.5	106	237	343	24	58	21	3	0	148	37	1	51	17.5	42.5	27.9	12.1
MV20.8	20.8	206	89	295	25	156	19	6	0	78	1	0	10				
MV20.5	20.5	138	49	187	19	66	52	1	0	35	6	2	6	11.5	45.3	27.2	16.2
MV19.5	19.5	38	47	85	16	4	18	1	0	14	3	1	29	15.4	46.2	28.0	10.5
MV18.5	18.5	167	34	201	30	45	91	1	0	21	4	8	1	4.4	45.1	40.6	9.9
MV17.5	17.5	194	105	299	67	45	80	2	0	102	2	0	1	8.3	36.7	38.3	16.7
MV16.5	16.5	260	82	342	42	4	203	9	2	48	8	26	0	16.7	26.7	41.0	15.7
MV15.5	15.5	239	67	306	85	17	132	5	0	26	5	18	18	8.3	30.6	61.6	0.0
MV14.5	14.5	67	131	198	13	14	40	0	0	55	19	16	41	9.7	32.3	58.0	0.0
MV13.5	13.5	204	69	273	53	93	55	3	0	44	19	0	6	6.6	30.3	63.1	0.0
MV12.5	12.5	9	46	55	4	4	5	0	0	31	7	1	7	7.7	38.3	44.9	8.9

APPENDIX C. Raw data from Rock Canyon (RC). Foraminiferal data represent numbers of specimens. Clay mineral data are expressed as percentage of total clays. Additional benthic specimens were picked for samples BC1-BC16 (Leckie, 1985).

Sample	Meters Above base Bdg Crk	Total Planktics	Total Benthics	Total Foraminifers	Biserial Heterohelix	Triserial Guembelitria	Trochospiral planktics	Planispiral planktics	Keeled planktics	Neobulimina	Gavelinella	Other Calcareous	Agglutinated	Mixed-Layer Illite-Smectite	Discrete Illite	Kaolinite	Chlorite
BC22	6.65	333	1	334	167	1	165	0	0	1	0	0	0	54.8	25.8	19.4	0.0
BC21	6.30	207	0	207	98	0	109	0	0	0	0	0	0	68.6	17.1	14.3	0.0
BC20	5.85	306	0	306	128	0	178	0	0	0	0	0	0	89.3	7.1	3.6	0.0
BC19	5.30	288	45	333	170	10	108	0	0	37	7	1	0	87.2	7.7	5.1	0.0
BC18	4.95	284	50	334	128	0	153	3	0	39	10	1	0	77.1	14.3	8.6	0.0
BC17	4.85	248	53	301	100	0	148	0	0	36	17	0	0	69.0	14.3	16.7	0.0
BC16	4.45	449	172	621	288	2	159	0	0	261	71	12	0	88.5	7.7	3.8	0.0
BC15	4.35	381	68	449	164	2	215	0	0	279	42	6	0	93.8	2.8	3.4	0.0
BC14	4.20	411	94	505	225	4	182	0	0	191	102	11	0	91.5	4.2	4.3	0.0
BC13	4.10	345	74	419	217	7	120	1	0	239	34	68	0	87.5	8.3	4.2	0.0
BC12	3.65	502	13	515	291	0	210	1	0	363	12	8	0	91.5	3.4	5.1	0.0
BC11	3.45	512	48	560	284	5	222	1	0	275	33	2	0	96.9	2.1	1.0	0.0
BC10	3.20	485	57	542	249	6	230	0	0	257	63	28	0	92.5	4.5	3.0	0.0
BC9	2.70	335	22	357	122	0	211	2	0	52	195	17	0	85.4	7.3	7.3	0.0
BC8	2.55	522	53	575	404	1	106	11	0	140	154	4	0	38.9	27.8	33.3	0.0
BC7	2.10	299	27	326	61	3	226	9	0	114	12	186	0	96.5	1.7	1.8	0.0
BC6	1.75	429	8	437	45	1	374	9	0	156	38	22	0	80.3	4.2	15.5	0.0
BC5	1.45	593	17	610	73	13	456	51	0	15	7	279	0	98.3	0.0	1.7	0.0
BC4	1.10	323	95	418	81	28	180	33	1	20	0	297	0	98.7	0.0	1.3	0.0
BC3	0.95	421	166	587	70	31	262	58	0	7	5	296	2	98.3	0.0	1.7	0.0
BC2	0.65	477	22	499	1	2	428	46	0	6	1	335	0	92.1	3.9	4.0	0.0
BC1	0.50	663	72	735	80	24	519	36	4	118	5	211	0	87.5	6.2	6.3	0.0
HL1	-0.25	352	1	353	44	0	282	11	15	0	0	1	0	66.7	16.6	16.7	0.0

REFERENCES

ARMSTRONG, R. L., 1968, Sevier orogenic belt in Nevada and Utah: Geological Society of America Bulletin, v. 79, p. 429-458.

ARMSTRONG, R. L., AND WARD, P. L., 1993, Late Triassic to earliest Eocene magmatism in the North American cordillera: Implications for the western interior basin, in Caldwell, W. G. E., and Kauffman, E. G., eds., Evolution of the Western Interior Basin: St. John's, Geological Association of Canada, Special Paper 39, p. 49-72.

ARTHUR, M. A., BRUMSACK, H. -J., JENKYNS, H. C., AND SCHLANGER, S. O., 1990, Stratigraphy, geochemistry, and paleoceanography of organic carbon-rich Cretaceous sequences, in Ginsburg, R. N., and Beaudoin, B., eds., Cretaceous Resources, Events, and Rhythms: Netherlands, Kluwer Academic Publishers, p. 75-119.

ARTHUR, M. A., DEAN, W. E., AND PRATT, L. M., 1988, Geochemical and climatic effects of increased marine organic carbon burial at the Cenomanian/Turonian boundary: Nature, v. 335, p. 714-717.

ARTHUR, M. A., DEAN, W. E., AND SCHLANGER, S. O., 1985, Variations in the global carbon cycle during the Cretaceous related to climate, volcanism, and changes in atmospheric CO_2, in Sundquist, E. T., and Broecker, W. S., eds., The Carbon Cycle and Atmospheric CO_2: Natural Variations Archean to Present: Washington, D.C., American Geophysical Union, Monograph 32, p. 504-529.

ARTHUR, M. A., AND SAGEMAN, B. B., 1994, Marine black shales: Depositional mechanisms and environments of ancient deposits: Annual Reviews of Earth and Planetary Sciences, v. 22, p. 499-551.

ARTHUR, M. A., SCHLANGER, S. O., AND JENKYNS, H. C., 1987, The Cenomanian-Turonian oceanic anoxic event, II. Paleoceanographic controls on organic-matter production and preservation, in Brooks, J., and Fleet, A. J., eds., Marine Petroleum Source Rocks: London, Geological Society, Special Publication no. 26, p. 401-420.

ARTHUR, M. A., SLINGERLAND, R., AND KUMP, L. R., 1996, A new hypothesis for the origin of limestone/marlstone couplets in deposits of the Cretaceous Western Interior Sea of North America (abs.): Geological Society of America Abstracts with Programs, v. 28, p. A-65.

BANDY, O. L., 1956, Ecology of foraminifera in the northeastern Gulf of Mexico: Washington, D.C., U.S. Geological Survey Professional Paper 274G, p. 179-204.

BARRON, E. J., ARTHUR, M. A., AND KAUFFMAN, E. G., 1985, Cretaceous rhythmic bedding sequences: A plausible link between orbital variations and climate: Earth and Planetary Science Letters, v. 72, p. 327-340.

BE, A. W. H., 1977, An ecological, zoogeographic, and taxonomic review of Recent planktonic foraminifera, in Ramsay, A. T. S., ed., Oceanic Micropaleontology, Vol. 1: New York, Academic Press, p. 1-100.

BEAUMONT, C., 1981, Foreland basins: Geophysical Journal of the Royal Astronomical Society, v. 65, p. 291-329.

BERGER, W. H., AND DIESTER-HAAS, L., 1988, Paleoproductivity: The benthic/planktonic ratio in foraminifera as a productivity index: Marine Geology, v. 81, p. 15-25.

BERGER, W. H., AND VON RAD, U., 1972, Cretaceous and Cenozoic sediments from the Atlantic Ocean, in Hayes, D. E., Pimm, A. C., et al., Initial Reports of the Deep Sea Drilling Project, Vol. 14: Washington, D.C., U.S. Government Printing Office, p. 787-954.

BERNHARD, J. M., 1986, Characteristic assemblages and morphologies of benthic foraminifera from anoxic, organic-rich deposits: Jurassic through Holocene: Journal of Foraminiferal Research, v. 19, p. 207-215.

BERNHARD, J. M., 1996, Microaerophilic and facultative anaerobic benthic foraminifera: A review of experimental and ultrastructural evidence: Revue de Paleobiologie, v. 15, p. 261-275.

BLOCH, J., SCHROEDER-ADAMS, C. J., LECKIE, D. A., MCINTYRE, D. J., CRAIG, J., AND STANILAND, M., 1993, Revised stratigraphy of the lower Colorado Group (Albian to Turonian), western Canada: Bulletin of Canadian Petroleum Geology, v. 41, p. 325-348.

BOERSMA, A., AND PREMOLI-SILVA, I., 1989, Atlantic Paleogene biserial heterohelicid foraminifera and oxygen minima: Paleoceanography, v. 4, p. 271-286.

BOTTJER, D. J., ARTHUR, M. A., DEAN, W. E., HATTIN, D. E., AND SAVRDA, C. E., 1986, Rhythmic bedding produced in Cretaceous pelagic carbonate environments: Sensitive recorders of climatic cycles: Paleoceanography, v. 1, p. 467-481.

BRALOWER, T. J., 1988, Calcareous nannofossil biostratigraphy and assemblages of the Cenomanian-Turonian boundary interval: Implications for the origin and timing of oceanic anoxia: Paleoceanography, v. 3, p. 275-316.

BRINDLEY, G. W., AND BROWN, G., 1980, Crystal Structures of Clay Minerals and Their X-ray Identification: London, Mineralogical Society, 495 p.

CALDWELL, W. G. E., DINER, R., EICHER, E. L., FOWLER, S. P., NORTH, B. R., STELCK, C. R., AND VON HOLDT WILHELM, L., 1993, Foraminiferal biostratigraphy of Cretaceous marine cyclothems, in Caldwell, W. G. E., and Kauffman, E. G., eds., Evolution of the Western Interior Basin: St. John's, Geological Association of Canada Special Paper 39, p. 477-520.

CALDWELL, W. G. E., NORTH, B. R., STELCK, C. R., AND WALL, J. H., 1978, A foraminiferal zonal scheme for the Cretaceous System in the interior plains of Canada, in Stelck, C. R., and Chatterton, B. D. E., eds., Western and Arctic Canadian Biostratigraphy: St. John's, Geological Association of Canada Special Paper 18, p. 495-575.

CARON, M., AND HOMEWOOD, P., 1982, Evolution of early planktonic foraminifers: Marine Micropaleontology, v. 7, p. 453-462.

CHAMLEY, H., 1989, Clay Sedimentology: New York, Springer-Verlag, 623 p.

CHAMNEY, T. P., 1976, Foraminiferal morphogroup symbol for paleoenvironmental interpretation of drill cutting samples: Arctic America, in Schafer, C. T., and Pelletier, B. R., eds., First International Symposium on Benthonic Foraminifera of Continental Margins: Halifax, Maritime Sediments Special Publication No. 1, p. 585-624.

CHRISTIE-BLICK, N., AND DRISCOLL, N. W., 1995, Sequence stratigraphy: Annual Reviews of Earth and Planetary Science, v. 23, p. 451-478.

CIFELLI, R., AND BENIER, C. S., 1976, Planktonic foraminifera from near the west African coast and a consideration of faunal parceling in the North Atlantic: Journal of Foraminiferal Research, v. 6, p. 258-273.

COBBAN, W. A., AND HOOK, S. C., 1984, Mid-Cretaceous molluscan biostratigraphy and paleogeography of southwestern part of western interior, United States, in Westermann, G. E. G., ed., Jurassic-Cretaceous Biochronology and Paleogeography of North America: St. John's, Geological Association of Canada Special Paper 27, p. 257-271.

COBBAN, W. A., AND SCOTT, G. R., 1972, Stratigraphy and Ammonite Fauna of the Graneros Shale and Greenhorn Limestone Near Pueblo, Colorado: Washington, D.C., U.S. Geological Survey Professional Paper 645, 108 p.

COBBAN, W. A., AND REESIDE, J. B., JR., 1952, Correlation of the Cretaceous formations of the western interior of the United States: American Association of Petroleum Geologists Bulletin, v. 63, p. 1011-1044.

CORFIELD, R. M., HALL, M. A., AND BRASIER, M. D., 1990, Stable isotope evidence for foraminiferal habitats during the development of the Cenomanian/Turonian oceanic anoxic event: Geology, v. 18, p. 175-178.

CORLISS, B. H., 1985, Microhabitats of benthic foraminifera within deep sea sediments: Nature, v. 314, p. 435-438.

CORLISS, B. H., AND CHEN, C., 1988, Morphotype patterns of Norwegian Sea deep-sea benthic foraminifera and ecological implications: Geology, v. 16, p. 716-719.

CORLISS, B. H., AND EMERSON, S., 1990, Distribution of Rose Bengal stained deep-sea benthic foraminifera from the Nova Scotian continental margin and Gulf of Maine: Deep Sea Research, v. 97, p. 381-400.

CORLISS, B. H., AND FOIS, E., 1991, Morphotype analysis of deep-sea foraminifera from the northwest Gulf of Mexico: Palaios, v. 6, p. 589-605.

CROSS, T. A., AND PILGER, R. H., JR., 1978, Tectonic controls of Late Cretaceous sedimentation, western interior, USA: Nature, v. 274, p. 653-657.

CULVER, S. J., 1988, New foraminiferal depth zonation of the northwestern Gulf of Mexico: Palaios, v. 3, p. 69-85.

CUSHMAN, J. A., 1946, Upper Cretaceous foraminifera of the Gulf Coastal region of the United States and adjacent areas: Washington, D.C., U.S. Geological Survey Professional Paper 206, 241 p.

DOUGLAS, R. G., AND SAVIN, S. M., 1978, Oxygen isotopic evidence for the depth stratification of Tertiary and Cretaceous planktonic foraminifera: Marine Micropaleontology, v. 3, p. 175-196.

DOUGLAS, R. G., AND WOODRUFF, F., 1981, Deep sea benthic foraminifera, in Emiliani, C., ed., The Oceanic Lithosphere, The Sea, Volume 7: New York, Wiley-Interscience, p. 1233-1327.

EATON, J. G., KIRKLAND, J. I., AND KAUFFMAN, E. G., 1990, Evidence and dating of mid-Cretaceous tectonic activity in the San Rafael Swell, Emery County, Utah: The Mountain Geologist, v. 27, p. 39-45.

EATON, J. G., AND NATIONS, J. D., 1991, Introduction; Tectonic setting along the margin of the Cretaceous Western Interior Seaway, southwestern Utah and northern Arizona, in Nations, J. D., and Eaton, J. G., eds., Stratigraphy, Depositional Environments, and Sedimentary Tectonics of the Western Margin, Cretaceous Western Interior Seaway: Boulder, Geological Society of America Special Paper 260, p. 1-8.

EICHER, D. L., 1965, Foraminifera and biostratigraphy of the Graneros Shale: Journal of Paleontology, v. 39, p. 875-909.

EICHER, D. L., 1966, Foraminifera from the Cretaceous Carlile Shale of Colorado: Cushman Foundation for Foraminiferal Research Contributions, v. 17, p. 16-31.

EICHER, D. L., 1967, Foraminifera from Belle Fourche Shale and equivalents, Wyoming and Montana: Journal of Paleontology, v. 41, p. 167-188.

EICHER, D. L., 1969a, Cenomanian and Turonian planktonic foraminifera from the western interior of the United States, in Bronnimann, P., and Renz, H. H., eds., Proceedings of the first international conference on planktonic microfossils: Leiden, E. J. Brill, v. 2, p. 163-174.

EICHER, D. L., 1969b, Paleobathymetry of the Cretaceous Greenhorn Sea in eastern

Colorado: American Association of Petroleum Geologists Bulletin, v. 53, p. 1075-1090.

EICHER, D. L., AND DINER, R., 1985, Foraminifera as indicators of water mass in the Cretaceous Greenhorn Sea, western interior, *in* Pratt, L. M., Kauffman, E. G., and Zelt, F. B., eds., Fine-grained Deposits and Biofacies of the Cretaceous Western Interior Seaway: Evidence of Cyclic Sedimentary Processes, Field Trip Guidebook no. 4: Tulsa, Society of Economic Paleontologists and Mineralogists, p. 60-71.

EICHER, D. L., AND DINER, R., 1989, Origin of the Cretaceous Bridge Creek cycles in the western interior, United States: Palaeogeography, Palaeoclimatology, Palaeoecology, v. 74, p. 127-146.

EICHER, D. L., AND WORSTELL, P., 1970, Cenomanian and Turonian foraminifera from the Great Plains, United States: Micropaleontology, v. 16, p. 269-324.

ELDER, W. P., 1985, Biotic patterns across the Cenomanian-Turonian boundary near Pueblo, Colorado, *in* Pratt, L. M., Kauffman, E. G., and Zelt, F. B., eds., Fine-grained Deposits and Biofacies of the Cretaceous Western Interior Seaway: Evidence of Cyclic Sedimentary Processes, Field Trip Guidebook No. 4: Tulsa, Society of Economic Paleontologists and Mineralogists, p. 157-169.

ELDER, W. P., 1987, The paleoecology of the Cenomanian-Turonian (Cretaceous) stage boundary extinctions at Black Mesa, Arizona: Palaios, v. 2, p. 24-40.

ELDER, W. P., 1990, Soft-bottom paleocommunity dynamics in the Cenomanian-Turonian boundary interval of the western interior, United States, *in* Miller, W., III, ed., Paleocommunity Temporal Dynamics: The Long-Term Development of Multispecies Assemblages: Ithica, Paleontological Society Special Paper No. 5, p. 210-235.

ELDER, W. P., 1991, Molluscan paleoecology and sedimentation patterns of the Cenomanian-Turonian extinction interval in the southern Colorado Plateau region, *in* Nations, J. D., and Eaton, J. G., eds., Stratigraphy, Depositional Environments, and Sedimentary Tectonics of the Western Margin, Cretaceous Western Interior Seaway: Boulder, Geological Society of America Special Paper 260, p. 113-137.

ELDER, W. P., GUSTASON, E. R., AND SAGEMAN, B. B., 1994, Correlation of basinal carbonate cycles to nearshore parasequences in the Late Cretaceous Greenhorn seaway, western interior U.S.A.: Geological Society of America Bulletin, v. 106, p. 892-902.

ELDER, W.P., AND KIRKLAND, J.I., 1985, Stratigraphy and depositional environments of the Bridge Creek Limestone Member of the Greenhorn Formation at Rock Canyon Anticline near Pueblo, Colorado, *in* Pratt, L. M., Kauffman, E. G., and Zelt, F. B., eds., Fine-grained Deposits and Biofacies of the Cretaceous Western Interior Seaway: Evidence of Cyclic Sedimentary Processes, Field Trip Guidebook No. 4: Tulsa, Society of Economic Paleontologists and Mineralogists, p. 122-134.

ELDER, W. P., AND KIRKLAND, J. I., 1993a, Cretaceous paleogeography of the Colorado Plateau and adjacent areas, *in* Morales, M., ed., Aspects of Mesozoic Geology and Paleontology of the Colorado Plateau: Flagstaff, Museum of Northern Arizona Bulletin 59, p. 129-152.

ELDER, W. P., AND KIRKLAND, J. I., 1993b, Cretaceous paleogeography of the southern western interior region, *in* Caputo, M. V., Peterson, J. A., and Franczyk, K. J., eds., Mesozoic Systems of the Rocky Mountain Region, USA: Denver, Society of Economic Mineralogists and Paleontologists, Rocky Mountain Section, p. 415-440.

ERICKSEN, M. C., AND SLINGERLAND, R., 1990, Numerical simulations of tidal and wind-driven circulation in the Cretaceous interior seaway of North America: Geological Society of America Bulletin, v. 102, p. 1499-1516.

FINKELSTEIN, D. B., 1991, The Clay Mineralogy of the Upper Cretaceous Greenhorn Cyclothem of the Mancos Shale: unpublished M.S. thesis, University of Massachusetts, Amherst, 117 p.

FISCHER, A. G., 1980, Gilbert-bedding rhythms and geochronology, *in* Yochelson, E. I., ed., The Scientific Ideas of G.K. Gilbert: Boulder, Geological Society of America Special Paper 183, p. 93-104.

FISCHER, A. G., HERBERT, T., AND PREMOLI-SILVA, I., 1985, Carbonate bedding cycles in Cretaceous pelagic and hemipelagic sequences, *in* Pratt, L. M., Kauffman, E. G., and Zelt, F. B., eds., Fine-grained Deposits and Biofacies of the Cretaceous Western Interior Seaway: Evidence of Cyclic Sedimentary Processes, Field Trip Guidebook No. 4: Tulsa, Society of Economic Paleontologists and Mineralogists, p. 1-10.

FISHER, C. G., HAY, W. W., AND EICHER, D. L., 1994, Oceanic front in the Greenhorn Sea (late middle through late Cenomanian): Paleoceanography, v. 9, p. 879-892.

FOX, S. K., JR., 1954, Cretaceous Foraminifera from the Greenhorn, Carlile, and Cody Formations, South Dakota, Wyoming: Washington, D.C., U.S. Geological Survey Professional Paper, No. 254-E, p. 97-124.

FRANCZYK, K. J., FOUCH, T. D., JOHNSON, R. C., MOLENAAR, C. M., AND COBBAN, W. A., 1992, Cretaceous and Tertiary paleogeographic reconstructions for the Uinta-Piceance basin study area, Colorado and Utah: Washington, D.C., U.S. Geological Survey Bulletin 1787-Q, 37 p.

FRIZZELL, D. L., 1954, Handbook of Cretaceous Foraminifera of Texas: Austin, University of Texas, Bureau of Economic Geology, Report of Investigations No. 22, 232 p.

FRUSH, M. P., AND EICHER, D. L., 1975, Cenomanian and Turonian foraminifera and paleoenvironments in the Big Bend region of Texas and Mexico, *in* Caldwell, W. G. E., ed., The Cretaceous System in the Western Interior of North America: St. John's, Geological Association of Canada Special Paper 13, p. 277-301.

GARDNER, M. H., AND CROSS, T. A., 1994, Middle Cretaceous paleogeography of Utah, *in* Caputo, M. V., Peterson, J. A., and Franczyk, K. J., eds., Mesozoic Systems of the Rocky Mountain Region, USA: Tulsa, Society of Economic Mineralogists and Paleontologists, Rocky Mountain Section, p. 471-502.

GIBBS, R. J., 1971, X-ray diffraction mounts, *in* Carver, R. J., ed., Procedures in Sedimentary Petrology: New York, Wiley Interscience, p. 531-539.

GIBBS, R. J., 1977, Clay mineral segregation in the marine environment: Journal of Sedimentary Petrology, v. 47, p. 237-243.

GIBSON, T. G., 1989, Planktonic benthonic foraminiferal ratios: Modern patterns and Tertiary applicability: Marine Micropaleontology, v. 15, p. 29-52.

GILBERT, G. K., 1895, Sedimentary measurement of geologic time: Journal of Geology, v. 3, p. 121-127.

GLANCY, T. J., JR., ARTHUR, M. A., BARRON, E. J., AND KAUFFMAN, E. G., 1993, A paleoclimate model for the North American Cretaceous (Cenomanian-Turonian) epicontinental sea, *in* Caldwell, W.G.E., and Kauffman, E.G., eds., Evolution of the Western Interior Basin: St. John's, Geological Association of Canada Special Paper 39, p. 219-242.

GLANCY, T. J., JR., BARRON, E. J., AND ARTHUR, M. A., 1986, An initial study of the sensitivity of modeled Cretaceous climate to cyclical insolation forcing: Paleoceanography, v. 1, p. 523-537.

GOODAY, A. J., 1993, Deep-sea benthic foraminiferal species which exploit phytodetritus: Characteristic features and controls on distribution: Marine Micropaleontology, v. 22, p. 187-206.

GRIFFIN, G. M., 1971, Interpretation of x-ray diffraction data, *in* Carver, R. J., ed., Procedures in Sedimentary Petrology: New York, Wiley Interscience, p. 541-569.

HALLOCK, P., PREMOLI-SILVA, I., AND BOERSMA, A., 1991, Similarities between planktonic and larger foraminiferal evolutionary trends through Paleogene paleoceanographic changes: Palaeogeography, Palaeoclimatology, Palaeoecology, v. 83, p. 49-64.

HANCOCK, J. M., AND KAUFFMAN, E. G., 1979, The great transgressions of the Late Cretaceous: Journal of the Geological Society, v. 136, p. 175-186.

HAQ, B. U., HARDENBOL, J., AND VAIL, P. R., 1988, Mesozoic and Cenozoic Chronostratigraphy and Cycles of Sea-Level Change: Tulsa, Society of Economic Paleontologists and Mineralogists Special Publication 42, p. 71-108.

HARDY, R., AND TUCKER, M., 1988, X-ray diffraction of sediments, *in* Tucker, M., ed., Techniques in Sedimentology: Boston, Blackwell Scientific Publications, p. 191-228.

HART, M. B., 1980, A water depth model for the evolution of the planktonic Foraminiferida: Nature, v. 286, p. 252-254.

HATTIN, D. E., 1971, Widespread, synchronously deposited, burrow-mottled limestone beds in Greenhorn Limestone (Upper Cretaceous) of Kansas and southeastern Colorado: American Association of Petroleum Geologists Bulletin, v. 55, p. 412-431.

HATTIN, D. E., 1985, Distribution and significance of widespread, time-parallel pelagic limestone beds in Greenhorn Limestone (Upper Cretaceous) of the central Great Plains and southern Rocky Mountains, *in* Pratt, L. M., Kauffman, E. G., and Zelt, F. B., eds., Fine-grained Deposits and Biofacies of the Cretaceous Western Interior Seaway: Evidence of Cyclic Sedimentary Processes, Field Trip Guidebook No. 4: Tulsa, Society of Economic Paleontologists and Mineralogists, p. 28-37.

HATTIN, D. E., 1986a, Carbonate substrates of the Late Cretaceous sea, central Great Plains and southern Rocky Mountains: Palaios, v. 1, p. 347-367.

HATTIN, D. E., 1986b, Interregional model for deposition of Upper Cretaceous pelagic rhythmites, U.S. western interior: Paleoceanography, v. 1, p. 483-494.

HATTIN, D. E., 1987, Pelagic/hemipelagic rhythmites of the Greenhorn Limestone (Upper Cretaceous) of northeastern New Mexico and southeastern Colorado: Albuquerque, New Mexico Geological Society Guidebook, 38th Field Conference, p. 237-247.

HAY, W. W., EICHER, D. L., AND DINER, R., 1993, Physical oceanography and water masses in the Cretaceous Western Interior Seaway, *in* Caldwell, W. G. E., and Kauffman, E. G., eds., Evolution of the Western Interior Basin: St. John's, Geological Association of Canada Special Paper 39, p. 297-318.

HAYDEN SCOTT, C.C., 1992, Clay Mineralogy of the Upper Cretaceous Mancos Shale near Mesa Verde National Park, Southwestern Colorado: Clues to the Paleoceanography of the Western Interior Seaway: unpublished M.S. thesis, University of Massachusetts, Amherst, 225 p.

HAZENBUSH, G. C., 1973, Stratigraphy and depositional environments of the Mancos Shale (Cretaceous), Black Mesa, Arizona, *in* Fassett, J. E., ed., Cretaceous and

Tertiary Rocks of the Southern Colorado Plateau: Durango, Four Corners Geological Society Memoir, p. 57-71.

HELLER, P. L., BEEKMAN, F., ANGEVINE, C. L., AND CLOETINGH, S. A. P. L., 1993, Cause of tectonic reactivation and subtle uplifts in the Rocky Mountain region and its effect on the stratigraphic record: Geology, v. 21, p. 1003-1006.

HERGUERA, J. C., AND BERGER, W. H., 1991, Paleoproductivity from benthic foraminifera abundance: Glacial to postglacial change in the west-equatorial Pacific: Geology, v. 19, p. 1173-1176.

HILBRECHT, H., AND HOEFS, J., 1986, Geochemical and paleontological studies of the δ^{13}C anomaly in boreal and north Tethyan Cenomanian-Turonian sediments in Germany and adjacent areas: Palaeogeography, Palaeoclimatology, Palaeoecology, v. 53, p. 169-189.

HOWER, J., ESLINGER, E., HOWER, M., AND PERRY, E., 1976, Mechanism of burial metamorphism of argillaceous sediment: 1) mineralogic and chemical evidence: Geological Society of America Bulletin, v. 87, p. 725-737.

JARVIS, I., CARSON, G. A., COOPER, M. K. E., HART, M. B., LEARY, P. N., TOCHER, B. A., HORNE, D., AND ROSENFELD, A., 1988, Microfossil assemblages and the Cenomanian-Turonian (Late Cretaceous) oceanic anoxic event: Cretaceous Research, v. 9, p. 3-103.

JEWELL, P. W., 1993, Water-column stability, residence times, and anoxia in the Cretaceous North American seaway: Geology, v. 21, p. 579-582.

JONES, D. J., ed., 1953, Microfossils of the Upper Cretaceous of northeastern Utah and southwestern Wyoming: Salt Lake, Utah Geological and Mineralogical Survey Bulletin 47 (Contributions to Micropaleontology No. 1), 158 p.

JORDAN, T. E., 1981, Thrust loads and foreland basin evolution, Cretaceous, western United States: American Association of Petroleum Geologists Bulletin, v. 65, p. 2506-2520.

JORISSEN, F. J., DE STIGTER, H. C., AND WIDMARK, J. G. V., 1995, A conceptual model explaining benthic foraminiferal microhabitats: Marine Micropaleontology, v. 26, p. 3-15.

KAIHO, K., AND HASEGAWA, T., 1994, End-Cenomanian benthic foraminiferal extinctions and oceanic dysoxic events in the northwestern Pacific Ocean: Palaeogeography, Palaeoclimatology, Palaeoecology, v. 111, p. 29-43.

KAUFFMAN, E. G., 1975, Dispersal and biostratigraphic potential of Cretaceous benthonic Bivalvia in the western interior, in Caldwell, W. G. E., ed., The Cretaceous System in the Western Interior of North America: St. John's, Geological Association of Canada Special Paper 13, p. 163-194.

KAUFFMAN, E. G., 1977, Geological and biological overview: western interior Cretaceous basin: The Mountain Geologist, v. 14, p. 75-100.

KAUFFMAN, E. G., 1984, Paleobiogeography and evolutionary response dynamic in the Cretaceous Western Interior Seaway of North America, in Westermann, G. E. G., ed., Jurassic-Cretaceous Biochronology and Paleogeography of North America: St. John's, Geological Association of Canada Special Paper 27, p. 273-306.

KAUFFMAN, E.G., 1985, Cretaceous evolution of the western interior basin of the United States, in Pratt, L. M., Kauffman, E. G., and Zelt, F. B., eds., Fine-grained Deposits and Biofacies of the Cretaceous Western Interior Seaway: Evidence of Cyclic Sedimentary Processes, Field Trip Guidebook No. 4: Tulsa, Society of Economic Paleontologists and Mineralogists, p. iv-xiii.

KAUFFMAN, E. G., AND CALDWELL, W. G. E., 1993, The western interior basin in space and time, in Caldwell, W. G. E., and Kauffman, E. G., eds., Evolution of the Western Interior Basin: St. John's, Geological Association of Canada Special Paper 39, p. 1-30.

KAUFFMAN, E. G., PRATT, L. M., AND OTHERS, 1985, A field guide to the stratigraphy, geochemistry, and depositional environments of the Kiowa-Skull Creek, Greenhorn, and Niobrara marine cycles in the Pueblo-Canon City area, Colorado, in Pratt, L. M., Kauffman, E. G., and Zelt, F. B., eds., Fine-grained Deposits and Biofacies of the Cretaceous Western Interior Seaway: Evidence of Cyclic Sedimentary Processes, Field Trip Guidebook No. 4: Tulsa, Society of Economic Paleontologists and Mineralogists, p. FRS1-26.

KIRKLAND, J. I., 1991, Lithostratigraphic and biostratigraphic framework for the Mancos Shale (late Cenomanian to middle Turonian) at Black Mesa, northeastern Arizona, in Nations, J. D., and Eaton, J. G., eds., Stratigraphy, Depositional Environments, and Sedimentary Tectonics of the Western Margin, Cretaceous Western Interior Seaway: Boulder, Geological Society of America Special Paper 260, p. 85-111.

KIRKLAND, J. I., LECKIE, R. M., AND ELDER, W. P., 1995, A new principal reference section for the Mancos Shale (Late Cretaceous) at Mesa Verde National Park, in Santucci, V. L., and McClelland, L., eds., National Park Service Paleontological Research: Denver, National Park Service Technical Report NPS/NRPO/NRTR-95/16, p. 77-81.

KOUTSOUKOS, E. A. M., AND HART, M. B., 1990, Cretaceous foraminiferal morphogroup distribution patterns, paleocommunities and trophic structures: A case study from the Sergipe Basin, Brazil: Transactions of the Royal Society of Edinburgh, Earth Sciences, v. 81, p. 221-246.

KROON, D., AND NEDERBRAGT, A. J., 1990, Ecology and paleoecology of triserial

planktic foraminifera: Marine Micropaleontology, v. 16, p. 25-38.

LAMB, G. M., 1968, Stratigraphy of the lower Mancos Shale in the San Juan Basin: Geological Society of America Bulletin, v. 79, p. 827-854.

LARSON, R. L., 1991a, Geological consequences of superplumes: Geology, v. 19, p. 963-966.

LARSON, R. L., 1991b, Latest pulse of Earth: Evidence for a mid-Cretaceous superplume: Geology, v. 19, p. 547-550.

LAWTON, T. F., 1985, Style and timing of frontal structures, thrust belt, central Utah: American Association of Petroleum Geologists Bulletin, v. 69, p. 1145-1159.

LECKIE, R. M., 1985, Foraminifera of the Cenomanian-Turonian boundary interval, Greenhorn Formation, Rock Canyon Anticline, Pueblo, Colorado, in Pratt, L. M., Kauffman, E. G., and Zelt, F. B., eds., Fine-grained Deposits and Biofacies of the Cretaceous Western Interior Seaway: Evidence of Cyclic Sedimentary Processes, Field Trip Guidebook No. 4: Tulsa, Society of Economic Paleontologists and Mineralogists, p. 139-149.

LECKIE, R. M., 1987, Paleoecology of mid-Cretaceous planktonic foraminifera: A comparison of open ocean and epicontinental sea assemblages: Micropaleontology, v. 33, p. 164-176.

LECKIE, R. M., 1989, A paleoceanographic model for the early evolutionary history of planktonic foraminifera: Palaeogeography, Palaeoclimatology, Palaeoecology, v. 73, p. 107-138.

LECKIE, R. M., KIRKLAND, J. I., AND ELDER, W. P., 1997, Stratigraphic framework and correlation of a principal reference section of the Mancos Shale (Upper Cretaceous), Mesa Verde, Colorado, in Anderson, O. J., Kues, B. S., and Lucas, S. G., eds., Mesozoic Geology and Paleontology of the Four Corners Region, Field Conference Guidebook No. 48: Albuquerque, New Mexico Geological Society Guidebook, p. 163-216.

LECKIE, R. M., SCHMIDT, M. G., FINKELSTEIN, D., AND YURETICH, R., 1991, Paleoceanographic and paleoclimatic interpretations of the Mancos Shale (Upper Cretaceous), Black Mesa Basin, Arizona, in Nations, J. D., and Eaton, J. G., eds., Stratigraphy, Depositional Environments, and Sedimentary Tectonics of the Western Margin, Cretaceous Western Interior Seaway: Boulder, Geological Society of America Special Paper 260, p. 139-152.

LEITHOLD, E. L., 1993, Preservation of laminated shale in ancient clinoforms; comparison to modern subaqueous deltas: Geology, v. 21, p. 359-362.

LEITHOLD, E. L., 1994, Stratigraphical architecture at the muddy margin of the Cretaceous Western Interior Seaway, southern Utah: Sedimentology, v. 41, p. 521-542.

LESSARD, R. H., 1973, Micropaleontology and paleoecology of the Tununk Member of the Mancos Shale: Salt Lake, Utah Geological and Mineral Survey, Special Studies 45, 28 p.

LI, H., AND HABIB, D., 1996, Dinoflagellate stratigraphy and its response to sea level change in Cenomanian-Turonian sections of the western interior of the United States: Palaios, v. 11, p. 15-30.

LIPPS, J. H., 1979, Ecology and paleoecology of planktonic foraminifera, in Lipps, J. H., et al., eds., Foraminiferal Ecology and Paleoecology: Tulsa, Society of Economic Paleontologists and Mineralogists Short Course No. 6 (Houston), p. 62-104.

LORENZ, J. C., 1982, Lithospheric flexture and the history of the Sweetgrass arch, northwestern Montana, in Powers, R. B., ed., Geologic Studies of the Cordilleran Thrust Belt: Denver, Rocky Mountain Association of Geologists, p. 77-89.

MANCINI, E. A., 1982, Foraminiferal population changes in a shallow, epicontinental marine carbonate-claystone sequence: Main Street-Grayson interval (Cretaceous), north-central Texas: Third North American Paleontological Convention Proceedings, Vol. 2, p. 353-358.

MANN, K. H., AND LAZIER, J. R. N., 1991, Dynamics of Marine Ecosystems, Biological-Physical Interactions in the Oceans: Boston, Blackwell Scientific Publications, 466 p.

McGOOKEY, D. P., HAUN, J. D., HALE, L. A., GOODELL, H. G., McCUBBIN, D. G., WEIMER, R. J., AND WULF, G. R., 1972, Cretaceous System, in Mallory, W. W., ed., Geologic Atlas of Rocky Mountain Region: Denver, Rocky Mountain Association of Geologists, p. 190-228.

McNEIL, D. H., AND CALDWELL, W. G. E., 1981, Cretaceous Rocks and Their Foraminifera in the Manitoba Escarpment: St. John's, Geological Association of Canada Special Paper 21, 438 p.

MEREWETHER, E. A., AND COBBAN, W. A., 1986, Biostratigraphic units and tectonism in the mid-Cretaceous foreland of Wyoming, Colorado, and adjoining areas, in Peterson, J. A., ed., Paleotectonics and Sedimentation in the Rocky Mountain Region, United States: Tulsa, American Association of Petroleum Geologists, Memoir 41, p. 443-468.

MOLENAAR, C. M., 1983, Major depositional cycles and regional correlations of Upper Cretaceous rocks, southern Colorado Plateau and adjacent areas, in Reynolds, M. W., and Dolly, E. B., eds., Mesozoic Paleogeography of West-Central United States: Denver, Rocky Mountain Section, Society of Economic Mineralogists and Paleontologists, p. 201-224.

MOLENAAR, C. M., AND COBBAN, W. A., 1991, Middle Cretaceous stratigraphy on the south and east sides of the Uinta Basin, northeastern Utah and northwestern

Colorado: Washington, D.C., U.S. Geological Survey Bulletin 1787-P, 34 p.

MOORE, D. M., AND REYNOLDS, R. C., JR., 1989, X-ray Diffraction and the Identification and Analysis of Clay Minerals: New York, Oxford University Press, 332 p.

MURRAY, J. W., 1976, A method of determining proximity of marginal seas to an ocean: Marine Geology, v. 22, p. 103-119.

MURRAY, J. W., 1991, Ecology and Paleoecology of Benthic Foraminifera: Essex, England, Longman Scientific and Technical, 397 p.

NEDERBRAGT, A. J., 1990, Biostratigraphy and Paleoceanographic Potential of the Cretaceous Planktic Foraminifera Heterohelicidae: Amsterdam, Vrije Universiteit, 204 p.

NORTH, B. R., AND CALDWELL, W. G. E., 1975, Foraminiferal faunas in the Cretaceous System of Saskatchewan, in Caldwell, W. G. E., ed., The Cretaceous System in the Western Interior of North America: St. John's, Geological Association of Canada Special Paper 13, p. 303-331.

OLESEN, J., 1991, Foraminiferal biostratigraphy and paleoecology of the Mancos Shale (Upper Cretaceous), southwestern Black Mesa Basin, Arizona, in Nations, J. D., and Eaton, J. G., eds., Stratigraphy, Depositional Environments, and Sedimentary Tectonics of the Western Margin, Cretaceous Western Interior Seaway: Boulder, Geological Society of America Special Paper 260, p. 153-166.

ORTH, C. J., ATTREP, M., JR., QUINTANA, L. R., ELDER, W. P., KAUFFMAN, E. G., DINER, R., AND VILLAMIL, T., 1993, Elemental abundance anomalies in the late Cenomanian extinction interval: A search for the source(s): Earth and Planetary Science Letters, v. 117, p. 189-204.

PARRISH, J. T., 1993, Mesozoic climates of the Colorado Plateau, in Morales, M., ed., Aspects of Mesozoic Geology and Paleontology of the Colorado Plateau: Flagstaff, Museum of Northern Arizona Bulletin 59, p. 1-11.

PARRISH, J. T., GAYNOR, G. C., AND SWIFT, D. J. P., 1984, Circulation in the Cretaceous Western Interior Seaway of North America, a review, in Stott, D. F, and Glass, D. J., eds., The Mesozoic of Middle North America: Calgary, Canadian Society of Petroleum Geologists Memoir 9, p. 221-231.

PHLEGER, F. B., 1951, Ecology of Foraminifera, Northwest Gulf of Mexico, Part I. Foraminifera Distribution: Boulder, Geological Society of America Memoir 46, p. 1-88.

POLLASTRO, R. M., 1985, Mineralogical and morphological evidence for the formation of illite at the expense of illite/smectite: Clay and Clay Minerals, v. 33, p. 265-274.

PRATT, L. M., 1984, Influence of paleoenvironmental factors on preservation of organic matter in middle Cretaceous Greenhorn Formation, Pueblo, Colorado: American Association of Petroleum Geologists Bulletin, v. 68, p. 1146-1159.

PRATT, L. M., 1985, Isotopic studies of organic matter and carbonate in rocks of the Greenhorn marine cycle, in Pratt, L. M., Kauffman, E. G., and Zelt, F. B., eds., Fine-grained Deposits and Biofacies of the Cretaceous Western Interior Seaway: Evidence of Cyclic Sedimentary Processes, Field Trip Guidebook No. 4: Tulsa, Society of Economic Paleontologists and Mineralogists, p. 38-48.

PRATT, L. M., ARTHUR, M. A., DEAN, W. E., AND SCHOLLE, P. A., 1993, Paleo-oceanographic cycles and events during the Late Cretaceous in the Western Interior Seaway of North America, in Caldwell, W.G.E., and Kauffman, E.G., eds., Evolution of the Western Interior Basin: St. John's, Geological Association of Canada Special Paper 39, p. 333-353.

PRATT, L. M., AND THRELKELD, C. N., 1984, Stratigraphic significance of $\delta^{13}C/^{12}C$ ratios in mid-Cretaceous rocks of the western interior, U.S.A., in Stott, D.F, and Glass, D.J., eds., The Mesozoic of Middle North America: Calgary, Canadian Society of Petroleum Geologists Memoir 9, p. 305-312.

PRICE, R. A., 1973, Large-scale gravitational flow of supracrustal rocks, southern Canadian Rockies, in Delong, K. A., and Scholten, R., eds., Gravity and Tectonics: New York, John Wiley and Sons, p. 491-502.

RESIG, J. M., 1993, Cenozoic stratigraphy and paleoceanography of biserial planktonic foraminifera, Ontong Java Plateau, in Berger, W. H., Kroenke, L.W., Mayer, L. A., et al., Proceedings of the Ocean Drilling Program, Scientific Results, Volume 130: College Station, Ocean Drilling Program, p. 231-244.

REYNOLDS, R. C., 1989, Principles and techniques of quantitative analysis of clay minerals by X-ray powder diffraction, in Pevear, D. R., and Mumpton, eds., Quantitative Mineral Analysis of Clays: Evergreen, Colorado, Clay Mineral Society, p. 4-37.

REESIDE, J. B., 1957, Paleoecology of the Cretaceous Seas of the Western Interior of the United States: Boulder, Geological Society of America Memoir 67, p. 505-542.

RYER, T. A., AND LOVEKIN, J. R., 1986, The Upper Cretaceous Vernal Delta of Utah-depositional or paleotectonic feature? in Peterson, J. A., ed., Paleotectonics and Sedimentation in the Rocky Mountain Region, United States: Tulsa, American Association of Petroleum Geologists Memoir 41, p. 497-510.

RYER, T. A., 1993, Speculations on the origins of mid-Cretaceous clastic wedges, central Rocky Mountain region, United States, in Caldwell, W. G. E., and Kauffman, E. G., eds., Evolution of the Western Interior Basin: St. John's, Geological Association of Canada Special Paper 39, p. 189-198.

SAGEMAN, B. B., 1985, High-resolution stratigraphy and paleobiology of the Hartland

Shale Member: Analysis of an oxygen-deficient epicontinental sea, in Pratt, L. M., Kauffman, E. G., and Zelt, F. B., eds., Fine-grained Deposits and Biofacies of the Cretaceous Western Interior Seaway: Evidence of Cyclic Sedimentary Processes, Field Trip Guidebook No. 4: Tulsa, Society of Economic Paleontologists and Mineralogists, p. 110-121.

SAGEMAN, B. B., AND ARTHUR, M. A., 1994, Early Turonian paleogeographic/ paleobathymetric map, western interior, U.S., in Caputo, M. V., Peterson, J. A., and Franczyk, K. J., eds., Mesozoic Systems of the Rocky Mountain Region, USA: Tulsa, Society of Economic Paleontologists and Mineralogists, Rocky Mountain Section, p. 457-469.

SAVRDA, C. E., AND BOTTJER, D. J., 1993, Trace fossil assemblages in fine-grained strata of the Cretaceous western interior, in Caldwell, W. G. E., and Kauffman, E. G., eds., Evolution of the Western Interior Basin: St. John's, Geological Association of Canada Special Paper 39, p. 621-639.

SCHEDL, A., AND WILTSCHKO, D. V., 1984, Sedimentological effects of a moving terrain: Journal of Geology, v. 92, p. 273-287.

SCHLANGER, S. O., ARTHUR, M. A., JENKYNS, H. C., AND SCHOLLE, P. A., 1987, The Cenomanian-Turonian oceanic anoxic event; 1, stratigraphy and distribution of organic carbon-rich beds and the marine C excursion, in Brooks, J., and Fleet, A. J., eds., Marine Petroleum Source Rocks: London, Geological Society of London Special Publication 26, p. 371-399.

SCHLANGER, S. O., AND JENKYNS, H. C., 1976, Cretaceous oceanic anoxic events: Causes and consequences: Geol. en Mijnbouw, v. 55, p. 179-184.

SCHROEDER-ADAMS, C. J., LECKIE, D. A., BLOCH, J., CRAIG, J., McINTYRE, D. J., AND ADAMS, P. J., 1996, Paleoenvironmental changes in the Cretaceous (Albian to Turonian) Colorado Group of western Canada: Microfossil, sedimentological and geochemical evidence: Cretaceous Research, v. 17, p. 311-365.

SCHWARTZ, R. K., AND DeCELLES, P. G., 1988, Cordilleran foreland basin evolution in response to interactive Cretaceous thrusting and foreland partitioning, southwestern Montana, in Schmidt, C. J., and Perry, W. J., Jr., eds., Interaction of the Rocky Mountain Foreland and the Cordilleran Thrust Belt: Boulder, Geological Society of America Memoir 171, p. 489-513.

SETHI, P. S., AND LEITHOLD, E. L., 1994, Climatic cyclicity and terrigenous sediment influx to the early Turonian Greenhorn Sea, southern Utah: Journal of Sedimentary Research, v. B64, p. 26-39.

SINGER, A., 1980, Paleoclimatic interpretation of clay minerals in soils and weathering profiles: Earth-Science Reviews, v. 15, p. 303-327.

SINGER, A., 1984, The paleoclimatic interpretation of clay minerals in sediment - a review: Earth-Science Reviews, v. 21, p. 251-293.

SLINGERLAND, R., KUMP, L. R., ARTHUR, M. A., FAWCETT, P. J., SAGEMAN, B. B., AND BARRON, E. J., 1996, Estuarine circulation in the Turonian Western Interior seaway of North America: Geological Society of America Bulletin, v. 108, p. 941-952.

SLITER, W. V., 1972, Upper Cretaceous planktonic foraminiferal zoogeography and ecology-eastern Pacific margin: Palaeogeography, Palaeoclimatology, Palaeoecology, v. 12, p. 15-31.

SOHL, N. F., MARTINEZ, R. E., SALMERON-URENA, P., AND SOTO-JARAMILLO, F., 1991, Chapter 10: Upper Cretaceous, in Salvador, A., ed., The Gulf of Mexico Basin: Boulder, Colorado, Geological Society of America, The Geology of North America, Vol. J, p. 205-244.

STELCK, C. R., AND WALL, J. H., 1954, Kaskapau foraminifera from Peace River area of western Canada: Calgary, Alberta Research Council Report No. 68, p. 2-38.

STELCK, C. R., AND WALL, J. H., 1955, Foraminifera of the Cenomanian Dunveganoceras Zone from the Peace River area of western Canada: Calgary, Alberta Research Council Report No. 70, p. 1-81.

TAPPAN, H., 1940, Foraminifera from the Grayson Formation of northern Texas: Journal of Paleontology, v. 14, p. 93-126.

TAPPAN, H., 1962, Foraminifera from the Arctic Slope of Alaska; Part 3-Cretaceous foraminifera: U.S. Geological Survey, Professional Paper no. 236-C, p. 91-209.

THOMAS, E., AND GOODAY, A. J., 1996, Cenozoic deep-sea foraminifers: Tracers for changes in oceanic productivity?: Geology, v. 24, p. 355-358.

THUROW, J., BRUMSACK, H. -J., RULLKOTTER, J., LITTKE, R., AND MEYERS, P., 1992, The Cenomanian/Turonian boundary event in the Indian Ocean - a key to understand the global picture, in Synthesis of Results from Scientific Drilling in the Indian Ocean: Washington, D.C., American Geophysical Union, Geophysical Monograph 70, p. 253-273.

TUCHOLKE, B. E., AND VOGT, P. R., 1979, Western North Atlantic: Sedimentary evolution and aspects of tectonic history, in Tucholke, B. E., Vogt, P. R., et al., Initial Reports of the Deep Sea Drilling Project, Volume 43: Washington, D.C., U.S. Government Printing Office, p. 791-825.

VILLIEN, A., AND KLIGFIELD, R. M., 1986, Thrusting and synorogenic sedimentation in central Utah, in Peterson, J. A., ed., Paleotectonics and Sedimentation in the Rocky Mountain Region, United States: Tulsa, American Association of Petroleum Geologists Memoir 41, p. 281-308.

WALL, J. H., 1960, Upper Cretaceous foraminifera from the Smoky River area, Alberta: Calgary, Research Council of Alberta Bulletin 6, 43 p.

WALL, J. H., 1967, Cretaceous foraminifera of the Rocky Mountain foothills, Alberta: Calgary, Research Council of Alberta Bulletin 20, 185 p.

WATKINS, D. K., BRALOWER, T. J., COVINGTON, J. M., AND FISHER, C. G., 1993, Biostratigraphy and paleoecology of the Upper Cretaceous calcareous nannofossils in the western interior basin, North America, in Caldwell, W. G. E., and Kauffman, E. G., eds., Evolution of the Western Interior Basin: St. John's, Geological Association of Canada Special Paper 39, p. 521-537.

WEAVER, C. E., 1989, Clays, Muds and Shales: New York, Elsevier, 819 p.

WEST, O. L., 1993, Do foraminifera in sulfide-rich marine environments harbor symbiotic chemautotrophic bacteria?(abs.): Fourth East Coast Conference on Protozoa, Program with Abstracts, p. 9.

WEST, O. L., LECKIE, R. M., AND SCHMIDT, M., 1990, Cenomanian-Turonian foraminiferal changes in a transect across the Cretaceous Western Interior Seaway: Boulder, Geological Society of America, Abstracts with Programs, v. 22(7), p. A234-235.

WILTSCHKO, D. V., AND DORR, J. A., JR., 1983, Timing of deformation in Overthrust Belt and foreland of Idaho, Wyoming, and Utah: American Association of Petroleum Geologists Bulletin, v. 67, p. 1304-1322.

YOUNG, K., 1951, Foraminifera and stratigraphy of the Frontier Formation (Upper Cretaceous), southern Montana: Journal of Paleontology, v. 25, p. 35-68.

YURETICH, R. F., 1979, Modern sediments and sedimentary processes in Lake Rudolf (Lake Turkana), eastern Rift Valley, Kenya: Sedimentology, v. 26, p. 313-331.

ICHNOLOGY OF THE BRIDGE CREEK LIMESTONE: EVIDENCE FOR TEMPORAL AND SPATIAL VARIATIONS IN PALEO-OXYGENATION IN THE WESTERN INTERIOR SEAWAY

CHARLES E. SAVRDA

Department of Geology, Auburn University, Auburn, AL 36849-5305

ABSTRACT: The Upper Cretaceous (Cenomanian-Turonian) Bridge Creek Limestone is characterized by decimeter-scale alternation of pelagic limestones and marly shales that were deposited under variably oxygenated waters. Vertical stacking patterns of laminites and four oxygen-related ichnocoenoses in the Bridge Creek in two cores—USGS #1 Portland from east-central Colorado and Amoco #1 Rebecca K. Bounds from western Kansas—provide a record of both temporal and spatial changes in benthic oxygenation within the distal offshore parts of the Western Interior Seaway.

Paleo-oxygenation histories reconstructed for the Portland core reflect: (1) a broad trend towards decreased benthic oxygenation through the entire Bridge Creek interval; (2) a high-amplitude redox cyclicity that corresponds to limestone/marly shale couplets; and (3) a higher-frequency, lower-amplitude redox cyclicity expressed within the marly shale intervals.

Trends (1) and (2) are well expressed in the Bounds core. However, bioturbated horizons in marly shale intervals are less common, thinner, of lower ichnocoenosis rank, or absent altogether. This pattern indicates that paleo-oxygenation levels were lower at the Bounds locality, at least during clastic-dominated phases of depositional cycles, and may reflect higher productivity and oxygen demand in the eastern part of the basin.

INTRODUCTION

The rhythmically bedded Bridge Creek Limestone Member of the Greenhorn Formation represents one of two Upper Cretaceous intervals targeted for detailed study as part of the Cretaceous Western Interior Seaway Drilling Project. Among many other objectives (Dean and Arthur, this volume), this multidisciplinary research program was undertaken to better establish the mechanisms responsible for depositional cyclicity and to determine controls on accumulation rate, character, and preservational state of organic matter in the Western Interior basin. Attainment of these and related goals depends, in part, on our ability to evaluate how benthic oxygenation levels varied through time and space.

Benthic oxygenation is an important oceanographic parameter that may reflect aspects of oceanic circulation, some of which may be mediated by climate (e.g., upwelling intensity, degree of thermohaline density stratification, etc.), and exerts significant control over the quantity and preservational state of organic matter in marine deposits. Oxygenation is also one of the most crucial factors influencing the character and activities of benthic organisms in quiet marine settings. Fortunately, the response of bioturbating infauna to variations in benthic oxygenation are preserved and reflected by ichnological parameters. Hence, by serving as proxy indicators of paleo-redox conditions, ichnofabrics of hemipelagic and pelagic strata can play a significant role in paleoceanographic reconstructions and source-rock studies.

With this as incentive, I analyzed ichnofabrics and component ichnofossils at the centimeter scale throughout the Bridge Creek interval of two cores: USGS #1 Portland in east-central Colorado and Amoco #1 Rebecca K. Bounds in western Kansas (Dean and Arthur, this volume). The objectives of this paper are to (1) summarize the ichnologic data collected during these studies and (2) discuss their general implications for both temporal patterns (long-term trends and higher-frequency cycles) and spatial variations in benthic oxygenation in the offshore part of the Western Interior Seaway.

BRIDGE CREEK LIMESTONE

General Framework

The Cenomanian-Turonian Bridge Creek Limestone Member of the Greenhorn Formation records deposition in distal off-shore settings of the Western Interior Seaway during a major transgressive episode (Elder and Kirkland, 1985; Elder et al., 1994). The Bridge Creek Limestone, which is approximately 15 m thick on average (Eicher and Diner, 1989), overlies the Hartland Shale Member of the Greenhorn Formation and is overlain by the Fairport Member of the Carlile Shale (Elder and Kirkland, 1985; Hattin, 1985). In Colorado, the Bridge Creek Limestone is subdivided into three units (lower, middle, and upper Bridge Creek; Pratt, 1984; Elder and Kirkland, 1985), which are generally equivalent to the upper Hartland, Jetmore, and Pfiefer Members of the Greenhorn Limestone as defined in Kansas (Hattin, 1985). Delineation of these units is based primarily upon bedding characteristics and minor lithological variability (see Elder and Kirkland, 1985).

As a whole, the Bridge Creek is characterized by well-defined, decimeter-scale rhythmic alternation of highly bioturbated, organic-poor, micritic limestones and laminated to bioturbated, organic-rich marlstones and marly shales (Pratt, 1984; Arthur et al., 1985; Elder and Kirkland, 1985; Arthur and Dean, 1991; Pratt et al., 1993). Subordinate lithological components include numerous bentonites, derived from the volcanic arc to the west, and calcarenites of variable thickness (Pratt, 1984; Elder, 1985). Bentonites, as well as prominent limestone beds, can be traced over a large area of the Western Interior and define lithochronozones within which spatial changes in sedimentation patterns can be assessed (Hattin, 1985; Elder and Kirkland, 1985; Elder et al., 1994). Calcarenites, which are attributed to bottom-current winnowing, are relatively minor components of the lower and middle units, but are extremely abundant in and diagnostic of the upper Bridge Creek (Pratt, 1984).

Carbonate Cyclicity

Decimeter-scale rhythmicity in the Bridge Creek traditionally has been interpreted to be the product of orbitally driven climate cycles (Gilbert, 1895; Hattin, 1975; Kauffman, 1977; Fischer, 1980; Barron et al., 1985). However, there is considerable disagreement regarding the primary paleoceanographic mechanisms responsible for these carbonate rhythms. Bridge Creek carbonate oscillations are attributed to combined productivity/redox cycles that may be linked to climate-driven changes in density stratification of the Tethys to the south (Eicher and Diner, 1989, 1991; Ricken, 1991, 1994). Alternatively, carbonate rhythms may record combined dilution-redox cycles associated with changes in the flux of freshwater and clastic sediment to the basin in response to alternating periods of wet and dry climate (e.g., Hattin, 1971; Pratt, 1984; Arthur et al., 1985; Arthur and Dean, 1991; Pratt et al., 1993). The recently established link between limestones of the Bridge Creek and parasequence-bounding flooding surfaces in nearshore siliclastic facies to the east (Elder et al., 1994) provides strong evidence in support of the latter mechanism. However, even if dilution cycles are the primary cause of carbonate cycles in the Bridge Creek, coincident variations in productivity cannot be excluded altogether (Arthur and Dean, 1991). In fact, limited nannofossil evidence suggests that carbonate productivity may have been higher during the clastic-dominated parts of cycles (Watkins, 1989).

Owing to poorly constrained ages for the Bridge Creek interval (Obradovich, 1993), estimates of cyclic periods vary. Hence, there are also differences of opinion regarding the specific orbital parameters responsible. Fischer et al. (1985; Fischer, 1986) inferred a period of 40 ky for Bridge Creek carbonate cycles and, hence, implicated obliquity cycles. Their estimate was made on the basis of the relatively even spacing of prominent limestones and the apparent lack of the well-expressed bundling that typically characterizes precession cycles (~20 ky) modulated by eccentricity cycles (~100 ky). Employing an updated time scale and an estimated average sedimentation rate for the member (0.5-1.0 cm/ky), Elder (1985) estimated a period of roughly 100 ka for cycles defined by the most prominent and laterally continuous limestones, suggesting that eccentricity played a role. Elder's (1985) calculation implies that shorter cycles expressed by intervening less prominent limestones and marlstones may reflect a precession signal. Ichnological evidence presented by Savrda and Bottjer (1994) supports Elder's interpretations. However, as demonstrated by Sageman et al. (this volume), the relationships between Bridge Creek bedding rhythms and Milankovitch parameters may be more complex than previously thought.

Diagenesis

Interpretations of Bridge Creek bedding rhythms are hindered to some degree by differential diagenesis and associated cycle enhancement (Ricken, 1994). Limestones apparently experienced more diagenetic alteration than less carbonate-rich intervals (marlstones and calcareous shales) (Arthur et al., 1985), and overall diagenetic intensity increases westward across the basin as a function of increasing burial depth (Arthur and Dean, 1991; Pratt et al., 1993).

ICHNOFOSSILS AS INDICATORS OF BENTHIC PALEO-OXYGENATION

Active bioturbation in fine-grained marine substrates typically occurs in one of two layers: (1) a surface mixed layer characterized by rapid and complete biohomogenization; and (2) an underlying transition layer characterized by heterogeneous bioturbation by deeper dwelling and/or feeding organisms (Ekdale et al., 1984). The general response of these ichnological components to changes in benthic oxygenation are addressed in detail in earlier papers (e.g., Savrda and Bottjer, 1987, 1989, 1991). Summarily, as oxygen availability decreases, either along lateral sea-floor redox gradients or through time, the thickness of the surface mixed layer decreases and the diversity, diameters, and depths of penetration of transition-layer burrows generally decline. Eventually, when oxygen concentrations drop below a critical threshold, bioturbation virtually ceases and laminated, organic-rich sediments accumulate.

Hence, oxygen-related changes in endobenthic communities result in the production of different ichnocoenoses that are manifested in either simple or compound ichnofabrics in the stratigraphic record (Savrda, 1992; Savrda and Bottjer, 1994). Oxygen-related ichnocoenoses (ORI) representing the low end of the benthic oxygenation spectrum are characterized by low-diversity assemblages of small-diameter, shallowly penetrating burrows. ORI produced under successively higher oxygen concentrations are characterized by progressively higher-diversity ichnofossil assemblages that include larger and more vertically extensive burrows. Stratigraphic intervals deposited under temporally variable levels of oxygenation are characterized by compound ichnofabrics produced by the vertical stacking and superposition of different ichnocoenoses. When deciphered through detailed vertical sequence analyses, ORI stacking patterns can be translated to paleo-oxygenation curves that reflect more or less time-averaged redox histories of marine basins (Savrda, 1992; Savrda and Bottjer, 1994).

Variability of bioturbation in the Bridge Creek Limestone previously was recognized and linked to changes in benthic paleo-oxygenation (e.g., Hattin, 1971, 1975; Pratt, 1984; Elder, 1985; Elder and Kirkland, 1985). Most recently, Savrda and Bottjer (1994) attempted to reconstruct Bridge Creek paleo-oxygenation histories using the ORI approach. However, this study was restricted to the lowermost Bridge Creek and was based solely on the reexamination of acetate-peel replicates originally prepared by Pratt (1984) from a core drilled near Pueblo, Colorado. The Cretaceous Western Interior Seaway Drilling Project provided an opportunity to assess temporal oxygenation histories for the entire Bridge Creek interval via direct observation of core. Moreover, by allowing a comparison of ichnofabrics in well-defined lithochronozones between two cores, this project provided the unique opportunity to assess, in detail, *spatial* gradients in basin paleo-oxygenation.

GENERAL APPROACH

Ichnological studies addressed herein focused on two cores: (1) the Portland core drilled near Florence, central Colorado; and (2) the Bounds core drilled in Greeley County, western Kansas (for locations see Dean and Arthur, this volume). All observations were made at the USGS Core Research Center in Denver. Although the Bridge Creek interval was emphasized, the uppermost parts of the subjacent Hartland Shale Member were also described. Core recovery of the target interval in the Portland core was 100%. Recovery was also very good for the Bounds core, although some core was missing from the basal part of the lower Bridge Creek.

Core surfaces were wetted to enhance visibility and examined to document vertical changes in lithology and general sedimentary fabric (e.g., laminated vs. bioturbated) at the centimeter scale. Analyses of individual bioturbated intervals involved identification of burrow types, measurement of maximum burrow diameters, and, where possible, assessment of burrow-penetration depths. An attempt was made to identify all discrete biogenic structures to the ichnogeneric level. However, owing to locally poor contrast between burrows and surrounding sediment and/or restricted two-dimensional views provided by core surfaces, ichnotaxonomic assignments of some burrow types could not be made with confidence. Ichnocoenoses were identified on the ba-

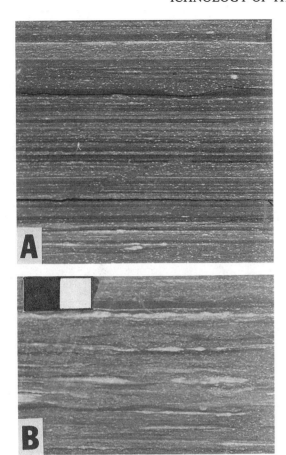

FIG. 1.—Laminated fabrics of marly shales in Bridge Creek Limestone. (A) Pelagic laminite characterized by relatively continuous fine lamination. Light specks are pelagic fecal pellets and foraminifera. Note localized disruption of laminae by small burrows. Burrows likely are piped from overlying bioturbated beds, although the precise sources often cannot be determined. (B) Calcarenitic laminite characterized by thin, irregular, and discontinuous laminae and pods of fine skeletal debris. Note disruption of laminae by small, poorly defined burrows piped from overlying bioturbated interval. Both examples are from the Bounds core. Bar scale is 2 cm long.

sis of recurring associations of burrow types among all bioturbated intervals. Vertical stacking patterns of ichnocoenoses and associated laminites were used to construct paleo-oxygenation curves for the Bridge Creek intervals of both cores.

Visibility of primary fabrics and discrete trace fossils is better overall in the Bounds core. Differences in contrast presumably are related to the aforementioned cross-basin diagenetic gradient. However, it is doubtful that differential diagenesis significantly affected or biased ichnological comparisons.

BRIDGE CREEK SEDIMENTARY FABRICS

Laminites

The term "laminite" is used here to describe all intervals with background fabrics characterized by primary sedimentary structures. For the purposes of this paper, laminites in the Bridge Creek are divided into two broad groups: pelagic laminites and calcarenitic laminites.

Pelagic laminites are most commonly represented in organic-rich, marly shale intervals of both cores, although some marlstones and limestones in the upper Bridge Creek may also qualify. Those in calcareous shales are typically characterized by alternating light and dark laminae on the scale of 1 to 5 mm (Fig. 1A). Lighter laminae appear to be enriched in foraminifera and/or small, elliptical, coccolith-rich fecal pellets of presumed planktonic origin (Hattin, 1975), while darker laminae presumably are more clastic-rich. These laminites probably reflect slow pelagic deposition. However, some thicker (5-20 mm), homogeneous bands that are intercalated with finer lamination may record more rapid, distal storm deposition. Although some pelagic laminites are virtually undisturbed, most intervals are weakly to strongly disrupted by burrows that pipe downward from superjacent bioturbated intervals (Fig. 1A). Intensity of this disruption increases upward towards the bases of these bioturbated horizons but varies with ichnocoenosis rank. Based on the absence of bioturbation by resident infauna, pelagic laminites are ascribed to deposition under anoxic or nearly anoxic conditions (Savrda and Bottjer, 1991).

Calcarenitic intervals, which are most common in the upper part of the Hartland Shale and upper Bridge Creek, are almost invariably characterized by plane to wavy lamination and/or ripple-cross lamination and, hence, are collectively referred to here as calcarenitic laminites (Fig. 1B). Like the pelagic laminites with which they typically are intercalated, these intervals may be partly disrupted by burrows piped from overlying bioturbated units. Less commonly, calcarenitic laminites are disrupted by small, poorly defined, horizontal burrows that cannot be easily attributed to piping. However, in many cases, fabrics produced by diffuse, shallow biogenic disruption could not be distinguished from other deformational fabrics, such as those produced by compaction of starved ripples. Calcarenites reflect periods of enhanced bottom-current energy and associated winnowing and traction transport. Nonetheless, the general lack of resident bioturbation in these and associated pelagic laminites suggests that, despite intensified currents, benthic oxygenation levels were generally too low to permit establishment of a benthic infauna.

Bioturbated Beds

Bioturbated intervals are strictly defined here as strata that are characterized by bioturbated, homogeneous to diffusely burrow-mottled background fabrics as well as discrete burrows. Individual bioturbated intervals in the Bridge Creek Limestone, which range in thickness from 5 mm to over 90 cm, are characterized by one or more of four ichnocoenoses defined on the basis of recurring assemblages of one or more of eight burrow types. The general characteristics of ichnocoenoses are summarized below and in Figure 2.

Ichnocoenosis 1.—

This ichnocoenosis, which is most often associated with marly shales and marlstones, is represented by an extremely low-diversity burrow assemblage. With the exception of a few occurrences where *Chondrites* is also present, ichnocoenosis 1 is characterized only by relatively narrow, horizontal to subhorizontal, unlined burrows that are collectively assigned to the ichnogenus *Planolites* (Fig. 2A, B). Maximum burrow diameters in all occurrences of ichnocoenosis 1 range from 2 to 6

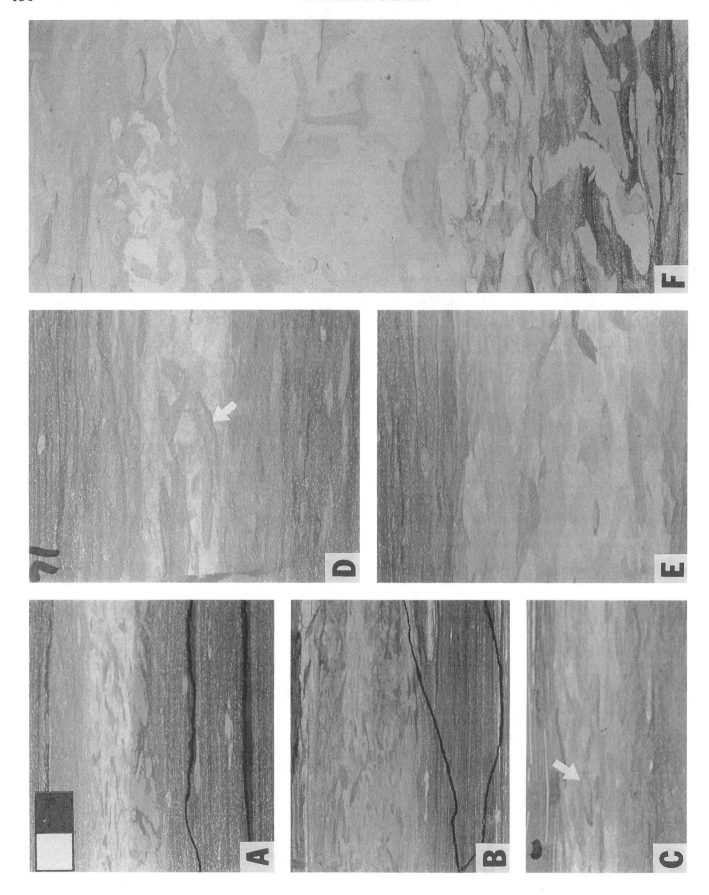

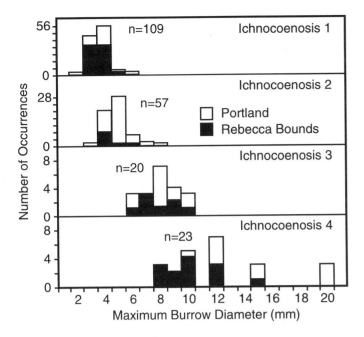

FIG. 3.— Maximum burrow diameters recorded for all occurrences of Bridge Creek ichnocoenoses in both cores.

mm, but are most typically 3 to 4 mm (Fig. 3). Based on depths of dense piping into subjacent laminites or other ichnocoenoses, burrow penetration depths of these small *Planolites* are generally less than 3 cm.

Ichnocoenosis 2.—

Ichnocoenosis 2, which is also generally restricted to marly shale and marlstone intervals, is slightly more diverse. In addition to the small *Planolites* noted above, this ichnocoenosis also contains one or both of two additional burrow types. The first of these is tentatively assigned to the ichnogenus *Taenidium*. These are horizontal to subhorizontal, apparently unbranched and unlined, backfilled tubes (Fig. 2C). Burrows of the second form are also horizontal to subhorizontal structures. These, however, are more elongate on two-dimensional core surfaces (Fig. 2D) and appear sheet-like in limited and poor three-dimensional views. These structures are likely those that were referred to as *Zoophycos* in earlier acetate peel studies (Savrda and Bottjer, 1994). However, spreite are either not observed or are very poorly defined (Fig. 2D). For the purpose of this paper, these are referred to descriptively as *Zoophycos*-like sheet structures. Maximum burrow diameter for all occurrences of ichnocoenosis 2

ranges from 3 to 8 mm but is most commonly 4 to 5 mm (Fig. 3). Where determined from the vertical extent of piping, typical burrow-penetration depths range from 2 to 5 cm.

Ichnocoenosis 3.—

This ichnocoenosis is most commonly associated with marlstones and limestones. Although all recurring burrow types associated with ichnocoenosis 2 are typically present, ichnocoenosis 3 also includes larger-diameter, horizontal to subhorizontal, unlined and apparently unbranched burrows referred to here as large *Planolites* (Fig. 2E). Diameters of these burrows are typically on the order of 8 mm but range from 6 to 10 mm (Fig. 3). Limited information on piping from this ichnocoenosis indicates that burrow penetration depths were typically on the order of 3 to 6 cm.

Ichnocoenosis 4.—

Ichnocoenosis 4, which occurs only in limestones, differs from ichnocoenosis 3 primarily by the addition of branching systems of large burrows assigned to the ichnogenus *Thalassinoides* (Fig. 2F). However, this comparatively diverse ichnocoenosis also may include vertically extensive *Teichnichnus* and/or, less commonly, *Zoophycos* (Fig. 2F). Maximum burrow diameters, in all cases recorded from *Thalassinoides*, range from 8 to 20 mm with an average of 12 mm (Fig. 3). Minimum burrow-penetration depths, estimated from burrow piping and/or lengths of vertical burrow segments, range from 4 to 8 cm.

Trace fossil diversity, burrow diameters, and general burrow-penetration depths progressively increase from ichnocoenosis 1 through ichnocoenosis 4. On this basis, the four Bridge Creek ichnocoenoses are interpreted to record progressively higher levels of benthic oxygenation.

PALEO-OXYGENATION HISTORIES

Vertical stacking patterns of laminites and oxygen-related ichnocoenoses were used to reconstruct paleo-oxygenation histories for both cores. Interpreted paleo-oxygenation curves, constructed in the manner outlined by Savrda and Bottjer (1994; Savrda, 1992) for the entire Bridge Creek interval in the Portland and Bounds cores, are shown in Figures 4 and 5, respectively.

Portland Core

The paleo-oxygenation curve for the Portland core reflects the following patterns: (1) a broad trend towards decreased benthic oxygenation through the Bridge Creek interval; (2) a high-amplitude redox cyclicity defined by the recurrence of ichnocoenoses 3 or 4; and (3) a higher-frequency, lower-amplitude redox cyclicity generally defined by the recurrence of ichnocoenoses 1 and 2.

The trend of decreasing oxygenation through the Bridge Creek is reflected by changes in extent of bioturbation in general (Fig. 4). Approximately 87% of the lower Bridge Creek is completely bioturbated. This includes all of the prominent limestones (labeled LS1 through LS 10 in Fig. 4, following the scheme of Elder, 1985) and most of the intervening marlstone/marly shale intervals, although the frequency of pelagic laminites in the latter increases up section. In contrast, only ~43% of the middle Bridge Creek is bioturbated. Reduction of bioturbation here is accounted for in the marlstone/marly shale intervals, which are

FIG. 2.— Ichnocoenoses represented in the Bridge Creek Limestone. (A) and (B) Ichnocoenosis 1 characterized by small *Planolites*, rare *Chondrites*, and nondescript burrow mottling. Note small burrow diameters and minimal burrow penetration depths into subjacent pelagic laminites. Bar scale in (A) is 2 cm long. (C) and (D) Ichnocoenoses 2 characterized by small *Planolites*, *Taenidium* (arrow in C), *Zoophycos*-like sheet structures (arrow in D), and diffuse burrow mottling. (E) Ichnocoenosis 3 in limestone. Small *Planolites*, *Taenidium*, and *Zoophycos*-like sheet structures cross cut a background fabric that includes large *Planolites*. (F) Ichnocoenosis 4 characterized by a more diverse assemblage of burrows that includes *Thalassinoides* and *Teichichnus*. Note larger burrow diameters and vertically extensive piping into subjacent sediment. Scales in B through F are the same as in A. All examples are from the Bounds core.

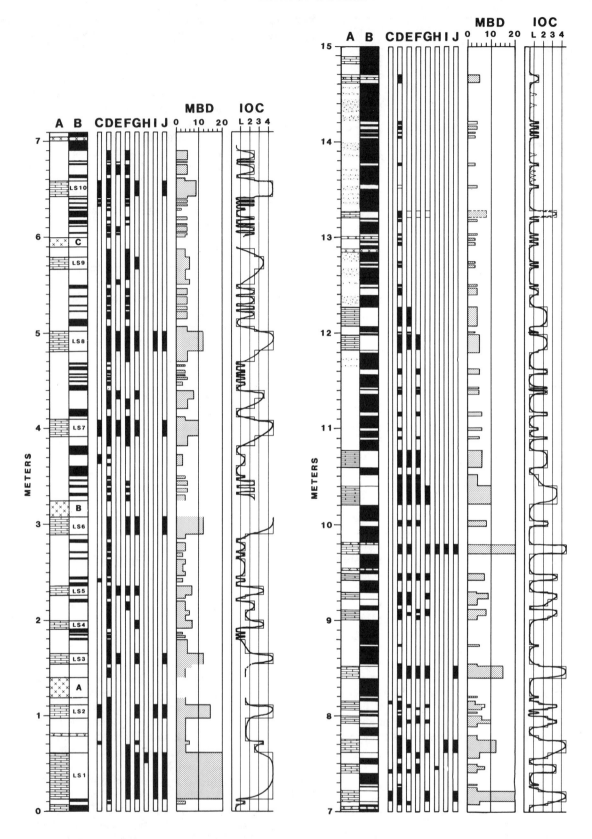

FIG. 4.—Distribution of general lithologies (A; limestones and bentonites are patterned, calcarenites are stippled, and marlstones/marly shales are white), general sedimentary fabrics (B; bioturbated intervals are white, laminites are black), and recurring biogenic structure types (C, *Chondrites*; D, small *Planolites*; E, *Taenidium*; F, *Zoophycos*-like sheet structures; G, large *Planolites*; H, *Zoophycos*; I, *Teichnichnus*; and J, *Thalassinoides*), maximum burrow diameter (MBD, in mm), and interpreted oxygenation curve (IOC) for the Bridge Creek interval in the Portland core. IOC is constructed using four reference lines. Line L represents oxygen levels below which laminites are preserved and above which producers of ichnocoenosis 1 can survive. Lines 2, 3, and 4 represent threshold oxygen levels required for development of ichnocoenoses 2, 3, and 4, respectively. Bed designations for limestones (LS1-LS10) and thicker bentonites (A, B, and C) in lower Bridge Creek (at left) are after Elder (1985). The boundary between middle and upper Bridge Creek Limestone is inferred to be at ~10.8 m.

dominated by pelagic laminites. Interbedded limestones are completely bioturbated and of high ichnocoenosis rank. The upper Bridge Creek is dominated by pelagic and calcarenitic laminites. Bioturbated horizons comprise only 33% of this interval overall and are characterized by ichnocoenoses 1 or 2, which reflect short, low-amplitude oxygenation events only. Most of these events occurred during deposition of marlstones or marly shales. However, diffuse bioturbation also may be recorded in some calcarenitic intervals.

High-amplitude redox cycles reflected by the recurrence of ichnocoeneoses 3 or 4 are expressed only in the lower and middle Bridge Creek (Fig. 4). These cycles generally correspond to limestone-marlstone/marly shale couplets. Spacing between high-amplitude oxygenation events, which are represented by the limestones, varies considerably from ~10 cm to 1 m, but is generally lower in the middle Bridge Creek. It is currently unclear whether this trend reflects true differences in actual frequency of events or temporal variations in sediment accumulation rates.

High-frequency, lower-amplitude oxygenation cycles defined by the interbedding of laminites and ichnocoenoses 1 or 2 are expressed within marlstone/marly shale intervals between limestones. Although comparatively rare in the middle Bridge Creek, low-amplitude redox events are recorded throughout the member. The cyclical nature of these events is particularly well expressed in the upper part of the lower Bridge Creek (Fig. 4). There, low- and high-amplitude cycles together define a bundling pattern that is reminiscent of sedimentary rhythms mediated by orbital eccentricity and axial precession (Fischer et al., 1985; Fischer, 1986). However, this pattern may be deceiving; results of spectral analyses of ichnologic data (ichnocoenosis rank and maximum burrow diameter) from the Portland core suggest a more complex link between redox conditions and Milankovitch orbital parameters (Sageman et al., this volume).

Bounds Core

The interpreted paleo-oxygenation curve for the Bounds core also reflects a progressive decrease in benthic oxygenation through Bridge Creek time (Fig. 5). However, comparison with the record for the Portland core indicates that oxygenation was comparatively lower overall at the western Kansas site. In the Bounds core, bioturbated intervals comprise only 52%, 34%, and, at most, 13% of the lower, middle, and upper Bridge Creek, respectively (compared to 87%, 43%, and 33%, respectively, in the Portland core).

Eastward reduction in oxygenation is most obvious in the lower Bridge Creek, particularly in the upper part where prominent limestones and bentonites provide the basis for confident correlation and comparison of narrow lithochronozones between the two cores. Although the high-amplitude oxygenation events associated with limestone deposition are similarly expressed in both cores, bioturbated horizons in marlstone-marly shale intervals are clearly less common, thinner, and/or of lower ichnocoenosis rank in the Bounds core (Fig. 6). Generally, intervals characterized by ichnocoenosis 2 in central Colorado are replaced in western Kansas by ichnocoenosis 1 or pelagic laminites.

Differences in oxygenation between the two sites are less obvious in the middle Bridge Creek. As in the Portland core, this interval in the Bounds core is characterized by a high-amplitude redox cyclicity reflected by alternating, predominantly laminated marly shales and limestones of high ichnocoenosis rank (ichnoceoenosis 3 and 4) (Fig. 5). Oxygenation levels during clastic-dominated phases of deposition were apparently very low at both sites. Direct comparison of high-amplitude events in this interval is hindered somewhat by current uncertainties regarding correlation of discrete limestones between the two cores and questionable ichnocoenosis-rank assignments of certain limestones. However, cross-basin redox variations during limestone deposition, if they existed at all, were likely minor.

Eastward reduction in benthic oxygenation is again more evident in the upper Bridge Creek interval. Unlike that of the Portland core, where numerous low-amplitude oxygenation events are recorded, this interval in the Bounds core contains no fabrics that can be ascribed unequivocally to bioturbation. Possible biogenic disruption is suspected in some thin calcarenitic layers (Fig. 5). However, as previously noted, these cannot be easily distinguished from abiogenic, compactionally deformed fabrics.

Comparison of oxygenation histories for the two distal offshore sites addressed here suggest the following: (1) during limestone deposition, benthic oxygenation levels were generally high and varied little, if any at all, between the two sites; while, (2) during clastic-dominated phases of carbonate cycles, oxygenation levels were generally depleted but lower to the east at the Bounds locale. The cause of the latter trend cannot be adequately assessed on the basis of ichnologic data. If the western Kansas sequence accumulated in deeper water, this trend could reflect a depth-related oxygenation gradient associated with basinal thermohaline stratification. Alternatively, the eastward decrease in oxygenation may reflect a lateral redox gradient possibly associated with increased upwelling, higher productivity, and oxygen demand in the eastern part of the seaway. Various sedimentologic, geochemical, and paleontologic data reported elsewhere in this volume will likely help constrain the mechanisms responsible for spatial redox variations. Whatever the cause, spatial trends in oxygenation recognized from ichnological analysis are consistent with and help to explain parallel trends in organic geochemical data (Dean and Arthur, this volume).

SUMMARY

(1) On the basis of ichnofossil diversity, burrow size, and bioturbation depth, four recurring oxygen-related ichnocoenoses are recognized in bioturbated intervals of the rhythmically bedded Bridge Creek Limestone in two cores: the Portland core from east-central Colorado and the Bounds core from western Kansas.

(2) Paleo-oxygenation curves, constructed on the basis of the vertical disposition of ichnocoenoses and intervening laminites, document a broad, progressive decrease in benthic oxygenation through Bridge Creek time; a high-amplitude redox cyclicity corresponding to limestone-marly shale couplets; and a higher-frequency, low-amplitude redox cyclicity expressed within marlstone/marly shale intervals. Together, low- and high-amplitude cycles in the lower Bridge Creek define a bundling pattern that suggests, although perhaps deceptively, control by orbital eccentricity and axial precession.

(3) Comparison of paleo-oxygenation histories between central Colorado and western Kansas demonstrate lower benthic oxygen availability towards the east during clastic-dominated phases of carbonate cycles. Provided that the positions of these two sites with respect to water-mass stratification were not significantly different, this spatial pattern may reflect lateral basin gradients

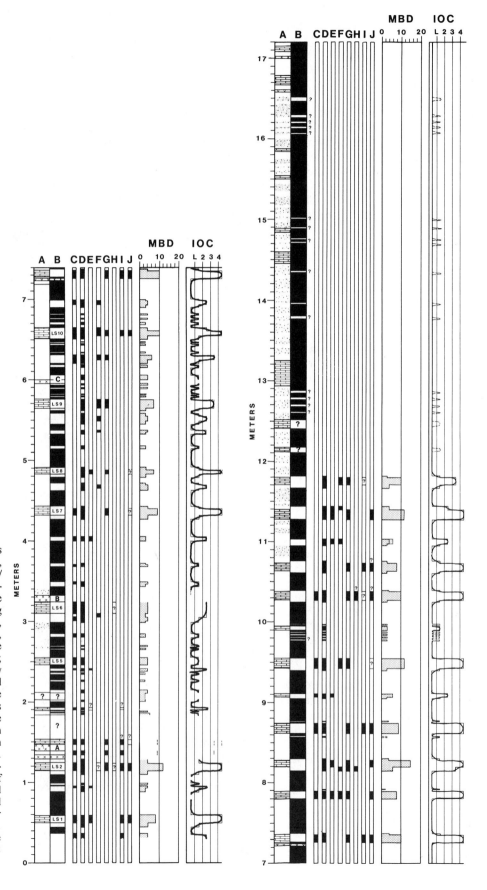

FIG. 5.— Distribution of general lithologies (A; limestones and bentonites are patterned, calcarenites are stippled, and marlstones/marly shales are white), general sedimentary fabrics (B; bioturbated intervals are white, laminites are black), and recurring biogenic structure types (C, *Chondrites*; D, small *Planolites*; E, *Taenidium*; F, *Zoophycos*-like sheet structures; G, large *Planolites*; H, *Zoophycos*; I, *Teichnichnus*; and J, *Thalassinoides*), maximum burrow diameter (MBD, in mm), and interpreted oxygenation curve (IOC) for the Bridge Creek interval of the Bounds core. IOC is constructed using four reference lines. Line L represents oxygen levels below which laminites are preserved and above which producers of ichnocoenosis 1 can survive. Lines 2, 3, and 4 represent threshold oxygen levels required for development of ichnocoenoses 2, 3, and 4, respectively. Bed designations for limestones (LS1-LS10) and thicker bentonites (A, B, and C) in lower Bridge Creek (at left) are after Elder (1985). Boundary between middle and upper Bridge Creek Limestone is inferred to be at ~11.8 m.

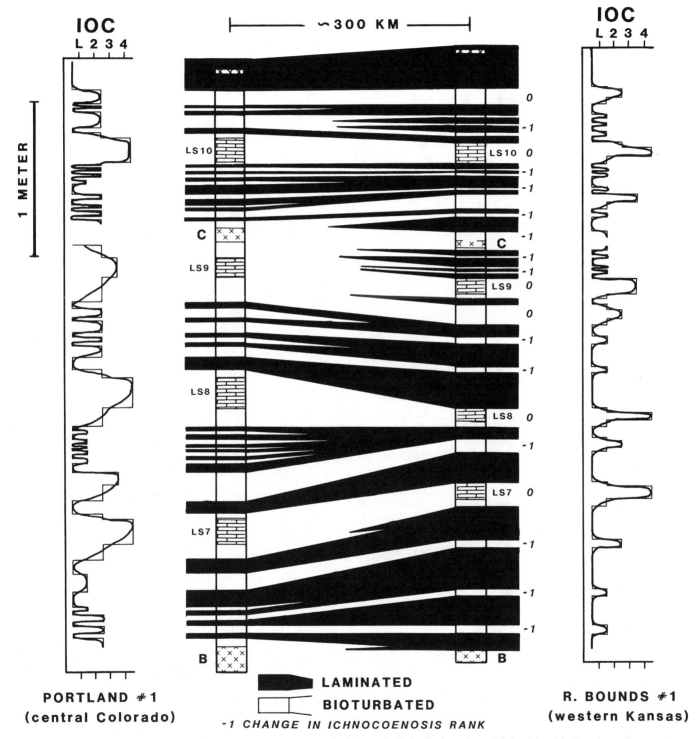

FIG. 6. — Oxygenation curves (IOC) for the upper part of the lower Bridge Creek Limestone in the Portland core (central Colorado) and the Bounds core (western Kansas). Comparison of oxygenation histories, facilitated by limestone (LS7-LS10) and bentonite (B and C) marker beds, indicates eastward increase in laminated fabrics and general reduction in both thickness and ichnocoenosis rank of bioturbated beds in marlstone/marly shale intervals.

in productivity and oxygen demand across the distal offshore part of the Western Interior Seaway.

(4) Further integration of ichnological observations with geochemical, sedimentologic, and paleontologic data from the Bridge Creek Limestone reported elsewhere in this volume will provide a clearer understanding of paleoceanography, cyclic sedimentation, and organic matter accumulation in the basin.

ACKNOWLEDGMENTS

Ichnological aspects of the Western Interior Seaway Drilling Project (WISDP) were supported by a grant from the Department of Energy to Michael A. Arthur and Walter E. Dean. The principal investigators and other members of the WISDP team, particularly Bradley B. Sageman, are acknowledged for helpful discussion. Walter E. Dean, David J. Bottjer, and Allan A. Ekdale provided helpful comments on an earlier version of this manuscript.

REFERENCES

ARTHUR, M. A., AND DEAN, W. E., 1991, A holistic geochemical approach to cyclomania: Examples from Cretaceous pelagic limestone sequences, in Einsele, G., Ricken, W., and Seilacher, A., eds., Cycles and Events in Stratigraphy, New York, Springer-Verlag, p. 126-166.

ARTHUR, M. A., DEAN, W. E., POLLASTRO, R. M., CLAYPOOL, G. E., AND SCHOLLE, P. A., 1985, Comparative geochemical and mineralogical studies of two cyclic transgressive pelagic limestone units, Cretaceous Western Interior Basin, in Pratt, L. M., Kauffman, E. G., and Zelt, F. B., eds., Fine-grained Deposits and Biofacies of the Cretaceous Western Interior Seaway: Evidence for Cyclic Sedimentary Processes, Field Trip Guidebook No. 4: Tulsa, Society of Economic Paleontologists and Mineralogists, p. 16-27.

BARRON, E. J., ARTHUR, M. A., AND KAUFFMAN, E. G., 1985, Cretaceous rhythmic bedding sequences: A plausible link between orbital variations and climate: Earth and Planetary Science Letters, v. 72, p. 327-340.

EICHER, D. L., AND DINER, R., 1989, Origin of the Cretaceous Bridge Creek cycles in the Western Interior, United States: Palaeogeography, Palaeoclimatology, Palaeoecology, v. 74, p. 127-146.

EICHER, D. L., AND DINER, R., 1991, Environmental factors controlling Cretaceous limestone-marlstone rhythms, in Einsele, G., Ricken, W., and Seilacher, A., eds., Cycles and Events in Stratigraphy, New York, Springer-Verlag, p. 79-93.

EKDALE, A. A., BROMLEY, R. G., AND PEMBERTON, S. G., 1984, Ichnology: Tulsa, Society of Economic Paleontologists and Mineralogists Short Course Notes 15, 317 p.

ELDER, W. P., 1985, Biotic patterns across Cenomanian-Turonian extinction boundary near Pueblo, Colorado, in Pratt, L. M., Kauffman, E. G., and Zelt, F. B., eds., Fine-grained Deposits and Biofacies of the Cretaceous Western Interior Seaway: Evidence for Cyclic Sedimentary Processes, Field Trip Guidebook No. 4: Tulsa, Society of Economic Paleontologists and Mineralogists, p. 157-169.

ELDER, W. P., AND KIRKLAND, J. I., 1985, Stratigraphy and depositional environments of the Bridge Creek Limestone Member of the Greenhorn Limestone at Rock Canyon Anticline near Pueblo, Colorado, in Pratt, L. M., Kauffman, E. G., and Zelt, F. B., eds., Fine-grained Deposits and Biofacies of the Cretaceous Western Interior Seaway: Evidence for Cyclic Sedimentary Processes, Field Trip Guidebook No. 4: Tulsa, Society of Economic Paleontologists and Mineralogists, p. 122-134.

ELDER, W. P., GUSTASON, E. R., AND SAGEMAN, B. B., 1994, Correlation of basinal carbonate cycles to nearshore parasequences in the Late Cretaceous Greenhorn seaway, Western Interior, U.S.A.: Geological Society of America Bulletin, v. 106, p. 892-902.

FISCHER, A. G., 1980, Gilbert-bedding rhythms and geochronology: Boulder, Geological Society of America Special Paper 183, p. 93-104.

FISCHER, A. G., 1986, Climatic rhythms recorded in strata: Review of Earth and Planetary Science, v. 14, p. 351-376.

FISCHER, A. G., HERBERT, T., AND PREMOLI-SILVA, I., 1985, Carbonate bedding cycles in Cretaceous pelagic and hemipelagic sequences, in Pratt, L. M., Kauffman, E. G., and Zelt, F. B., eds., Fine-grained Deposits and Biofacies of the Cretaceous Western Interior Seaway: Evidence for Cyclic Sedimentary Processes, Field Trip Guidebook No. 4: Tulsa, Society of Economic Paleontologists and Mineralogists, p. 1-10.

GILBERT, G. K., 1895, Sedimentary measurement of geologic time: Journal of Geology, v. 3, p. 121-127.

HATTIN, D. E., 1971, Widespread, synchronously deposited, burrow-mottled limestone beds in Greenhorn Limestone (Upper Cretaceous) of Kansas and southeastern Colorado: American Association of Petroleum Geologists Bulletin, v. 55, p. 412-431.

HATTIN, D. E., 1975, Petrology and origin of fecal pellets in Upper Cretaceous strata of Kansas and Saskatchewan: Journal of Sedimentary Petrology, v. 45, p. 686-696.

HATTIN, D. E., 1985, Distribution and significance of widespread, time-parallel pelagic limestone beds in Greenhorn Limestone (Upper Cretaceous) of the central Great Plains and southern Rocky Mountains, in Pratt, L. M., Kauffman, E. G., and Zelt, F. B., eds., Fine-grained Deposits and Biofacies of the Cretaceous Western Interior Seaway: Evidence for Cyclic Sedimentary Processes, Field Trip Guidebook No. 4: Tulsa, Society of Economic Paleontologists and Mineralogists, p. 28-37.

KAUFFMAN, E. G., 1977, Geological and biological overview: Western Interior Cretaceous basin: The Mountain Geologist, v. 14, p. 75-99.

OBRADOVICH, J. D., 1993, A Cretaceous time scale, in Caldwell, W. G. E., and Kauffman, E. G., eds., Evolution of the Western Interior basin: St. Johns, Geological Association of Canada Special Paper 39, p. 379-396.

PRATT, L. M., 1984, Influence of paleoenvironmental factors on preservation of organic matter in middle Cretaceous Greenhorn Formation: American Association Petroleum Geologists Bulletin, v. 68, p. 1146-1159.

PRATT, L. M., ARTHUR, M. A., DEAN, W. E., AND SCHOLLE, P. A., 1993, Paleo-oceanographic cycles and events during the Late Cretaceous in the Western Interior seaway of North America, in Caldwell, W. G. E., and Kauffman, E. G., eds., Evolution of the Western Interior basin: St. Johns, Geological Association of Canada Special Paper 39, p. 333-353.

RICKEN, W., 1991, Variation of sedimentation rates in rhythmically bedded sediments. Distinction between depositional types, in Einsele, G., Ricken, W., and Seilacher, A., eds., Cycles and Events in Stratigraphy: New York, Springer-Verlag, p. 167-187.

RICKEN, W., 1994, Complex rhythmic sedimentation related to third-order sea-level variations: Upper Cretaceous, Western Interior basin, U.S.A., in de Boer, P., and Smith, D. G., eds., Orbital Forcing and Cyclic Sequences, International Association of Sedimentologists Special Publication 19, Oxford, Blackwell Scientific Publications, p. 167-193.

SAVRDA, C. E., 1992, Trace fossils and benthic oxygenation, in Maples, C., and West, R., Trace Fossils: Knoxville, Paleontological Society Short Course Notes 5, p. 172-196.

SAVRDA, C. E., AND BOTTJER, D. J., 1987, Trace fossils as indicators of bottom-water redox conditions in ancient marine environments, in Bottjer, D. J., ed., New concepts in the use of biogenic sedimentary structures for paleoenvironmental interpretation: Los Angeles, Society of Economic Paleontologists and Mineralogists, Pacific Section, Volume and Guidebook 52, p. 3-26.

SAVRDA, C. E., AND BOTTJER, D. J., 1989, Trace fossil model for reconstructing oxygenation histories of ancient marine bottom waters: Application to Upper Cretaceous Niobrara Formation, Colorado: Palaeogeography, Palaeoclimatology, Palaeoecology, v. 74, p. 49-74.

SAVRDA, C. E., AND BOTTJER, D. J., 1991, Oxygen-related biofacies in marine strata: An overview and update, in Tyson, R. V., and Pearson, T. H., eds., Modern and ancient continental shelf anoxia: London, Geological Society of London Special Publication 58, p. 201-219.

SAVRDA, C. E., AND BOTTJER, D. J., 1994, Ichnofossils and ichnofabrics in rhythmically bedded pelagic/hemipelagic carbonates: Recognition and evaluation of benthic redox and scour cycles, in de Boer, P., and Smith, D. G., eds., Orbital Forcing and Cyclic Sequences, International Association of Sedimentologists Special Publication 19: Oxford, Blackwell Scientific Publications, p. 195-210.

WATKINS, D. K., 1989, Nannoplankton productivity fluctuations and rhythmically-bedded carbonates of the Greenhorn Limestone (Upper Cretaceous): Palaeogeography, Palaeoclimatology, Palaeoecology, v. 74, p. 75-85.

ICHNOCOENOSES IN THE NIOBRARA FORMATION: IMPLICATIONS
FOR BENTHIC OXYGENATION HISTORIES

CHARLES E. SAVRDA

Department of Geology, Auburn University, Auburn, AL 36849-5305

ABSTRACT: The vertical stacking patterns of laminites and six oxygen-related ichnocoenoses were used to reconstruct paleo-oxygenation histories for the Niobrara Formation (Fort Hays Member and lower parts of the Smoky Hill Member) as expressed in two cores—USGS #1 Portland from east-central Colorado and Amoco #1 Rebecca Bounds from western Kansas. Oxygenation records for the Fort Hays reflect high-amplitude redox cycles that correspond to decimeter-scale limestone/shale couplets. Ichnofabrics of the lowermost Smoky Hill record a complex history of intermediate- to low-amplitude redox fluctuations that are superimposed upon a general deoxygenation trend. The remainder of the Smoky Hill is characterized by low- and high-frequency, low- to intermediate-amplitude redox cycles defined by the decimeter-scale alternation between relatively thick laminite sequences and intervals containing clusters of closely spaced, thin bioturbated beds separated by laminites. Decimeter-scale redox cyclicity throughout the study interval generally corresponds to carbonate rhythms and is tentatively atttributed to astronomically forced climate cycles mediated by axial precession. Lower-frequency cycles (for example, those reflected by varying cluster spacing in the Smoky Hill) may reflect a control by orbital eccentricity. High-frequency cycles, such as those reflected by centimeter-scale ichnofabric variations within Smoky Hill clusters, record periodic or episodic processes that operated at time scales shorter than Milankovitch orbital cycles. Comparison of paleo-oxygenation histories for the two cores indicate that benthic oxygenation levels were lower toward the east at least during deposition of the lowermost Smoky Hill.

INTRODUCTION

As part of the Cretaceous Western Interior Drilling Project, detailed ichnologic investigations were undertaken of two rhythmically bedded, pelagic-hemipelagic carbonate sequences deposited during peak transgressive phases in the distal parts of the Western Interior Seaway—the Bridge Creek Limestone Member of the Greenhorn Formation (Cenomanian-Turonian) and the Niobrara Formation (Upper Turonian-Lower Campanian). The immediate goals of these studies were to reconstruct benthic oxygenation histories for both sequences in two cores using ichnocoenosis stacking patterns, and to evaluate the nature of temporal cycles and spatial trends in redox conditions within the basin. Results of the Bridge Creek study are reported elsewhere (Savrda, this volume). This paper (1) describes recurring ichnofossils and oxygen-related ichnocoenoses identified in the Niobrara Formation, (2) presents paleo-oxygenation curves for the lower parts of the Niobrara in both cores, and (3) summarizes their general implications for depositional cyclicity and paleoceanographic conditions.

NIOBRARA FORMATION

General Stratigraphy

The Niobrara Formation comprises the bulk of the Niobrara cyclothem as represented in the distal offshore parts of the Western Interior Seaway (Kauffman, 1977, 1984). The Niobrara is divided into two members (Hattin, 1982; Barlow and Kauffman, 1985). The basal Fort Hays Member (Turonian-Coniacian) varies from 5 to 40 m thick and is characterized by decimeter- to meter-scale alternation of relatively thick, resistant, light gray micritic limestones and thin, dark gray to black calcareous shale or marlstone seams. The overlying Smoky Hill Member, which is typically on the order of 200 m thick, is subdivided into seven informal units based on broad trends in carbonate content (Scott and Cobban, 1964; Hattin, 1982). In ascending order, these are the shale and limestone, lower shale, lower limestone, middle shale, middle limestone, upper shale, and upper limestone. Like the Fort Hays, each of these units is characterized by decimeter-scale carbonate rhythms, in this case couplets of marlstone and limestone.

Carbonate Cyclicity

Carbonate rhythms in the Niobrara generally are attributed to astronomically forced paleoclimatic cycles mediated by Milankovitch orbital parameters (e.g., Gilbert, 1895; Fischer, 1980, 1986, 1993; Hattin, 1982; Arthur et al., 1984, 1986; Barlow and Kauffman, 1985, Barron et al., 1985; Scholle and Pollastro, 1985; Bottjer et al., 1986; Savrda and Bottjer, 1989; Pratt et al., 1993). Several workers suggested that the decimeter-scale carbonate rhythms of the Niobrara have periodicities of approximately 20 ky and, therefore, implicated axial precession as a forcing mechanism (e.g., Fischer, 1986, 1993; Pratt et al., 1993). Bundles of bedding couplets in both the Fort Hays and Smoky Hill Members apparently define longer-term cycles of approximately 100 and 400 ky and, hence, have been attributed to the short and long cycles of orbital eccentricity (Fischer et al., 1985; Laferriere et al., 1987; Fischer, 1993). The broader alternation of carbonate-rich and carbonate-poor packages expressed by the Fort Hays Member and the informal units of the Smoky Hill has been attributed to regionally synchronous sea-level cycles (e.g., Barlow and Kauffman, 1985). Fischer (1991) proposed that these long-term carbonate variations may be related to a lower-frequency (~1300 ka) eccentricity cycle.

The specific paleoceanographic mechanisms responsible for the decimeter-scale carbonate rhythmicity in the Niobrara are not well understood. Some workers suggest that these rhythms record dilution cycles that are associated with wet-dry climate cycles and their impact on clastic input and water-column stratification (e.g., Arthur et al., 1985; Savrda and Bottjer, 1989; Arthur and Dean, 1991; Pratt et al., 1993). Others attribute high-frequency carbonate variations to carbonate-productivity cycles (e.g., Fischer et al., 1985; Ricken, 1991; Fischer, 1993) or, for some parts of the Smoky Hill, to a combination of productivity and dilution mechanisms (Ricken, 1994).

Diagenesis

The Niobrara was affected by a variety of early and late diagenetic processes (Scholle, 1977; Hattin, 1981; Scholle and Pollastro, 1985; Laferriere, 1992; Ricken, 1994), of which dissolution compaction is the most pronouced. The extent to which dissolution affected the Niobrara varies both stratigraphically and

Copyright 1998 SEPM (Society for Sedimentary Geology), ISBN 1-56576-044-1, p. 137-151.

geographically (Scholle, 1977; Laferriere, 1992; Ricken, 1994). Dissolution compaction and associated diagenetic enhancement of bedding is most evident in the Fort Hays Member, where original sedimentary fabrics near limestone-shale contacts, and in some cases, in interior parts of limestone beds, were variably obscured by solution seams and microstylolites. Magnitude of dissolution compaction increases from east to west across the basin, in response to progressively deeper burial (Scholle, 1977). This diagenetic modification inhibited detailed ichnofabric descriptions of some Fort Hays intervals. However, its effect on interpretations of benthic oxygenation histories for the Niobrara overall is regarded as minimal.

Previous Ichnologic Studies

The Niobrara was the subject of several previous ichnologic investigations of varying scope. Earlier studies (e.g., Frey, 1970; Hattin, 1982), which emphasized the diverse trace fossil assemblages of the Fort Hays, addressed ichnologic implications for paleobathymetry, environmental energy, sedimentation rates, and substrate consistency. Later works emphasized the use of ichnofossils in evaluating basin oxygenation. Barlow and Kauffman (1985) employed degree of bioturbation to assess short-term redox cyclicity and long-term paleo-oxygenation trends recorded in the Niobrara. Subsequently, Savrda and Bottjer (1989, 1993) performed more detailed ichnofabric studies of the Fort Hays Member and basal shale and limestone unit of the Smoky Hill from the western Denver basin. In the latter investigation, four recurring oxygen-related ichnocoenoses were recognized and employed to reconstruct paleo-oxygenation curves. The current study represents an extension of this ichnocoenosis approach to a thicker interval of the Niobrara at two localities.

ANALYTICAL APPROACH

Studies of Niobrara ichnofossils, all of which were performed at the USGS Core Research Center in Denver, focused on the same two cores addressed in comparable studies of the Bridge Creek Limestone (Savrda, this volume): (1) the USGS #1 Portland core drilled near Florence, central Colorado; and (2) Amoco #1 Rebecca K. Bounds core drilled in Greeley County, western Kansas (for locations, see Dean and Arthur, this volume). The interval of study includes the Fort Hays Member and the lower parts of the Smoky Hill Member. This encompassed approximately 48 and 43 m of section in the Portland (108-260 ft) and Bounds (530-674 ft) cores, respectively. Core recovery of the target interval in the Portland core was 100%. Recovery was almost 100% for the Bounds core, but several short core segments were missing due to earlier sampling, and some minor core misplacement was recognized in the basal part of the Smoky Hill.

Sedimentologic and ichnologic observations were made in much the same manner as described for the Bridge Creek Limestone (Savrda, this volume). Wetted core surfaces were examined to document vertical changes in lithology and sedimentary fabric (laminated vs. bioturbated) at the centimeter scale. Ichnocoenoses were defined and delineated within bioturbated intervals based on recurring associations of burrow types. For each ichnocoenosis occurrence, a census of burrow forms was taken, apparent maximum burrow diameter was measured, and, where possible, burrow-penetration depth was estimated from piped-zone thickness. Vertical stacking patterns of ichnocoenoses and associated laminites were used to construct paleo-oxygenation

TABLE 1. — NIOBRARA UNITS IN PORTLAND AND BOUNDS CORES

Unit[1]	Core	Depth in Core (ft)	Height in Column[2] (m)	Thickness (m)
Unit 1- FH	Portland	212.5-258.2	0.92-14.90	13.98
	Bounds	604.5-671.8	0.55-20.47	19.92
Unit 2- SH	Portland	181.4-212.5	14.90-24.45	9.55
	Bounds	574.7-604.5	20.47-29.27	8.80
Unit 3- SH	Portland	163.3-181.4	24.45-29.90	5.45
	Bounds	548.6-574.7	29.27-37.03	7.76
Unit 4- SH	Portland	151.6-163.3	29.90-34.0	4.10
	Bounds	534.2-548.6	37.03-41.7	4.67
Unit 5- SH	Portland	105.4-151.6	34.0-48.0	14.0
	Bounds	529.6-534.2	41.7-43.06	1.36

[1]Units recognized in Fort Hays (FH) and Smoky Hill (SH) Members
[2]Height depicted in stratigraphic columns (Figs. 2, 3)

curves for the core intervals in the manner described by Savrda and Bottjer (1991, 1994), and to evaluate the nature of temporal redox cycles. Spatial variations in benthic oxygenation were assessed by comparing paleo-redox histories for age-equivalent intervals in the cores.

SUBDIVISION AND PRELIMINARY CORRELATION OF CORES

Five lithologic units are recognized in both the Portland and Bounds cores (Table 1). Unit 1 corresponds to the Fort Hays Member, while units 2 through 5 comprise the lower parts of the Smoky Hill. Placement of unit boundaries was somewhat subjective, and the degree of synchroneity of these boundaries between the two cores is not well established. Nonetheless, the five units are believed to represent crudely defined lithochronozones and are used herein as the basis for core comparison.

The Fort Hays is approximately 14 m thick in the Portland core and 20 m thick in the Bounds cores. This unit is characterized by rhythmic alternation of thick, well-cemented, light- to medium-gray, micritic limestone beds and relatively thin, poorly indurated, dark gray calcareous shale/marlstone beds with minor bentonites. Limestone beds range from 5 to 120 cm thick (average = 25-30 cm). Intervening calcareous shale/marlstone beds typically are thin (1 to 3 cm), but locally are up to 26 cm thick. Contacts between limestone and carbonate-poor beds are variably obscured by solution seams produced by chemical compaction. Approximately 54 and 48 limestone and calcareous shale/marlstone couplets are recognized in the Bounds and Portland cores, respectively.

The four stratigraphic units in the Smoky Hill were delineated based on apparent differences in carbonate content (as indicated by color variations) and, to a lesser extent, on tentative

TABLE 2.—LAMINITES AND OXYGEN-RELATED ICHNOCOENOSES (ORI) IN NIOBRARA FORMATION

PORTLAND CORE % of occurrences containing eight recurring traces[1]

Fabric/ Ichnocoen.	No. of Occurrences	Range in Thickness (cm)	Ave. Thickness (cm)	Chs	Chl	Ps	Ta	Zo	Te	Pl	Th	Range in MBD (mm)	Ave. MBD (mm)
Lam	283	0.2-70	7.0	-	-	-	-	-	-	-	-	-	-
ORI 1	47	0.2-1	0.5	100	-	-	-	-	-	-	-	1-2	1.0
ORI 2	155	1-12	2.5	5	32	100	-	-	-	-	-	2-5	3.0
ORI 3	~60	1-11	5.0	5	44	100	100	-	-	-	-	3-6	4.8
ORI 4	~37	2-45	12.0	8	83	94	39	58	85	-	-	3-8	5.2
ORI 5	29	5-40	12.0	-	50	81	22	41	63	100	-	7-12	8.5
ORI 6	36	6-90	28.0	4	48	66	14	34	74	84	100	9-20	15.0

BOUNDS CORE % of occurrences containing eight recurring traces[1]

Fabric/ Ichnocoen.	No. of Occurrences	Range in Thickness (cm)	Ave. Thickness (cm)	Chs	Chl	Ps	Ta	Zo	Te	Pl	Th	Range in MBD (mm)	Ave. MBD (mm)
Lam	230	0.2-92	8.0	-	-	-	-	-	-	-	-	-	-
ORI 1	104	0.2-1	0.5	100	-	-	-	-	-	-	-	1-2	1.0
ORI 2	~58	1-9	2.5	49	51	100	-	-	-	-	-	1-4	2.6
ORI 3	~59	1-12	4.0	50	35	93	100	-	-	-	-	2-7	3.9
ORI 4	33	2-30	7.5	24	45	91	54	70	58	-	-	3-8	4.9
ORI 5	39	5-26	11.0	17	60	88	25	55	73	100	-	6-10	7.6
ORI 6	37	6-80	31.0	2	53	80	4	62	78	78	100	10-20	13.6

[1]Chs = *small Chondrites*, Chl = large *Chondrites*, Ps = small *Planolites*, Ta = *Taenidium*, Zo = *Zoophycos*, Te = *Teichichnus*, Pl = large *Planolites*, Th = *Thalassinoides*

correlation of bentonitic intervals. Unit 2 is characterized by decimeter-scale alternation of beds of limestone, marly limestone, marlstone, and calcareous/marly shale and contains several potentially correlative bentonites. Unit 3 is relatively carbonate-poor and is characterized by calcareous/marly shale and subordinate marlstone. Unit 4 appears more calcareous and contains alternating beds of calcareous to marly shale and marly limestone. Unit 5 resembles unit 3. Thicknesses of units 2, 3, and 4, are approximately 9.5 m and 8.8 m, 5.5 m and 7.7 m, and 4.1 and 4.7 m in the Portland and Bounds cores, respectively (Table 1). Unit 5 extends for approximately 14 m to the top of the studied interval in the Portland core. Only the lowermost 1.4 m of unit 5 is represented in the Bounds section.

Given the lack of carbonate data, the relationships between units 2 through 5 and the informal subunits of the Smoky Hill defined by Scott and Cobban (1964) are questionable. Unit 2 is likely equivalent to the basal shale and limestone subunit. The placement of the upper three units is less certain. All three of these units may correspond to the lower shale subunit. Alternatively, units 3, 4, and 5 may roughly correspond to the lower shale, lower limestone, and middle shale subunits, respectively.

NIOBRARA SEDIMENTARY FABRICS

Laminites

The term "laminite" used here describes all intervals within which background fabrics are characterized by primary parallel lamination. Approximately 283 and 230 of these intervals, ranging from a few millimeters to 92 cm in thickness (average=7 cm), were recognized in the Portland and Bounds cores, respectively (Table 2). Niobrara laminites, which are largely restricted to calcareous or marly shales, are characterized by alternating diffuse yet continuous thin (1 to 5 mm), light, and dark laminae

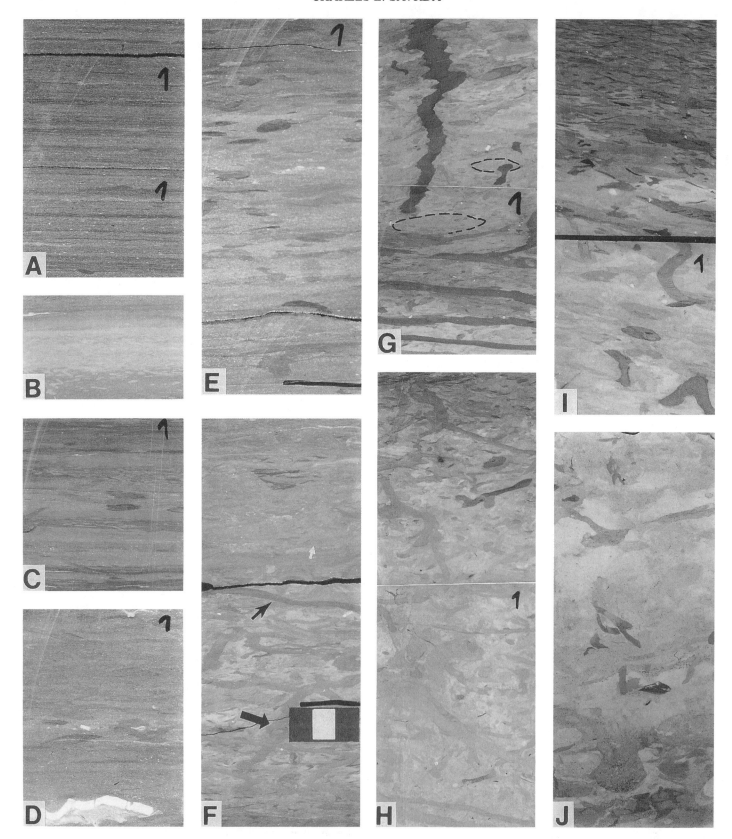

(Fig. 1A). They typically appear speckled, owing to the presence of abundant small (<1 mm), elliptical fecal pellets of presumed planktonic origin (see Hattin, 1975). Whole or partly fragmented inoceramids and/or other bivalves are locally abundant. However, calcarenitic horizons like those in the Bridge Creek Limestone are absent or very rare.

Although some thicker Niobrara laminites contain no vestige of macroscopic bioturbation, most laminites are variably disrupted by burrows extending down from overlying bioturbated intervals. Lamina disruption generally increases upward toward the bases of bioturbated beds, and disruption is nearly complete in intervals that are bounded by closely spaced bioturbated beds (Fig. 1C).

Bioturbated Beds

General Character.—

Bioturbated intervals are defined here as strata that are characterized by bioturbated, homogeneous to diffusely burrow-mottled background fabrics as well as discrete burrows; vestiges of primary sedimentary fabric are absent altogether. These intervals typically are underlain by piped zones, in which burrows derived from bioturbated intervals variably disrupt primary laminae in subjacent laminite beds. Bioturbated intervals in the Niobrara range from less than 1 cm to over 2.5 m in thickness, and they make up approximately 60% and 57% of the investigated parts of the Niobrara in the Portland and Bounds cores, respectively.

Recurring Ichnofossils.—

Eight common ichnofossil types, assigned to six different ichnogenera, are recognized within bioturbated beds and associated piped zones of the Niobrara in both the Portland and Bounds cores. These include: two forms of *Chondrites*, which are distinguished primarily on the basis of burrow diameter (large and small *Chondrites*); two forms of *Planolites*, also distinguished on the basis of burrow size (large and small *Planolites*); *Taenidium*; *Zoophycos*; *Teichichnus*; and *Thalassinoides*.

Chondrites is a complex burrow system of regularly branching tunnels of relatively small, yet uniform diameter. On Niobrara core surfaces, both small and large *Chondrites* appear most commonly as clusters of circular to elliptical blebs that lack any internal structure. Tunnel branching is observed locally within clusters. Specimens of small *Chondrites* have tunnel diameters ranging from 0.5 to 1.0 mm and, in most cases, appear to be relatively short; burrow penetration depths typically are only 0.5 to 2.0 cm (Figs. 1B, C). Burrow systems referred to as large *Chondrites* have tunnel diameters of 2 to 3 mm (Fig. 1F) and generally are more vertically extensive.

Planolites is an unlined, typically unbranched, straight to curved, horizontal to subhorizontal cylindrical burrow with structureless fill. On vertical core faces, both small and large *Planolites* appear as circular to highly elliptical, horizontal to subhorizontal pods, depending on degree of compaction and/or relations between burrow axes and core surfaces (Figs. 1D, G-I). Although burrow diameters vary over the complete range from 3 to 12 cm, two distinct size populations are recognized. The diameters of small and large *Planolites* average 3 mm (range = 1 to 5 mm) and 8 mm (range = 6 to 12 mm), respectively.

Taenidium refers to typically branched, unlined and unmantled, cylindrical burrows with meniscate backfill. In the Niobrara, *Taenidium* typically appears as circular to elliptical, concentrically laminated pods (Fig. 1E). Clear evidence of branching is rare. Burrow diameters range from 2 to 7 mm (average = 4.4 mm). As inferred from piped-zone thicknesses, the vertical extent of *Taenidium* varies highly but typically exceeds 35 cm. In some intervals, particularly those where color contrast between burrow fills and ambient sediment is poor, the differentiation between *Taenidium* and small *Planolites* is problematic.

Zoophycos is a complex spreiten structure. On core surfaces, *Zoophycos* appears as horizontal to shallowly inclined structures with straight to highly irregular axes and chevron-shaped or meniscate internal lamination (e.g., Figs. 1F, G). Morphological variability reflects the presence of both planar and helicoidal forms. Heights of spreiten, which reflect causative burrow diameters, generally range from 3 to 8 mm (average = 5 mm). Although recognized in both cores, well-defined *Zoophycos* is more common in the Bounds core.

Teichichnus is a vertical spreiten structure formed by a series of vertically stacked, longitudinally nested, horizontal to gently inclined burrows. As seen in cross sections of Niobrara cores, *Teichichnus* is expressed as vertical to subvertical structures with typically concave-up spreite laminae (e.g., Figs. 1F-I). Some specimens exhibit relatively straight vertical axes and, hence, are allied with *Teichichnus rectus*. However, most have highly irregular axes and are more appropriately assigned to the ichnospecies *Teichichnus zigzag* (Frey and Bromley, 1985). Height of spreite range from 2 to 20 cm, while spreite width, taken as a measure of causative burrow diameter, varies from 3 to 8 mm (average=5 mm).

Thalassinoides refers to three-dimensional branching systems of relatively large, unlined cylindrical burrows. In the Niobrara, structures assigned to this ichnotaxon are generally restricted to limestones, where they are variably expressed as circular to elliptical pods and vertical to steeply inclined shafts with localized branch junctions (Fig. 1J). Although certain identification is problematic in some cases, *Thalassinoides* generally could be distinguished from large *Planolites* based on the presence of vertical burrow elements or obvious branching and/or larger burrow diameters in the former. Burrow diameters of Niobrara *Thalassinoides* range from 9 to 20 mm (average=14.4 mm).

FIG. 1.—Examples of laminites and ichnocoenoses in the Niobrara Formation. (A) Pelagic laminite characterized by relatively continuous, fine lamination. Light specks are pelagic fecal pellets and foraminifera(?). Note localized disruption of laminae by small burrows piped from superjacent beds. (B) Thin oxygenation-event bed (lighter layer) characterized by ichnocoenosis 1 (small *Chondrites* only). (C) Ichnocoenosis 1 in multiple thin beds (lighter layers) separated by remnant laminites. Larger burrows near center are *Taenidium* piped from above. (D)- Ichnocoenosis 2, dominated here by small *Planolites*, overlying partially disrupted laminite with inoceramid(?). (E) Ichnocoenosis 3 dominated by *Taenidium* and small *Planolites*. Note disrupted laminite at base, and transition to ichnocoenosis 2 near top. (F) Ichnocoenosis 4 characterized by *Teichichnus* (e.g., large black arrow), *Zoophycos* (e.g., small black arrow), large *Chondrites* (e.g., small white arrow), small *Chondrites*, and small *Planolites* and/or *Taenidium*. Bar scale is in centimeters. (G) Ichnocoenosis 5 with vague large *Planolites* (dashed outlines) overprinted by *Zoophycos*, *Teichichnus*, small *Planolites* and/or *Taenidium*, and large *Chondrites*, some of which are piped from an overlying interval dominated by ichnocoenosis 4. (H) Ichnocoenosis 5 grading to ichnocoenosis 4 near top. (I) Ichnocoenosis 5 (in lighter limestone) grading upwards to ichnocoenoses 4 and 2 (darker marlstone). (J) Ichnocoenosis 6 characterized by *Thalassinoides*. Scales in (A) through (E) and (G) through (J) are the same as in (F). All examples are from the Portland core.

Chondrites, *Zoophycos*, and *Teichichnus* all are regarded as combined feeding/dwelling structures (fodinichnia) produced by various infaunal, nonvagile, worm-like organisms. *Planolites* and *Taenidium* generally are interpreted as endogenic pascichnia produced by vagile, deposit-feeding, vermiform animals. *Thalassinoides* is attributed to the feeding/dwelling activities of decapod crustaceans.

Recurring Ichnocoenoses.—

Ichnocoenoses are ecologically pure assemblages of ichnofossils that are inferred to derive from the activities of a single endobenthic community (Bromley, 1990). Six ichnocoenoses (ichnocoenoses 1 through 6) are recognized within bioturbated intervals of the Niobrara. Data pertaining to relative abundance, composition and diversity of ichnofossil assemblages, and burrow-size parameters for each of these ichnocoenoses are given in Table 2.

Stratigraphic intervals assigned to ichnocoenosis 1 are characterized by the presence of small *Chondrites* only (Figs. 1B, C); all other traces are absent. Approximately 47 and 104 such intervals are recognized in the Portland and Bounds cores, respectively. Thickness of these intervals and the thickness of their subjacent piped zones generally range from 2 to 10 mm and 0.5 to 2 cm, respectively. Thicker (2-3 cm) intervals are rare and apparently represent amalgamated beds produced by two or more closely spaced events.

Ichnocoenosis 2 refers to intervals characterized by small *Planolites* with or without associated small and/or large *Chondrites* (Fig. 1D). Small and large *Chondrites* characterize 5% and 32%, respectively, of the 155 occurrences of this ichnocoenosis in the Portland core, and 49% and 51%, respectively, of the 58 occurrences recognized in the Bounds core. Intervals assigned to ichnocoenosis 2 range from 1 to 12 cm in thickness. Associated piped zones are typically 1-3 cm deep.

Intervals assigned to ichnocoenosis 3 are those that contain *Taenidium* and other relatively small-diameter ichnofossils (Fig. 1E). Roughly equal numbers (~60) of these intervals are observed in the Portland and Bounds sections. *Taenidium* is ubiquitous. Small *Planolites*, large *Chondrites*, and small *Chondrites* co-occur with *Taenidium* in 100%, 44%, and 5%, respectively, of all ichnocoenosis 3 intervals in the Portland core, and in 93%, 35%, and 50%, respectively, of all ichnocoenosis 3 intervals in the Bounds core (Table 2). Thicknesses of these intervals range from 1 to 12 cm (average = 4 to 5 cm). Depths of associated piped zones most commonly are on the order of 15 cm, although many exceed 35 cm.

Ichnocoenosis 4 refers to intervals that contain *Teichichnus* and/or *Zoophycos* and one or more of all other ichnofossil types except for large *Planolites* and *Thalassinoides* (Figs. 1F, H). Approximately 37 and 33 intervals of ichnocoenosis 4 are recognized in the Portland and Bounds cores, respectively. *Teichichnus* is present in 85% and 58% of occurrences in the Portland and Bounds cores. *Zoophycos* is present in only 58% of occurrences in the Portland core but is more important in the Bounds core where it is observed in 70% of ichnocoenosis 4 intervals. Ichnofossil types co-occurring with *Zoophycos* and/or *Teichichnus* include, in order of decreasing importance: small *Planolites*, large *Chondrites*, *Taenidium* and small *Chondrites* (see Table 2). Intervals assigned to this ichnocoenosis range in thickness from 2 to 45 cm, but most commonly are 7-12 cm thick. Maximum burrow-penetration depths vary broadly from 2 to 80 cm, but most commonly are 25 to 30 cm.

Ichnocoenosis 5 refers to all intervals that contain large *Planolites* but lack obvious *Thalassinoides* (Figs. 1G-I). Approximately 29 and 39 of these intervals are recognized in the Portland and Bounds cores, respectively. Burrow types that concur with large *Planolites* in this ichnocoenosis include, in general order of decreasing importance: small *Planolites*, *Teichichnus*, large *Chondrites*, *Zoophycos*, *Taenidium*, and small *Chondrites* (Table 2). Intervals assigned to ichnocoenosis 5 range from 5 to 40 cm thick (average=11 to 12 cm). Although data are sparse, maximum burrow-penetration depths apparently range from 8 to 40 cm.

Ichnocoenosis 6 is distinguished by the presence of *Thalassinoides* (Fig. 1J), but it also may include, in order of decreasing importance, large and small *Planolites*, *Teichichnus* and/ or *Zoophycos*, large *Chondrites*, *Taenidium*, and small *Chondrites* (Table 2). This most diverse ichnocoenosis is recognized in 36 and 37 intervals in the Portland and Bounds cores, respectively, ranging from 6 to 90 cm in thickness (average=28-31 cm). Burrow- penetration depths are comparable to those of ichnocoenosis 5.

Occurrences of ichnocoenoses roughly correlate with rock type. Ichnocoenosis 1 generally is restricted to calcareous or marly shale intervals; ichnocoenoses 2 and 3 are prevalent in marly shale and marlstone intervals; ichnocoenosis 4 occurs in marlstone, marly limestone, and limestone beds; and ichnocoenoses 5 and 6 are largely restricted to limestones and marly limestones.

Implications for Benthic Oxygenation

Laminites of the Niobrara resemble the pelagic laminites described for the Bridge Creek Limestone (Savrda, this volume), and are similarly interpreted here to record deposition under relatively quiet, oxygen-deficient conditions. Laminites lacking macrobenthic body fossils likely reflect anaerobic conditions, while those containing inoceramids and/or other bivalve remains may reflect the exaerobic (Savrda and Bottjer, 1991) or the benthic boundary biofacies (Sageman, 1989).

Bioturbated intervals obviously reflect periods of improved benthic conditions. Following the ichnologic models of Savrda and Bottjer (1991, 1994) and Savrda (1992), the systematic increase in ichnofossil diversities, maximum burrow diameters, and burrow-penetration depths from ichnocoenosis 1 through ichnocoenosis 6 (Table 2) indicate progressively higher levels of paleo-oxygenation; that is, ichnocoenosis rank (1-6) is interpreted to be proportional to benthic oxygen availability.

Based on these relationships, the vertical disposition of laminites and oxygen-related ichnocoenoses were employed to construct paleo-oxygenation curves for the study intervals (for a summary of curve construction, see Savrda and Bottjer, 1991, 1994). However, confident assessment of oxygenation histories was precluded for some intervals (for example, the lower parts of unit 2; see below), particularly those wherein poor burrow visibility and/or complex overprinting of piped burrows made it difficult to decipher ichnocoenosis stacking patterns.

PALEO-OXYGENATION HISTORIES

Paleo-oxygenation curves for the Portland and Bounds cores are shown in Figures 2 and 3, respectively. Long-term trends in paleo-oxygenation and shorter-term redox cycles reflected by these curves are summarized below.

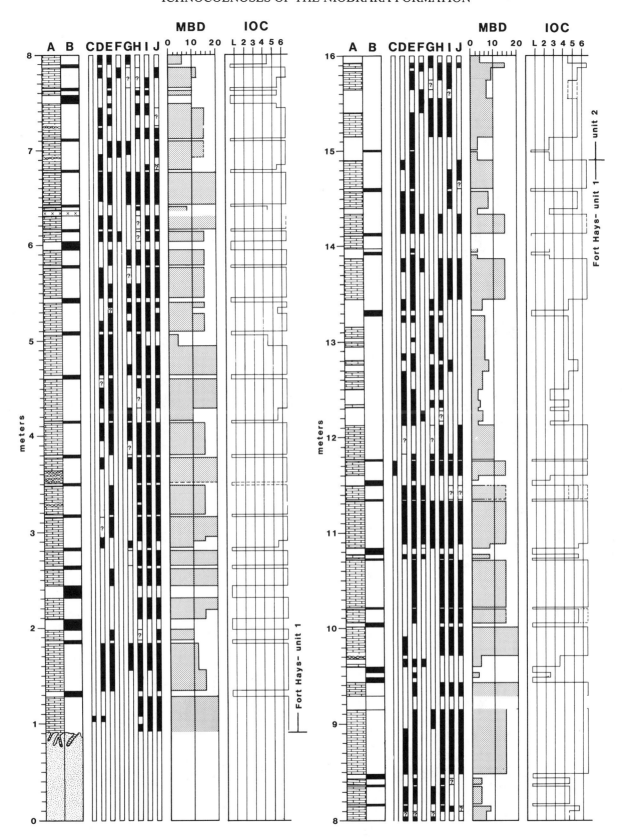

FIG. 2.— Distribution of general lithologies (A; limestones, marly limestones, and bentonites are patterned, and marlstones and marly shales are white), general sedimentary fabrics (B; bioturbated intervals are white, laminites are black), and recurring biogenic structure types (C, small *Chondrites*; D, large *Chondrites*; E, small *Planolites*; F, *Taenidium*; G, *Zoophycos*; H, *Teichichnus*; I, large *Planolites*; and J, *Thalassinoides*), maximum burrow diameter (MBD, in mm), and interpreted oxygenation curve (IOC) for the Fort Hays and lower Smoky Hill intervals in the Portland core. IOC is constructed using six reference lines. Line L represents oxygenation levels below which laminites are preserved and above which producers of ichnocoenosis 1 can survive. Lines 2, 3, 4, 5, and 6 represent threshold oxygen levels required for development of ichnocoenoses 2, 3, 4, 5, and 6, respectively.

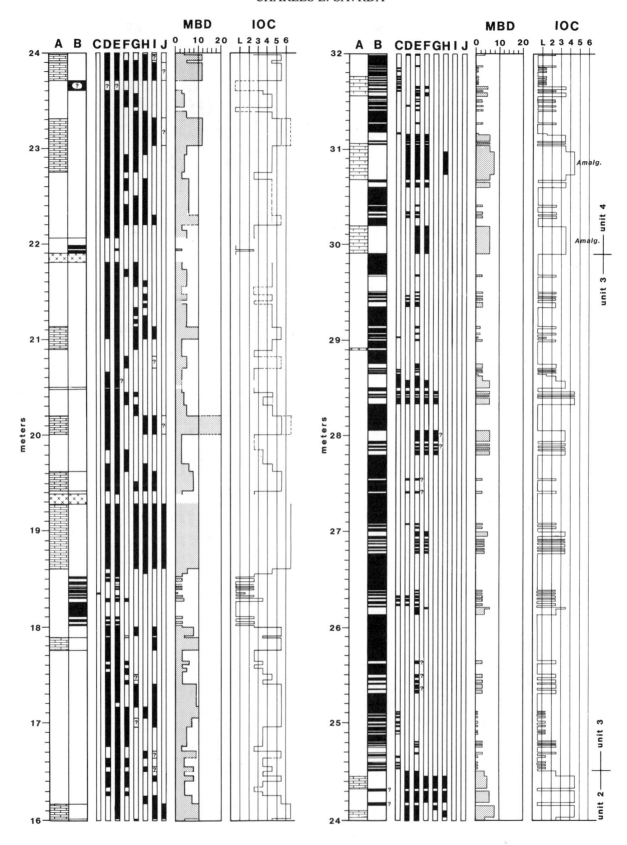

Fig. 2.— Continued.

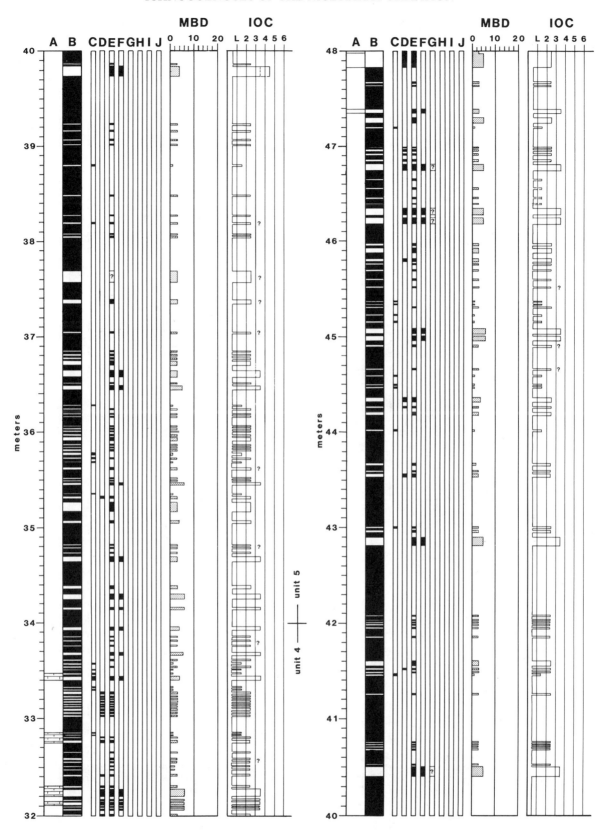

FIG. 2.— Continued.

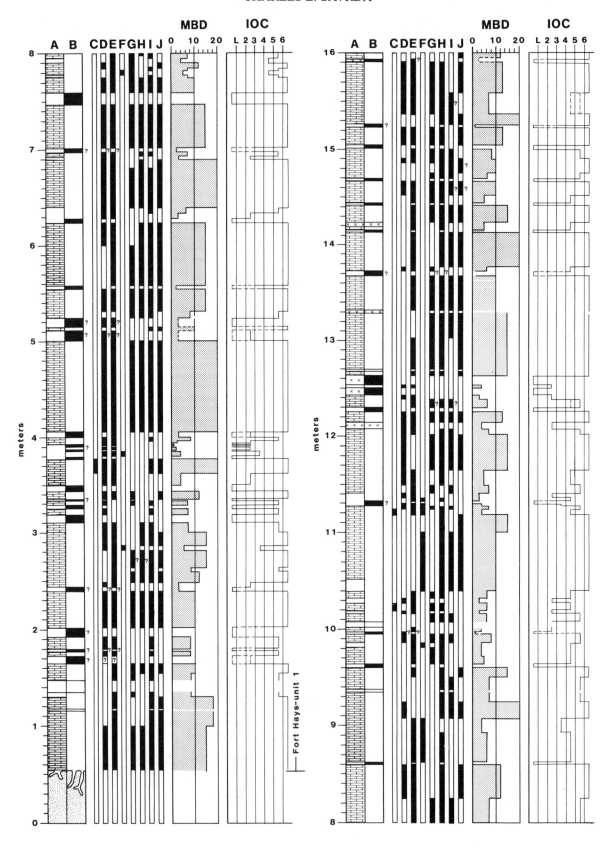

FIG. 3.— Distribution of general lithologies (A; limestones, marly limestones, and bentonites are patterned, and marlstones and marly shales are white), general sedimentary fabrics (B; bioturbated intervals are white, laminites are black), and recurring biogenic structure types (C, small *Chondrites*; D, large *Chondrites*; E, small *Planolites*; F, *Taenidium*; G, *Zoophycos*; H, *Teichichnus*; I, large *Planolites*; and J, *Thalassinoides*), maximum burrow diameter (MBD, in mm), and interpreted oxygenation curve (IOC) for the Fort Hays and lower Smoky Hill intervals in the Bounds core. IOC was constructed as in Figure 2.

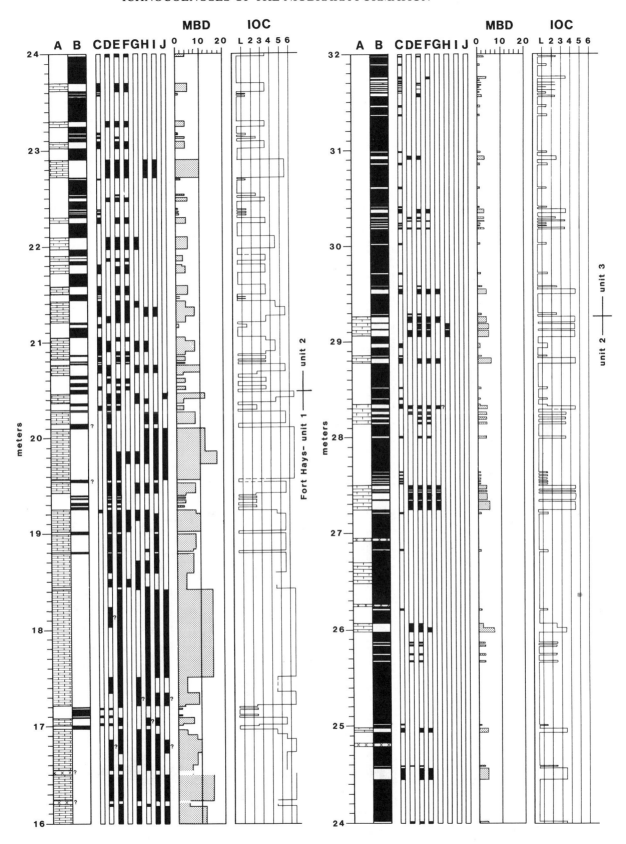

Fig. 3.— Continued.

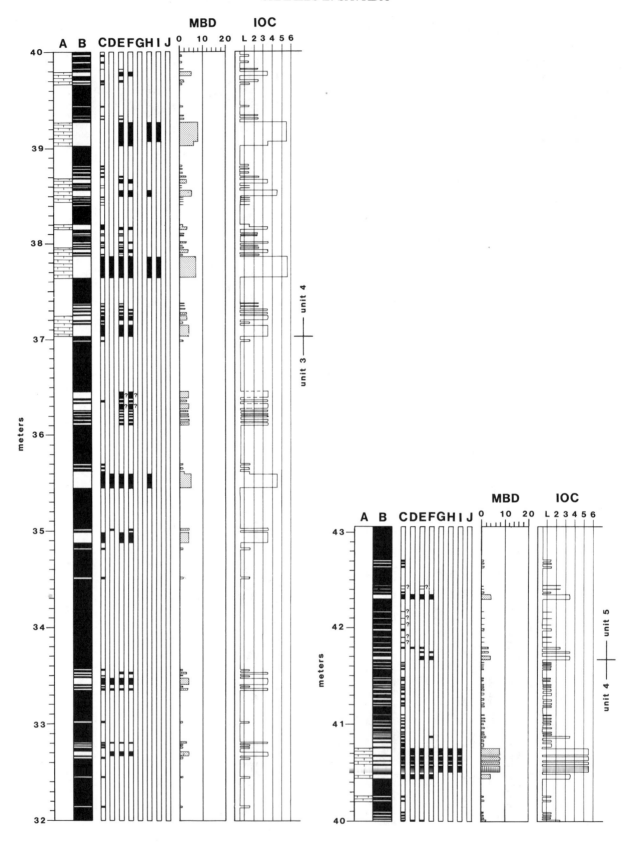

Fig. 3.— Continued.

Portland Core

The paleo-oxygenation curve for the Portland core (Fig. 2) reflects a general trend towards decreased benthic oxygenation from the Fort Hays through the Smoky Hill. Despite this trend, all five of the units record shorter-term redox cycles of variable frequency and amplitude.

The Fort Hays Member (unit 1) is characterized by decimeter- to meter-scale, high-amplitude cycles that generally correspond to interbedded limestone and calcareous shale/marlstone couplets. In the lower half of the member, limestones are dominated by ichnocoenosis 6 and, therefore, reflect very well oxygenated substrates. Although intervening calcareous shale/marlstone seams were influenced significantly by dissolution compaction, they appear to be characterized by primary laminated fabrics. Limestones in the upper half of the Fort Hays also are dominated by ichnocoenoses 6. However, all or parts of some limestones are characterized by ichnocoenoses 4 or 5, which indicates slightly lower oxygenation levels than are inferred for the lower part of the Fort Hays. Calcareous shale/marlstone seams in the upper Fort Hays are either completely laminated or bioturbated (ichnocoenoses 2 3, and/or 4) or alternately laminated and bioturbated. Where bioturbated, ichnocoenoses 2 through 4 are prevalent. Bundling, reflected by variations in the amplitude and/or thickness of cycles in maximum burrow diameter or the interpreted oxygenation curve, is not readily apparent in the Fort Hays (Fig. 2).

Unit 2 in the Smoky Hill is almost completely bioturbated; laminites make up only 6% of this interval by stratigraphic thickness. In some bioturbated intervals low color contrast between burrows and surrounding sediments hindered positive identification of burrow types and ichnocoenoses, whereas in other intervals intense overprinting by piped burrows complicates ichnofabric interpretations. Hence, parts of the oxygenation curve for unit 2 were constructed with limited confidence. Nonetheless, some generalizations regarding redox cyclicity can be made. Low-frequency, intermediate-amplitude cycles are crudely defined by a decimeter-scale alternation between intervals characterized by ichnocoenosis 5 and, less commonly, ichnocoenoses 4 or 6, and intervals characterized by ichnocoenosis 2 or 3 and/or laminites. The latter intervals generally correspond to relatively carbonate-poor beds or bedsets. Occurrences of laminites define three bundles of 4 to 6 of these decimeter-scale cycles. Higher-frequency, lower-amplitude redox variations are also evident, particularly in intervals that represent overall lower levels of oxygenation (for example, in the 18.0-18.5-meter interval, Fig. 2).

Unit 3, which is dominated by laminites (73% by stratigraphic height), is characterized by compound low- to intermediate-amplitude redox cycles. Nine low-frequency cycles, averaging ~60 cm thick, are defined by alternating thick laminite sequences with intervals that contain clusters of 2 to 15 closely spaced, thin bioturbated beds. These beds typically are characterized by ichnocoenosis 1, 2, or 3, although ichnocoenosis 4 occurs locally. A higher-frequency cycle is expressed within the clusters by centimeter-scale alternation of bioturbated beds and laminites.

The oxygenation record for Unit 4 reflects a compound cyclicity similar to that of unit 3. Numerous high-frequency, low- to intermediate-amplitude cycles are reflected by the interbedding of laminites and thin bioturbated beds represented by ichnocoenoses 1, 2, 3, and 4. Clustering of bioturbated beds defines 7 lower-frequency cycles that roughly correspond to the carbonate rhythms in the unit. Unit 4 differs from unit 3 in that clusters are more closely spaced and intervening laminite intervals are less pronounced. This may reflect lower sedimentation rates (condensation) and/or an increase in frequency of oxygenation episodes.

Paleo-oxygenation histories for unit 5 also resemble those for unit 3. Numerous low- to intermediate-amplitude redox cycles are arranged in clusters that define a low-frequency rhythm. The spacing of clusters and the number of bioturbated beds within vary inversely upwards through the unit, and they may reflect an even lower-frequency redox cycle.

Bounds Core

The interpreted oxygenation record for the Bounds section (Fig. 3) is similar in most respects to that for the Portland core. The Bounds oxygenation curve reflects a long-term deoxygenation trend upon which redox cycles varying in frequency and amplitude are superimposed. The general patterns of redox cyclicity recorded in units 1 (Fort Hays) and 3 through 5 are essentially identical to those observed in the Portland core. The Fort Hays Member (unit 1) primarily records decimeter- to meter-scale, high-amplitude redox cycles that generally correspond to limestone and calcareous shale/marlstone couplets. Units 3 and 4 are characterized by compound cycles of low to intermediate amplitude. Both units are characterized by alternating laminites and bioturbated beds. Most bioturbated intervals are thin and characterized by ichnocoenoses 1, 2, or 3, although ichnocoenoses 4 and 5 are locally important, particularly in unit 4. Clustering of bioturbated beds defines nine lower-frequency redox cycles in unit 3, and six or seven such cycles in unit 4. As in the Portland core, clusters in unit 4 generally are more closely spaced and contain greater numbers of bioturbated beds.

The most obvious difference between the two cores is in the paleo-oxygenation curves for unit 2. The oxygenation record for unit 2 in the Bounds core reflects significantly lower oxygen levels compared to those indicated for the Portland section. Laminites comprise 72% of unit 2 by stratigraphic height in the Bounds core, compared to only 6% in the Portland core (compare Figs. 2 and 3). In the Bounds core, the lowermost part of unit 2 (~2 m thick) is characterized by a complex pattern of intermediate- to high-amplitude cycles reflected by alternating intervals dominated by ichnocoenoses 3, 4, and/or 5 with intervals dominated by laminites and/or ichnocoenoses 1 or 2. The remainder of unit 2 is characterized by more clearly expressed low- and high-frequency redox cycles of low to intermediate amplitude. Similar to the compound cycles of units 3 through 5, these are reflected by clustering within predominately laminated sequences of closely spaced, thin bioturbated beds characterized by ichnocoenosis 1, 2, 3, or 4.

DISCUSSION

Redox Cyclicity

Given the current lack of age constraints for the core intervals studied, it is difficult to accurately determine periodicities for the various redox cycles recognized in the Niobrara paleo-oxygenation records. The high- and intermediate-amplitude redox cycles recorded in the Fort Hays Member and in Portland's unit 2, and the low-frequency, low- to intermediate-amplitude cycles reflected by bed clustering in units 3 through 5 and Bounds' unit 2, generally coincide with decimeter-scale carbonate cycles. These are interpreted here to reflect the 20-ky precession cycle, which is the same periodicity calculated by Pratt et al. (1993) and Dean and Arthur (this volume). The spacing of bioturbated bed clusters, and the number and amplitude of short-term oxygenation events recorded therein, appear to vary cyclically through units 3-5 (and unit 2 of the Bounds core); that is, clusters appear to be crudely bundled. This bundling pattern may reflect a control by orbital eccentricity. However, further quantitative treatment of ichnologic data (e.g., Sageman et al., this volume) will be required in order to more confidently establish a link between very low-frequency redox rhythms and the various eccentricity cycles.

If the aforementioned decimeter-scale, low-frequency rhythms reflect the the precession cycle, then the high-frequency cycles manifested within clusters of bioturbated beds in the Smoky Hill likely have periods of several hundreds to several thousands of years. This would indicate that benthic oxygenation in the Western Interior Seaway was not controlled solely by Milankovitch-scale climatic and paleoceanographic changes. Rather, other periodic or episodic processes operating over shorter-time scales (e.g., storms, etc.) also must have influenced redox dynamics. High-frequency redox cyclicity is clearly expressed in unit 2 of the Bounds core and units 3 through 5 in both cores. In contrast, it is only weakly expressed, if at all, in the Fort Hays and lowermost parts of unit 2. This could be taken to indicate that short-term redox variations capable of forcing ichnocoenosis change did not occur during the earliest phases of Niobrara deposition. However, poor expression of high-frequency cycles could also be related to preservational biases. High-frequency redox variations may have occurred during early Niobrara time, but deeper bioturbation associated with overall higher oxygenation levels may have destroyed the evidence for short-lived, low-amplitude redox events and, hence, resulted in a more time-averaged paleo-oxygenation record (Savrda and Bottjer, 1991, 1994).

Spatial Variations in Redox Conditions

As previously noted, the precise unit-age relationships between the two cores are currently poorly established. However, there is no evidence to indicate that the gradational Fort Hays/ Smoky Hill contact is significantly time-transgressive (unlike the base of the Fort Hays). Hence, differences in the oxygenation records for the lowermost parts of the Smoky Hill likely reflect a true spatial redox gradient. Lower oxygen levels at the Bounds locality indicate an apparent east-west gradient analogous to that recognized in the Bridge Creek Limestone (Savrda, this volume). This cross-basin gradient is not evident in other parts of the sequence. However, several additional ichnologic trends may provide at least subtle indications of an eastward reduction in benthic oxygen availability. First, *Chondrites* and *Zoophycos*, which are generally considered to belong to an oxygen-deficient ichnoguild (see Bromley, 1990; Savrda, 1992), are more common in the Bounds core. Second, among the numerous low-magnitude oxygenation-event beds in units 3 through 5, ichnocoenosis 1 is most common in the Bounds core, whereas ichnocoenosis 2 is most common in the Portland core. Finally, the average maximum burrow diameters for ichnocoenoses 2 through 6 is slightly higher in the Portland core (Table 2).

Mechanisms responsible for the apparent spatial redox gradient are not currently known. This gradient may be related to differences in paleobathymetry and position of the seafloor relative to depth-stratified water masses. Alternatively, it may reflect greater upwelling, productivity, and oxygen demand towards the eastern side of the seaway during Niobrara deposition. The latter might be expected if, as previously alleged (Laferriere, 1992), deeper Tethyan waters had a greater influence on the eastern side of the seaway. A better understanding of spatial paleoceanographic gradients, as well as depositional cycles, will undoubtedly come from the synthesis of ichnologic observations with other sedimentologic, geochemical, and paleontologic data collected as part of the Cretaceous Western Interior Drilling Project.

ACKNOWLEDGMENTS

Ichnological studies of the Niobrara Formation were supported by the USGS and a grant from the Department of Energy to Michael A. Arthur and Walter E. Dean. These principal investigators and other members of the Cretaceous Western Interior Drilling Project team, particularly Tim White and Brad Sageman, are acknowledged for helpful discussions regarding the Niobrara. Walter E. Dean, David J. Bottjer, and Allan A. Ekdale provided helpful comments on an earlier version of this manuscript.

REFERENCES

Arthur, M. A., and Dean, W. E., 1991, A holistic geochemical approach to cyclomania: Examples from Cretaceous pelagic limestone sequences, in Einsele, G., Ricken, W., and Seilacher, A., eds., Cycles and Events in Stratigraphy: New York, Springer-Verlag, p. 126-166.

Arthur, M. A., Dean, W. E., Bottjer, D. J., and Scholle, P. A., 1984, Rhythmic bedding in Mesozoic-Cenozoic pelagic carbonate sequences: the primary and diagenetic origin of Milankovitch-like cycles, in Berger, A., et al., eds., Milankovitch and Climate, Part 1: Dordrecht, D. Riedel, p. 191-222.

Arthur, M. A., Dean, W. E., Pollastro, R. M., Claypool, G. E., and Scholle, P. A., 1985, Comparative geochemical and mineralogical studies of two cyclic transgressive pelagic limestone units, Cretaceous Western Interior Basin, in Pratt, L. M., Kauffman, E. G., and Zelt, F. B., eds., Fine-grained Deposits and Biofacies of the Cretaceous Western Interior Seaway: Evidence for Cyclic Sedimentary Processes, Field Trip Guidebook No. 4: Tulsa, Society of Economic Paleontologists and Mineralogists, p. 16-27.

Arthur, M. A., Bottjer, D. J., Dean, W. E., Fischer, A. G., Hattin, D. E., and Kauffman, E. G., 1986, Rhythmic bedding in Upper Cretaceous pelagic carbonate sequences— Varying sedimentary response to climatic forcing: Geology, v. 14, p. 153-156.

Barlow, L. K., and Kauffman, E. G., 1985, Depositional cycles in the Niobrara Formation, Colorado Front Range, in Pratt, L. M., Kauffman, E. G., and Zelt, F. B., eds., Fine-grained Deposits and Biofacies of the Cretaceous Western Interior Seaway: Evidence for Cyclic Sedimentary Processes, Field Trip Guidebook No. 4: Tulsa, Society of Economic Paleontologists and Mineralogists, p. 199-208.

Barron, E.J., Arthur, M. A., and Kauffman, E. G., 1985, Cretaceous rhythmic bedding sequences: A plausible link between orbital variations and climate: Earth and Planetary Science Letters, v. 72, p. 327-340.

Bottjer, D. J., Arthur, M. A., Dean, W. E., Hattin, D. E., and Savrda, C. E., 1986, Rhythmic bedding produced in Cretaceous pelagic carbonate environments: Sensitive recorders of climatic cycles: Paleoceanography, v. 1, p. 467-481.

Bromley, R.G., 1990, Trace Fossils: Biology and Taphonomy: London, Unwin-Hyman, 280 p.

Fischer, A. G., 1980, Gilbert-bedding Rhythms and Geochronology: Boulder, Geological Society of America Special Paper 183, p. 93-104.

Fischer, A. G., 1986, Climatic rhythms recorded in strata: Annual Review of Earth and Planetary Sciences, v. 14, p. 351-376.

Fischer, A. G., 1991, Orbital cyclicity in Mesozoic strata, in Einsele, G., Ricken, W., and Seilacher, A., eds., Cycles and Events in Stratigraphy: New York, Springer-Verlag, p. 48-62.

Fischer, A. G., 1993, Cyclostratigraphy of Cretaceous chalk-marl sequences, in Caldwell, W. G. E., and Kauffman, E. G., eds., Evolution of the Western Interior basin: St. Johns, Geological Association of Canada Special Paper 39, p. 283-295.

FISCHER, A. G., HERBERT, T., AND PREMOLI-SILVA, I., 1985, Carbonate bedding cycles in Cretaceous pelagic and hemipelagic sequences, in Pratt, L. M., Kauffman, E. G., and Zelt, F. B., eds., Fine-grained Deposits and Biofacies of the Cretaceous Western Interior Seaway: Evidence for Cyclic Sedimentary Processes, Field Trip Guidebook No. 4: Tulsa, Society of Economic Paleontologists and Mineralogists, p. 1-10.

FREY, R. W., 1970, Trace fossils of the Fort Hays Limestone Member of the Niobrara Chalk (Upper Cretaceous) of west-central Kansas: Lawrence, University of Kansas, Paleontological Contributions, Article 53, 41 p.

FREY, R. W., AND BROMLEY, R. G., 1985, Ichnology of American chalks: The Selma Group (Upper Cretaceous), western Alabama: Canadian Journal of Earth Sciences, v. 22, p. 801-828.

GILBERT, G. K., 1895, Sedimentary measurement of geologic time: Journal of Geology, v. 3, p. 121-127.

HATTIN, D. E., 1975, Petrology and origin of fecal pellets in Upper Cretaceous strata of Kansas and Saskatchewan: Journal of Sedimentary Petrology, v. 45, p. 686-696.

HATTIN, D. E., 1981, Petrology of the Smoky Hill Chalk Member, Niobrara Chalk (Upper Cretaceous), in the type area, western Kansas: American Association of Petroleum Geologists Bulletin, v. 65, p. 831-849.

HATTIN, D. E., 1982, Stratigraphy and depositional environment of the Smoky Hill Chalk Member, Niobrara Chalk (Upper Cretaceous) of the type area, western Kansas: Kansas State Geological Survey Bulletin, v. 225, 108 p.

KAUFFMAN, E. G., 1977, Geological and biological overview: Western Interior Cretaceous basin: The Mountain Geologist, v. 14, p. 75-99.

KAUFFMAN, E. G., 1984, Paleobiogeography and evolutionary response dynamic in the Cretaceous Western Interior Seaway of North America, in Westerman, G. E. G., ed., Jurassic-Cretaceous Biochronology and Paleogeography of North America: Geological Association of Canada Special Paper 27, p. 273-306.

LAFERRIERE, A. P., 1992, Regional isotopic variations in the Fort Hays Member of the Niobrara Formation, United States Western Interior: Primary signals and diagenetic overprinting in a Cretaceous pelagic rhythmite: Geological Society of America Bulletin, v. 104, p. 980-992.

LAFERRIERE, A. P., HATTIN, D. E., AND ARCHER, A. W., 1987, Effects of climate, tectonics, and sea-level changes on rhythmic bedding patterns in the Niobrara Formation (Upper Cretaceous), U.S. Western Interior: Geology, v. 15, p. 233-236.

PRATT, L. M., ARTHUR, M. A., DEAN, W. E., and SCHOLLE, P. A., 1993, Paleo-oceanographic cycles and events during the Late Cretaceous in the Western Interior seaway of North America, in Caldwell, W. G. E., and Kauffman, E. G., eds., Evolution of the Western Interior basin: St. Johns, Geological Association of Canada Special Paper 39, p. 333-353.

RICKEN, W., 1991, Variation of sedimentation rates in rhythmically bedded sediments: Distinction between depositional types, in Einsele, G., Ricken, W., and Seilacher, A., eds., Cycles and Events in Stratigraphy: New York, Springer-Verlag, p. 167-187.

RICKEN, W., 1994, Complex rhythmic sedimentation related to third-order sea-level variations: Upper Cretaceous, Western Interior, U.S.A., in de Boer, P., and Smith, D. G., eds., Orbital Forcing and Cyclic Sequences, International Association of Sedimentologists Special Publication 19: Oxford, Blackwell Scientific Publications, p. 167-193.

SAGEMAN, B. B., 1989, The benthic boundary layer biofacies model: Hartland Shale Member, Greenhorn Formation (Cenomanian), Western Interior, North America: Palaeogeography, Palaeoclimatology, Palaeoecology, v. 74, p. 87-110.

SAVRDA, C. E., 1992, Trace fossils and benthic oxygenation, in Maples, C., and West, R., eds., Trace Fossils: Knoxville, Paleontological Society Short Course Notes 5, p. 172-196.

SAVRDA, C. E., AND BOTTJER, D. J., 1989, Trace fossil model for reconstructing oxygenation histories of ancient marine bottom waters: Application to Upper Cretaceous Niobrara Formation, Colorado: Palaeogeography, Palaeoclimatology, Palaeoecology, v. 74, p. 49-74.

SAVRDA, C. E., AND BOTTJER, D. J., 1991, Oxygen-related biofacies in marine strata: An overview and update, in Tyson, R. V., and Pearson, T. H., eds., Modern and Ancient Continental Shelf Anoxia: London, Geological Society of London, Special Publication 58, p. 201-219.

SAVRDA, C. E., AND BOTTJER, D. J., 1993, Trace fossil assemblages in fine-grained strata of the Cretaceous Western Interior, in Caldwell, W. G. E., and Kauffman, E. G., eds., Evolution of the Western Interior basin: St. Johns, Geological Association of Canada Special Paper 39, p. 621-639.

SAVRDA, C. E., AND BOTTJER, D. J., 1994, Ichnofossils and ichnofabrics in rhythmically bedded pelagic/hemipelagic carbonates: Recognition and evaluation of benthic redox and scour cycles, in de Boer, P., and Smith, D. G., eds., Orbital Forcing and Cyclic Sequences, International Association of Sedimentologists Special Publication 19: Oxford, Blackwell Scientific Publications, p. 195-210.

SCOTT, R. W., AND COBBAN, W. A., 1964, Stratigraphy of the Niobrara Formation at Pueblo, Colorado: Washington, U.S. Geological Survey Professional Paper 454-L, 30 p.

SCHOLLE, P. A., 1977, Chalk diagenesis and its relation to petroleum exploration—Oil from chalks, a modern miracle?: American Association of Petroleum Geologists Bulletin, v. 61, p. 982-1009.

SCHOLLE, P. A., AND POLLASTRO, R. M., 1985, Sedimentology and reservoir characteristics of the Niobrara Formation (Upper Cretaceous), Kansas and Colorado, in Longman, M. W., Shanley, K. W., Lindsay, R. F., and Eby, D. E., eds., Rocky Mountain carbonate reservoirs— A core workshop: Tulsa, Society of Economic Paleontologists and Mineralogists Core Workshop No. 7, p. 447-482.

MULTIPLE MILANKOVITCH CYCLES IN THE BRIDGE CREEK LIMESTONE (CENOMANIAN-TURONIAN), WESTERN INTERIOR BASIN

B.B. SAGEMAN[1], J. RICH[1], M.A. ARTHUR[2], W.E. DEAN[3], C.E. SAVRDA[4], AND T.J. BRALOWER[5]

[1]*Department of Geological Sciences, Northwestern University, Evanston, Illinois, 60208*

[2]*Department of Geosciences, Penn State University, University Park, Pennsylvania, 16802*

[3]*U.S. Geological Survey, MS 980, Federal Center, Denver, Colorado, 80225*

[4]*Department of Geology, Auburn University, Auburn, Alabama, 36849*

[5]*Department of Geology, University of North Carolina, Chapel Hill, North Carolina, 27599*

ABSTRACT: Spectral analyses of complementary lithologic and paleoecologic data sets from the Cenomanian-Turonian Bridge Creek Limestone are used to test for Milankovitch periodicities. Because the analyses quantitatively assess variance in the data sets through the study interval, they also offer a new method for evaluating relationships between different components of the depositional system. The analysis was made possible by high-resolution sampling for geochemistry, ichnology, and microfossils from a complete section of the Bridge Creek Limestone Member in the USGS #1 Portland core, coupled with a detailed chronostratigraphic framework established for the interval in a recently published compilation of new radiometric dates and biozones. The analyzed data included weight percent carbonate (% $CaCO_3$) and organic carbon (% OC), grayscale pixel values, ichnologic measures such as maximum burrow diameter and ichnocoenosis rank, and relative abundance values for selected nannofossil taxa. The lithologic parameters produced significant spectral responses for all three major orbital cycles (eccentricity, obliquity, and precession) in the upper 6 meters of the study interval. Spectra for ichnologic data are similar to those for % OC, possibly because both of these variables are dominantly controlled by benthic redox conditions. Spectra for some nannofossil taxa show results similar to % OC or % $CaCO_3$ but are less definitive due to preservational effects. A new model to explain the Bridge Creek cycles is developed based on the spectral results. The model combines dilution and productivity mechanisms by suggesting that obliquity predominantly forces dilution through its effect on high-latitude precipitation, whereas precession predominantly forces carbonate productivity through its effect on evaporation and nutrient upwelling in the Tethyan realm to the south. The two influences mix in the shallow Western Interior epicontinental basin, where they result in constructive and destructive interference (because of the different frequencies of obliquity and precession) to produce the complex bedding pattern observed in the Bridge Creek Limestone.

INTRODUCTION

In recent years Fourier techniques have been successfully employed to quantify evidence for Milankovitch forcing in sedimentary records of pre-Pleistocene strata (e.g., Park and Herbert, 1987). However, they have not been applied to the rhythmically bedded Cretaceous strata of the Greenhorn Formation, Western Interior basin, from which Gilbert (1895) first proposed the theory of climatic forcing of sedimentary rhythms. In this paper we report evidence of Milankovitch periodicities in the Bridge Creek Limestone Member of the Greenhorn Formation, obtained by applying Fourier techniques to lithologic and paleontologic data. The study was made possible by recent refinements in chronostratigraphy (Obradovich, 1993; Kauffman et al., 1993), and the collection of continuously sampled, high-resolution stratigraphic, paleontologic, and geochemical data sets from a core drilled in central Colorado (Dean and Arthur, this volume). These data include lithologic variations quantified in measured sections and core photographs, variations in measured contents of carbonate (% $CaCO_3$) and organic carbon (% OC) of the rocks (Sageman et al., 1997), variations in ichnologic characteristics observed in the core (Savrda, this volume), and changes in nannofossil assemblages analyzed in core samples (Burns and Bralower, this volume).

Variations in % $CaCO_3$ and % OC of pelagic and hemipelagic strata commonly are interpreted to reflect the interplay between clastic dilution, primary production, carbonate dissolution, and bottom-water oxygen content, factors that may be influenced or controlled by climate cycles (Arthur et al., 1984). These data were successfully employed in spectral estimates of Milankovitch forcing in Italian limestone/marlstone cycles (Park and Herbert, 1987). Variations in trace-fossil content provide an independent assessment of changes in benthic redox and substrate conditions (Savrda and Bottjer, 1991), and were successfully employed recently in spectral analyses of Cretaceous limestone/marlstone sequences (Erba and Premoli-Silva, 1994). Changes in the relative abundance of nannofossil species are interpreted to reflect water-mass conditions in the Western Interior basin, including surface-water temperature, salinity, and nutrient supply (Watkins, 1985; Bralower, 1988; Roth and Krumbach, 1988; Watkins et al., 1993). These parameters are also potentially sensitive to climate cycles (Watkins, 1989), but relative abundance trends in Cretaceous nannofossils of the Western Interior have not been tested with spectral techniques.

In this study we analyzed independent lithologic and paleontologic data sets of the Bridge Creek Limestone for Milankovitch periodicities. The results represent the first spectral evaluation of periodic oscillations corresponding to the Milankovitch precession, obliquity, and (or) eccentricity frequencies in the stratal patterns of the Bridge Creek Limestone. What is most significant about the study, however, is the simultaneous spectral analysis of three independent data records that are interpreted as largely reflecting sedimentation processes, benthic redox and substrate conditions, and surface-water conditions. Comparing results among the different data sets allows independent assessments of the mechanisms for sedimentary expression of orbitally influenced climate cycles. These results are helpful in evaluating different models for the linkage between climate and depositional systems in the Northern Hemisphere during Cenomanian-Turonian time.

Stratigraphy and Paleoenvironments of the Cretaceous Western Interior Seaway, USA, SEPM Concepts in Sedimentology and Paleontology No. 6
Copyright 1998 SEPM (Society for Sedimentary Geology), ISBN 1-56576-044-1, p. 153-171.

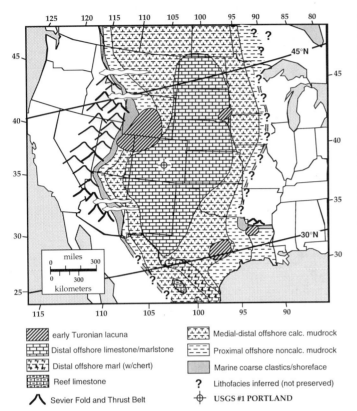

early Turonian lacuna

Distal offshore limestone/marlstone

Distal offshore marl (w/chert)

Reef limestone

Sevier Fold and Thrust Belt

Medial-distal offshore calc. mudrock

Proximal offshore noncalc. mudrock

Marine coarse clastics/shoreface

? Lithofacies inferred (not preserved)

⊕ USGS #1 PORTLAND

FIG. 1. Locality map shows the site of the Portland core in relation to paleogeography and the distribution of lithofacies for the Western Interior Basin during Early Turonian time (*Watinoceras devonense* Biozone).

GEOLOGICAL BACKGROUND

The history of the Western Interior basin of North America (Fig. 1) is well documented in several recent compilations (Pratt et al., 1985; Caldwell and Kauffman, 1993; Caputo et al., 1994). Aspects of this history relevant to our study include the paleogeography, climatic history, sedimentology, and oceanography of the seaway during deposition of the Bridge Creek Limestone. At this time, (during the Late Cenomanian to Early Turonian), a combination of foreland-basin subsidence and tectono-eustatic highstand resulted in maximum flooding of the Western Interior basin (Jordan, 1981; Kauffman, 1984). As a result, a meridional seaway formed that extended across more than 30° of latitude, connecting the circumpolar ocean to the western Tethys (Kauffman, 1977, 1984). Thus, the basin spanned both the lower latitude zone of high evaporation as well as the higher latitude precipitation belt (inferred from generalized modern atmospheric circulation patterns). Climatic interpretations from geologic data suggest warm temperate to subtropical climates with humid to subhumid conditions for the central Western Interior region (Kauffman, 1984; Pratt, 1984; Upchurch and Wolfe, 1993; Ludvigson et al., 1994; Witzke and Ludvigson, 1994). In addition, general circulation model experiments to analyze the effects of orbitally forced changes in insolation suggest that monsoonal circulation intensified in northern mid-latitudes and led to periodic fluctuations in the hydrologic cycle (Glancy et al., 1993; Park and Oglesby, 1991, 1994).

Sedimentation was dominated by abundant siliciclastic input from the uplifted fold and thrust belt to the west, but com-

paratively little clastic sediment from the east (Kauffman, 1984). Deposition of the Bridge Creek Limestone began just prior to and spanned one of the peak highstand events in the basin. Sedimentation was characterized by oscillations between mud-rich and carbonate-rich facies that ultimately formed limestone/marlstone bedding couplets. These couplets are characterized by fluctuations in biofacies and organic carbon content that are interpreted by many (e.g., Barron et al., 1985) to reflect changes in bottom-water oxygen content. The most common oceanographic model for circulation in the Western Interior basin during Late Cenomanian to Early Turonian time proposes fluctuations between a stratified and mixed water column to explain the benthic redox cycles (Barron et al., 1985; Pratt et al., 1993). Proponents of this hypothesis contend that the lithologic and interpreted redox characteristics of the Bridge Creek couplets are accounted for by changes in clastic dilution and water column stratification, respectively, due to Milankovitch forcing of precipitation and runoff. Alternate hypotheses were proposed for productivity as the dominant mechanism in forming the carbonate-rich units (Eicher and Diner, 1985, 1989, and 1991; Ricken, 1991, 1994) and for the combined influence of dilution and productivity (Arthur and Dean, 1991).

Previous investigators evaluated the Milankovitch hypothesis for the Bridge Creek Limestone, typically using bed counts and established time scales to estimate the periodicity of the bedding couplets. Kauffman (1977) used radiometric data from Obradovich and Cobban (1975) to estimate bedding periodicity at 60-80 ky. Fischer (1980) calculated periodicities of 27 to 22 ky and later revised that estimate to 41 ky (Fischer et al., 1985) based on additional data collected by Pratt (1984). Elder (1985) identified prominent limestone beds in the Bridge Creek sequence, and using Kauffman's (1977) time scale, estimated periods of 100 ky for them. Many authors cite a range between 20 and 40 ky for the lithologic couplets (e.g., Arthur et al., 1984; Arthur and Dean, 1991), or simply offer 40 ky as the average between minimum and maximum published estimates (Barron et al., 1985; Pratt, 1985; Pratt et al., 1993). Two recent studies illustrate the continued lack of resolution in periodicity estimates of the Bridge Creek bedding cycles. Based on analysis of ichnofossils in a core of the Bridge Creek Limestone drilled near Pueblo, Colorado, Savrda and Bottjer (1994) found that the thickest limestone beds in the lower part of the Bridge Creek, which correspond to the limestone marker beds of Cobban and Scott (1972), contain the largest and deepest burrows, and assigned these the highest values for their oxygen-related ichnocoenoses (ORI) index. However, between these bioturbated beds (presumably representing the highest oxygen levels), they found bundles of four to six ORI cycles of lower degrees of bioturbation (presumably representing lower oxygenation levels) that were not manifested as obvious lithologic variations. Savrda and Bottjer interpreted these cycles as evidence of 20-ky precessional cycles (smaller ORI cycles) within the predominant 100-ky eccentricity cycles (major limestone beds). Kauffman (1995) analyzed lithologic couplets defined in the Rock Canyon outcrop section and using Obradovich's (1993) new radiometric dates, calculated a period of 41 ky for the bedding couplets and 100 ky for intervals bounded by "prominent" (i.e., thicker) limestones. These varied interpretations illustrate the difficulties inherent in defining couplets based on different parameters as well as the problems that stem from age uncertainties. With the recent improvements in radiometric dating of the Late Cenomanian to Early Turonian interval (Obradovich, 1993), time series analysis of data from a con-

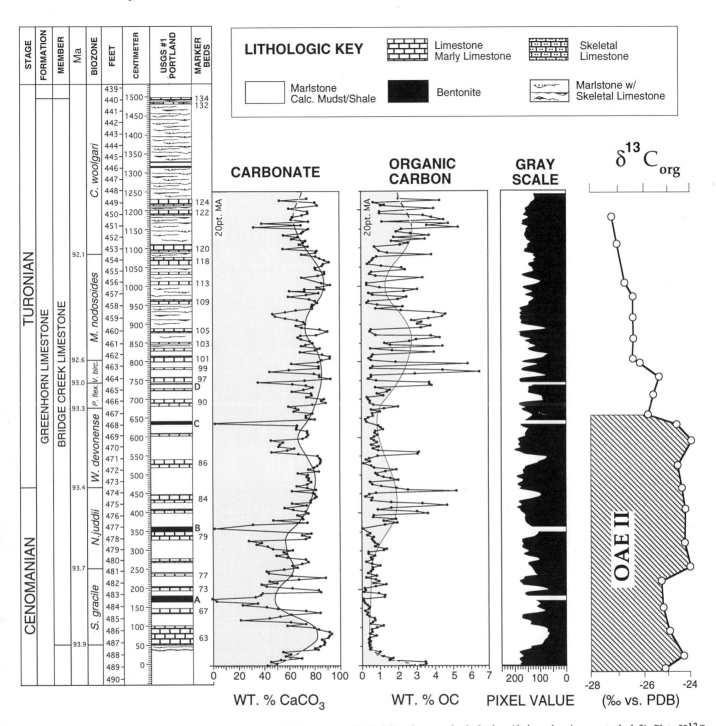

FIG. 2. Lithologic section of the Portland core plotted with weight % CaCO3, weight % OC, and gray-scale pixel values (darker values increase to the left). Plot of δ13C organic matter (Pratt, 1985) defines OAE II. Depths are given in feet below the top of the core, and in centimeters above a core break in the Hartland Shale Member just below the base of the Bridge Creek Limestone Member (487.5 ft). Shaded regions in CaCO3 and OC plots show 20-point moving averages (MA). Marker bed numbers of Cobban and Scott (1972) and bentonites A-D of Elder (1985) indicate correlations to the Rock Canyon Anticline reference section. Radiometric ages are based on the Kauffman et al. (1993) time scale. Biozonation from Kennedy and Cobban (1991).

tinuous core offers a means to more objectively and quantitatively test for periodicities in the Bridge Creek Limestone Member.

METHODS

The Portland core was acquired as part of the Cretaceous Western Interior Seaway Drilling Project (Dean and Arthur, 1994,

and this volume). At the drill site Cretaceous strata of interest are horizontal and structurally undisturbed, which allowed total recovery of the core (Dean and Arthur, this volume). High-resolution lithologic description and measurement of the Portland core was completed by Arthur and Sageman following collection and curation by the USGS Core Research Center in Denver, CO (Dean and Arthur, this volume). The Bridge Creek Limestone interval is illustrated in Figure 2. Marker beds defined by

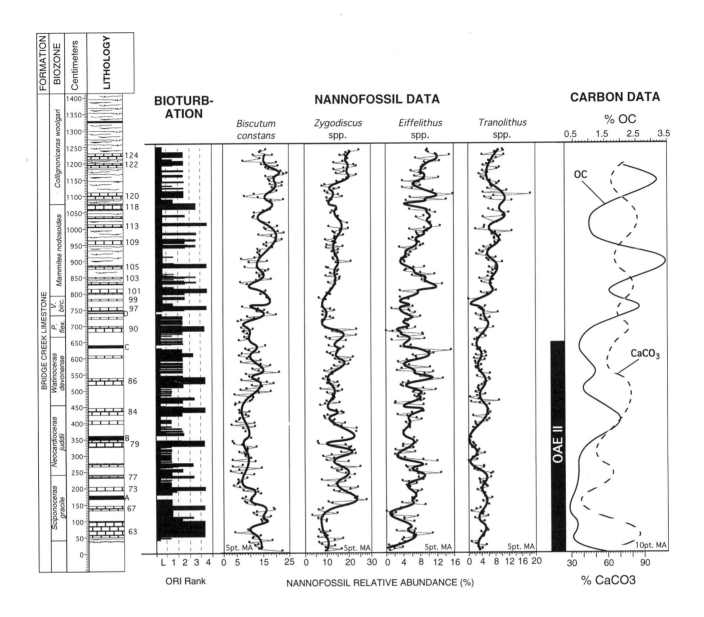

FIG. 3. Lithologic section of USGS #1 Portland core plotted with oxygen-related ichnocoenosis rank values (ORI) and relative abundance data (percent of assemblage) for four nannofossil taxa. Shown for comparison are 10-point moving average (MA) curves of % CaCO₃ and % OC data, as well as a black bar indicating OAE II.

Cobban and Scott (1972) and Elder (1985) were identified in the core by correlating to the Cretaceous reference section at Rock Canyon Anticline (which is approx. 40 km east of the drill site) where these beds were first recognized. The boundaries of standard Western Interior ammonite biozones (Cobban, 1984, 1985; Elder, 1985; Kennedy and Cobban, 1991) were defined for the core based on the lithologic correlations (Fig. 2).

Lithologic Data

To quantify lithologic variations through the 12.5 m of well-developed limestone-marlstone bedding couplets in the study interval, 1-cm-thick samples were taken at 5-cm intervals and analyzed for carbon content. The samples were crushed to <200 mesh and analyzed by coulometry (Engleman et al., 1985). The carbonate in the untreated whole samples was acidified with perchloric acid to liberate CO_2, which was titrated in the coulometer cell to measure carbonate carbon. Total carbon was measured by titrating CO_2 that was liberated during sample combustion at $950°$ C in a stream of oxygen. The technique has a precision of better than ± 0.5% for both carbonate and total carbon. Organic carbon (OC) was calculated as the difference between total and carbonate carbon, and % $CaCO_3$ was calculated based on stoichiometry. Samples within altered volcanic ash (bentonite) beds were removed from the time series prior to spectral analysis. Because of slight irregularities in lithology and core recovery, some of the sample intervals varied up to 1 or 2 cm. A cubic spline was applied to the slightly unevenly spaced

TABLE 1.- *Effective sedimentation rates for selected biozones of the Greenhorn Limestone.*

Biozone	Age at base	Duration	Thickness	S	
	Ma	My	m	My/m	cm/ky
base Fairport Chalky Shale	91.90				
C. woolgari (part)	92.10	0.20	4.20	0.05	2.10
M. nodosoides	92.60	0.50	2.80	0.18	0.56
W. devonense-V. birchbyi	93.40	0.80	3.40	0.24	0.42
N. juddii	93.70	0.30	2.10	0.14	0.70
S. gracile	93.90	0.20	2.05	0.10	1.02
D. conditum	94.18	0.27	5.50	0.05	2.00
Average for Bridge Creek Mbr.		1.80	10.35	0.17	0.57

1. Table shows interpolated ages of biozone boundaries from Kauffman et al. 1993 time scale. A weighted average value for S is calculated using the base of the Bridge Creek Limestone and the top of the M. nodosoides biozone (note dashed lines) as datums.

"raw" geochemical data to obtain the necessary regularity for spectral analysis.

To complement the geochemical data set, black and white photographs of the slabbed archive half of the Portland core were scanned to create 8-bit TIFF files and analyzed for variations in gray-scale values using Image 1.49, a public domain image analysis software package produced by the National Institutes of Health. Such programs allow for scans of optical density in which pixel gray-scale values ranging from 0 (white) to 255 (black), are averaged over a user-selected area. To produce a pixel data set as closely comparable to the geochemical data as possible, 1-2 cm areas were scanned at 5-cm intervals that corresponded to the geochemical samples. Several scans of each sample site were averaged, edited to remove spurious values (dark cracks, bright reflections, etc.), and spliced into a data series for spectral analysis (Fig. 2). As with the geochemical data, a cubic spline was applied to ensure regular spacing of data points, and values representing altered volcanic ash (bentonite) beds were edited out prior to spectral analysis.

Paleoecologic Data

In collaboration with the description and geochemical sampling of the Portland core, C. Savrda analyzed trace fossil content at 2-cm intervals through the Bridge Creek Limestone (for methods see Savrda, this volume). Data for maximum burrow diameter (MBD) and oxygen-related ichnocoenoses (ORI) were chosen for spectral analysis. These represent ordinal and rank data series, respectively, and show very similar trends because ORI is partly based on MBD. For simplicity, only the ORI rank data are plotted in Figure 3.

Burns and Bralower (this volume) analyzed abundance trends of major nannofossil taxa through the bedding couplets of the Bridge Creek in the Portland core. The resulting relative abundance data (% of assemblage) are based on counts of 220 specimens in smear slides made from the same samples used in CaCO₃ and OC determinations (Burns and Bralower, this volume). In their study, Burns and Bralower analyzed abundant taxa at the

species level, but grouped less common species by genera, assuming similar paleoecological affinities. Four taxonomic groups that fluctuate between limestone and shale hemicycles were chosen for spectral analysis. These include two groups commonly interpreted to indicate fertility (*Biscutum constans, Zygodiscus* spp.) and two groups whose paleoecological affinities are less certain (*Eiffellithus* spp., *Tranolithus* spp.) (Burns and Bralower, this volume). Because nannofossil recovery was incomplete in some samples, gaps of up to 10 or 15 cm occur in a few segments of the data series. A smoothed cubic spline was applied to interpolate between existing data points.

Spectral Analysis

The fundamental challenge in spectral determination of Milankovitch forcing from geological data stems from uncertainties in age-depth relationships. These result from time-scale uncertainties, sampling problems, interference by nonperiodic climatic or geologic factors, and nonlinearity in the sedimentary expression of periodic forcing (Park and Herbert, 1987). Given these difficulties, the most reliable test for Milankovitch forcing in pre-Pleistocene geological time series is the detection of sinusoids with ratios approximating the modern orbital periods (or periods corrected to pre-Pleistocene values: Berger et al., 1989, 1992) (Park and Herbert, 1987).

In analyzing the Bridge Creek data sets we employed the multitaper method (MTM) of Thomson (1982), which is believed to provide the optimal spectral estimate in noisy geologic time series (Park and Herbert, 1987). In addition to the familiar amplitude spectra, MTM also provides a statistical test (F variance-ratio test) of the likelihood that the data contain a monochromatic signal at each discrete frequency. In our illustrations, the F-test results are superimposed on the amplitude spectra and the 90, 95, and 99% confidence levels for the F-tests are shown. All data tables include only those frequencies with F-tests above the 90% confidence level. MTM was used to test for dominant periods matching those most commonly reported in studies of Cretaceous rhythmic bedding (e.g., Fischer et al., 1985), including

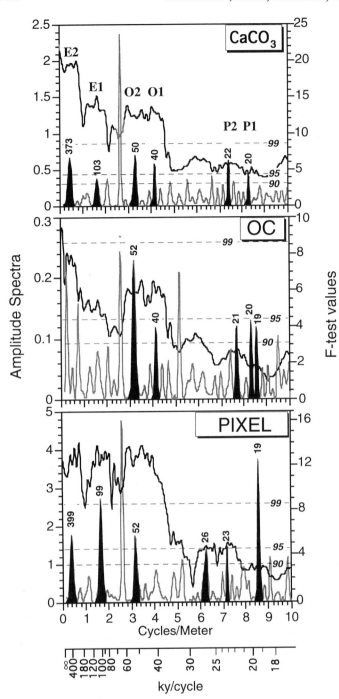

FIG.4. Plots of multitaper method (MTM) spectra for % CaCO3, % OC, and gray-scale pixel data from the upper Bridge Creek Limestone. F-test peaks (gray lines) are superimposed on amplitude spectra (black lines). The amplitude scales are on the left, the F-test scales on the right. The 90%, 95%, and 99% confidence limits for F-tests are indicated by dashed lines in each plot; they correspond to values of 3.11, 4.46, and 8.65, respectively. The blackened F-test peaks represent frequencies with F-test values >90% confidence that correspond to amplitude maxima; bold numbers show cycle periods assuming S=0.6 cm/ky (note ky/cycle scale below figure).

Methodological steps used in the spectral analysis are described in detail in another paper (Sageman et al., 1997) and will be only briefly reviewed. The major tasks included determining appropriate MTM parameters and a reasonable age-depth relationship. MTM parameters were optimized by analyzing synthetic time series of the modern orbital parameters (Berger, 1978) constructed to parallel the Bridge Creek data sets in terms of series length. The best results were obtained with five 3-pi prolate tapers and maximum data series padding. All subsequent MTM estimates used these parameters.

Determining an appropriate age-depth relationship for the core was facilitated by new high-precision $^{40}Ar/^{39}Ar$ dates from volcanic ash beds analyzed within the study interval (Obradovich, 1993), which were used to develop an interpolated time scale for Cenomanian-Turonian biozone boundaries (Kauffman et al., 1993) (Fig. 2). Because of the scale on which sedimentation rates change in the Greenhorn Formation, and the positions of dated bentonites relative to these changes, biozone interpolation offered a better approach than a linear fit between dated horizons and stratigraphy. Based on the interpolated time scale the entire study interval represents about 1.9 Ma. Effective sedimentation rates (S), which are not corrected for compaction (Park and Herbert, 1987), can be calculated for each biozone (Table 1). An average S value of 0.57 cm/ky is determined for the study interval that contains bedding cycles, and this value is rounded to 0.6 cm/ky for analysis of the spectral results. At this rate, one meter of section represents an average time interval of 167 ky.

RESULTS

Lithologic/Geochemical Data

The lithologic characteristics of the limestone-marlstone bedding couplets in the Bridge Creek were described in detail by Pratt (1984), Arthur et al. (1985), Elder (1985), Ricken (1994), Savrda and Bottjer (1994), Sageman et al. (1997), and many others. We will first summarize those characteristics that helped guide the application of spectral methods.

The major lithotypes include a series of light-colored (light gray to medium-gray), carbonate-rich rocks such as limestone, marly limestone, and burrowed marlstones ($CaCO_3$ from >60%) and a series of dark-colored (medium light-gray or olive-gray to olive-black) marlstones and calcareous mudstones ($CaCO_3$ <60%). Fluctuations in carbonate content appear to be a dominant variable and account for the conspicuous weathering profile in outcrop as well as the light-dark color variations obvious in the core. A concomitant (but inverse) fluctuation in % OC usually occurs in the bedding couplets and may contribute to the color variation. Variations in clay content and mixing across bed boundaries due to burrowing can also influence color patterns. The idealized bedding couplet grades from a dark-colored, OC-rich, laminated marlstone upward into a light-colored, carbonate-rich, OC-poor, burrowed limestone.

Detailed plots of lithologies and variations in $CaCO_3$ and OC contents for the Bridge Creek are shown in Figure 2. Couplets in the lower half (below 600 cm in Fig. 2) generally consist of dark-colored calcareous shale or mudstone overlain by lighter-colored marlstone, marly limestone, or micritic limestone. The clay-rich hemicycles of the lower interval are anomalously low in % OC, with the exception of beds that bracket the Cenomanian-Turonian boundary, and there are numerous conspicuous bento-

the 21-ky peak for precession, the 41-ky peak for obliquity, and the 100- and 413-ky peaks for eccentricity (E1 and E2). In addition, because the MTM is successful in resolving additional components of the orbital quasi-periods (Berger, 1984; Park and Herbert, 1987), we were alert to the presence of 97- and 127-ky eccentricity signals (E1a and E1b), 39 and 53 ky obliquity signals (O1 and O2), and 19- and 23-ky precession signals (P1 and P2).

TABLE 2.- *Results from MTM analysis of upper Bridge Creek Limestone.*

Frequency (cycle/m)	Amplitude (wt.%, pix)	Phase (degrees)	F-test (>90% only)	Period-kyr (0.6cm/kyr)	
Weight %CaCO$_3$					
0.4468	0.0287	36.91	6.7	**373.04**	E2
1.6235	0.0197	-110.17	3.88	**102.66**	E1
2.085	0.013	64.04	3.88	79.94	
2.6465	0.0167	69.86	24.52	62.98	
3.3081	0.0183	157.36	7.11	**50.38**	O2
4.1602	0.0185	62.48	5.81	**40.06**	O1
4.8462	0.0077	175.2	3.4	34.39	
5.5957	0.007	-82.45	3.63	29.78	
6.6724	0.0063	19.59	4.16	24.98	
7.4097	0.0083	42.55	6.47	**22.49**	P2
7.644	0.0073	-65.02	3.43	21.8	
8.3203	0.0064	42.63	4.3	**20.03**	P1
Weight %OC					
0.2759	0.0034	135.72	7.1	604.13	
0.7666	0.0024	-41.9	5.43	217.41	
2.6172	0.0016	-67.62	8.15	63.68	
3.208	0.0026	59.14	7.77	**51.95**	O2
4.1577	0.0023	-126.84	3.96	**40.09**	O1
5.1831	0.0011	-13.52	7.21	32.16	
7.7075	0.0011	45.77	4	**21.62**	P2
8.3569	0.0007	172.75	4.3	19.94	
8.5962	0.0006	31.55	3.97	**19.39**	P1
9.5044	0.0007	61.79	3.47	17.54	
Grayscale Pixel Value					
0.4175	0.0544	-73.94	5.75	**399.22**	E2
1.6773	0.0561	39.84	8.69	**99.37**	E1
2.6221	0.0495	-138.14	15.54	63.56	
3.1812	0.0528	53.26	5.63	**52.39**	O2
5.2368	0.02	-113.28	3.59	31.83	
6.2842	0.02	5.48	4.57	26.52	
7.1997	0.022	-167.85	4.62	**23.15**	P2
8.6206	0.0134	-21.09	12.71	**19.33**	P1
9.0942	0.0089	30.05	3.73	18.33	
9.8706	0.011	105.54	3.86	16.88	

1. *Only frequencies with F-tests above 90% confidence level are shown.*

2. *Frequency values close to the dominant orbital cycles are designated as E1, E2, etc. on the right, and their period values are printed in bold.*

nites. In addition, some limestone beds are anomalously thick; based on regional stratigraphic data these beds were interpreted to represent amalgamation of several thinner limestones (Sageman et al., 1997). In contrast, most of the bedding couplets in the upper Bridge Creek Member (above 600 cm in Fig. 2) consist of dark-colored marlstone overlain by light-colored marly or micritic limestone and show an overall increase in CaCO3 and OC content. Bioturbation trends closely follow the lithologic oscillations (Fig. 3), and there is a decrease in bed thicknesses and a shift to more regular oscillations between the two end-member depositional phases of the upper Bridge Creek. In fact, the hemicycles show a coordinated (but inverse) oscilla-

tion in CaCO3 and OC content at both shorter (bedding cycles) and longer (meter-scale) wavelengths (Fig. 2). These data suggest that a major change occurred in the nature of deposition from the lower part to the upper part of the Bridge Creek Limestone. A quantifiable record of orbital influence on the depositional system, if it is preserved at all, should be most clearly recorded in the upper part of the study interval. Spectral analyses supported this conclusion, and thus all spectral results reported herein are limited to the data series from 600 to 1230 cm, which represents about 0.8 - 0.9 Ma (Fig. 2).

The results of MTM analysis of the % CaCO3, % OC and gray-scale pixel data from the upper part of the Bridge Creek are

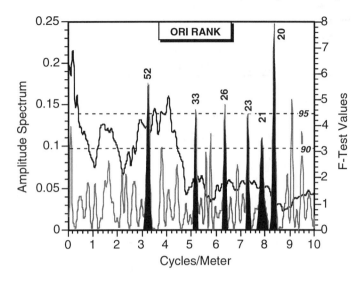

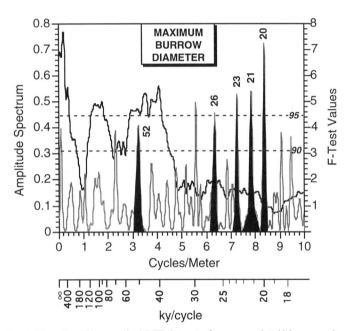

FIG. 5. Plots of multitaper method (MTM) spectra for oxygen-related ichnocoenosis rank (ORI) and maximum burrow diameter (MBD) from the upper Bridge Creek Limestone. F-test peaks (gray lines) are superimposed on amplitude spectra (black lines). The blackened F-test peaks represent frequencies with F-test values >90% confidence that correspond to amplitude maxima; bold numbers show cycle periods assuming S=0.6 cm/ky (note ky/cycle scale below figure).

shown in Figure 4 and tabulated in Table 1. The F-test output indicates scattered noise below the 90% confidence level, but also shows a series of peaks with F-tests >90% within the ranges of about 0.4, 1.6, 3.3, 4.1, and 7.0-8.5 cycles/m (darkened in Fig. 4). At a sedimentation rate of 0.6 cm/ky, which was estimated for the upper Bridge Creek Member based on the interpolated biozone boundary ages, these peaks correspond closely to the frequencies for the E2, E1, O2, O1, and P2-P1 orbital cycles, respectively (Table 2). Variations in the spectral results among the different data sets are summarized as follows:

1. % CaCO3 shows distinct plateaus in the amplitude spec-

trum corresponding to E and O signals, and a weaker and broader plateau at frequencies corresponding to P. However, significant F-test results occur for E1, E2, O1, O2, and P1 - P2. All but one of these F-test peaks is above the 95% confidence level.

2. % OC shows a dominant plateau in the amplitude spectrum that corresponds to the obliquity signal. F-test results confirm this with significant peaks for O1 and O2. The O2 peak is above the 95% confidence level. Several peaks with F-tests above the 90% confidence level also occurred at frequencies suggesting that there might be a precessional signal.

3. Gray-scale pixel data show strong amplitude plateaus at frequencies corresponding to E and O, and significant F-test responses for E1, E2, O2, and P1/P2. The F-test peaks for E1 and P1 were above the 99% confidence level and strongly suggest the precessional index.

4. Peaks with F-tests >90% confidence that do not match the dominant Milankovitch periods were also observed at 2.6 cycles/m in all three data sets and at 5.2 cycles/m in % OC. Note that these peaks correlate with troughs in the amplitude spectrum; as such they only indicate a well-defined lack of strength in the signal at that frequency.

In summary, spectral analysis of the lithologic/geochemical data suggest that there is evidence for each of the major orbital parameters. The amplitude records show conspicuous plateaus at 3-4 cycles/m corresponding to the obliquity cycle, and significant F-tests appear for O1 and O2 in spectra for % CaCO3 and % OC. Interestingly, O2 produces a stronger signal, which is especially pronounced in the % OC data. Although a number of F-tests with >90% confidence could be produced purely by chance, the likelihood that they would conform so closely to the ratio of the Cretaceous orbital periods (Table 3) is extremely small. One possible explanation for anomalous peaks in the F-test record is the influence of halftones or overtones. For example, the normally weak 52-ky peak may have been amplified in the record by the beat frequency of another quasiperiod.

Ichnologic Data

Systematic variations in the diversity, diameter, and penetration depth of burrows are common features of limestone-marlstone cycles in pelagic and hemipelagic strata (Savrda and Bottjer, 1994). In the Bridge Creek Limestone, for example, a change from laminated, organic carbon-rich shale (OC>6%) to densely burrowed limestone (OC<1%) is typical of many bedding couplets. Savrda and Bottjer (1994) analyzed these variations in the lower part of the study interval using acetate peels of a core drilled at Rock Canyon Anticline near Pueblo, Colorado (Pratt, 1984). They identified four recurrent ichnocoenoses defined by their trace fossil assemblages: (1) *Chondrites* ichnocoenosis; (2) *Planolites* ichnocoenosis; (3) *Zoophycos/Teichichnus* ichnocoenosis; and (4) *Thalassinoides* ichnocoenosis. Using measurements of trace fossil diversity, maximum burrow diameter, penetration depth, and vertical stacking patterns, Savrda and Bottjer (1994) interpreted a history of changes in benthic paleo-oxygenation that correspond to the lithologic changes of the bedding couplets. Savrda (this volume) employed the same techniques to analyze the Bridge Creek study interval in the Portland core, and Figure 3 includes a plot of oxygen-related ichnocoenosis (ORI) rank values from his study.

Several features of the ORI plot are relevant to our investi-

TABLE 3.- *Analysis of MTM frequency data.*

	E2	E1	O2	O1	P2	P1
CaCO$_3$	>95	>90	>95	>95	>95	>90
OC	-	(<90)	>95	>90	>90	>90
Pixel	>95	>99	>95	(<90)	>90	>99
f (cycles/m)	0.432	1.65	3.23	4.16	7.44	8.51
σ (cycles/m)	0.02	0.04	0.07	0.002	0.26	0.17
Period (kyr)	385.8	101	51.6	40.06	22.4	19.58
Scaled ratio	3.79	0.99	0.507	0.39	0.22	0.19
Predicted ratio	4.09	1	0.507	0.388	0.229	0.185

1. *Only spectral peaks with at least two F-test values >90% confidence among the three data series (CaCO3, OC, and gray-scale) were used.*

2. *F-test confidence levels for the peaks corresponding approximately to frequencies for E2, E1, O2, O1, P2, and P1 are shown at top for CaCO3, OC and Pixel data sets. Calculated values include: average of the frequency (f), standard deviation (σ), the corresponding period using a sedimentation rate of 0.6 cm/kyr, and the ratio between the resulting periods. The ratio is calculated by setting peaks with the strongest response (e.g., O2 > 95% confidence in all three data sets) to 0.507, which is equivalent to its estimated Cretaceous value, and scaling the other values accordingly. Based on the standard deviations shown, the error in period estimates among the different data records is E2±18.7 kyr; E1±2.38 kyr; O2±1.09 kyr; O1±0.02 kyr; P2±0.75 kyr; and P1±0.38 kyr. Finally, the scaled ratio is compared to that predicted for Cretaceous orbital cycles by Berger et al. (1989).*

gation. For example, burrowing intensity appears to correlate positively with the thickness and carbonate content of limestone beds. Thicker beds are characterized by ORI Rank 4. The thinner limestone or marlstone beds between them show lower rank values, and these are interbedded with laminated units (Fig. 3). In the lower part of the Bridge Creek Limestone, Savrda and Bottjer (1994) identified bundling in these burrowing trends that suggests the precessional index. Analysis of the Portland core (Savrda, this volume) showed that the same bundling does not occur in the upper part of the member. In fact, there is an overall decrease in burrowing levels within the clay-rich hemicycles from the lower to the upper part of the Bridge Creek, and fluctuations from laminated to highly burrowed hemicycles become much more pronounced (Fig. 3).

The multitaper method was applied to the data series for maximum burrow diameter (MBD) and ORI rank from the upper Bridge Creek Limestone. The analysis parameters were the same as those used for the lithologic and geochemical data sets. The results of the analyses were nearly identical for both types of ichnological measurements, which is to be expected given that the two measures represent strongly correlated dependent variables (Savrda and Bottjer, 1994). Results of MTM analyses are summarized in Figure 5 and Table 4.

The amplitude curves in Figure 5 show pronounced plateaus at 1.6 and 3-4 cycles/m. F-test results do not support the 1.6 cycle/m peak but include a >95% confidence peak at 3.3 cycles/m. In addition, multiple peaks above the 95% confidence level occur in the 7-8.5 cycles/m range. Using S = 0.6 cm/ky, these results correspond to O2 and P1/2 periodicities. Peaks below the 90% limit for E1 and O1 that correlate to significant signals in the lithologic/geochemical data were also observed but are statistically indistinguishable from signal noise.

Given our estimated sedimentation rate, the strongest sig-

nals in the ichnologic data correlate with frequencies suggesting obliquity and precession. The similarity between amplitude spectra for MBD and ORI Rank and those for CaCO3, OC, and pixel values suggests that variations in bottom-water oxygenation (MBD and ORI), clastic influx (CaCO3 dilution), and preservation of organic matter are linked and are collectively driven by orbital forcing. The ichnologic spectra are most similar to spectra for % OC.

Nannofossil Data

A detailed analysis of relative abundance trends in nannofossils through the entire sequence of bedding couplets in the Bridge Creek Limestone was conducted on the Portland core by Burns and Bralower (this volume). Although a large number of nannofossil taxa were counted, Burns and Bralower (this volume) focus on seven species or species groups that comprise the bulk of species abundance in the assemblages. Several of these taxa have been widely used for interpretations of taphonomy and paleoenvironmental conditions (Burns and Bralower, this volume). For example, the abundance of *Watznaueria barnesae*, a solution-resistant taxon, suggests that nannofossil preservation in the Portland core is comparatively poor. A decrease in abundance of *W. barnesae* and an increase in nannofossil species richness (Burn and Bralower, this volume) in the upper Bridge Creek Member, however, suggests preservation is better there and that valid paleoecological interpretations might be made for this part of the study interval. Thus, a decision to focus spectral analysis on the upper interval for comparative purposes is supported by preservational arguments as well.

Species or species groups commonly interpreted as fertility indicators include *Biscutum constans* and *Zygodiscus* spp. (e.g., Roth and Bowdler, 1981; Roth and Krumbach, 1988; Watkins,

TABLE 4.- *MTM analysis of ichnologic data from upper Bridge Creek Limestone.*

Frequency (cycles/m)	Amplitude (MBD/ORI)	Phase (degrees)	F-test (>90% only)	Period-kyr (0.6cm/kyr)	
Maximum Burrow Diameter					
0.1038	0.0029	-55.21	3.96	1606.27	
2.3193	0.0013	17.27	3.88	71.86	
3.2532	0.002	65.45	4.07	**51.23**	*O2*
5.5847	0.0007	94.68	4.95	29.84	
6.3477	0.0009	-87.24	4.57	26.26	
7.2693	0.0007	-136.24	5.29	**22.93**	*P2*
7.8491	0.0007	-160.94	5.42	21.23	
8.3801	0.0005	-71.17	7.22	**19.89**	*P1*
9.0759	0.0004	-26.98	3.79	18.36	
9.4727	0.0005	-173.72	3.64	17.59	
ORI Rank					
0.1038	0.0008	-61.38	3.98	1606.27	
3.2532	0.0006	62.82	5.54	**51.23**	*O2*
3.7903	0.0006	-35.69	3.17	**43.97**	*O1*
5.1636	0.0002	-152.75	4.61	32.28	
5.7678	0.0002	55.13	3.7	28.9	
6.3416	0.0002	-71.12	4.83	26.28	
7.2815	0.0002	-154.01	4.45	**22.89**	*P2*
7.8613	0.0002	-167.84	3.53	**21.2**	
8.3557	0.0002	-47.31	7.91	**19.95**	*P1*
9.0759	0.0001	-19.94	5.01	18.36	
9.491	0.0002	173.23	3.76	17.56	

1. *Only frequencies with F-tests above the 90% confidence level are shown.*

2. *Period values were calculated using S = 0.6 cm/kyr. Frequency values close to the dominant orbital cycles are designated as E1, E2, etc. on the right, and their period values are printed in bold.*

1989; Erba et al., 1992). Watkins (1989) observed that these taxa were relatively abundant in OC-rich or clay-rich hemicycles and less abundant in carbonate-rich beds. Although this relationship holds for many of the hemicycles in the Portland core, it is not entirely consistent from one hemicycle to the next, and the two taxa do not always covary closely (Fig. 3). A 5-point moving average curve through the data for *B. constans* and *Zygodiscus* spp. also does not indicate a common trend. In fact, whereas *Zygodiscus* spp. is most abundant in the lower part of the study interval, *B. constans* is least abundant there and reaches a maximum in the upper interval (Fig. 3). Nonetheless, because these taxa are established fertility proxies, they were analyzed with MTM. Two other taxa, *Eiffellithus* spp. and *Tranolithus* spp., were also chosen for analysis from among those studied by Burns and Bralower (this volume). Although Watkins (1989) correlated *Eiffellithus* spp. with limestone deposition and suggested that it reflected low fertility levels, Burns and Bralower (this volume) found no significant correlation between *Eiffellithus* spp. and CaCO3 content. However, *Eiffellithus* spp. and *Tranolithus* spp. show a range of behaviors in terms of fluctuations between limestone and shale hemicycles as well as some longer-term fluctuations in relative abundance (Fig. 3) and thus were considered appropriate for spectral analysis.

As in prior MTM analyses, only nannofossil data points above 600 cm (Fig. 3) were analyzed. This limits the effect of changes in nannofossil populations associated with OAE II (Schlanger et al., 1987) and the Cenomanian-Turonian extinction event, and focuses on the best preserved assemblages. Identical MTM parameters were used for the nannofossil analyses as employed in analyses of previous data sets. Like the results for MBD and ORI Rank, output from the nannofossil analyses includes various F-test peaks above the 90% confidence level (Fig. 6). Although some of these peaks match lows in the amplitude spectrum (e.g., peaks at 2.6 and 4.8 cycles/m in the *B. constans* plot), others do not. These may reflect either beat frequencies as suggested earlier, complicated and/or noisy signals, or nonlinear effects in the spectrum.

Distinct amplitude plateaus occur in the *B. constans* output from 1.5 to 1.8 cycles/m and from 3.0 to 4.6 cycles/m (Fig. 6). Although less pronounced than the amplitude plateaus of lithologic and ichnologic data sets, they are similar in frequency range. The *B. constans* F-test output is also similar to previous spectral estimates. It shows peaks above 95% confidence at 1.8, 3.4, and 6.8 cycles/m, which suggests frequencies within the range of E1, O2, and P2, respectively. Although some peaks are above the 90% confidence level, as would be predicted, those that correlate to orbital frequencies are the most significant. Among the nannofossil data tested, *B. constans* is the only taxon that produced spectra reasonably similar to those previously discussed.

The results for *Zygodiscus* spp. include several significant

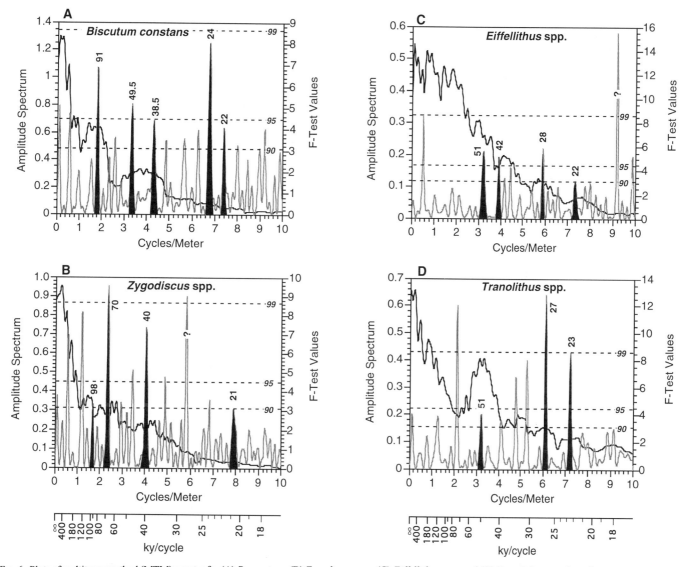

FIG. 6. Plots of multitaper method (MTM) spectra for (A) *B. constans*, (B) *Zygodiscus* spp., (C) *Eiffellithus* spp., and (D) *Tranolithus* spp. from the upper Bridge Creek Limestone. F-test peaks (gray lines) are superimposed on amplitude spectra (black lines). The blackened F-test peaks represent frequencies with F-test values >90% confidence that correspond to amplitude maxima; bold numbers show cycle periods assuming S=0.6 cm/ky (note ky/cycle scales below plots).

F-test peaks, but the amplitude record shows very weak response above a red noise spectrum. Thus, although certain peaks suggest the E1, O1, and P1 signals (at 1.7, 4.0, and 7.9 cycles/m, respectively), there are several other highly significant F-tests for frequencies that do not match orbital signals or reflect low points in the amplitude spectrum. These results suggest that there is considerable noise in the record and fail to clearly support an hypothesis of orbital influence. Although the *Eiffellithus* spp. results are similarly inconclusive, the amplitude spectrum for *Tranolithus* spp. reflects a possible O2 signal (note F-test >90% confidence for 51 ky), as well as a possible P2 signal (note F-test >99% confidence for 23 ky). The peak at 27 ky (F-test >99% confidence) has a corresponding amplitude response and while it is not correlated to an orbital quasi-period, may represent an interference tone.

DISCUSSION

Using sedimentologic, paleontologic, and geochemical evidence, many authors (e.g., Arthur et al., 1984, 1985; Barron et al., 1985; Pratt et al., 1993) have argued for a coordinated dilution-redox mechanism to explain the development of the Bridge Creek bedding couplets. In this model, increased rainfall in the highlands to the west increases freshwater input to the seaway which, in turn, increases the delivery of fine-grained siliciclastics throughout the basin (causing dilution of the carbonate flux to the sediment) and stratification of the water column (leading to lower benthic oxygen levels and increased OC preservation). The result is deposition of laminated, OC-rich shale or marlstone. Alternately, drier periods of low fresh-water input are characterized by decreased supply of fine-grained sediment, a clearer, better-mixed water column, increase in benthic oxygen levels, possible improvement in conditions for calcareous phytoplankton production, and deposition of a burrowed limestone. The driving force in this model is the modulation of climate by orbital influences to produce variations in precipitation over the land areas draining into the basin.

Other authors have argued that the planktonic microfossil record of the Western Interior basin does not support the interpretation of "freshened" surface waters during deposition of clay-

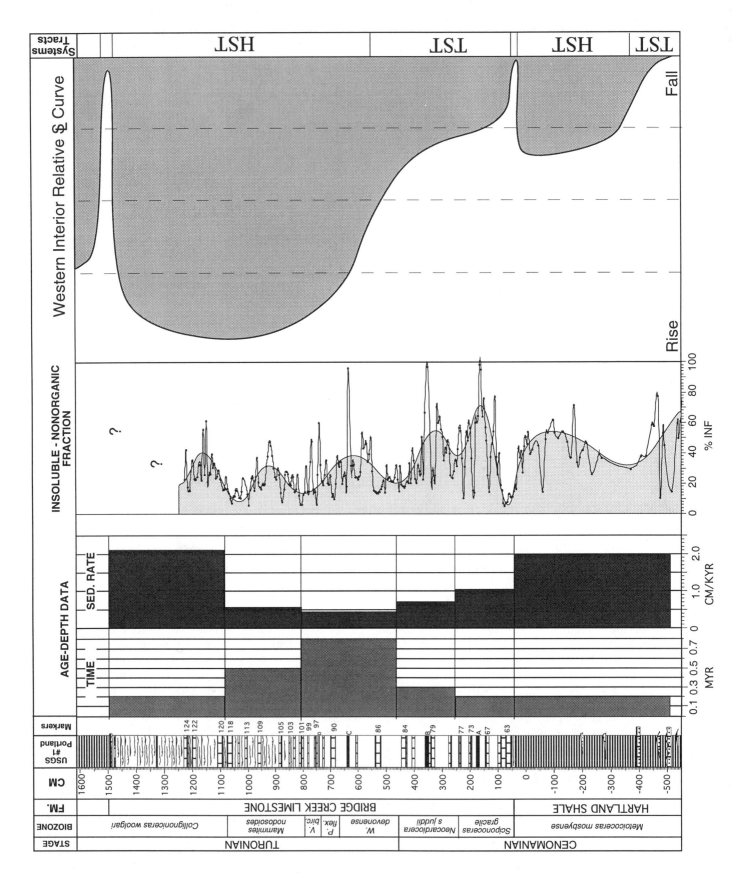

FIG. 7. Measured section of the Portland core with histograms showing duration of selected biozones and corresponding sedimentation rates, a plot of weight % insoluble-nonorganic fraction (INF) with superimposed 20-point moving average (MA) curve (see text), and an interpretation of relative sea level change based on regional stratigraphy and facies analysis (modified from Sageman, 1996). Equivalent systems tract designations (sensu Haq et al., 1987) are also shown.

rich hemicycles and suggest instead that changes in oceanic primary productivity due to cycles of nutrient upwelling could account for the bedding couplets (Eicher and Diner, 1985, 1989). In this model, orbital forcing leads to changes in evaporation at low latitudes sufficient to cause warm, saline bottom waters to form, which in turn displace nutrient-rich oceanic deep waters to the surface. Nutrient-driven blooms of phytoplankton would occur along the northern margin of the Tethys Sea and would have influenced the Western Interior basin through its southern aperture. A problem for this model, however, is suggested by the results of Watkins (1989) which indicate that high productivity characterizes the clay-rich (OC-rich) hemicycles rather than the carbonate-rich phases. Watkins (1989) suggests that the cycles do indeed reflect dilution, but in this case, dilution of the carbonate flux by organic matter.

Arthur and Dean (1991) and Ricken (1991 and 1994) argued that patterns in bedding cycles may be understood in terms of the mixing of three primary fluxes, detrital clay, $CaCO_3$ and OC, as well as the degree of subsequent modification due to carbonate dissolution, OC oxidation from biogenic activity, and other diagenetic factors. Analyzing elemental, organic, and isotopic geochemical data, Arthur and Dean (1991) concluded that the Bridge Creek bedding couplets were dominantly controlled by dilution of carbonate through orbitally forced changes in the detrital flux, but that changes in the productivity of calcareous plankton probably also played an important but secondary role. Arthur and Dean (1991) pointed out the need to understand the nature of the different fluxes to the sediment as well as the degree to which they represent dependent or independent variables. For example, the observed correlation between aluminum and OC in the Bridge Creek Member was interpreted to reflect a linkage of independent variables by a common forcing factor (higher freshwater input causes increase in detrital flux as well as enhanced water-column stratification leading to better preservation of OC), rather than a direct relationship (OC is adsorbed on clays, thus higher detrital flux leads to enhanced OC burial). However, because OC content in marine strata is ultimately controlled by a combination of organic production, bulk sedimentation rate, and redox conditions (that influence benthic biogenic activity) (Arthur and Sageman, 1994), the implications of linkage are not so clear.

Thus, the most important unanswered questions concerning Western Interior bedding cycles relate to the nature and origin of linkages: (1) Is OC production necessarily linked to carbonate production, and can changes in nannofossil and planktonic foraminiferal assemblages occur independently of changes in the noncalcareous phytoplankton and vice versa? (2) Are changes in OC and carbonate production linked to oceanic nutrient cycles, to nutrient input from fluvial sources, or to a combination of both, and do they respond independently to these forcing factors? (3) Is OC content primarily controlled by organic production, by preservation due to water column stratification and resulting oxygen deficiency, or to a combination of both? and (4) Does the influence of orbitally forced changes in climate impart an overall coordination to multiple independent processes, and, if so, how is this accomplished?

Our purpose was to first address whether or not the bedding couplets could be quantitatively related to orbital cyclicity using spectral techniques, and then to employ spectral results from independent data records as a tool to interpret links between different components of the depositional system. The bedding couplets of the Bridge Creek Limestone are complex and do not consistently show a 1:5 ratio like other Cretaceous limestone-

marlstone units (Schwarzacher and Fischer, 1982). Nor do they show the uniform bedding characteristic of couplets that are interpreted to result from obliquity (Fischer at al., 1985). Comparing the spectral signatures of % OC, % $CaCO_3$, and gray-scale data with those for burrow diameter and nannofossil abundance in the Bridge Creek Member, Portland core, provides an opportunity to test the Milankovitch hypothesis more rigorously and a way to evaluate the relationships between lithology, benthic oxygen, and planktic productivity during Cenomanian-Turonian time.

Sources of Error

Sources of error that could influence the interpretations described herein include variability in sedimentation rates within and between bedding couplets, time-scale errors, and errors introduced in the process of spectral analysis. For example, it is estimated that given a series of 100 data points, there may be up to 10 F-test peaks from the MTM estimate that are above the 90% confidence level completely by random chance (Thomson, 1982). However, because most of our data sets showed consistent and significant F-test responses at frequencies that closely approximate the ratio of orbital cycles (Table 3), they were judged to reflect Milankovitch cyclicity.

To objectively assess variation in sedimentation rates one must analyze time-scale errors. Although the radiometric dates used in this study are the best available for the Upper Cretaceous (Obradovich, 1993), they nonetheless have an average error of ±0.6 Ma. This means that bulk sedimentation rates could have varied by as much as 1 cm/ky from the average value used here (Table 1). Clearly this would significantly affect the interpretation of MTM spectra. However, the average sedimentation rate of 0.6 cm/ky is close to that estimated for other Cretaceous hemipelagic facies (e.g., Park and Herbert, 1987), and is within the range of bulk sedimentation rates estimated in previous studies of the Bridge Creek Limestone (Pratt, 1985).

Although the question of variations in sedimentation rate within individual bedding couplets is difficult if not impossible to resolve, Ricken's (1994) calculations of changes in relative sedimentation rate from clay-rich to carbonate-rich hemicycles of the lower Bridge Creek Member suggest that they are minor (<0.10 cm/ky). Such differences as indicated by Ricken (1994) would be least in the upper part of the study interval, where the bedding rhythms are most uniform and sedimentation rates are lowest (Fig. 7). Anomalies in trends of bed thickness, $CaCO_3$, and OC, and regional stratigraphic relationships (Sageman, 1991, Sageman et al., 1997), as well as direct calculation of age-depth relationships in the core (Fig. 7) show that sedimentation rates changed significantly from the lower to the upper part of the study interval. The change in sedimentation rate can be correlated to an increase in the weight percent of the insoluble, non-organic fraction [100-(% $CaCO_3$ + % OC)], called the INF, and reflects increases in the contribution of detrital and/or volcanic clay to the sediment. Variations in the INF upsection are consistent with predictions based on interpretation of relative sea-level change for the Greenhorn Formation (Sageman, 1996). The lower Bridge Creek records a transgressive episode during which sediment delivery to the basin changed rapidly. Volcanic ash may have diluted detrital, carbonate, and organic fractions due to decreased background detrital sediment flux or intensified eruptive activity. The decrease in detrital clastic flux from the lower to upper Bridge Creek corresponds well with a decrease in calculated sedimentation rate, and suggests that the maximum rate of relative sea level rise occurred just after the Cenomanian-Turonian boundary event and resulted in a condensed section (sensu Van Wagoner et al., 1988) in the earliest Turonian.

Significance of Paleoecologic Data

Addressing the dominant influences on preserved ORI's, Savrda and Bottjer (1994) discussed a complex set of control mechanisms and feedbacks related to OC preservation. Bottom-water oxygen levels may be most strongly controlled by pore-water oxygen, which in turn is influenced by the OC content of sediments and bacterial activity (e.g., Tyson, 1987). Sediment OC content in a shallow epicontinental sea is mainly controlled by organic production, rate and mechanism of OC transport through the water column, burial rate, sediment grain size, and perhaps most important, the rate of consumption by bacteria and macrofauna in the benthic transition layer. Because biotic activity is largely controlled by O_2 levels, the rate of OC supply vs. the rate of oxygen advection to the benthic zone is the key factor limiting biotas, and thus organic richness of the sediment. Savrda and Bottjer (1994) noted, however, that additional factors, such as the consistency of substrates, could influence infaunas and impart some control on burrowing patterns. Based on these factors, three hypotheses may be formed: (1) if ORI's mainly depend on physical changes in the substrate due to dilution or productivity cycles, they should correlate with changes in % $CaCO_3$ (and have similar spectra); (2) if ORI's mainly depend on changes in organic richness and pore water O_2, they should correlate with changes in % OC (and have similar spectra); and (3) if ORI's correlate with either % $CaCO_3$ or % OC because of a common external forcing mechanism, they should have similar spectra.

Comparison of Figures 4 and 5 shows that the spectra for ORI Rank and MBD are most similar to that for OC; each indicates dominance of obliquity and precession, but there is no statistically significant record of eccentricity. This result suggests stronger coupling between bioturbation and OC content of substrates than between bioturbation and $CaCO_3$ content (and presumed substrate consistency). However, it does not resolve whether changes in OC content reflect increased OC flux or decreased O_2 advection. The spectral result also indicates that the records of OC and $CaCO_3$ may be decoupled, and perhaps slightly out of phase, which suggests that the mechanisms responsible for OC production or water column stratification in the basin are different from those that control carbonate production (or siliciclastic dilution).

Although nannofossil assemblages from strata of the Western Interior basin have been described as unusual and enigmatic compared to assemblages from oceanic settings, levels of diversity nonetheless reach oceanic values in the peak transgressive Bridge Creek Limestone of the Greenhorn Cyclothem (Watkins et al., 1993). Watkins (1989) analyzed nannofossil assemblages in Bridge Creek limestone/marlstone bedding couplets and found that changes in assemblage diversity correlated significantly with changes in % $CaCO_3$. Based on analysis of two couplets at localities in Kansas, Nebraska, and South Dakota, Watkins (1989) found that limestone hemicycles tend to be characterized by high diversity and high equitability levels whereas marlstone hemicycles show lower diversity and dominance of species that reflect high fertility (e.g., Roth and Krumbach, 1986). Certain species showed strong correlations in relative abundance to the carbonate-rich hemicycles (e.g., *Helicolithus* or *Eiffellithus* spp.) while others correlated significantly with the organic carbon-rich laminated marls (e.g., *Biscutum constans, Zeugrhabdotus* or *Zygodiscus* spp.).

Watkins (1989) interpreted these trends as reflective of changes in surface water conditions related to climate cycles. He suggested that the deposition of organic carbon-rich marlstone hemicycles was characterized by unstable conditions in the water column, increased nutrient levels, and high fertility among a few opportunistic taxa. Carbonate-rich hemicycles, on the other hand, were characterized by more stable conditions leading to higher diversity and equitability levels. Watkin's (1989) results contradict the suggestion of Eicher and Diner (1985 and 1989) that high productivity levels accounted for the deposition of Bridge Creek limestone beds.

Comparison of trends in the nannofossil record with % $CaCO_3$, % OC, and bioturbation offers an opportunity to assess linkages between surface-water conditions, sedimentation processes, and redox history. Burns and Bralower (this volume) indicate that several taxa have specific interpretive value: W. *barneseae* reflects nannofossil preservation and *B. constans* indicates fertility. Poor results in nannofossil spectra for the Portland core may reflect poor preservation, as suggested by trends in *W. barnesae* (Burns and Bralower, this volume), or they may indicate that changes in the relative abundance of some taxa (e.g., *Zygodiscus* spp., *Eiffellithus* spp.) are unrelated to the forcing factors that influence the depositional system. The spectrum for *B. constans*, however, showed a result most similar to that for % $CaCO_3$, and the spectrum for *Tranolithus* spp. was similar to that for % OC. Given that *B. constans* is regarded as a fertility indicator and generally is most abundant in organic carbon-rich hemicycles, we might expect it to more closely reflect fluctuations in % OC if organic productivity were the driving mechanism. However, it is possible that slight dissolution of *B. constans* in association with the minor diagenesis that attended formation of the Bridge Creek limestone beds (Ricken, 1994) modified relative abundance values to more closely follow variations in $CaCO_3$ content. Unfortunately the ecological affinities of *Tranolithus* spp. are not well know (Burns and Bralower, this volume).

In addition to the characteristics discussed above, each of the data sets shows variations in expression through the Bridge Creek that bear on the origin of bedding cyclicity. For example, Sageman et al. (1997) described variations in carbonate content and bed thickness trends from the lower to the upper Bridge Creek. Burns and Bralower (this volume) indicated that preservation was better in the upper part. Savrda (this volume) illustrated that the ORI cycles recognized in the lower part of Bridge Creek Member of the Pueblo core (and interpreted to reflect the precessional index) do not occur in the upper part. Thus, the lowest part of the Bridge Creek is characterized by a record that may reflect both amalgamation and dilution of cycles. The middle part (upper *N. juddi* and *W. devononse* Biozones) includes bedding couplets that suggest a 1:5 pattern (note two to three E1 cycles marked with *'s in Fig. 8), but this could not be confirmed by spectral analysis. Although E and P signals were detected in the spectral results from the upper part of the Bridge Creek, the 1:5 pattern is not apparent. Instead, a 1:2 or 1:3 bedding pattern becomes predominant.

Analysis of Cycle Hierarchy

Based on the spectral results we hypothesize that P, O, and

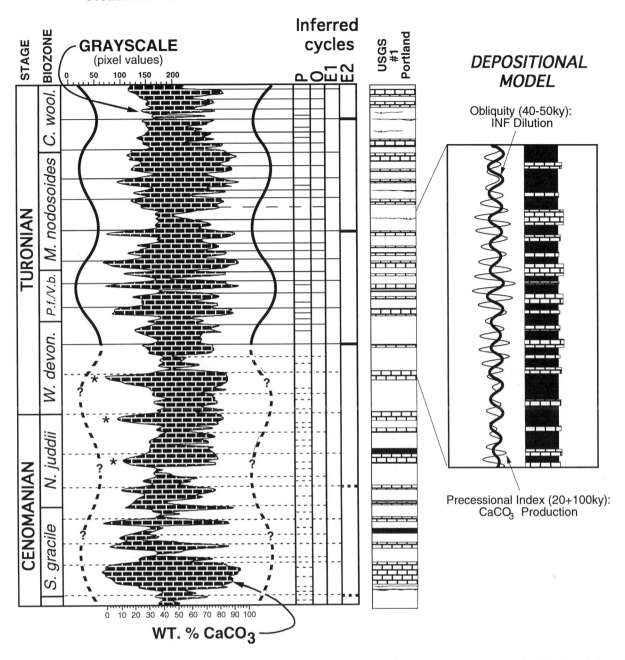

FIG. 8. Interpretation of Bridge Creek bedding cycles in the #1 Portland core using mirror plot (Herbert and Fischer, 1986) of grayscale pixel values (left curve) and % CaCO3 (right curve). Inferred P, O, E1, and E2 cycles are summarized in columns. Heavy lines in mirror plot define inferred E2 cycles. Dashed lines are used in the lower Bridge Creek to indicate uncertainty. Hypothesized depositional model relating orbital cycles to bedding pattern is illustrated in panel on right (see text).

E cycles can be identified in the upper Bridge Creek Member. To present our analysis of the pattern of bedding we employ the "mirror plot" method (Fig. 8) introduced by Herbert and Fischer (1986). First, the long eccentricity cycle (E2) is chosen based on its recognition in the 20-point MA curve of Figure 2 (a bold line marks the E2 "envelop" and boundaries are placed in the E2 column for inferred cycles). Then, assuming that obliquity comprises a dominant signal, every limestone bed greater than about 10 cm is marked by a line on the right side of the mirror plot, extending from the CaCO3 curve and ending in the O column. If the O signal represents about 40-50 ky, then approximately every other cycle would correspond to E1. The E1 cycles are thus delineated on the left side of Figure 8 (with lines extending from the gray-scale curve) and marked in the E1 column. Higher

frequency fluctuations are assumed to reflect precession and marked in the P column. This method of cycle identification is extended to the lower Bridge Creek but illustrated with dashed lines to reflect uncertainty.

This method attempts to superimpose a hierarchy of cycles on the record of bedding. In the upper Bridge Creek this hierarchy explains the data fairly well. Although there are irregularities in the pattern (e.g., some E1 cycles correspond to two O cycles whereas others have three), this is to be expected given the variations in the orbital quasi-periods and resulting phase relationships among cycles. We extended this cycle hierarchy to the lower Bridge Creek using the same principles, but the lower interval does not follow the same pattern as clearly (Fig. 8). The characteristics of the lower Bridge Creek suggest more than

TABLE 5.- *MTM analysis of nannofossil data from the Bridge Creek Limestone.*

Frequency (cycles/m)	Amplitude (Rel. Abund.)	Phase (degrees)	F-test (>90% only)	Period-kyr (0.6cm/kyr)	
Biscutum constans					
0.1196	0.0185	154.49	5.1	1393.19	
0.5908	0.0119	37.4	4.8	282.09	
1.8237	0.0098	109.56	7.05	**91.39**	*E1*
2.6099	0.0026	-146.9	3.65	63.86	
3.3667	0.0041	157.16	5.29	49.5	
4.3311	0.0038	74.02	4.42	**38.48**	*O1*
4.8535	0.002	-137.86	3.57	34.34	
5.6616	0.0014	-42.47	3.59	29.44	
6.2671	0.0011	145.5	3.98	26.59	
6.7676	0.0011	-43.77	8.2	**24.63**	*P2*
7.4048	0.0006	136.12	4.09	22.51	
7.9981	0.0008	-93.57	3.26	**20.84**	*P1*
8.9893	0.0003	-167.72	3.18	18.54	
9.209	0.0003	-110.11	4.06	18.1	
Zygodiscus spp.					
0.1147	0.012	135.48	3.76	1452.48	
0.5957	0.0096	-63.5	7.19	279.78	
1.1987	0.0051	-86.76	8.41	139.04	
1.6992	0.004	138.9	3.3	**98.08**	*E1*
2.3657	0.0052	30.27	9.61	70.45	
2.9663	0.0031	-71.45	3.42	56.19	
3.4717	0.0028	116.26	5.11	48.01	
4.082	0.0034	-74.5	7.34	**40.83**	*O1*
4.8999	0.002	-80.63	4.76	34.01	
5.8228	0.0014	161.57	9.14	28.62	
6.8384	0.0007	-67.92	3.59	**24.37**	*P2*
7.9004	0.0005	130.2	3.17	**21.1**	*P1*
Eiffellithus spp.					
0.5396	0.0063	21.29	8.73	308.9	
3.2178	0.0038	85.14	5.63	**51.8**	*O2*
3.9111	0.0023	14.95	5.27	**42.61**	*O1*
4.1748	0.0025	-51.28	3.39	39.92	
4.4385	0.0024	-176.5	4.38	37.55	
5.9106	0.0018	61.72	5.87	28.2	
7.3584	0.0009	-88.77	3.13	**22.65**	*P2*
9.187	0.0003	-149.49	16.52	18.14	
9.8731	0.0003	-147.91	5.35	16.88	
Tranolithus spp.					
0.1367	0.0085	144.08	4.06	1219.04	
1.2647	0.0045	-17.48	3.87	131.79	
2.124	0.0033	-95.14	12.08	78.47	
3.2129	0.0052	85.13	4.07	**51.87**	*O2*
4.1113	0.0023	13.8	3.65	**40.54**	*O1*
4.773	0.0025	146.45	6.91	34.92	
5.2979	0.0023	-13.45	8.12	31.46	
6.1255	0.0025	53.98	13.02	27.21	
7.2144	0.0017	96.26	8.73	**23.1**	*P2*

1. Only frequencies with F-tests above the 90% confidence level are shown.

2. Period values were calculated using S = 0.6 cm/kyr. Frequency values close to the dominant orbital cycles are designated as E1, E2, etc., and their period values are printed in bold.

simple variance in sedimentation rates. Some limestone beds may be amalgamated and others missing entirely. Sageman et al. (1997) concluded that changes in the production and (or) preservation of OC related to OAE II, the influence of aperiodic ash fall, and changes in detrital flux related to a transgressive event, altered the pattern of bedding thicknesses in the lower Bridge Creek sufficiently to modify the preservation of orbital influence. Extending the cycle hierarchy from the upper to the lower Bridge Creek thus provides a framework within which variations in the linkage between climate and the depositional system can be better understood.

Origin Of Bedding Cyclicity

As pointed out by Ricken (1994), LaFerriere et al. (1987), Arthur and Dean (1991), and Elder et al. (1994), the development of bedding cycles in Cretaceous rocks of the Western Interior basin reflects a highly sensitive depositional system. This system was influenced by minor changes in relative sea level and climate that periodically modulated the fluxes of organic matter, carbonate, and detrital material to the basin. At times it was also sensitive to the aperiodic input of volcanic ash. The preservation of bedding cycles that record orbital influence sufficiently to allow spectral analysis is limited to the highstand of a relative sea-level cycle that occurred near peak marine flooding of the basin (Fig. 7).

The results of spectral analyses for this highstand interval suggest that the complex pattern of the Bridge Creek bedding couplets may reflect constructive and destructive interference of separate orbital cycles that preferentially influenced different parts of the depositional system (Sageman et al., 1997). As a result of the meridional configuration of the seaway, the basin spanned multiple climate zones, and was thus subject to latitudinal variations in climate forcing. The major control of bedding cycles is thought to be dilution by detrital input (e.g., Arthur et al., 1985; Barron et al., 1985; Pratt et al., 1993). Variations in precipitation in the northern part of the seaway, which probably played a significant role in the dilution process, would have been particularly sensitive to obliquity (Park and Oglesby, 1991, 1994). However, if precession produced the dominant effect on climate at lower latitudes, as is predicted, it may have influenced carbonate productivity in the manner suggested by Eicher and Diner (1985, 1989) and superimposed variations in carbonate flux on dilution cycles within the basin. The relative influence of the northern and southern climate belts in the seaway would determine the degree of cycle mixing. Phase relations would cause positive and negative interference during cycle development where mixing was strong, resulting in a complex bedding pattern (this interpretation may explain some of the F-test results interpreted to represent interference beats).

This depositional model is illustrated on the right side of Figure 8. The formation of limestone beds is modeled as the product of clastic dilution (represented by the bold "O" or obliquity curve, which increases to the right) and carbonate input (represented by the thin "P" or precessional index curve, which increases to the right). Only where O is low can P produce thick limestone beds (in some cases by amalgamating two P cycles). When O is high, P may produce a small bed or no bed at all depending on the modulation of E1 and E2 (Fig. 8). This model produces a 1:2 or 1:3 pattern of bedding couplets like that observed in the upper Bridge Creek, and the long term variation it creates (E2 scale) also matches the observations from the Port-

land core (Fig. 8). This model implies that the onset of bedding cyclicity in the lower Bridge Creek was dominated by a southern (precessional) influence, but that through time the northern (obliquity) influence became more dominant. Sageman et al. (1997) hypothesized that this shift was caused by an Early Turonian cooling event precipitated by the global burial of carbon during OAE II, as was suggested by Arthur et al. (1988).

The depositional model is supported by the interpretation of decoupling between the % CaCO3 record and % OC record, which suggests that the dominant control on carbonate production was unrelated or not entirely related to the controls on OC burial. One possibility is that OC preservation solely depended on stratification intensity, and thus was closely tied to runoff volume. But carbonate content, being the net product of dilution (due to run-off) and carbonate productivity (controlled by Tethyan surface waters), had a more complex origin. Another possibility is that OC burial was strongly influenced by production of noncalcareous plankton (e.g., diatoms), and these taxa were more sensitive to fluctuations in nutrient supply within the seaway. In this scenario the correlation between ORI and %OC in the upper Bridge Creek would reflect increases in river-borne nutrients, productivity among noncalcareous plankton, and organic loading of substrates leading to lower oxygen levels in bottom waters. The low % OC levels of clay-rich hemicycles in the lower Bridge Creek may indicate that nutrient supply was too low to stimulate excess productivity among the noncalcareous plankton. As a result, OC flux to the sediment would have been insufficient to drive pore waters anoxic. This allowed more infauna to colonize, further reducing preserved OC levels. The increase in relative abundance of the fertility indicator *B. constans* from the lower to the upper Bridge Creek provides support for this interpretation (see also discussion in Burns and Bralower, this volume).

CONCLUSIONS

The Bridge Creek Limestone has long been interpreted as a product of orbital influences (Fischer, 1980), although determining of the periodicity and depositional mechanisms responsible for the cycles has been difficult. Detailed quantitative analysis of complementary paleoecological and geochemical data, however, helps to reveal the complexities of orbital expression in the Bridge Creek record. Understanding these complexities illuminates the paleoenvironmental history of the Western Interior basin.

Our study was made possible by new high-resolution [40]Ar/[39]Ar radiometric dates, which improve estimates of age-depth relationships for the Bridge Creek Limestone. These estimates allow meaningful application of spectral analyses. Spectral analysis of three complementary analytical records from the Portland core, including lithologic data (weight % CaCO3, weight % OC, and optical densitometry of core photographs), ichnologic data (maximum burrow diameter, oxygen-related ichnocoenoses), and nannofossil data (species abundance trends), indicate the presence of obliquity and precessional index signals in the upper Bridge Creek.

Analysis of the spectral results from multiple complementary data sets suggests a depositional model for the bedding couplets that combines the dilution and productivity mechanisms. Obliquity, through its effect on high-latitude precipitation, is thought to predominantly influence dilution processes whereas the precessional index mainly controls changes in carbonate productivity through its effect on evaporation and nutrient upwelling in the Tethyan realm. The two influences mix in the shallow Western Interior basin and result in constructive and destructive interference, producing a complex bedding pattern. In addition, variations in cycle expression through the Bridge Creek provided a high-resolution record of changes in the depositional system, revealing intervals of unusual condensation and aperiodic dilution. Some of these corroborate and help refine interpretations of relative sea level history for the study interval.

Comparison of %OC data with trends in trace fossils and nannofossils, which act as proxies for benthic redox conditions and primary production, respectively, suggest that organic matter production and preservation were decoupled from carbonate production. The recognition that $CaCO_3$ production and OC production may fluctuate in phase, out of phase, or antithetically probably reflects changes in the ecological structure of the primary producer community and the different responses of taxonomic groups to changing climatic and oceanographic conditions.

The interpretative model presented herein is based on (1) determination of patterns in Bridge Creek bedding cycles using a quantitative method (spectral analysis), and extension of that pattern to parts of the section where spectral analysis is not possible, and (2) analysis of multiple data sets reflecting different components of the depositional system. The interpretative model provides a reasonable explanation for the bedding patterns observed throughout the Bridge Creek, and attempts to synthesize the observations of past workers (Pratt, 1984; Arthur et al., 1985; Barron et al., 1985; Eicher and Diner, 1985, 1989; Watkins, 1989; Pratt et al., 1993) into a comprehensive scheme. Finally, the model highlights the relative importance of sea level, paleogeography and latitudinal climate belts, the hydrologic cycle, nutrient flux, and marine primary productivity in preserving Milankovitch orbital cycles in the Western Interior rock record. The development of these cycles within the highstand phase of the Greenhorn Cyclothem represented a stratigraphic "window

REFERENCES

Arthur, M. A., Dean, W. E., Bottjer, D., and Scholle, P. A., 1984, Rhythmic bedding in Mesozoic-Cenozoic pelagic carbonate sequences; the primary and diagenetic origin of Milankovitch-like cycles, in Berger, A., Imbrie, J., Hays, J. D, Kukla, G., and Saltzman, B., eds., Milankovitch and Climate, pt. 1: Amsterdam, Reidel Publishing Company, p. 191-222.

Arthur, M. A., Dean, W. E., Pollastro, R., Scholle, P. A., and Claypool, G. E., 1985, A comparative geochemical study of two transgressive pelagic limestone units, Cretaceous Western Interior basin, U.S., in Pratt, L. M., Kauffman, E. G., and Zelt, F. B., eds., Fine-grained Deposits and Biofacies of the Cretaceous Western Interior Seaway: Evidence of Cyclic Sedimentary Processes, Field Trip Guidebook No. 4: Tulsa, Society of Economic Paleontologists and Mineralogists, p. 16-27.

Arthur, M. A., Dean, W. E., and Pratt, L. M., 1988, Geochemical and climatic effects of increased marine organic carbon burial at the Cenomanian/Turonian boundary: Nature, v. 335, p. 714-717.

Arthur, M. A., and Dean, W. E., 1991, An holistic geochemical approach to cyclomania: Examples from Cretaceous pelagic limestone sequences, in Einsele, G., Ricken, W., and Seilacher, A., eds., Cycles and Events in Stratigraphy: Berlin, Springer Verlag, p. 126-166.

Arthur, M.A. and Sageman, B.B., 1994, Marine Black Shales: A Review of Depositional Mechanisms and Environments of Ancient Deposits: Annual Reviews of Earth & Planetary Science, v. 22, pp. 499-552.

Barron, E. J., Arthur, M. A., and Kauffman, E. G., 1985, Cretaceous rhythmic bedding sequences: A plausible link between orbital variations and climate: Earth Planetary Science Letters, v. 72, p. 327-340.

Berger, A., 1978, Long-term variations of daily insolation and Quaternary climate change: Journal of Atmospheric Science, v. 35, p. 2362-2367.

Berger, A., 1984, Accuracy and frequency stability of the Earth's orbital elements during the Quaternary, in Berger, A., Imbrie, J., Hays, J. D., Kukla, G., and Saltzman, B., eds., Milankovitch and Climate, pt. 1: Amsterdam, Reidel Publishing Company, p. 3-40.

Berger, A., Loutre, M. F., and Dehant, V., 1989, Astronomical frequencies for pre-Quaternary palaeoclimate studies: Terra Nova, v. 1, p. 474-479.

Berger, A., Loutre, M. F., and Laskar, J., 1992, Stability of the astronomical frequencies over the Earth's history for paleoclimate studies: Science, v. 255, p. 560-566.

Bralower, T. J., 1988, Calcareous nannofossil biostratigraphy and assemblages of the Cenomanian-Turonian boundary interval: Implications for the origin and timing of anoxia: Paleoceanography, v. 3, p. 275-316.

Caldwell, W. G. E. and Kauffman, E. G., eds., 1993, Evolution of the Western Interior Basin: St. John's, Geological Association of Canada Special Paper 39, 680 p.

Caputo, M., Peterson, J. and Franczyk, K., eds., 1994, Mesozoic Systems of the Rocky Mountain Region: Tulsa, Society of Economic Paleontologists and Mineralogists Special Publication, 536 p.

Cobban, W. A., 1984, Mid-Cretaceous ammonite zones, Western Interior, United States: Bulletin Geological Society Denmark, v. 33, p. 71-89.

Cobban, W. A., 1985, Ammonite record from Bridge Creek Member of Greenhorn Limestone at Pueblo Reservoir State Recreation Area, Colorado, in Pratt, L. M. Kauffman, E. G., and Zelt, F. B., eds., Fine-grained Deposits and Biofacies of the Cretaceous Western Interior Seaway: Evidence of Cyclic Sedimentary Processes, Field Trip Guidebook No. 4: Tulsa, Society of Economic Paleontologists and Mineralogists, p. 135-138.

Cobban, W. A., and Scott, R. W., 1972, Stratigraphy and Ammonite Fauna of the Graneros Shale and Greenhorn Limestone Near Pueblo, Colorado: Washington, D.C., U.S. Geological Survey Professional Paper 645, p. 1-108.

Dean, W. E., and Arthur, M. A., 1994, The Cretaceous Western Interior Seaway Drilling Project, An overview (abs.): American Association of Petroleum Geologists Annual Meeting, Abstracts with Program, v. 3, p. 134.

Eicher, D. L., and Diner, R., 1985, Foraminifera as indicators of water mass in the Cretaceous Greenhorn Sea, Western Interior, in Pratt, L. M., Kauffman, E. G., and Zelt, F. B., eds., Fine-grained Deposits and Biofacies of the Cretaceous Western Interior Seaway: Evidence of Cyclic Sedimentary Processes, Field Trip Guidebook No. 4: Tulsa, Society of Economic Paleontologists and Mineralogists, p. 60-71.

Eicher, D. L., and Diner, R., 1989, Origin of the Cretaceous Bridge Creek cycles in the Western Interior, United States: Palaeogeography, Palaeoclimatology, and Palaeoecology, v. 74, p. 127-146.

Eicher, D. L., and Diner, R., 1991, Environmental factors controlling Cretaceous limestone-marlstone rhythms, in Einsele, G., Ricken, W., and Seilacher, A., eds., Cycles and Events in Stratigraphy: Berlin, Springer Verlag, p. 79-93.

Elder, W. P., 1985, Biotic patterns across the Cenomanian-Turonian extinction boundary near Pueblo, Colorado, in Pratt, L. M., Kauffman, E. G., and Zelt, F. B., eds., Fine-grained Deposits and Biofacies of the Cretaceous Western Interior Seaway: Evidence of Cyclic Sedimentary Processes, Field Trip Guidebook No. 4: Tulsa, Society of Economic Paleontologists and Mineralogists, p. 157-169.

Elder, W. P., Gustason, E. R., and Sageman, B. B., 1994, Basinwide correlation of parasequences in the Greenhorn Cyclothem, Western Interior, U.S.: Geological Society of America Bulletin, v. 106, p. 892-902.

Engleman, E. E., Jackson, L. L., Norton, D. R., and Fischer, A. G., 1985, Determination of carbonate carbon in geological materials by coulometric titration: Chemical Geology, v. 53, p. 125-128.

Erba, E., Castradori, D., Guasti, G., and Ripepe, M., 1992, Calcareous nannofossils and Milankovitch cycles: The example of the Albian Gault Clay Formation (southern England): Palaeogeography, Palaeoclimatology, and Palaeoecology, v. 93, p. 47-69.

Erba, E., and Premoli-Silva, I., 1994, Orbitally driven cycles in trace fossil distribution from the Piobbico core (late Albian, central Italy), in deBoer, P. L., and Smith, D. G., eds., Orbital Forcing and Cyclic Sequences: International Association of Sedimentologists Special Publication 19, Oxford, Blackwell Scientific Publications, p. 219-225.

Fischer, A. G., 1980, Gilbert-bedding rhythms and geochronology, in Yochelson, E. I., ed., The Scientific Ideas of G. K. Gilbert: Boulder, Geological Society of America Special Paper 183, p. 93-104.

Fischer, A. G., Herbert, T., and Premoli-Silva, I., 1985, Carbonate bedding cycles in Cretaceous pelagic and hemipelagic sediments, in Pratt, L. M., Kauffman, E. G., and Zelt, F. B., eds., Fine-grained Deposits and Biofacies of the Cretaceous Western Interior Seaway: Evidence of Cyclic Sedimentary Processes, Field Trip Guidebook No. 4: Tulsa, Society of Economic Paleontologists and Mineralogists, p. 1-10.

Gilbert, G. K., 1895, Sedimentary measurement of geologic time: Geology, v. 3, p. 121-127.

Glancy, T. J., Jr., Arthur, M. A., Barron, E. J., and Kauffman, E. G., 1993, A paleoclimate model for the North American Cretaceous (Cenomanian-

Turonian) epicontinental sea, *in* Caldwell, W. G. E., and Kauffman, E. G., eds., Evolution of the Western Interior Basin: St. John's, Geological Association of Canada Special Paper 39, p. 219-242.

HAQ, B. V., HARDENBOL, J., AND VAIL, P. R., 1987, Chronology of fluctuating sea levels since the Triassic (250 million years ago to present): Science, v. 235, p. 1159-1167.

HERBERT, T. D., AND FISCHER, A. G., 1986, Milankovitch climatic origin of mid-Cretaceous black shale rhythms in central Italy: Nature, v. 321, p. 739-743.

JORDAN, T. E., 1981, Thrust loads and foreland basin evolution, Cretaceous, western United States: American Association of Petroleum Geologists Bulletin, v. 65, p. 2506-2520.

KAUFFMAN, E. G., 1977, Geological and biological overview: Western Interior Cretaceous Basin: Mountain Geologist, v. 13, p. 75-99.

KAUFFMAN, E. G., 1984, Paleobiogeography and evolutionary response dynamic in the Cretaceous Western Interior Seaway of North America, *in* Westermann, G. E. G., ed., Jurassic-Cretaceous Biochronology and Paleogeography of North America: St. John's, Geological Association of Canada Special Paper 27, p. 273-306.

KAUFFMAN, E. G., 1995, Global change leading to biodiversity crisis in a greenhouse world: The Cenomanian-Turonian (Cretaceous) mass extinction. Effects of past global change on life, *in* Studies in Geophysics: Washington, D.C., National Academy Press, p. 47-71.

KAUFFMAN, E. G., SAGEMAN, B. B., ELDER, W. P., KIRKLAND, J. I., AND.VILLAMIL, T., 1993, Cretaceous molluscan biostratigraphy and biogeography, Western Interior Basin, North America, *in* Caldwell, W. G. E., and Kauffman, E. G., eds., Evolution of the Western Interior Basin: St. John's, Geological Association of Canada Special Paper 39, p. 397-434.

KENNEDY, W. J., AND COBBAN, W. A., 1991, Stratigraphy and interregional correlation of the Cenomanian-Turonian transition in the Western Interior of the United States near Pueblo, Colorado, a potential boundary stratotype for the base of the Turonian stage: Newsletter of Stratigraphy, v. 24(1/2), p. 1-33.

LAFERRIERE, A. P., HATTIN, D., AND ARCHER, A. W., 1987, Effects of climate, tectonics and sea-level changes on rhythmic bedding patterns in the Niobrara Formation (Upper Cretaceous), U.S., Western Interior: Geology, v. 15, p. 233-236.

LUDVIGSON, G. A., WITZKE, B. J., GONZALEZ, L. A., HAMMOND, R. H., AND PLOCHER, O. W., 1994, Sedimentology and carbonate geochemistry of concretions from the Greenhorn Marine Cycle (Cenomanian-Turonian), eastern margin of the Western Interior Seaway, *in* Shurr, G. W., Ludvigson, G. A., and Hammond, R. H., eds., Perspectives on the Eastern Margin of the Cretaceous Western Interior Basin: Boulder, Geological Society of America Special Paper 287, p. 145-174.

OBRADOVICH, J., 1993, A Cretaceous time scale, *in* Caldwell, W. G. E., and Kauffman, E. G., eds., Evolution of the Western Interior Basin: St. John's, Geological Association of Canada Special Paper 39, p. 379-396.

OBRADOVICH, J. D., AND COBBAN, W. A., 1975, A time-scale for the Late Cretaceous of the Western Interior of North America, *in* Caldwell, W. G. E., ed., The Cretaceous System in the Western Interior of North America: St. John's, Geological Association of Canada Special Paper 13, p. 31-54.

PARK, J., AND HERBERT, T. D., 1987, Hunting for paleoclimatic periodicities in a geologic time series with an uncertain time scale: Journal of Geophysical Research, v. 92, p. 14027-14040.

PARK, J., AND OGLESBY, R. J., 1991, Milankovitch rhythms in the Cretaceous: A GCM modelling study: Palaeogeography, Palaeoclimatology, and Palaeoecology (Global and Planetary Change), v. 90, p. 329-355.

PARK, J., AND OGLESBY, R. J., 1994, The effect of orbital cycles on Late and Middle Cretaceous climate: A comparative general circulation model study, *in* deBoer, P. L., and Smith, D. G., eds., Orbital Forcing and Cyclic Sequences: International Association of Sedimentologists Special Publication 19, Oxford, Blackwell Scientific Publications, p. 509-529.

PRATT, L. M., 1984, Influence of paleoenvironmental factors on the preservation of organic matter in middle Cretaceous Greenhorn Formation near Pueblo, Colorado: American Association of Petroleum Geologists Bulletin, v. 68, p. 1146-1159.

PRATT, L. M., 1985, Isotopic studies of organic matter and carbonate in rocks of the Greenhorn marine cycles, *in* Pratt, L. M., Kauffman, E. G., and Zelt, F. B., eds., Fine-grained Deposits and Biofacies of the Cretaceous Western Interior Seaway: Evidence of Cyclic Sedimentary Processes, Field Trip Guidebook No. 4: Tulsa, Society of Economic Paleontologists and Mineralogists, p. 38-48.

PRATT, L. M., KAUFFMAN, E. G., AND ZELT, F. B., eds., 1985, Fine-grained Deposits and Biofacies of the Cretaceous Western Interior Seaway: Evidence of Cyclic Sedimentary Processes, Field Trip Guidebook No. 4: Tulsa, Society of Economic Paleontologists and Mineralogists, 249 p.

PRATT, L. M., ARTHUR, M. A., DEAN, W. E., AND SCHOLLE, P. A., 1993, Paleoceanographic cycles and events during the Late Cretaceous in the Western Interior Seaway of North America, *in* Caldwell, W. G. E., and Kauffman, E. G., eds., Evolution of the Western Interior Basin: St. John's, Geological Association of Canada

Special Paper 39, p. 333-354.

RICKEN, W., 1991, Variation of sedimentation rates in rhythmically bedded sediments - distinction between depositional types, *in* Einsele, G., Ricken, W., and Seilacher, A., eds., Cycles and Events in Stratigraphy: Berlin, Springer-Verlag, p. 167-187.

RICKEN, W., 1994, Complex rhythmic sedimentation related to third-order sea-level variations: Upper Cretaceous, Western Interior Basin, USA, *in* deBoer, P. L., and Smith, D. G., eds., Orbital Forcing and Cyclic Sequences: International Association of Sedimentologists Special Publication 19, Oxford, Blackwell Scientific Publications, p. 167-193.

ROTH, P. H., AND BOWDLER, J. L., 1981, Middle Cretaceous calcareous nannoplankton biogeography and oceanography of the Atlantic Ocean: Tulsa, Society of Economic Paleontologists and Mineralogists Special Publication 32, p. 517-546.

ROTH, P. H. AND KRUMBACH, K. R., 1986, Middle Cretaceous nannofossil biogeography and preservation in the Atlantic and Indian Oceans: Implications for paleoceanography: Marine Micropaleontology, v. 10, p. 235-266.

SAGEMAN, B. B., 1996, Lowstand Tempestites: Depositional model for Cretaceous skeletal limestones, Western Interior, U.S.: Geology, v. 24, p. 888-892.

SAGEMAN, B. B., 1991, High-resolution event stratigraphy, carbon geochemistry, and paleobiology of the Upper Cenomanian Hartland Shale Member (Cretaceous), Greenhorn Formation, Western Interior, U.S.: Unpublished Ph.D. Dissertation, University of Colorado, Boulder, 572 p.

SAGEMAN, B. B., RICH, J., ARTHUR, M. A., BIRCHFIELD, G. E., AND DEAN, W. E., 1997, Evidence For Milankovitch periodicities in Cenomanian-Turonian lithologic and geochemical cycles, Western Interior U.S., Journal of Sedimentary Research, v. 67, p. 285-301.

SAVRDA, C. E., AND BOTTJER, D. J., 1991, Oxygen-related biofacies in marine strata: An overview and update, *in* Tyson, R. V., and Pearson, T. H., eds., Modern and Ancient Continental Shelf Anoxia: London, Geological Society of London Special Publication 58, p. 201-220.

SAVRDA, C. E., AND BOTTJER, D. J., 1994, Ichnofossils and ichnofabrics in rhythmically bedded pelagic/hemi-pelagic carbonates: Recognition and evaluation of benthic redox and scour cycles, *in* deBoer, P. L., and Smith, D. G., eds., Orbital Forcing and Cyclic Sequences: International Association of Sedimentologists Special Publication 19, Oxford, Blackwell Scientific Publications, p. 195-210.

SCHLANGER, S. O., ARTHUR, M. A., JENKYNS, H. C., AND SCOLLE, P. A., 1987, The Cenomanian-Turonian Oceanic Anoxic Event, I. Stratigraphy and distribution of organic carbon-rich beds and the marine δ^{13}C excursion, *in* Brooks, J., and Fleet, A. J., eds., Marine Petroleum Source Rocks: London, Geological Society of London Special Publication 26, p. 371-399.

SCHWARZACHER, W., AND FISCHER, A. G., 1982, Limestone-shale bedding and perturbations of the earth's orbit, *in* Einsele, G., Ricken, W., and Seilacher, A., eds., Cyclic and Event Stratification: Berlin, Springer-Verlag, p. 72-95.

THOMSON, D. J., 1982, Spectrum estimation and harmonic analysis: Institute of Electrical and Electronics Engineers (IEEE), Proceedings, v. 70, p. 1055-1096.

TYSON, R. V., 1987, The genesis and palynofacies characteristics of marine pertroleum source rocks, *in* Brooks, J., and Fleet, A. J., eds., Marine Petroleum Source Rocks: London, Geological Society of London Special Publication 26, p. 47-67.

UPCHURCH, G.R., AND WOLFE, J.A., 1993, Cretaceous vegetation of the Western Interior and adjacent regions of North America, *in* Caldwell, W. G. E., and Kauffman, E. G., eds., Evolution of the Western Interior Basin: St. John's, Geological Association of Canada Special Paper 39, p. 243-282.

VAN WAGONER, J. C., POSIMENTIER, H. W., MITCHUM JR., R. M., VAIL, P. R., SARG, J. F., LOUTTIT, T. S., AND HARDENBOL, J., 1988, An overview of the fundamentals of sequence stratigraphy and key definitions, *in* Wilgus et al., eds., Sea-Level Changes: An Integrated Approach: Tulsa, Society of Economic Paleontologists and Mineralogists Special Publication 42, p. 39-46.

WATKINS, D. K., 1985, Biostratigraphy and paleoecology of nannofossils in the Greenhorn marine cycle, *in* Pratt, L. M., Kauffman, E. G., and Zelt, F. B., eds Fine-grained Deposits and Biofacies of the Cretaceous Western Interior Seaway: Evidence of Cyclic Sedimentary Processes, Field Trip Guidebook No. 4: Tulsa, Society of Economic Paleontologists and Mineralogists, p. 151-156.

WATKINS, D. K., 1989, Nannoplankton productivity fluctuations and rhythmically bedded pelagic carbonates of the Greenhorn Limestone (Upper Creatceous): Palaeogeography, Palaeoclimatology, and Palaeoecology, v. 74, p. 75-86.

WATKINS, D. K., BRALOWER, T. J., COVINGTON, J. M., AND FISHER, C. G., 1993, Biostratigraphy and paleoecology of the upper Cretaceous calcareous nannofossils in the Western Interior Basin, North America, *in* Caldwell, W. G. E., and Kauffman, E. G., eds., Evolution of the Western Interior Basin: St. John's, Geological Association of Canada Special Paper 39, p. 521-538.

WITZKE, B. J. AND LUDVIGSON, G. A., 1994, The Dakota Formation in Iowa and the type area, *in* Shurr, G. W., Ludvigson, G. A., and Hammond, R. H., eds., Perspectives on the Eastern Margin of the Cretaceous Western Interior Basin: Boulder, Geological Society of America Special Paper 287, p. 43-78.

ORGANIC GEOCHEMISTRY OF THE CRETACEOUS WESTERN INTERIOR SEAWAY: A TRANS-BASINAL EVALUATION

RICHARD D. PANCOST, KATHERINE H. FREEMAN, AND MICHAEL A. ARTHUR

Department of Geosciences, The Pennsylvania State University, University Park, PA 16802

ABSTRACT: We used hydrocarbon distributions to establish the dominant sources and controls on organic matter deposition and preservation in rocks from the latest Cenomanian-early Turonian interval in cores representing a three-site transect across the Cretaceous Western Interior Seaway. Here we present the results of free *n*-alkane and isoprenoid hydrocarbon analyses. Maturity varies slightly between the three cores: organic matter in the central basin core is most mature (equivalent vitrinite reflectane ~0.55), while in the western and east-central basin cores organic matter is less mature (equivalent vitrinite reflectance ~0.4). Geographic controls on compound abundances were dominant over lithologic controls, with terrestrial sources dominating on the western margin proximal to the Sevier highlands and a mixture of marine and terrestrial sources present in the central basin. Pristane and phytane in samples from the east-central basin exhibit ^{13}C-enrichment relative to the other sites; it is proposed that this difference results from a greater Tethyan contribution to surface waters at that site than at the others. Interpretation of pristane/phytane ratios is complicated by maturity differences but indicates that depositional conditions were most reducing at the basin margins.

At a given location, variations in compound abundances through lithologic cycles primarily record increased degradation of organic matter during deposition of bioturbated limestones. In those same beds, selective degradation of marine organic matter and/or lower fluxes of marine organic matter also resulted in preferential preservation of terrestrial organic matter. These processes affected the compound-specific record of the Cenomanian-Turonian positive carbon isotope excursion, which is expressed in the $\delta^{13}C$ values of pristane and phytane but is less clear in $\delta^{13}C$ values of other compounds. This is likely a function of multiple sources for individual compounds, whose relative contributions are influenced by both surface-ocean and early diagenetic processes.

INTRODUCTION

The Cretaceous Western Interior Seaway was a unique environment for organic matter deposition. In contrast to the modern "ice-house" world, high global sea level (Haq et al., 1987) persisted throughout the Cretaceous, and is associated with warm, equable climates that could have resulted from volcanic outgassing of CO_2 (Lasaga, 1985; Arthur et al., 1985c, 1991). Concentrated accumulation of organic matter occurred periodically during presumed oceanic anoxic events (Schlanger and Jenkyns, 1976; Arthur and Schlanger, 1979; Jenkyns, 1980; Arthur et al., 1987; Schlanger et al., 1987; Arthur et al., 1990), which possibly influenced the global carbon budget. Deposition in the seaway responded to those influences as well as to possible Milankovitch-type forcing of the local environment which resulted in the deposition of lithologic couplets (Fischer, 1980; Pratt, 1981; Arthur et al., 1984; Barron et al., 1985; Barlow and Kauffman, 1985; Fischer et al., 1985; Bottjer et al., 1986; ROCC Group, 1986; Pratt et al., 1993; Arthur et al., 1994; Sageman et al., 1997).

Theories that explain the circulation and dominant oceanographic processes in the Cretaceous Western Interior Seaway are abundant and controversial. Considerable confusion over these processes exists because there is no true modern analogue to the Cretaceous Western Interior Seaway. The seaway was open to oceans at two locations, the Boreal ocean to the north and the Tethys ocean to the south. Important oceanographic processes could include: (1) the mixture of Tethys and Boreal ocean waters, typically with either northward expansion of the former as sea level rises (Eicher and Worstell, 1970; Frush and Eicher, 1975; Kauffman, 1985; Pratt, 1985; Eicher and Diner, 1985; Hay et al., 1993) or caballing (Fisher, 1991; Hay et al., 1993; Fisher et al., 1995) suggested as outcomes; (2) formation of a fresh-water lid and anoxia in bottom waters due to an excess of precipitation over evaporation over the seaway and adjacent drainage basins (Kauffman, 1975; Pratt, 1981, 1984, 1985; Barron et al., 1985; Hattin, 1985; ROCC Group, 1986; Watkins, 1986; Kauffman, 1988; Glancy et al., 1993; Kyser et al., 1993; Pratt et al., 1993; Jewell, 1993; Sethi and Leithold, 1994); and (3) Ekman-transport driven upwelling, that enhanced productivity and organic carbon burial (Parrish et al., 1984). Recently Slingerland et al. (1996) studied circulation in the seaway using a three-di-

mensional coastal ocean model using climatic and wind conditions from the GENESIS general atmospheric circulation model developed at the National Center for Atmospheric Research (Pollard and Thompson, 1992). Their results suggest that during the Early Turonian, the seaway operated as a large estuary, exporting runoff-freshened water both to the north (runoff from eastern drainages) and to the south (runoff from western drainages). This counter-clockwise circulation pattern simultaneously drew in Tethys and Boreal waters from the Southeast and Northwest, respectively. These models and hypotheses serve as a framework for interpreting organic matter distributions and isotopic compositions in the seaway.

The abundances and distributions of hydrocarbon biomarkers, when coupled with compound-specific isotope analyses, can be very useful in the study of paleoenvironmental conditions. Other organic geochemical and carbon-isotopic investigations of the Cretaceous Western Interior Seaway were reported by Pratt (1984), Hayes et al. (1989, 1990) and Curiale (1994). Here, we build on those previous efforts by examining a transect of three cores across the basin; we present the results of the free *n*-alkane, pristane, and phytane analyses, and evaluate how the composition and distribution of each reflects paleoceanographic processes in the Cretaceous Western Interior Seaway. *N*-alkane distributions can be used as indicators of terrestrial *vs.* marine inputs; thus, we expect to see variations in *n*-alkane abundances with lithology and with proximity to the Sevier highlands. Isotopic analyses of the *n*-alkanes further assist in deconvolving the complex mixture of bacterial, algal, and terrestrial plant sources. Pristane and phytane are degradation products of phytol, a chlorophyll side chain, and thus could record algal productivity. Thus, these analyses are used to evaluate diagenesis and organic matter inputs across the basin and the impact of local oceanographic conditions on the record of the Cenomanian-Turonian event. Results from other compounds, including cyclic hydrocarbons, are used here only to assess organic matter thermal maturity (Table 1).

EXPERIMENTAL TECHNIQUES

Bulk Geochemical Data

CO_2 from carbonate and total carbon was quantified by coulometry upon liberation by immersion in phophoric acid and com-

TABLE 1. – BIOMARKER INDICATORS OF THERMAL MATURITY

Indicator	Escalante Core	Portland Core	BoundsCore
C_{29} 20S / (20S+20R)[a]	0.1 ± 0.02	0.3 ± 0.04	0.1 ± 0.03[1]
C_{27} 20S / (20S+20R)[b]	0.27 ± 0.02	0.37 ± 0.02	0.26 ± 0.03
C_{29} $\alpha\beta\beta$/ $(\alpha\alpha\beta + \alpha\alpha\alpha)$[c]	0.36 ± 0.03	0.26 ± 0.04	0.36 ± 0.03
C_{31} Hopane 22S / (22S+22R)[d]	0.3 ± 0.1	0.58 ± 0.01	0.29 ± 0.03
C_{32} Hopane 22S \ (22S+22R)[e]	0.4 ± 0.1	0.56 ± 0.01	0.23 ± 0.03
C_{30} Moretane / (C_{30} Moretane +C_{30} Hopane)[f]	0.27 ± 0.06	0.17 ± 0.02	0.27 ± 0.04
C_{32} Moretane 22S / (22S+22R)[g]	0.3 (n=1)	0.40 ± 0.02	0.25 ± 0.04
C_{33} Moretane 22S / (22S+22R)[h]	-	0.54 ± 0.02	0.3 ± 0.1
C_{32} Moretane / (C_{32} Moretane + C_{32} Hopane)[i]	0.6 (n=1)	0.23 ± 0.02	0.61 ± 0.05
C_{33} Moretane / (C_{33} Moretane + C_{33} Hopane)[j]	-	0.3 ± 0.06	0.83 ± 0.22

[1] Values reported are averages of measurements of all samples from a given core; n = 10 (Portland), = 3 (Escalante), = 4 (Bounds).
[a] 5α(H),14α(H),17α(H) 24-ethylcholestane 22S / 5α(H),14α(H),17α(H) 24-ethylcholestane (22S+22R)
[b] 5α(H),14α(H),17α(H) cholestane 22S / 5α(H),14α(H),17α(H) cholestane (22S+22R)
[c] 5α(H),14β(H),17β(H) 24-ethylcholestane (22S+22R) / (5α(H),14α(H),17α(H) + 5α(H),14β(H),17β(H) 24-ethylcholestane (22S+22R))
[d] C_{31} 17α(H),21β(H)-hopane 22S / C_{31} 17α(H),21β(H)-hopane (22S+22R)
[e] C_{32} 17α(H),21β(H)-hopane 22S / C_{32} 17α(H),21β(H)-hopane (22S+22R)
[f] 17β(H),21α(H)-C_{30} moretane / (17β(H),21α(H)-C_{30} moretane + 17α(H),21β(H)-C_{30} hopane)
[g] 17β(H),21α(H)-C_{32} moretane 22S / 17β(H),21α(H)-C_{32} moretane (22S+22R)
[jh] 17β(H),21α(H)-C_{33} moretane 22S / 17β(H),21α(H)-C_{33} moretane (22S+22R)
[i] 17β(H),21α(H)-C_{32} moretane / (17β(H),21α(H)-C_{32} moretane + 17α(H),21β(H)-C_{32} hopane (22S+22R))
[j] 17β(H),21α(H)-C_{33} moretane / (17β(H),21α(H)-C_{33} moretane + 17α(H),21β(H)-C_{33} hopane (22S+22R))

bustion at 1200°C, respectively. Percent total organic carbon (%TOC) was determined by difference. Accuracy is typically ± 5% of the measured value, and values are given in Tables 2-4. Carbonate δ^{13}C values were determined by exposing samples to phosphoric acid and measuring δ^{13}C of liberated CO_2 on a stable-isotope mass spectrometer. Prior to measurement, samples were heated for 1 hour under vacuum at 380° to remove volatile organics. TOC δ^{13}C values were determined by heating decalcified samples in the presence of cupric oxide at 800° for 5 hours; liberated CO_2 was cryogenically purified in a glass vacuum system and introduced to the stable-isotope mass spectrometer via the dual inlet system. Carbon-isotopic compositions of carbonate and TOC are provided in Tables 4-7.

Extraction and Separation

Samples were washed with methanol to remove handling and storage contamination, ground with mortar and pestle, and powdered with a ball-mill device. The powdered samples were Soxhlet extracted with a 2:1 dichloromethane:methanol azeotrope for 24 hours. The total solvent extract was separated by column chromatography into hydrocarbon, aromatic/ketone, and polar fractions using 20 mL each of hexane, toluene, and methanol passed through 8 g of activated silica gel.

The hydrocarbon fraction was further divided into n-alkanes and branched/cyclic hydrocarbons by adduction with urea (Michalczyk, 1985). Samples were dissolved in 200 µL each of methanol saturated with urea, pentane, and acetone, and then were refrigerated. The samples were evaporated under N_2 and urea crystals were extracted with hexane to yield the branched/cyclic fraction. Urea crystals were then dissolved by adding 500 uL of extracted H_2O, and the solution was extracted with hexane to yield the n-alkane fraction.

Compound Identification and Quantification

Structural Gas Chromatography/Mass Spectrometry was performed at the Energy and Geosciences Institute (EGI) in Salt Lake City, Utah. Samples were injected via autosampler into a gas chromatograph and passed through a DB-1 column (60 m x 0.25 mm x 0.25 µm) programmed from 35°C (2 min) to 310°C (held at 70.5 min) at 2°C/min. Eluting compounds then passed into a Hewlett Packard mass spectrometer operating in selective ion monitoring (SIM) mode. Additional full scan analyses were made and compared to published spectra to confirm compound identifications. Compound abundances were determined by a flame ionization detector following gas chromatography, in which androstane was used as an internal standard (Tables 2-4).

Frequently molecular ratios, rather than absolute abundances, are used to evaluate organic matter source and quality. The carbon preference index (CPI) is a measure of the relative predominance of n-alkanes with an odd number of carbon atoms and was calculated using the following equation based on Bray and Evans, (1961):

$$CPI = \frac{2 \times \sum oddC_{21} - C_{31}}{\sum evenC_{20} - C_{30} + \sum evenC_{22} - C_{32}}$$

Isotope-Ratio Monitoring Gas Chromatography/ Mass Spectrometry

Isotope ratios were determined by isotope-ratio monitoring Gas Chromatography/Mass Spectrometry (Hayes et al., 1990); compounds eluting from the gas chromatograph are converted to CO_2 by combustion at 850°C over cupric oxide and water is removed by selectively permeable membrane flushed externally by helium. Isotopic compositions were then measured on a stable-

TABLE 2. – LITHOLOGIC AND ORGANIC MATTER ABUNDANCE DATA: ESCALANTE CORE

	784.9'	786.8'	795.8'	799.9'
C_{15}	-	-	-	0.001
C_{16}	0.02[1]	0.06	0.07	0.04
C_{17}	0.02	0.01	0.01	0.02
C_{18}	0.03	0.02	0.01	0.03
C_{19}	0.06	0.03	0.07	0.03
C_{20}	0.10	0.06	0.03	0.04
C_{21}	0.21	0.09	0.08	0.05
C_{22}	0.22	0.06	0.07	0.04
C_{23}	0.38	0.04	0.15	0.06
C_{24}	0.67	0.04	0.28	0.06
C_{25}	1.15	0.05	0.50	0.10
C_{26}	1.57	0.05	0.68	0.11
C_{27}	1.78	0.06	0.77	0.14
C_{28}	1.56	0.04	0.67	0.11
C_{29}	1.37	0.05	0.60	0.11
C_{30}	0.95	0.02	0.41	0.07
C_{31}	0.79	0.04	0.31	0.08
C_{32}	0.42	0.01	0.17	0.03
C_{33}	0.29	0.02	0.12	0.03
C_{34}	0.15	0.01	0.06	0.01
C_{35}	0.11	-	0.04	0.01
C_{36}	0.04	-	0.02	0.01
C_{37}	0.04	-	-	0.01
CPI	1.09	1.33	1.09	1.23
Pristane	0.26	0.043	0.047	0.12
Phytane	0.28	0.043	0.117	0.15
Pr/Ph	0.96	1.0	0.4	0.8
%TOC	0.66	0.6	0.51	0.59
%CaCO₃	47	49	11	10.5

[1] Values reported in units of µg/g whole rock

TABLE 3. – LITHOLOGIC AND ORGANIC MATTER ABUNDANCE DATA: BOUNDS CORE

	970.65'	971.55'	973.6'	974.1'
C_{13}	-	-	-	0.46
C_{14}	0.03[1]	-	-	0.26
C_{15}	0.14	0.08	-	0.65
C_{16}	0.17	0.39	0.03	0.33
C_{17}	0.31	0.44	0.01	0.67
C_{18}	0.25	0.41	0.01	0.51
C_{19}	0.18	0.38	0.02	0.42
C_{20}	0.11	0.33	0.02	0.33
C_{21}	0.13	0.36	0.03	0.34
C_{22}	0.17	0.39	0.03	0.33
C_{23}	0.19	0.39	0.03	0.34
C_{24}	0.15	0.36	0.02	0.34
C_{25}	0.13	0.30	0.02	0.25
C_{26}	0.10	0.31	0.02	0.28
C_{27}	0.08	0.21	0.01	0.17
C_{28}	0.05	0.16	0.01	0.14
C_{29}	0.05	0.27	0.01	0.14
C_{30}	0.03	0.12	0.01	0.10
C_{31}	0.08	0.14	0.01	0.12
C_{32}	0.02	0.05	0.01	0.06
C_{33}	0.01	-	-	0.06
C_{34}	0.01	-	-	0.02
C_{35}	-	-	-	0.03
C_{36}	-	-	-	0.01
C_{37}	-	-	-	0.03
CPI	1.17	1.10	1.19	0.99
Pristane	0.24	1.46	0.62	2.05
Phytane	0.17	0.92	0.52	0.82
Pr/Ph	1.4	1.59	1.21	2.5
%TOC	0.22	3.3	0.27	4.5
% CaCO₃	83	71	79	62

[1] Values reported in units of µg/g whole rock

TABLE 4. – LITHOLOGIC AND ORGANIC MATTER ABUNDANCE DATA: PORTLAND CORE

	468.9'	470.3'	471.4'	472.85	474.6'	475.3'	475.65	475.9'	477.7'	478.6'
C_{13}	-	0.04	-	0.03	-	0.28	-	0.01	-	-
C_{14}	-	0.24	-	0.17	-	1.13	-	0.08	-	-
C_{15}	0.05[1]	0.61	-	0.59	-	2.55	0.02	0.54	0.01	0.01
C_{16}	0.21	0.68	0.12	1.70	0.36	2.22	1.85	1.49	0.15	0.40
C_{17}	0.16	0.94	0.002	1.25	0.01	3.56	0.72	2.06	0.04	0.12
C_{18}	0.15	0.79	0.01	1.12	0.02	3.11	1.07	2.05	0.05	0.19
C_{19}	0.19	0.78	0.04	1.15	0.05	2.91	1.36	1.89	0.06	0.27
C_{20}	0.24	0.73	0.10	1.24	0.13	2.51	1.65	1.63	0.09	0.35
C_{21}	0.25	0.75	0.15	1.61	0.28	2.48	2.07	1.63	0.15	0.44
C_{22}	0.21	0.68	0.12	1.70	0.26	2.22	1.84	1.49	0.15	0.40
C_{23}	0.17	0.61	0.08	1.36	0.28	2.09	1.51	1.44	0.11	0.25
C_{24}	0.15	0.52	0.06	1.05	0.20	1.81	1.24	1.31	0.09	0.32
C_{25}	0.14	0.46	0.04	0.89	0.17	1.54	1.12	1.18	0.07	0.31
C_{26}	0.14	0.39	0.03	0.77	0.17	1.31	0.99	1.01	0.06	0.28
C_{27}	0.13	0.33	0.02	0.63	0.17	0.87	0.84	0.79	0.08	0.29
C_{28}	0.11	0.25	0.02	0.47	0.16	0.67	0.65	0.41	0.04	0.25
C_{29}	0.10	0.29	0.02	0.34	0.14	0.61	0.76	0.54	0.04	0.24
C_{30}	0.06	0.21	0.01	0.35	0.10	0.42	0.43	0.34	0.03	0.17
C_{31}	0.06	0.22	0.02	0.21	0.08	0.41	0.24	0.34	0.04	0.17
C_{32}	0.03	0.12	0.01	0.18	0.05	0.28	0.24	0.21	0.03	0.11
C_{33}	0.03	0.11	-	0.09	0.04	0.23	0.13	0.18	0.03	0.09
C_{34}	0.01	0.05	-	0.06	0.02	0.12	0.13	0.10	-	0.06
C_{35}	0.01	0.05	-	0.03	0.02	0.08	0.08	0.09	-	0.04
C_{36}	0.01	0.03	-	0.05	0.01	0.05	-	0.06	-	0.03
C_{37}	0.01	0.03	-	0.04	0.01	0.04	-	-	-	0.02
CPI	1.07	1.07	1.14	1.00	1.05	1.02	1.04	1.08	1.16	1.08
Pristane	0.16	0.61	0.006	3.84	0.15	1.50	3.06	1.77	0.12	0.44
Phytane	0.04	0.15	0.003	0.94	0.04	0.33	0.76	0.51	0.029	0.21
Pr/Ph	4.3	4	2	4.1	4	4.5	4	3.5	4	2.1
%TOC	0.75	3.1	0.25	1.6	0.3	4.7	1.6	3.6	0.47	0.9
%CaCO$_3$	75	59	82	71	81	57	75	68	95	35

[1] Values are in units of μg/g whole rock

isotope mass spectrometer, and delta values relative to Pee Dee Belemnite (PDB) were calculated by comparison against a calibrated CO_2 gas introduced by a second capillary: $\delta^{13}C = (^{13}R_{SA}/^{13}R_{PDB} - 1) \times 10^3$ where $^{13}R_{SA}$ represents the $^{13}C/^{12}C$ abundance ratio for the sample and PDB, respectively. The working standard was calibrated against NBS-19. Based on coinjected standards, accuracy in compound-specific isotope determinations for *n*-alkanes, pristane, and phytane was ±0.5‰. Precision, determined from multiple (typically duplicate) analyses, was typically better than ±0.3‰. $\delta^{13}C$ values and precision are reported in Tables 5-7.

Samples

Cores sampled in this study (Dean and Arthur, this volume) were drilled on the Brush Hollow Anticline near Portland, Colorado, (USGS Portland No. 1 core) in June, 1992 and in the Kaiparowits Basin near Escalante, Utah, (USGS Escalante No. 1 core) in June, 1991. The Cretaceous portion of a third core from Greeley County, Kansas was obtained from AMOCO (Amoco Rebecca K. Bounds No. 1 core). These three cores represent a transect across the Cretaceous Western Interior Seaway (Fig. 1). From the Portland core, approximately 50 g of

TABLE 5. – CARBON ISOTOPIC COMPOSITION OF ORGANIC MATTER AND WHOLE ROCK CARBONATE: ESCALANTE CORE

	784.9'		786.8'		795.8'		799.9'	
C_{17}	-		-28.3	0.2	-		-27.0	0.2
C_{18}	-		-29.6	0.4	-		-26.4	0.4
C_{19}	-28.8	0.4[1]	-29.1	0.2	-26.3	0.2	-26.3	0.2
C_{20}	-29.4	0.1	-29.3	0.2	-26.7	0.1	-25.7	0.2
C_{21}	-29.1	0.2	-29.5	0.1	-26.0	0.1	-24.2	0.1
C_{22}	-28.8	0.2	-29.4	0.1	-26.7	0.2	-25.5	0.2
C_{23}	-28.5	0.2	-28.9	0.5	-26.9	0.3	-24.8	0.1
C_{24}	-28.0	0.3	-28.9	0.2	-27.5	0.1	-26.7	0.3
C_{25}	-28.9	0.3	-28.5	0.3	-28.0	0.2	-26.9	0.2
C_{26}	-28.7	0.1	-29.2	0.1	-28.6	0.1	-28.0	0.2
C_{27}	-28.6	0.1	-29.4	0.3	-28.7	0.2	-27.6	0.2
C_{28}	-29.2	0.1	-31.3	0.8	-29.2	0.2	-28.4	0.2
C_{29}	-28.8	0.1	-29.8	0.1	-29.5	0.1	-28.6	0.1
C_{30}	-29.0	0.1	-30.2	0.5	-29.4		-28.4	0.2
C_{31}	-29.4	0.2	-30.4	0.5	-		-28.3	0.1
C_{32}	-29.5	0.3	-		-		-28.4	0.1
C_{33}	-		-		-		-28.8	0.1
C_{34}	-		-		-		-28.9	0.1
Pristane	-28.6	0.2	-29.2	0.6	-30.5	0.6	-29.4	0.2
Phytane	-27.8	0.6	-28.2	0.6	-28.4	0.1	-27.1	0.2
TOC	-23.4		-22.3		-22.6		-22.3	
$CaCO_3$	2.8		3.1		3.9		3.3	

[1] Standard deviation based on duplicate measurements.

TABLE 6. – CARBON ISOTOPIC COMPOSITION OF ORGANIC MATTER AND WHOLE ROCK CARBONATE IN THE BOUNDS CORE

	970.65'		971.55'		973.6'		974.1'	
C_{14}	-		-		-		-28.5	0.3
C_{15}	-28.9	0.2[1]	-29.5		-		-29.4	0.7
C_{16}	-28.4	0.7	-28.8	0.4	-		-29.8	0.7
C_{17}	-28.9	0.2	-29.0	0.2	-28.7	0.5	-29.3	0.7
C_{18}	-28.8	0.4	-28.9	0.1	-29.4	0.2	-28.9	0.5
C_{19}	-29.2	0.3	-28.8	0.1	-29.0	0.1	-28.6	0.4
C_{20}	-29.3	0.1	-28.6	0.2	-29.4	0.1	-28.3	0.2
C_{21}	-29.5	0.1	-29.3	0.1	-29.5	0.1	-28.3	0.3
C_{22}	-29.9	0.3	-29.3	0.4	-29.4	0.4	-27.8	0.2
C_{23}	-29.4	0.2	-28.6	0.7	-29.2	0.2	-29.1	0.4
C_{24}	-28.6	0.7	-29.5	0.5	-28.8	0.2	-28.2	0.3
C_{25}	-30.1	0.6	-28.9	0.2	-29.4	0.6	-28.4	0.6
C_{26}	-29.9	0.7	-29.9		-29.2	0.1	-28.2	
C_{27}	-31.4	0.5	-28.5		-28.2	0.8	-27.2	0.3
C_{28}	-		-28.3		-30.3	0.8	-26.4	
C_{29}	-30.7		-27.1		-29.2	1.0	-26.3	
C_{30}	-		-		-30.8	0.8	-	
C_{31}	-		-		-30.1	0.2	-	
Pristane	-30.1		-29.4	0.1	-29.4	0.4	-28.3	0.4
Phytane	-29.1	0.5	-28.7	0.4	-29.0	0.5	-28.4	0.7
TOC	-25.1		-24.0		-22.6		-23.1	
$CaCO_3$	2.0		2.6		2.2		2.9	

[1] Standard deviation based on duplicate measurements.

sample were collected from each bed of five limestone and marlstone couplets (Fig. 2; limestone units PBC-10, -13, -14, -15, and -16 [Elder and Kirkland, 1985] and subjacent marlstone beds for each) spanning the peak of the Cenomanian-Turonian oceanic anoxic event (Schlanger et al., 1987). Similar quantities of rock were collected from the PBC-13 and -15 couplets of the Bounds core. From the clastic-dominated Bridge Creek equivalent of the Escalante core (Tropic Shale), four samples were selected based on approximate correlation with samples chosen from the Bounds and Portland cores.

The samples analyzed in this study span the Cenomanian-Turonian boundary and the uppermost portion of the positive carbon-isotope excursion (Scholle and Arthur, 1980; Pratt, 1985). However, no attempt was made to sample the excursion in detail. The Cenomanian-Turonian boundary is located at the top of the *Neocardioceras juddii* ammonite zone just above limestone bed PBC-14 of Elder and Kirkland (1985). This corresponds to a depth of 778 ft to 782 ft in the Escalante core (E. Kauffman, pers. comm.). The peak of the bulk organic carbon-isotope excursion is located at 794 ft in the Escalante core (Dean et al., 1995), from 974 ft to 981 ft in the Bounds core (Dean et al., 1995), and from 478 ft to 474 ft in the Portland core as extrapolated from analyses on the nearby Pueblo-Princeton core (Arthur et al., 1988). Data collected for this project are consistent with these observations (Tables 5-7).

Carbonate in the study interval of the Portland core ranges from 35% to 85%. Most values in the section fall between 60% and 85% (Fig. 2), consistent with values reported by other workers (Arthur et al., 1985b; Sageman et al., 1997 and this volume). Samples from the Bounds core have comparable carbonate percentages, while values are much lower in the western Escalante core, probably because of clastic dilution resulting from the proximity of the Sevier highlands (Arthur et al., 1985; Fig. 1).

Thermal Maturity

Structural analyses using Gas Chromatography/Mass Spectrometry can be highly useful in assessing thermal maturity of the Cenomanian-Turonian interval of these cores. Increased exposure to thermal processes rearranges the stereochemical configurations of both terpanes and steranes. This process proceeds until the proportion of each configuration achieves a thermodynamically favored distribution. For instance, the biological configuration of hopanes at the C_{22} carbon is R (see Peters and Moldowan [1993] for a discussion of biomarker stereochemistry); with increased maturity 22R isomers are converted toward the more stable 22S configuration, until the ratio of 22S/(22S+22R) achieves an equilibrium value of 0.6 (Seifert and Moldowan, 1986). By considering a suite of these indicators, we can estimate equivalent vitrinite reflectance (eqR$_o$). Values

TABLE 7. – CARBON ISOTOPIC COMPOSITIONS OF HYDROCARBONS IN THE PORTLAND CORE

	468.9'	470.3'	471.4'	472.85'	474.6'	475.3'	475.65'	475.9'	477.7'	478.6'
C_{13}	-	-	-	-	-	-	-	-	-	-
C_{14}	-	-	-	-	-	-28.1 0.2[1]	-	-	-	-
C_{15}	-28.0	-29.0	-	-28.6 0.3	-	-28.4 0.3	-	-270	-	-
C_{16}	-29.6	-29.6	-	-29.1 0.1	-	-28.7 0.4	-	-28.4	-	-
C_{17}	-29.4 0.4	-29.5 0.2	-	-28.9 0.1	-	-28.8 0.4	-	-28.8 0.1	-28.6 0.3	-27.0 0.3
C_{18}	-29.0 0.1	-29.5 0.1	-	-28.8 0.2	-26.7 0.4	-28.8 0.3	-	-28.9 0.1	-28.7 0.3	-28.1 0.3
C_{19}	-29.4 0.4	-29.5 0.2	-29.1	-28.9 0.1	-27.6 0.1	-28.8 0.4	-28.4 0.3	-28.8 0.2	-29.0 0.1	-28.3 0.2
C_{20}	-29.0 0.1	-29.5 0.1	-29.7	-29.4 0.2	-28.8 0.3	-28.8 0.3	-28.5 0.1	-28.8 0.2	-29.0 0.1	-28.5 0.2
C_{21}	-29.2 0.4	-29.4 0.2	-29.7	-29.6 0.4	-29.2 0.8	-28.8 0.2	-28.8 0.3	-28.5 0.1	-29.3 0.1	-28.8 0.2
C_{22}	-28.9 0.3	-29.5 0.1	-29.8	-29.8 0.4	-29.6 0.7	-28.6 0.3	-28.8 0.3	-28.3 0.2	-29.5 0.1	-28.7 0.1
C_{23}	-28.7 0.1	-29.5 0.4	-29.6	-29.7 0.3	-29.7 1.0	-28.7 0.2	-27.9 0.2	-28.2 0.2	-29.7 0.1	-28.6 0.2
C_{24}	-28.7 0.3	-29.3 0.4	-29.1	-29.5 0.1	-29.2 0.6	-30.1	-	-28.1 0.1	-29.0 0.1	-28.4 0.2
C_{25}	-29.1 0.5	-29.3 0.3	-29.6	-29.1 0.1	-29.4 0.5	-28.5	-29.9 0.4	-28.6 0.1	-29.2 0.5	-28.3 0.1
C_{26}	-29.1 0.1	-29.7	-29.4	-29.1 0.1	-29.8 0.4	-27.8	-28.5 0.1	-28.1 0.2	-29.9 0.2	-28.9 0.2
C_{27}	-28.5	-30.1	-29.1	-28.9 0.2	-30.2 0.2	-28.6	-28.4 0.1	-28.3β 0.3	-30.0 0.4	-29.3 0.4
C_{28}	-29.3	-29.3	-	-29.0	-30.0 0.6	-26.7 0.4	-	-28.5 0.2	-30.1	-29.5 0.3
C_{29}	-30.8	-29.4	-30.1 0.7	-28.7	-30.7 0.1	-28.2 0.5	-29.5 0.2	-29.5 0.6	-29.7	-29.0 0.1
C_{30}	-	-	-	-28.4	-30.0 0.4	-	-29.7 0.4	-27.7 0.4	-	-29.1 0.4
C_{31}	-	-	-	-	-30.7 0.2	-	-27.6 0.1	-27.9 0.1	-	-29.0 0.1
C_{32}	-	-	-	-	-30.0 0.1	-	-	-27.9 0.2	-	-
C_{33}	-	-	-	-	-30.5 0.3	-	-	-	-	-
C_{34}	-	-	-	-	-29.2 0.1	-	-	-	-	-
Pristane	-31.8 0.4	-30.9 0.5	-32.8	-29.5 0.1	-29.0 0.1	-29.1 0.1	-30.3 0.2	-29.7 0.1	-29.3 0.1	-29.5 0.6
Phytane	-31.0 0.3	-30.3 0.3	-31.6 0.3	-29.6 0.3	-30.3 0.5	-29.4 0.3	-28.9 0.2	-29.6 0.3	-	-29.3 0.3
TOC	-25.2	-24.6	-25.2	-23.6	-26.7	-23.5	-23.2	-23.4	-26.1	-23.5
$CaCO_3$	1.9	2.7	1.4	2.8	1.3	3.0	3.0	3.0	1.3	3

[1] Represents one half the difference between duplicate measurements; all numbers are rounded up to the one-tenth digit

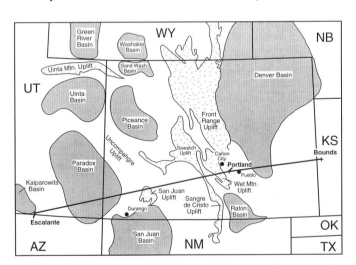

FIG. 1 – Map of the Utah, Colorado, and Kansas area showing the location of the Escalante, Portland, and Bounds cores.

for biomarker maturity indicators are tabulated in Table 1 and plotted in Figures 3 and 4.

All maturity indicators indicate that each of these cores is thermally immature with respect to oil generation. For C_{31} and C_{32} hopane isomerization (22S/(22R+22S)) (Fig. 3), samples from the Portland core plot near but just below the equilibrium value of $R_o \sim 0.6$ (Seifert and Moldowan, 1986), while samples from the Bounds and Escalante cores lie well below equilibrium values. The same general trend is observed for the C_{30} moretane/hopane ratio and the $5\alpha(H),14\beta(H),17\beta(H)$ C_{27} and C_{29} sterane (20S/(20S+20R)) ratio (Fig. 3). However, for all samples, including those from the Portland core, the moretane/hopane and sterane ratios are well below equilibrium (Fig. 3) (Seifert and Moldowan, 1986). Moreover, in all three cores extended moretanes persist and, in the Escalante and Bounds cores, are present in significant quantities relative to analogous hopanes (Table 1). 22S/(22S+22R) ratios calculated for extended moretanes are consistent with the previously described trends (Table 1). Thus, Portland core samples are somewhat immature with respect to oil generation (eq$R_o \sim 0.5$ to 0.55), while Escalante and Bounds samples are immature (eq$R_o \sim 0.4$).

CONTROLS ON ORGANIC MATTER PRODUCTION AND PRESERVATION

Sources of N-alkanes, Pristane, and Phytane

The interpretation of variations in compound distributions and carbon-isotopic composition is contingent upon an understanding of the sources of those compounds. This is a non-trivial

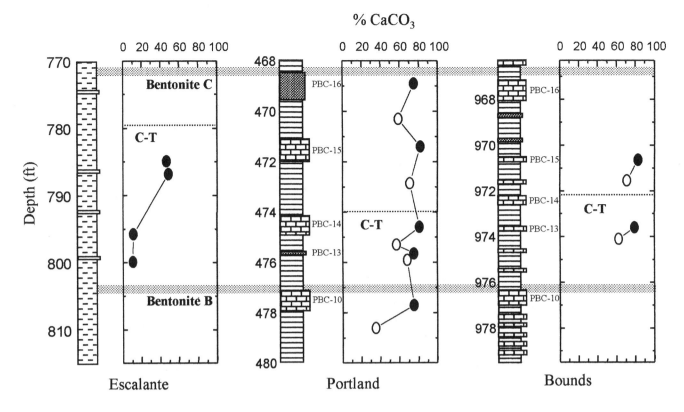

FIG. 2 – Lithologic logs and plots of percent carbonate against depth for Escalante, Portland, and Bounds cores. Open circles denote marlstone beds and closed circles denote limestone beds. The Cenomanian-Turonian boundary is defined in each core by the approximate location of the top of the *N. juddii* ammonite zone, and limestone and bentonite designations are those of Elder and Kirkland (1985).

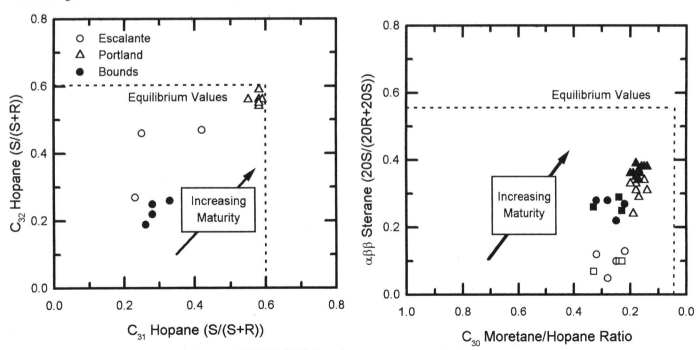

FIG. 3 – Plot of C_{32} and C_{31} hopane isomerization ratios (22S/(22S+22R)), thermal maturity indices, for the Escalante, Portland, and Bounds cores. Dashed lines represent the equilibrium values (Peters and Moldowan, 1993).

FIG. 4 – Plot of C_{30} moretane to C_{30} hopane ratio against C_{27} 5α, 14β, 17β cholestane (closed symbols) and ethylcholestane (open symbols) 20S:(20S+20R) ratios, thermal maturity indices, for the Escalante (■), Portland (▲) and Bounds (●) cores. Dashed lines represent the equilibrium values (Peters and Moldowan, 1993).

consideration, because many sources can contribute a given compound to the sediment. A combination of distribution patterns, sedimentological evidence, and isotopic evidence with a knowledge of dominant sources in modern environments will help mitigate these concerns. Typical sources of organic matter in a

marine basin such as the Cretaceous Western Interior Seaway include (but are not limited to) phytoplankton, terrestrial plant material, and bacteria, all of which can vary in importance spatially and temporally. Eglinton et al. (1962) proposed that odd-

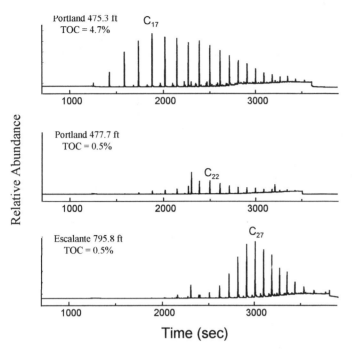

FIG. 5 – Gas chromatograms of urea-adductable saturate fractions showing n-alkane distributions for representative samples from the Escalante core and organic-poor and organic-rich samples from the Portland core.

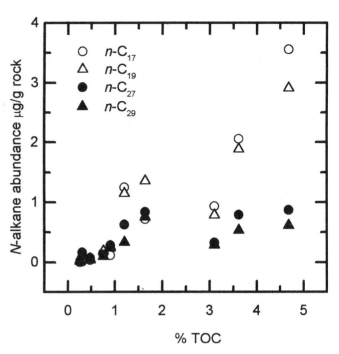

FIG. 6 – Plot of low molecular-weight n-alkane abundance (n-C_{17} and n-C_{19}) and high molecular weight n-alkane abundance (n-C_{27} and n-C_{29}) against %TOC for the Portland core.

molecular weight n-alkanes between C_{25} and C_{35} are generated predominantly by terrestrial plants. Although the carbon preference index (CPI) can be used to estimate the significance of terrestrial organic matter, its utility is limited by thermal maturation, which overprints the odd-over-even pattern. Furthermore, odd, high-molecular weight n-alkanes can be derived from algae (Gelpie et al., 1970; Kolattukudy, 1976) and should not be considered unambiguous markers of higher plants (Lichtfouse et al., 1994).

Odd carbon number and low-molecular weight n-alkanes (C_{15}-C_{19}) are considered markers for algae (Gelpie et al., 1970; Collister et al., 1994). Pristane and phytane can be derived from degradation of the phytol side-chain present in several chlorophylls, and thus, also can record algal inputs (Hayes et al., 1990). However, tocopherol, an isoprenoid not related to chlorophyll, is also a source of pristane (Goosens et al., 1984) and archaeobacterial lipids from methanogens can generate both pristane and phytane (Risatti, 1983; Volkman, 1986). Nonetheless, previous work in Cenomanian-Turonian strata of the Cretaceous Western Interior Seaway (Hayes et al., 1990) suggests a common source for pristane and phytane. Carbon-isotopic compositions of isoprenoids are 4.5‰ depleted relative to porphyrins, the same offset observed in modern plants, and the authors suggested that both pristane and phytane are derived from phytoplankton. Because the samples used by Hayes et al. (1990) were of the same age as our samples and from a location near the Portland and Amoco Bounds cores used in this study, we infer that this conclusion is valid for our samples as well.

Carbon-isotopic analyses are helpful in distinguishing the multiple sources of these compounds (Freeman et al., 1990, 1994; Lichtfouse et al., 1994; Collister et al., 1994). However, this tool is of limited use when applied to much of the Cretaceous record, because marine and terrestrial organic matter had overlapping carbon-isotopic compositions during that time (-26‰ to -28‰ for marine and –24‰ to -30‰ for mixed marine/terrestrial; Dean et al., 1986).

Paleoceanography

We expect to observe variations in n-alkane abundances and carbon-isotopic compositions of n-alkanes across the basin with decreased terrestrial input from the Sevier highlands. We also expect to see variations in these parameters with lithology. If marine productivity was higher during carbonate deposition as has been postulated (Eicher and Diner, 1989), and terrestrial inputs were relatively constant, then the abundance of marine markers relative to terrestrial markers should be lower in marls than in carbonates. However, Arthur et al. (1984) suggested that higher marine productivity prevailed during marlstone deposition, and under these conditions, the opposite biomarker pattern would be observed.

Paleoceanographic processes should also exert significant influence on both production of organic matter and the depositional conditions that control its preservation. The main controls on the carbon-isotopic composition of primary organic matter are the $\delta^{13}C$ values of $CO_{2(aq)}$, the concentrations of $CO_{2(aq)}$, which cause ^{13}C-depletion at higher concentrations, and growth rates, which cause ^{13}C-enrichment at higher values (Francois et al, 1993; Goericke and Fry, 1994; Laws et al., 1995; Bidigare et al., 1997). This understanding allows us to predict how oceanographic processes can affect the carbon-isotopic record of primary organic matter. Near sources of runoff, potentially high nutrient influx could elevate productivity and remove CO_2 from surface waters; both lower CO_2 concentrations and higher growth rates will result in enrichment of ^{13}C in marine organic matter. Upwelling of sub-thermocline waters can also supply nutrients to surface waters, enhance growth rates, and cause ^{13}C-enrichment in ma-

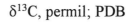

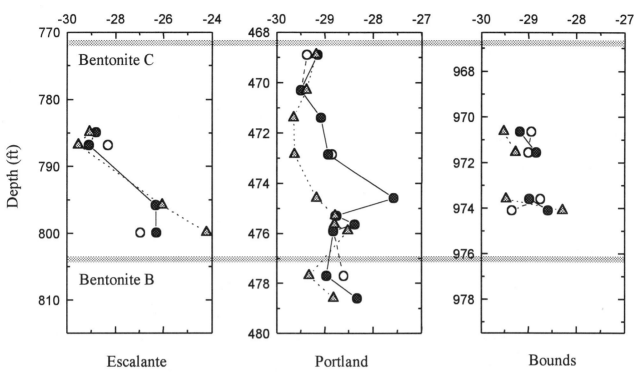

FIG. 7 – Plots of low molecular-weight n-alkane (n-C_{27} [O], n-C_{29} [●], and n-C_{31} [▲]) $\delta^{13}C$ values (‰) against depth for Escalante, Portland, and Bounds cores.

rine organic matter as seen in the modern Peru upwelling region (R. Pancost, K. Freeman, and S. Wakeham, unpublished results). These process, through enhanced production, also can deplete oxygen in bottom waters, and rocks from highly productive areas should be relatively rich in marine organic matter.

Biological compounds of marine origin can vary in both isotopic composition and distribution, depending on whether Boreal or Tethys water masses predominate. Tethyan water was significantly warmer than Boreal water (23°C versus 6°C, Slingerland et al., 1996); this will result in lowered values of surface-water CO_2 concentration and enrichment of ^{13}C in organic matter. Significant differences in salinity (estimated as 32‰ for the Boreal and 36‰ for the Tethys; Slingerland et al., 1996) and temperature should result in different phytoplankton assemblages. Density contrasts will impact upon physical processes such as stratification, caballing, and mixing, which can in turn affect bottom-water oxygenation and preservation of organic matter. Thus, we anticipate that both the distributions and carbon-isotopic compositions of compounds vary spatially and through lithologic cycles.

RESULTS

N-alkane Distributions and Carbon-Isotopic Compositions

N-alkane distributions vary widely depending on both lithology and sample location (Tables 2-4). Overall, the abundances of high-molecular weight homologs relative to lower-molecular weight homologs are greater in the western, proximal Escalante core than in the central-basin Portland and Bounds cores (Fig. 5). The distributions of low molecular-weight ho-

mologs are more complex, with relatively high absolute abundances in the central Portland core, lower values in the proximal Escalante core, and intermediate values in the distal Bounds core (Tables 2-4). Variation on n-alkane distributions with lithology is also strong. Within the Portland core, absolute abundances of low-molecular weight n-alkanes decrease significantly (from >3.5 to <0.01 µg/g whole rock) with decreasing %TOC (from ~4.5% to ~0.3%) and increasing %$CaCO_3$, while high-molecular weight n-alkanes exhibit a much more moderate decrease (Fig. 6). Thus, the ratio of high to low- molecular weight n-alkanes is greatest in organic-lean units. The highest molecular weight n-alkanes (carbon number greater than 27) exhibit a very slight odd-over-even predominance in all cores regardless of lithology or maturity, and carbon preference indices are slightly greater than 1 for most samples.

Carbon-isotopic values for all analyzed compounds are listed in Tables 5-7, results for n-C_{17}, n-C_{19}, and n-C_{21} are plotted in Figure 7, and those for n-C_{27}, n-C_{29}, and n-C_{31} are plotted in Figure 8. Carbon-isotopic values for low molecular weight n-alkanes ranged from −28‰ to −30 ‰ in all cores but the Escalante, where compound in the lower two samples showed enriched values of ~ -25‰ to -27‰. In general, these values are consistent with a marine origin (Dean et al., 1986). Expression of the Cenomanian-Turonian excursion by these compounds was minimal, with $\delta^{13}C$ values for n-C_{19} varying by only 1‰ for all but one sample. In the Portland core, n-C_{19} was consistently enriched relative to pristane by 1‰ to 2‰. Hayes et al. (1990) made similar observations for the Bridge Creek, and, based on metabolic considerations that suggest that n-alkanes are more depleted in ^{13}C than isoprenoids (Monson and Hayes, 1982), argued that the two compound classes emanated from separate sources.

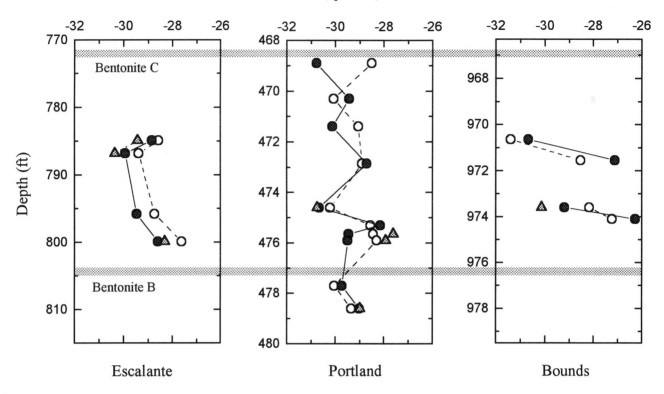

$\delta^{13}C$, permil; PDB

FIG. 8 – Plots of high molecular-weight n-alkane (n-C_{17} [O], n-C_{19} [●], and n-C_{21} [▲]) $\delta^{13}C$ values (‰) against depth for Escalante, Portland, and Bounds cores.

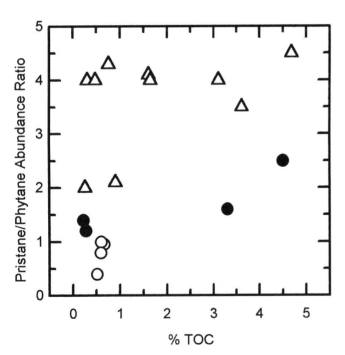

FIG. 9 – Plot of the pristane/phytane ratio against %TOC for the Escalante (O), Portland (△), and Bounds (●) cores.

Isotopic values for high molecular weight n-alkanes for the Portland core (-28‰ to -31‰) are also consistent with previously reported values for terrestrial and marine bulk carbon (Dean et al., 1986). $\delta^{13}C$ values for these compounds decrease slightly (~1‰) upsection and appear to express the Cenomanian-Turonian isotope excursion, but the record is masked by shifts related to

lithology throughout the section. These shifts are greater in the Bounds core, such that in correlative organic-rich units n-C_{29} $\delta^{13}C$ values in Bounds samples are enriched by 1.5‰ - 3‰ relative to the Portland core; in organic-poor units, n-C_{29} $\delta^{13}C$ values do not vary between the two cores.

Pristane and Phytane

The pristane/phytane ratios varied little between samples within a given core but are strongly related to basin location. Values are highest in the center of the basin, with ratios near 4 in the Portland core and lower at its edges (less than 1 in the Escalante core; between 1 and 2 in the Bounds core) (Fig. 9). The slightly enhanced maturity of the Portland core and associated release of pristane from other sources could be responsible for some of the difference between that core and the others (ten Haven et al., 1987), but this does not explain the difference between the immature Bounds and Escalante cores. Although other investigators have used this ratio as an anoxia indicator (Didyk et al., 1978), the pr/ph ratio does not systematically vary with %TOC in the Portland core. In fact, in both the Portland and Bounds cores, pr/ph ratios increase with %TOC, which is opposite of expectations that high %TOC is associated with relatively reducing conditions. Absolute abundances of pristane and phytane generally track each other, although pristane varies over a larger range both within cores and across the basin (pristane: 0.006 to 3.8 µg/g whole rock compared to phytane: 0.003 to 0.94 µg/g whole rock).

In the Portland core, $\delta^{13}C$ values of pristane and phytane track the bulk organic isotope excursion (previously shown by Hayes et al., 1989) (Fig. 10). In the Bounds core, $\delta^{13}C$ values are enriched by ~2 ‰ relative to correlative Portland samples

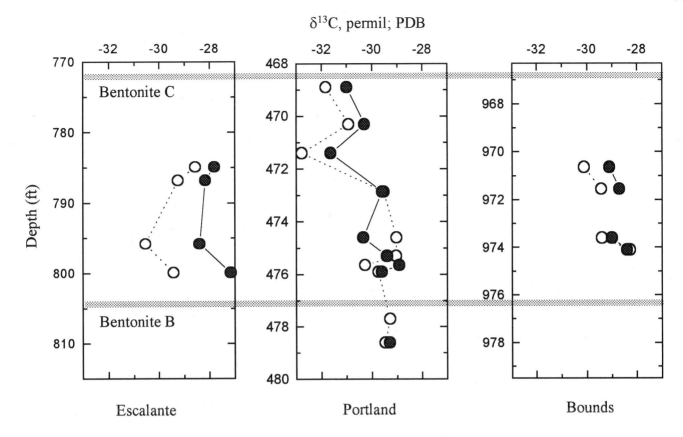

δ¹³C, permil; PDB

Escalante   Portland   Bounds

FIG. 10 – Plots of pristane (●) and phytane (○) δ¹³C values (‰) against depth for Escalante, Portland, and Bounds cores.

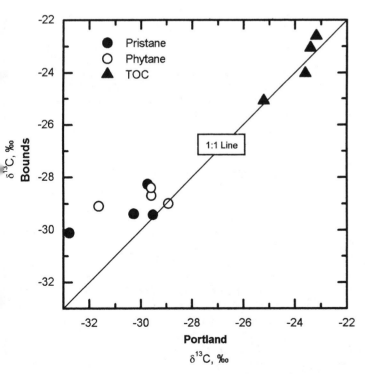

FIG. 11 – δ¹³C values of pristane, phytane, and TOC in the Bounds core plotted against those in correlative units of the Portland core. Correlation is based upon lithologic cyclicity; samples represent the PBC-13 and PBC-15 limestone beds (Elder and Kirkland, 1985) and subjacent marls in each core.

(correlations were made using lithologic couplets) (Fig. 11). This enrichment is not observed for $\delta^{13}C_{TOC}$ values, which indicates that this difference is limited to photosynthetic processes. Values from the Escalante core do not show shifts similar to those observed in the other cores, despite significant shifts in the $\delta^{13}C$ values of low-molecular weight n-alkanes.

DISCUSSION

Organic Matter Variation across The Basin

Organic matter in the Escalante core is dominated by relatively high quantities of terrestrial matter, with high molecular-weight n-alkanes present in much greater concentrations than low molecular-weight homologs. Although the sample set is limited, in the Escalante core carbon-isotope ratios of high molecular weight n-alkanes, pristane, and to a lesser degree, phytane exhibit little variation. In contrast, $\delta^{13}C$ values of low molecular weight n-alkanes are significantly enriched in the lower two samples relative to the uppermost samples, and we suggest these represent marine organic matter. Based on carbon-isotopic similarity to high molecular-weight n-alkanes, pristane and phytane could have terrestrial origins in the Escalante core. While these isoprenoids typically have marine sources, the dominance or selective preservation of terrestrial inputs to the western margin of the Cretaceous Western Interior Seaway apparently overwhelmed these contributions.

Overall, the Portland and Bounds cores exhibit similar compound distributions, and we suggest that both contain a mixture of terrestrial and marine organic matter. Moreover, the propor-

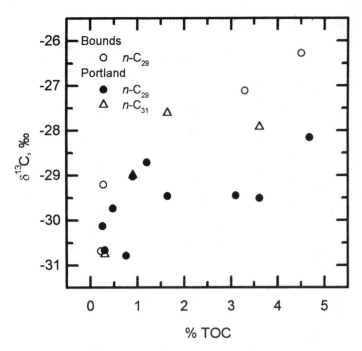

FIG. 12 – Plot of δ¹³C values for *n*-C$_{29}$ and *n*-C$_{31}$ against %TOC for samples from the Bounds and Portland cores.

tional contribution of marine organic matter to these cores is significantly greater than that in the Escalante core. The lower C$_{27}$, C$_{29}$, and C$_{31}$ *n*-alkane abundances relative to those in the Escalante core are consistent with diminished terrestrial input with greater distance to the Sevier orogen. The greater abundances of pristane and phytane relative to %TOC indicate greater contributions of marine organic matter in the central basin.

Organic Matter Variation with Lithologic Cyclicity

Small-scale lithologic cyclicity in the relative abundance of calcium carbonate, terrigenous siliciclastics, and organic matter characterize the Bridge Creek Limestone (Arthur et al., 1984; 1991; Sageman et al., 1997, this volume). Potential causes of this cyclicity include variations in calcium-carbonate production and/or terrigenous input. There is also evidence of decreased bottom-water oxygenation during marlstone deposition (Pratt, 1984; Arthur et al., 1985b; Savrda and Bottjer, 1994), and proposed mechanisms for cyclicity must account for such variations as well. Some workers have proposed that wet-dry climatic oscillations in sediment source regions were responsible for variable clastic input (Fischer, 1980; Pratt, 1981, 1984), and this theory has received wide support (Arthur et al., 1984, 1985, 1987; Barron et al., 1985; Elder and Kirkland, 1985; Hattin, 1985; Bottjer et al., 1986; Glancy et al., 1986; ROCC Group, 1986). An extension of this model would suggest that a high nutrient supply associated with terrigenous influx would increase production of organic matter, and density stratification caused by enhanced fresh-water runoff would produce the anoxic indicators observed in the marlstones. An alternative explanation is that enhanced vertical circulation in the seaway during carbonate deposition oxygenated the bottom waters and transported nutrients to the surface which increased carbonate productivity (Eicher and Diner, 1989). These are two contradictory hypotheses – in one, productivity is greatest during marl deposition and in the other productivity is greatest during carbonate deposition.

The most obvious change that cyclicity causes in organic matter character is compound abundance. In the Portland and Bounds cores, the absolute abundances of all hydrocarbon compounds increases with greater %TOC and decreases with %CaCO$_3$. However, the increase in low molecular-weight *n*-alkanes is much greater than in high molecular-weight homologs (Fig. 6). In the Escalante core, where clastic inputs dominate and cycles are not strongly expressed, there is no clear relationship between compound abundances and %TOC values.

If *n*-C$_{19}$ and *n*-C$_{29}$ are markers of predominantly marine (though not necessarily algal) and terrestrial input, respectively, then Figure 6 reveals either enhanced degradation of labile marine matter or decreased marine productivity during carbonate deposition. While directly contradicting the productivity-driven model of Eicher and Diner (1989), the productivity explanation is consistent with the abundance of nannofossil fertility indicators (Watkins, 1989; Burns and Bralower, this volume) as well as with the wet-dry model of cyclicity, in which higher productivity could be stimulated by increased runoff and elevated nutrient concentrations during marl deposition. It is also consistent with the δ¹³C values of low molecular weight *n*-alkanes in the Escalante core; ¹³C-enriched values occur in carbonate-poor samples which suggests lower values of surface-water CO$_2$ and/or higher growth rates at these times. This variation is ~3‰ - much greater than that recorded by carbonate carbon (~1‰). Variations of isotopic compositions with carbonate contents are not as clear in the central basin samples. *N*-C$_{19}$, *n*-C$_{21}$, pristane and phytane are enriched in carbonate-poor samples in the Bounds core, but not in the Portland core, and enrichment of ¹³C in high molecular-weight *n*-alkanes in carbonate-poor rocks of both cores probably does not reflect surface-water processes (see below).

While the isotopic evidence is consistent with a productivity-related control on the observed variations in low molecular weight *n*-alkane concentrations, the record varies among both compounds and cores and is complicated by isotopic variations related to the Cenomanian-Turonian excursion. Thus, the data are not conclusive regarding a productivity control versus a preservation control on *n*-alkane distributions; these mechanisms are not independent of each other in modern systems and it is possible that both played a role.

Compound-Specific Record of the Cenomanian-Turonian Positive Carbon-Isotope Excursion

In the Portland core, pristane and phytane exhibit a 3‰ shift to more negative carbon isotope values just above the Cenomanian-Turonian boundary, which suggests that the termination of the Cenomanian-Turonian positive isotope excursion is recorded by these compounds. This event is not recorded by the carbon-isotopic records of the *n*-alkanes. The δ¹³C values of low molecular weight *n*-alkanes show little variation through the either the Bounds or Portland cores, but a large (~3‰) shift to negative values occurs between the samples at 786.8 ft and 795.8 ft in the Escalante core. This shift spans the the Cenomanian-Turonian boundary. Hayes et al. (1990) argued that low-molecular weight *n*-alkanes were derived from different organisms than pristane and phytane; thus, their record of the Cenomanian-Turonian excursion in the Portland and Bounds cores could have been obscured by secondary processes (such as heterotrophy).

The high-molecular weight *n*-alkane isotopic record is complex. Although in the Portland core these compounds become depleted in ^{13}C upsection, the record is complicated by 2‰ - 3‰ shifts in $\delta^{13}C$ values. Similar shifts occur in the few samples gathered from the Bounds core. This variability appears related to changes in lithology and/or %TOC. Most positive $\delta^{13}C$ values are observed in the samples with the highest %$CaCO_3$ and the lowest %TOC (Fig. 12). A potential explanation is that multiple, isotopically distinct sources contributed high molecular weight, *n*-alkanes to the organic record. Variations in the relative contributions from those sources in concert with changing sediment inputs, such as preferential degradation or increased productivity of one pool over the other during carbonate deposition, would produce the observed shifts in $\delta^{13}C$ values.

Based upon *n*-alkane abundances, we conclude that during deposition of carbonates, refractory terrestrial matter was preferentially preserved relative to marine organic matter. This suggests that the sources of high-molecular weight *n*-alkanes consist of (1) a ^{13}C-depleted terrestrial source and (2) a ^{13}C-enriched marine source. This conclusion apparently contradicts the observation from analyses of bulk material that, in general, Cretaceous marine and terrestrial organic matter have similar carbon-isotopic compositions (Dean et al., 1986). During the Cenomanian-Turonian interval these relationships could have changed, explaining some of this discrepancy. In addition, we suggest that the observations for bulk organic carbon do not hold true for the molecular record, which requires that the isotopic relationships between individual compounds and bulk organic matter differ among biological sources. Indeed, Collister et al. (1994) observed that *n*-alkanes from terrestrial plants were depleted by 1.6‰ - 13.8‰ relative to total biomass, a considerably larger difference than reported by Park and Epstein (1961) for marine plants. We suggest that, although $\delta^{13}C$ values of bulk marine and terrestrial organic matter were similar durng this time, terrestrial *n*-alkanes were depleted in ^{13}C relative to marine *n*-alkanes.

Certainly other explanations are possible. For example, the shifts could record the isotopic composition of only one source, that responded to varying environmental conditions. Regardless of the mechanism, the records of both high- and low-molecular weight *n*-alkanes illustrate how processes influencing the properties of a certain compound class can obscure a significant geochemical signal preserved in the bulk organic matter and that $\delta^{13}C$ values of extractable compounds are not always representative of the bulk kerogen.

Ocean Circulation and Molecular and Isotopic Records

Compounds assumed to be dominantly of marine origin – pristane and phytane – are enriched in ^{13}C by ~ 2‰ in the Bounds core relative to the Portland core (Fig. 11). Possible explanations for enrichment at those locations include ^{13}C-enriched substrate CO_2, enhanced growth rates (as discussed earlier), or decreased surface water $CO_{2(aq)}$ concentrations due to higher temperatures, stratification, or higher productivity. Although $\delta^{13}C$ values of whole rock carbonate in these cores was compromised by diagenetic processes in organic carbon-rich units (Arthur et al., 1985b), inorganic $\delta^{13}C$ values exhibit little variability between the two locations (Tables 5-7) and we discount the first explanation. Surface-water circulation can explain these trends. The numerical circulation simulation of Slingerland et al. (1996)

clearly predicts a strong Tethyan supply of water to the east-central portion of the basin near the Bounds core. While the exact geographic boundary between Boreal and Tethyan influences can only be approximated, based on their simulations, it is likely that organisms producing the organic matter preserved in Bounds samples experienced more Tethyan conditions than those growing further west and preserved in Portland samples. Thus, ^{13}C-enriched marine organic matter in the Bounds core indicates the influence of warm, CO_2-poor, Tethyan water and/or that growth rates of organisms in Tethyan waters were greater.

The pristane/phytane ratio also exhibits significant variation across the basin. The elevated values expressed in the Portland core could be due, in part, to the core's slightly higher thermal maturity (Connan, 1974). Ichnofossil indicators of dissolved oxygen in bottom-waters indicate that organic matter in the Bounds core experienced more reducing conditions than that in the Portland core (Savrda, this volume). This is consistent with trends in pristane/phytane ratios, and we suggest the ratios record regional differences in redox processes. Isotopic trends discussed above suggest productivity was elevated at the Bounds location relative to the Portland core, and this could have led to the development of oxygen-poor bottom-waters. The decreased pr/ph ratio in Escalante samples relative to those from the Bounds core suggests even more reducing conditions on the western margin of the basin.

In summary, the molecular and isotopic characteristics of Cretaceous Western Interior Seaway sediments reveal variation through lithologic cycles and between Tethyan and Boreal water mass signatures. The carbon-isotopic composition of pristane and phytane indicate that Tethyan surface-waters were more productive than Boreal waters, and pristane/phytane ratios indicate that Tethyan bottom-waters were more reducing than Boreal waters. These interpretations can be tested by more detailed evaluation of molecular distributions. Trends in redox conditions and marine phytoplankton distribution have been evaluated by using a variety of biomarker proxies (reviewed in Peters and Moldowan, 1993) and will be discussed in future publications.

CONCLUSIONS

From our examination of *n*-alkanes and isoprenoids preserved in Cretaceous Western Interior Seaway rocks sampled from the Upper Cenomanian-Lower Turonian boundary, we reach several conclusions:

1. Inputs of terrestrial organic matter to basin sediments were greatest on the western margin and decreased towards the east, which indicates that the western margin was the dominant source of terrestrial matter to the basin. The relative marine contribution to preserved organic matter was significantly greater in the central and eastern parts of the basin, though terrestrial contributions persisted.

2. *N*-alkane distributions track variations in lithology and indicate that either marine inputs to the Cretaceous Western Interior Seaway were more labile than terrestrial inputs, marine phytoplanktic productivity was lower during times of carbonate deposition, or both. These trends are consistent with variations in the carbon-isotopic composition of high molecular-weight *n*-alkanes.

3. In the central basin, both pristane and phytane appear to record the termination of the Cenomanian-Turonian positive isotope excursion. However, that signal is masked for the *n*-alkanes: there is relatively little variation in $\delta^{13}C$ values of low-

molecular weight homologs, and there are large, lithologically related shifts in $\delta^{13}C$ values of high-molecular weight homologs. Thus, in the molecular record a major oceanographic signal recorded by primary organic matter was obscured by heterotrophic or mixing processes.

4. ^{13}C-enriched isotopic compositions of molecular markers for primary photosynthate in the Bounds core relative to those in the Portland core suggest that high growth rates and/or warm, CO_2-poor Tethyan waters dominated at the former location, while cold, CO_2-rich Boreal waters dominated at the latter. These conclusions are consistent with the modeled circulation of Slingerland et al. (1996).

5. Pristane/phytane ratios vary significantly across the seaway and record similar redox conditions as ichnofossils (Savrda, this volume). In particular, elevated pr/ph ratios in the Portland core suggest relatively oxidizing conditions at that location.

ACKNOWLEDGMENTS

Funding for travel and sample collection was provided by DOE grant DE-FG02-92ER14251 (PSU). We also thank Dennis Walizer for technical support in the Stable Isotope Geochemistry Lab at Penn State, Walter Dean at the United States Geological Survey in Denver, Colorado for assistance in procuring samples, and David Wavrek and James Collister at the Energy and Geoscience Institute (Dept. of Civil and Environmental Engineering, University of Utah; Salt Lake City, UT) for Gas Chromatography/Mass Spectrometry analyses. We also thank Lisa Pratt and David Hollander for valuable reviews of this manuscript.

REFERENCES

ARTHUR, M. A., DEAN, W. E., BOTTJER, D., AND SCHOLLE, P. A., 1984, Rhythmic bedding in Mesozoic-Cenozoic pelagic carbonate sequences: The primary and diagenetic origin of Milankovitch-like cycles In Berger, A., Imbrie, J., Hays, J., Kukla, G., and Saltzman B.,eds., Milankovitch and Climate: Dordrecht, Reidel, p. 191-222.

ARTHUR, M. A., DEAN, W. E., AND CLAYPOOL, G. E., 1985a, Anomalous ^{13}C enrichment in modern marine organic carbon: Nature, v. 315, p. 216-218.

ARTHUR, M. A., DEAN, W. E., POLLASTRO, R., CLAYPOOL, G. E., AND SCHOLLE, P. A., 1985b, Comparative geochemical and mineralogical studies of two cyclic transgressive pelagic limestone units, Cretaceous western interior basin, U.S., In Pratt, L. M., Kauffman, E. G., and Zelt, F. B.,eds., Field Trip Guidebook, 4: Tulsa, Society of Economic Paleontologists and Mineralogists, p. 16-27.

ARTHUR, M. A., DEAN, W. E., AND SCHLANGER, S. O., 1985c, Variations in the global carbon cycle during the Cretaceous related to climate, volcanism, and changes, In atmospheric CO_2, in Sundquist, E. T. and Broecker, W. S., eds., The Carbon Cycle and Atmospheric CO_2: Natural Variations Archean to Present: American Geophysical Union Monograph, v. 32, p. 504-529.

ARTHUR, M. A. AND SCHLANGER, S. O., 1979, Cretaceous 'oceanic anoxic events' as causal factors in development of reef-resevoired giant oil fields: Bulletin of the American Association of Petroleum Geologists, v. 63, p. 870-885.

ARTHUR, M. A., SCHLANGER, S. O. AND JENKYNS, H. C., 1987, The Cenomanian-Turonian Oceanic Anoxic Event, II, Paleoceanographic controls on organic matter production and preservation, In Brooks, J. and Fleet, A., eds., Marine Petroleum Source Rocks: Geological Society of London Special Publication, v. 26, p. 401-420.

ARTHUR, M. A., DEAN, W. E., AND PRATT, L. M., 1988, Geochemical and climatic effects of increased marine organic carbon burial at the Cenomanian/Turonian boundary: Nature, v. 335, p. 714-717.

ARTHUR, M. A., BRUMSACK, H. J., JENKYNS, H. C., AND SCHLANGER, S. O., 1990, Statigraphy, geochemistry, and paleoceanography of organic carbon-rich Cretaceous sequences, In Ginsburg, R. N. and Beaudoin, B.,eds., Cretaceous Resources, Events, Rhythms: Background and Plans for Research: Dordrecht, the Netherlands, Kluwer Academic Publishers, p.75-119.

ARTHUR, M. A., KUMP, L. R., DEAN, W. E., AND LARSON, L. R., 1991, Superplume,

supergreenhouse? (abs): EOS, Transactions of the American Geophysical Union, v. 72, p.301, 1991.

ARTHUR, M. A., SAGEMAN, B. B., RICH, J., DEAN, W. E., SAVRDA, C. E., BRALOWER, T. J., AND LECKIE, M., 1994, Cyclostratigraphy of the Cenomanian-Turonian Bridge Creek Limestone Member of the Greenhorn Formation in the U.S. Western Interior (abs.), Annual Meeting Abstracts, AAPG and SEPM, p. 95-96, 1994.

BARLOW, L. K. AND KAUFFMAN, E. G., 1985, Depositional cycles in the Niobrara Formation, Colorado Front Range, In Pratt, L. M., Kauffman, E. G., and Zelt, F. B., eds., Fine-grained Deposits and Biofacies of the Cretaceous Western Interior Seaway: Evidence of Cyclic Sedimentary Processes, Fieldtrip Guidebook 4: Tulsa, Society of Economic Paleontologists and Mineralogists, p. 199-208.

BARRON, E. J., ARTHUR, M. A., AND KAUFFMAN, E. G., 1985, Cretaceous rhythmic bedding sequences: a plausible link between orbital variations and climate: Earth Planetary Science Letters, v. 72, p. 327-340.

BIDIGARE, R. R., FLUEGG, A., FREEMAN, K. H., HANSON, K. L., HAYES, J. M., HOLLANDER, D., JASPER, J. P., KING, L. L., LAWS, E. A., MILLERO, F. J., PANCOST, R. D., POPP, B. N., STEINBERG, P. A., AND WAKEHAM, S. G., 1997, Consistent fractionation of ^{13}C in nature and in the laboratory: Growth-rate effects in some haptophyte algae: Global Biogeochemical Cycles, v. 11, p. 179-192.

BOTTJER, D. J., ARTHUR, M. A., DEAN, W. E., HATTIN, D. E., AND SAVRDA, C. E., 1986, Rhythmic bedding produced in Cretaceous pelagic carbonate environments: sensitive recorders of climate cycles: Paleoceanography, v. 1, p. 467-481.

BRAY, E. E., AND EVANS, D. E., 1961, Distributioin of n-paraffins as a clue to recognition of source beds: Geochimica et Cosmochimica Acta, v. 22, p. 2-15.

COLLISTER, J. W., RIELEY, G., STERN, B., EGLINTON, G., AND FRY, B., 1994, Compound-specific $\delta^{13}C$ analysis of leaf lipids from plants with different carbon dioxide metabolisms: Organic Geochemistry, v. 21, p. 619-627.

CONNAN, J., 1974, Diagenese naturelle et diagenese artificielle de la matiere organique a element vegetaux predominants, In Tissot, B.P. and Bienner, F., eds., Advances In Organic Geochemistry 1973: Paris, Editions Technip, p. 73-95.

CURIALE, J. A., 1994, High-resolution organic record of Bridge Creek deposition, northwest New Mexico: Organic Geochemistry, v. 21, p. 489-507.

DEAN, W. E., ARTHUR, M. A., AND CLAYPOOL, G. E., 1986, Depletion of ^{13}C in Cretaceous marine organic matter: Source, diagenetic, or environmental signal?: Marine Geology, v. 70, p. 119-157.

DEAN, W. E., ARTHUR, M. A., SAGEMAN, B. B., AND LEWAN, M. D., 1995, Core Descriptions And Preliminary Geochemical Data For The Amoco Production Company Rebecca K. Bounds #1 Well, Greeley County, Kansas: Washington, D. C., United States Geological Survey Open-File Report 95-209, 243 p.

DIDYK, B. M., SIMONEIT, B. R. T., BRASSEL, S. C., AND EGLINTON, G., 1979, Organic geochemical indicators of palaeoenvironmental conditions of sedimentation: Nature, v. 272, p. 216-222.

EGLINTON, G., GONZALEZ, A. G., HAMILTON, R. J., AND RAPHAEL, R. A., 1962, Hydrocarbon constituents of the wax coatings of plant leaves: A taxonomic survey: Phytochemistry, v. 1, p. 89-102.

EICHER, D. L. AND DINER, R., 1985, Foraminifera as indicators of water mass in the Cretaceous Greenhorn Sea, In Pratt, L. M., Kauffman, E. G., and Zelt, F. B., eds., Fine-grained Deposits and Biofacies of the Cretaceous Western Interior Seaway: Evidence of Cyclic Sedimentary Processes, Fieldtrip Guidebook 4: Tulsa, Society of Economic Paleontologists and Mineralogists, p. 60-71.

EICHER, D. L. AND DINER, R., 1989, Origin of the Cretaceous Bridge Creek cycles in the western interior, United States, Palaeogeography, Palaeoclimatology, Palaeoecology, v. 74, p. 127-146.

EICHER, D. L. AND WORSTELL, P., 1970, Cenomanian and Turonian foraminifera from the Great Plains, United States: Micropaleontology, v. 16, p. 269-324.

ELDER, W. P. AND KIRKLAND, J. I., 1985, Stratigraphy and depositional environments of the Bridge Creek Limestone Member of the Greenhorn Limestone at Rock Canyon Anticline near Pueblo, Colorado, In Pratt, L. M., Kauffman, E. G., and Zelt, F. B., eds., Fine-grained Deposits and Biofacies of the Cretaceous Western Interior Seaway: Evidence of Cyclic Sedimentary Processes, Fieldtrip Guidebook 4: Tulsa, Society of Economic Paleontologists and Mineralogists, p. 122-134.

FISCHER, A. G., 1980, Gilbert–bedding rhythms and geochronology: Geologcical Society of America Special Paper 183, p. 93-104.

FISCHER, A. G., HERBERT, T. D., PREMOLI-SILVA, I., 1985, Carbonate bedding cycles in Cretaceous pelagic and hemipelagic sequences, In Pratt, L. M., Kauffman, E. G., and Zelt, F. B., eds., Fine-grained Deposits and Biofacies of the Cretaceous Western Interior Seaway: Evidence of Cyclic Sedimentary Processes, Fieldtrip Guidebook 4: Tulsa, Society of Economic Paleontologists and Mineralogists, p. 156-169.

FISHER, C. G., 1991, Calcareous nannofossil and foraminifera definition of an oceanic front in the Greenhorn Sea (Late Middle through Late Cenomanian), northern Black Hills, Montana and Wyoming: paleoceanographic implictions, Unpublished Ph.D. Dissertation, Boulder, University of Colorado.

FRANCOIS, R., ALTABET, M. A., GOERICKE, R., MCCORKLE, D., BRUNET, C., AND POISSON, A., 1993, Changes in the $\delta^{13}C$ of surface water particulate organic matter across the subtropical convergence in the SW Indian Ocean: Global Biogeochemical Cycles, v. 7, p. 627-644.

FREEMAN, K. H., HAYES, J. M., TRENDEL, J. M., AND ALBRECHT, P., 1990, Evidence from carbon isotope measurements for diverse origins of sedimentary hydrocarbons: Nature, v. 343, p. 254-256.

FREEMAN, K. H., WAKEHAM, S. G., AND HAYES, J. M., 1994, Predictive isotopic biogeochemistry: hydrocarbons from anoxic marine basins: Organic Geochemistry, v. 21, p. 629-644.

FRUSH, M. P. AND EICHER, D. L., 1975, Cenomanian and Turonian foraminifera and paleoenvironments in the Big Bend region of Texas and Mexico, In Caldwell, W. G. E., ed., The Cretaceous System in North America: St. Johns, Geological Association of Canada Special Paper 13, p. 277-301.

GELPIE, E., SCHNEIDER, H., MANN, J., AND ORO, J., 1970, Hydrocarbons of geochemical significance in microscopic algae: Phytochemistry, v. 9, p. 603-612.

GLANCY, T. J., JR., ARTHUR, M. A., BARRON, E. J., AND KAUFFMAN, E. G., 1993, A paleoclimate model for the North American Cretaceous (Cenomanian-Turonian) epicontinental sea, In Caldwell, W. G. E. and Kauffman, E. G., eds., Evolution of the Western Interior Basin: St. Johns, Geological Association of Canada Special Paper 39, p. 219-241.

GOERICKE, R. AND FRY, B., 1994, Variations of marine plankton $\delta^{13}C$ with latitude, temperature, and dissolved CO_2 in the world ocean: Global Biogeochemical Cycles, v. 8, p. 85-90.

GOOSENS, H., DE LEEUW, J. W., SCHENCK, P. A., AND BRASSEL, S. C., 1984, Tocopherols as likely precursors of pristane in ancient sediments and crude oils: Nature, v. 312, p. 440-442.

HAQ, B., HARDENBOL, J., AND VAIL, P. R., 1987, Chronology of fluctuating sea levels since the Triassic (250 million years ago to present): Science, v. 235, p. 1156-1167.

HATTIN, D. E., 1985, Distribution and significance of widespread, time-parallel pelagic limestone beds in Greenhorn Limestone (Upper Cretaceous) of the central Great Plains and southern Rocky Mountains, In Pratt, L. M., Kauffman, E. G., and Zelt, F. B., eds., Fine-grained Deposits and Biofacies of the Cretaceous Western Interior Seaway: Evidence of Cyclic Sedimentary Processes, Fieldtrip Guidebook 4: Tulsa, Society of Economic Paleontologists and Mineralogists, p. 28-37.

TEN HAVEN, H. L., DE LEEUW, J. W., RULLKOTTER, J., AND SINNINGHE DAMSTE, J. S., Restricted utility of the pristane/phytane ratio as a paleoenvironmental indicator: Nature, v. 330, p. 641-643.

HAY, W. W., EICHER, D. L., AND DINER, R., 1993, Physical oceanography and water masses in the Cretaceous Western Interior Seaway, In Caldwell, W. G. E. and Kauffman, E. G., eds., Evolution of the Western Interior Basin: St. Johns, Geological Association of Canada Special Paper 39, p. 297-318.

HAYES, J. M., POPP, B. N., TAKIGIKU, R., AND JOHNSON, M. W., 1989, An isotopic study of biogeochemical relationships between carbonates and organic carbon in the Greenhorn Formation: Geochimica et Cosmochimica Acta, v. 53, p. 2961-2972.

HAYES, J. M., FREEMAN, K. H., POPP, B. N., AND HOHAM, C., 1990, Compound-specific isotopic analyses: A novel tool for the reconstruction of ancient biogeochemical processes, In Durand, B., and Behar, F., eds., Advances in Organic Geochemistry 1989, Organic Geochemistry, p. 1115-1128.

JENKYNS, H. C., 1980, Cretaceous anoxic events – from continents to oceans: Journal of the Geological Society of London, v. 137, p. 171-188.

JEWELL, P. W., 1993, Water-column stability, residence times, and anoxia in the Cretaceous North American seaway: Geology, v. 21, p. 579-582.

KAUFFMAN, E. G., 1975, Dispersal and biostratigraphic potential of Cretaceous benthonic Bivalvia in the Western Interior, In Caldwell, W. G. E., ed., The Cretaceous System in the Western Interior of North America: St. Johns, Geological Association of Canada Special Paper 13, p. 163-194.

KAUFFMAN, E. G., 1985, cretaceous evolution of the Western Interior basin of the United States, In Pratt, L. M., Kauffman, E. G., and Zelt, F. B., eds., Fine-grained Deposits and Biofacies of the Cretaceous Western Interior Seaway: Evidence of Cyclic Sedimentary Processes, Fieldtrip Guidebook 4: Tulsa, Society of Economic Paleontologists and Mineralogists, p. IV-XIII.

KAUFFMAN, E. G., 1988, Concepts and methods of high resolution event stratigraphy: Annual Reviews of Earth and Planetary Sciences, v. 16, p. 605-654.

KOCH, P. L., ZACHOS, J. C., AND GINGERICH, P. D., 1992, Correlation between isotope records in marine and continental carbon reservoirs near the Palaeocene/Eocene boundary: Nature, v. 358, p. 319-322.

KOHNEN, M. E. L., SINNINGHE DAMSTE, J. S., AND DE LEEUW, J. W., 1991, Biases from natural sulfurization in paleoenvironmental reconstruction based on hydrocarbon marker distributions: Nature, v. 349, p. 775-778.

KOLATTUKUDY, P. E. (Ed.) 1976, Chemistry and Biochemistry of Natural Waxes, Amsterdam, Elsevier.

KYSER, T. K., CALDWELL, W. G. E., WHITTAKER, S. G., AND CADRIN, A. J., 1993, Paleoenvironment and geochemistry of the northern portion of the Western Interior Seaway during Cretaceous Time, In Caldwell, W. G. E., and Kauffman, E.G., eds., Evolution of the Western Interior Basin: St. Johns, Geological Association of Canada Special Paper 39, p. 355-375.

LASAGA, A.C., BERNER, R. A., AND GARRELS, R. M., 1985, An improved geochemical model of atmospheric CO2 fluctuations over the past 100 million years, In Sundquist, E. T. and Broecker, W. S., eds., The Carbon Cycle and Atmospheric CO_2: Natural Variations Archean to Present: Washington D. C., American Geophysical Union, Monograph 32, p. 397- 411.

LAWS, E. A., POPP, B. N., BIDIGARE, R. R., KENNICUTT, M. C., AND MACKO, S. A., 1995, Dependence of phytoplankton carbon isotopic compositions on growth rate and $[CO_{2(aq)}]$: Theoretical considerations and experimental results: Geochimica Cosmochimica Acta, v. 59, p. 1131-1138.

LICHTFOUSE, E., DERENNE, S., MARIOTTI, A., AND LARGEAU, C., 1994, Possible algal origin of long chain odd n-alkanes in immature sediments as revealed by distributions and carbon isotope ratios: Organic Geochemistry, v. 22, p. 1023-1027.

MICHALCZYK, G., 1985, Determination of n- and iso-paraffins in hydrocarbon waxes - comparitive results of analyses by gas chromatography, urea adduction, and molecular sieve adsorption: Fette-Seifen-Anstrichmittel, v. 87, p. 481-486.

MONSON, K. D. AND HAYES, J. M., 1982, Carbon isotopic fractionation in the biosynthesis of bacterial fatty acids. Ozonolysis of unsaturated fatty acids as a means of determining the intramolecular distribution of carbon isotopes: Geochimica Cosmochimica Acta, v. 46, p. 139-149.

PARK, R., AND EPSTEIN, S., 1961, Metabolic fractionation of ^{13}C and ^{12}C in plants: Plant Physiology, v. 35, p. 133-138.

PARRISH, J. T., GAYNOR, G. C., AND SWIFT, D. J. P., 1984, Circulation in the Cretaceous Western Interior Seaway of North America, a review, In Stott D. F. and Glass, D. J., eds., The Mesozoic of Middle North America: Canadian Society of Petroleum Geologists Memoir, v. 9, p. 221-231.

PETERS, K. E., AND MOLDOWAN, J. M., 1993, The Biomarker Guide: Interpreting Molecular Fossils in Petroleum and Ancient Sediments: Englewood Cliffs, New Jersey, Prentice Hall, 363 p.

POLLARD, D. AND THOMPSON, S. L., 1992, User's guide to the GENESIS global climate model version 1.02, NCAR ICS unnumbered manuscript: Boulder, Colorado, National Center for Atmospheric Research, 58 p.

PRATT, L. M., 1981, A paleo-oceanographic interpretation of sedimentary structures, clay minerals, and organic matter in a core of the Middle Cretaceous Greenhorn Formation near Oueblo, Colorado, Unpublished Ph.D. Dissertation, Princeton University, Princeton, NJ, 176 p.

PRATT, L. M., 1984, Influence of paleoenvironmental factors on preservation of organic matter in Middle Cretaceous Greenhorn Formation, Pueblo, Colorado: Bulletin of the American Association of Petroleum Geologists, v. 68, p. 1146-1159.

PRATT, L. M., 1985, Isotopic studies of organic matter and carbonate in rocks of the Greenhorn Marine Cycle, In Pratt, L. M., Kauffman, E. G., and Zelt, F. B., eds., Fine-grained Deposits and Biofacies of the Cretaceous Western Interior Seaway: Evidence of Cyclic Sedimentary Processes, Fieldtrip Guidebook 4: Tulsa, Society of Economic Paleontologists and Mineralogists, p. 38-48.

PRATT, L. M., ARTHUR, M. A., DEAN, W. E., AND SCHOLLE, P. A., 1993, Paleoceanographic cycles and events during the Late Cretaceous in the Western Interior Seaway of North America, In Caldwell, W. G. E. and Kauffman, E. G., eds., Evolution of the Western Interior Basin: St. Johns, Geological Association of Canada Special Paper 39, p. 333-354.

RESEARCH ON CRETACEOUS CYCLES GROUP, 1986, Rhythmic bedding in Upper Cretaceous pelagic carbonate sequences: Varying sediment response to climatic forcing: Geology, v. 14, p. 153-156.

RISATTI, J. B., ROWLAND, S. J., YAN, D. A., AND MAXWELL, J.R., 1984, Stereochemical studies of acyclic isoprenoids - XII. Lipids of methanogenic bacteria and possible contributors to sediments: Organic Geochemistry, v. 6, p. 93-104.

SAGEMAN, B. B., RICH, J., ARTHUR, M. A., BIRCHFIELD, G. E., DEAN, W. E., 1997, Evidence for Milankovitch periodicities in Cenomanian-Turonian lithologic and geochemical cycles, Western Interior, U. S. A. Journal of Sedimentary Research, Section B: Stratigraphy and Global Studies, v. 67, p. 286-302.

SAVRDA, C. E., AND BOTTJER, D. J., 1994, Ichnofossils and ichnofabrics in rhythmically bedded pelagic/hemipelagic carbonates: Recognition and evaluation of benthic redox and scour cycles, Specical Publications International Association Sedimentology, v. 19, p. 195-210.

SCHLANGER, S. O. AND JENKYNS, H. C., 1976, Cretaceous oceanic anoxic events - Causes and consequences: Geolog.en. Mijnb., v. 55, p. 179-184.

SCHLANGER, S. O., ARTHUR, M. A., JENKYNS, H. C., AND SCHOLLE, P.A., 1987, The Cenomanian-Turonian oceanic anoxic event, 1. Stratigraphy and distribution of organic-carbon rich beds and the marine $\delta^{13}C$ excursion, In: Brooks, J. and Fleet, A., eds., Marine Petroleum Source Rocks: London, Geological Society of London Special Publication, No. 26, p. 371-399.

SCHOLLE, P. A. AND ARTHUR, M. A., 1980, Carbon isotope fluctuations in Cretaceous pelagic limestones: Potential stratigraphic and petroleum exploration tool: Bulletin of the American Association of Petroleum Geologists, v. 64, p. 67-87.

SEIFERT, W. K. AND MOLDOWAN, J. M., 1986, Use of biological markers in petroleum exploration, In Johns, R.B., ed., Methods in Geochemistry and Geophysics, v.

24, p. 261-290.

SETHI, P. S. AND LEITHOLD, E. L., 1994, Climatic cyclicity and terrigenous sediment influx to the Early Turonian Greenhorn Sea, Southern Utah: Journal of Sedimentary Research, v. B64, p. 26-39.

SLINGERLAND, R., KUMP, L. R., ARTHUR, M. A., FAWCETT, P. J., SAGEMAN, B. B., AND BARRON, E. J., 1996, Estuarine circulation in the Turonian Western Interior Seaway of North America: Geological Society of America Bulletin, v. 108, p. 941-952.

VOLKMAN, J. K., AND MAXWELL, J. R., 1986, Acyclic isoprenoids as biological markers, In: Johns, R. B., ed., Biological Markers in the Sedimentary Record, New York, Elsevier, p. 1-42.

WATKINS, D. K. 1986, Calcareous nannofossil paleoceanography of the Cretaceous Greenhorn Sea: Geological Society of America Bulletin, v. 97, p. 1239-1249.

WATKINS, D. K., 1989, Nannoplankton productivity fluctuations and rhythmically bedded pelagic carbonates of the Greenhorn limestone (Upper Cretaceous): Palaeogeography, Palaeoclimatology, Palaeoecology, v. 74, p. 75-86.

DEPOSITIONAL PROCESSES AND CARBON BURIAL ON A TURONIAN PRODELTA AT THE MARGIN OF THE WESTERN INTERIOR SEAWAY

ELANA L. LEITHOLD[1] AND WALTER E. DEAN[2]

[1] North Carolina State University, Department of Marine, Earth, and Atmospheric Sciences, Box 8208, Raleigh, North Carolina 27695

[2]U.S. Geological Survey, MS 980 Federal Center, Denver, Colorado 80225

ABSTRACT: Lower-middle Turonian strata of the Tropic Shale and correlative Tununk Shale Member of the Mancos Shale accumulated in muddy prodeltaic environments near the western margin of the Cretaceous Western Interior Seaway. These fine-grained rocks are well exposed in outcrops along the southern margin of the Kaiparowits Plateau and in the Henry Mountains region in southern Utah, and they have been sampled in a core drilled by the United States Geological Survey near the town of Escalante, Utah. The rocks consist of bottomset, foreset, and topset facies that accumulated at progressively increasing rates during the eastward progradation of subaqueous deltaic clinoforms. Bottomset facies accumulated in relatively distal settings where sedimentation was dominated by deposition of both terrigenous and biogenic particles from suspension. In these strata, the shapes of fecal pellets, thicknesses of event layers, and intensity of bioturbation point to relatively slow rates of sediment accumulation. Foreset and distal topset beds, in contrast, suggest more rapid sediment accumulation, primarily as the result of episodic turbidity currents and storm processes.

Within the progradational succession, upward increasing organic carbon content and hydrogen indices and decreasing $\delta^{13}C$ values suggest that the amount of marine organic carbon buried increased over time and with proximity to the shoreline. These trends parallel evidence for both increasing rates of sediment accumulation and decreasing levels of bottom water oxygenation and are interpreted to reflect increasing preservation of labile marine carbon. Studies of the Tropic and Tununk shales suggest that along the western margin of the seaway, proximal foreset facies are major repositories of marine organic carbon.

INTRODUCTION

During Cretaceous time, thick deposits of fine-grained sediment accumulated near the western margin of the North American Western Interior Seaway. These fine-grained marine facies contain an invaluable record of the dynamic history of structural, oceanographic, climatic, and ecologic changes in the seaway. Little is known, however, about the processes by which organic-carbon-rich, fine-grained sediments were dispersed along and across the seaway's margin. An understanding of these processes and of the modes and rates of sediment accumulation in muddy offshore environments is necessary to accurately reconstruct the seaway's history. The aim of this paper is to develop such an understanding of the setting in which the Tropic Shale accumulated in Cenomanian-Turonian time. During this part of the Cretaceous, oxygen deficiency was widespread in the global ocean (so-called Oceanic Anoxic Event) and abundant organic-carbon-rich sediments accumulated. In this paper we examine the onshore-offshore facies patterns in the Tropic Shale and their relationships to organic-matter accumulation and burial. Trends in the amount and character of organic carbon in the strata will be related to both local sea-level changes and global events.

The Tropic Shale and correlative Tununk Shale Member of the Mancos Shale of southern Utah accumulated within a few hundred kilometers offshore of the western margin of the Western Interior Seaway. The entire thickness of the Tropic Shale was sampled in an approximately 280-m-long core drilled in 1991 at the base of the Kaiparowits Plateau near the town of Escalante, Utah (Fig. 1). The core, called the USGS #1 Escalante core, provides an excellent opportunity for detailed reconstruction of processes of fine-grained sediment dispersal and accumulation of organic matter along the margin of the seaway.

The Tropic and Tununk shales record a part of the Greenhorn transgressive-regressive cycle, a sea-level oscillation of approximately 7 m.y. duration during which the Western Interior Seaway expanded to its maximum extent and then partially contracted (Kauffman, 1977; 1985; Kauffman and Caldwell, 1993;

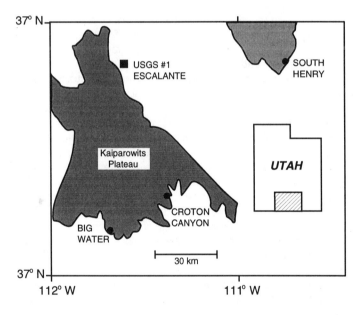

FIG. 1.—Location map, showing areas of outcrop of Upper Cretaceous rocks (shaded areas) in the Kaiparowits Plateau and Henry Mountains Region of southern Utah.

Kauffman et al., 1993). The Tropic and Tununk shales are composed primarily of calcareous shale and mudstone. As these fine-grained marine stata accumulated, more calcareous, basinal facies of the Bridge Creek Member of the Greenhorn Limestone and Fairport Chalky Shale Member of the Carlile Shale were deposited to the east in Colorado and Kansas. In southern Utah, the Tropic and Tununk shales overlie sandy, paralic facies of the Dakota Sandstone, which accumulated during the Greenhorn transgression, and they are overlain respectively by the Tibbet Canyon Member of the Straight Cliffs Sandstone and the Ferron Sandstone Member of the Mancos Shale, deposited during the regressive phase of the Greenhorn cycle.

Strata of the Tropic shales deposited during the early part of

Stratigraphy and Paleoenvironments of the Cretaceous Western Interior Seaway, USA, SEPM Concepts in Sedimentology and Paleontology No. 6
Copyright 1998 SEPM (Society for Sedimentary Geology), ISBN 1-56576-044-1, p. 189-200.

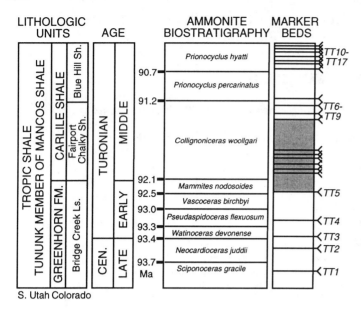

FIG. 2.—Biostratigraphy and bentonite marker beds in the Tropic Shale and correlative Tununk Shale Member of the Mancos Shale (Zelt, 1985; Kauffman et al., 1993; Obradovich, 1993). Study interval, between bentonites TT5 and TT6 is shaded.

the Greenhorn regression contain a particularly intriguing record of fine-grained sedimentation and are the focus of the present study. These strata record a transition between dominantly biogenic and dominantly terrigenous sources of sediment to the seafloor in southern Utah and a transition between processes of sediment transport and accumulation characteristic of the deeper basin and the shoreline. Leithold (1993, 1994) showed that during the early Greenhorn regression fine-grained prodeltaic environments in southern Utah were characterized by a clinoform morphology, with gently dipping topsets, more steeply dipping foresets, and gently dipping bottomsets. During the later regressive phase of the Greenhorn cycle, clinoforms prograded eastward, and the slowly deposited bottomset facies were buried by foreset facies that accumulated at increasingly rapid rates.

Fine-grained, regressive strata of the Tropic and Tununk shales thus accumulated at variable rates. In modern marine environments, sediment accumulation rates are known to be important controls on many seabed processes, including bioturbation, the formation and preservation of physical sedimentary structures, and the preservation of organic carbon (e.g., Moore and Scruton, 1957; Müller and Suess, 1979; Nittrouer and Sternberg, 1981; Nittrouer et al., 1984; Rhoads et al., 1985; Kuehl et al., 1986a and b; Henrichs and Reeburgh, 1987; Betts and Holland, 1991; Hedges and Keil, 1995). Few systematic studies have been conducted of the signatures and effects of rapid and slow sediment accumulation rates in ancient deposits. The paucity of such studies stems in part from the fact that sediment accumulation rates for ancient successions are generally known only in terms of very long-term averages, typically over time scales of several millions of years. Moreover, average sedimentation rates are typically calculated for successions that include a number of facies that likely accumulated at variable rates. The Tropic and Tununk shales are an exception to this rule. In these deposits, bentonite marker beds and facies patterns allow trends in relative accumulation rates to be defined in coeval prodeltaic facies that accumulated over an extensive region of the ancient seaway. These prodeltaic facies record a time span of a few hundreds of thousands of years.

In this investigation, strata of the Tropic Shale deposited during the early regressive phase of the Greenhorn cycle were studied in detail to address questions about sediment accumulation on an ancient prodelta. Specific objectives of this study included documenting the processes by which sediment was dispersed from the shoreline to deeper water settings of the Western Interior Seaway, delineating the sedimentological signatures of rapid versus slow sediment accumulation rates in the Tropic Shale, and evaluating the role that sediment accumulation rates played in organic carbon burial in this ancient prodeltaic system.

METHODS

Samples from the Escalante core were analyzed for total and inorganic carbon at the USGS in Denver by coulometry (Engleman et al., 1985; Dean and Arthur, this volume). Values of organic carbon (OC) were determined by difference between total carbon and inorganic carbon. Percent $CaCO_3$ was calculated by dividing percent carbonate carbon by 0.12, the fraction of carbon in $CaCO_3$. Carbonate content of samples from outcrops of the Tropic and Tununk Shale were analyzed at North Carolina State University (NCSU) by acidifying the sediment with 6N hydrochloric acid for 24 hours. The samples were then rinsed twice to remove chloride compounds and dried. Carbonate contents were calculated based on the weight lost by each sample as the result of acidification. Comparison of carbonate values obtained by the two different methods indicates a high degree of correlation ($r^2=0.96$), but reveals that systematically higher values were obtained by the acidification and weight loss method (Fig. 5). The offset probably is due to dissolution of other minerals during treatment with the acid.

Samples from the Escalante core were analyzed for hydrogen richness of organic matter at the USGS, Denver, using a Rock-Eval pyrolysis instrument, for comparison with similar analyses commonly used in the evaluation of hydrocarbon (HC) source rocks. The Rock-Eval pyrolysis method rapidly determines HC type, HC-generation potential, and degree of preservation of organic matter in sediments and sedimentary rocks (Dean and Arthur, this volume). Our data are expressed in terms of a hydrogen index (HI) as a measure of hydrogen richness of the organic matter, expressed as mg HC/g OC.

Stable-carbon isotope ratios in organic matter were determined on acidified whole-rock samples using standard techniques (Dean and Arthur, this). Results are reported in the standard per mil (‰) δ-notation based on the University of Chicago Pee Dee belemnite (PDB) marine-carbonate standard,

$$\delta =[(R_{sample}/R_{PDB})-1] \times 10^3,$$

where R is the ratio $^{13}C{\cdot}^{12}C$

The Escalante core was described jointly by E. L. Leithold and E. G. Kauffman at the USGS Core Research Center in Lakewood, Colorado. During description of the core, intensity of bioturbation was assessed on a bed-by-bed basis using a modified version of the ichnofabric indices proposed by Droser and Bottjer (1990). In this scheme, as in that of Droser and Bottjer, an ichnofabric index of 1 denotes a physically laminated fabric with no evidence for bioturbation, while an ichnofabric index of 5 denotes a fabric characterized by cross-cutting burrows with no physical structures preserved. In our scheme an ichnofabric

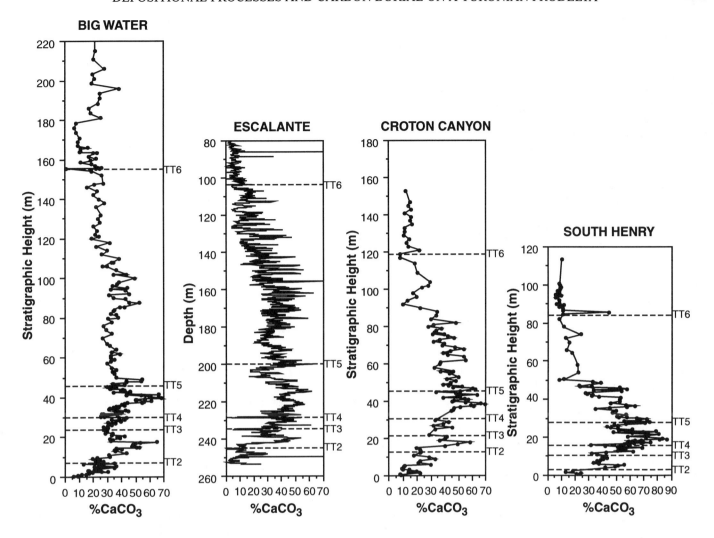

FIG. 3.—Vertical trends in CaCO₃ in the Tropic Shale and Tununk Shale Member of the Mancos Shale in the four study sections. Bentonite marker beds (TT2 through TT6), originally identified by Zelt (1985), are shown for reference.

index of 6 was assigned where strata were so completely homogenized by bioturbation that no discrete burrows were identifiable. To generalize the ichnofabric data, the weighted average ichnofabric index for each meter of core was calculated.

After generally describing of the Escalante core, Leithold studied the physical sedimentary structures in the interval between 189 and 100 m (620 and 300 ft). Within this interval, each physically stratified part of the core was carefully examined and sketched to a resolution of approximately 1 mm and then photographed. This attention to detail was particularly helpful for recognizing subtle discontinuity surfaces so that individual event layers could be identified. Normally graded beds were found throughout this interval, for example, and the thickness of the thickest discrete bed in each meter of core was tabulated.

Fecal pellets are present throughout most of the study interval. They are ovoid and typically about 100-200 μm long and 30-60 μm wide. The fecal pellets are white to tan in color, and study with Scanning Electron Microscopy by Leithold shows that they are composed primarily of coccoliths. Hattin (1975) described similar fecal pellets from the Greenhorn limestone and Carlile Shale in Kansas. He suggested that they were produced by selectively feeding zooplankton. Fecal pellets in the Tropic

and Tununk shale were used to test the hypothesis that the study interval in southern Utah accumulated at progressively decreasing rates with distance offshore. The length to width ratios of pellets were measured to evaluate compaction of sediments at the various geographic sites at various stratigraphic levels. Hand samples from outcrops of the Tropic Shale and Tununk Member and from the Escalante core were cut, polished, and examined under a microscope equipped with a video camera and connected to a personal computer. Two samples from an outcrop of the coeval Fairport Chalky Shale Member near Pueblo, 460 km (285 mi) east of Escalante in south-central Colorado, were also analyzed for comparison. In each of the samples, 10 1-mm square areas on the polished surface were chosen at random, and the long and short axes of the ten largest pellets in each of these areas were measured using image analysis software.

THE STUDY INTERVAL

The Tropic and Tununk shales are well exposed along the southern edge of the Kaiparowits Plateau and in the Henry Mountains region of southern Utah (Fig. 1). Zelt (1985) provided a preliminary biostratigraphy for the succession (Fig. 2) and identi-

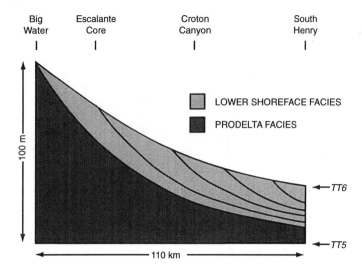

FIG. 4.—Facies and geometry of the interval between TT5 and TT6. Lower shoreface facies accumulated atop prodeltaic facies as the shoreline stepped seaward and downward during a sea-level fall in Late Middle Turonian time (see Leithold, 1994). Vertical exaggeration is approximately 750 x.

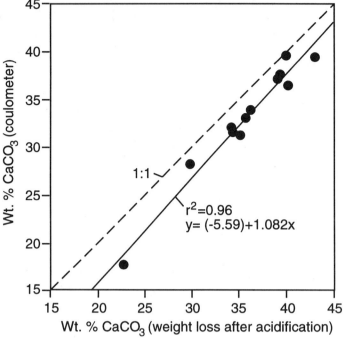

FIG. 5.—Comparison of carbonate data determined by coulometry at the U.S. Geological Survey and weight loss after acidification at North Carolina State University.

fied 17 bentonite layers in the Tropic and Tununk shales, which he traced throughout the region both in outcrops and in well logs. Leithold (1993, 1994) studied the Tropic and Tununk shales in detail in outcrops at Big Water, Croton Canyon, and in the southern Henry Mountains region (South Henry site; Fig. 1). She used nine of the bentonite layers (Fig. 2) identified by Zelt and designated them TT1 through TT9 to reconstruct stratal geometries and evolving depositional topography during accumulation of the fine-grained facies. With the aid of biostratigraphic data compiled by Kauffman (Kauffman, pers. commun., 1994), these bentonite layers can also be identified in the Escalante core. The lowermost four bentonite layers in Leithold's studies (TT1 through TT4) accumulated during the early, transgressive part of the Greenhorn cycle and are equivalent to bentonites A, B, C, and D of Elder (1991). Elder traced these layers from southern Utah to Arizona, New Mexico, and Colorado.

Deposits of the early Greenhorn regression lie between bentonites TT5 and TT6. Bentonite TT5 lies at the base of the *Mammites nodosoides* biozone. In the Escalante core and in the three outcrops studied by Leithold (1994), TT5 ranges from 2 to 5 cm thick. The tens of centimeters of calcareous shale that overlie TT5 are characterized by several thin beds of inoceramid shells and shell debris and by one thin but prominent siltstone bed. Bentonite TT6 is the lowermost of a cluster of four regionally traceable bentonite beds that lie near the top of the *Collignoniceras woollgari* biozone (Zelt, 1985; Leithold, 1994; Fig. 2). Bentonite TT6 ranges from 2 to 6 centimeters thick in outcrop. Commonly it is both underlain and overlain by a few centimeters of bentonitic mudstone.

Large-scale vertical trends in carbonate content are apparent within the Tropic and Tununk shales. The trends are very similar in outcrop sections and in the Escalante core (Fig. 3). Leithold (1994) interpreted vertical changes in the carbonate content of these strata as reflections of changes over time in the relative supplies of terrigenous detritus and of microfaunal and microfloral remains. Hence, vertical changes in carbonate content at various sites can be interpreted as a record of changing proximity to the shoreline during relative sea-level changes. Leithold (1994) showed that the most carbonate-rich strata in

the Tropic Shale and Tununk Member, which likely record the maximum Greenhorn transgression, lie between bentonites TT4 and TT5 (Fig. 3). In the mudrocks between TT5 and TT6, which are the focus of this study, carbonate content decreases progressively upward (Fig. 3), reflecting the Greenhorn regression. Within this interval, Leithold (1994) used smaller-scale fluctuations in carbonate and silt and sand content to delineate two higher-frequency sea-level cycles.

In southern Utah, the interval between bentonites TT5 and TT6 thins dramatically offshore from the paleoshoreline, from 109 m at Big Water to 97 m in the Escalante core to 73 m at Croton Canyon and 56.5 m at South Henry (Fig. 4). The geometry of the strata between TT5 and TT6 thus suggests that sediment accumulation rates decreased offshore by a factor of about two. Leithold (1994) documented evidence for a forced regression (Posamentier et al., 1992) in the interval. The forcing event occurred just prior to the deposition of bentonite TT6, and led to the rapid seaward and downward stepping of the shoreline, and deposition of lower shoreface facies on top of prodeltaic mudstones at Croton Canyon and South Henry. Careful study of the outcrops and core indicates that between the time that bentonite TT5 accumulated and this sea-level fall, at least 109 m and 97 m of muddy, prodeltaic sediment accumulated at Big Water and Escalante, respectively, while at the same time 40 m of sediment accumulated at Croton Canyon and 13 m of sediment accumulated at South Henry. These trends indicate that prodeltaic sediments accumulated about eight times faster at Big Water and Escalante than at South Henry.

The accumulation rates at the study sites can be approximated by noting that the interval between bentonites TT5 and TT6 encompasses most of the *Collignoceras woollgari* ammonite biozone. Interpolation of radiometric ages published by Obradovich (1993) suggests that this biozone represents approximately 0.9 Ma (Kauffman et al., 1993). Assuming that the forced

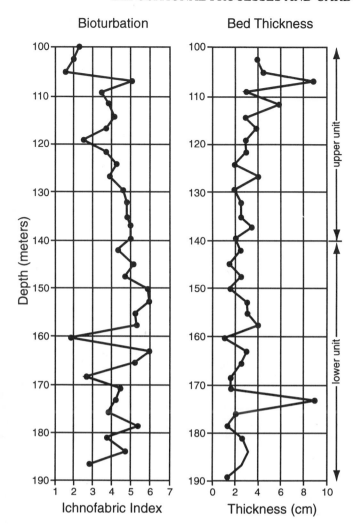

FIG. 6—Stratigraphic trends in the intensity of bioturbation and the thicknesses of the thickest discrete bed in each meter of the Escalante core.

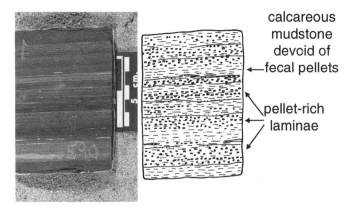

FIG. 7.—Photograph and sketch of a sample of laminated claystone and fecal pellets at 180 m (590 ft) depth in the Escalante core.

trast, sediment accumulation rates of steeply dipping foreset deposits are centimeters per year when averaged over 100-yr time scales and may be an order of magnitude higher on time scales of days to months (McKee et al., 1983; Kuehl et al., 1986a, 1986b; Nittrouer et al., 1986; Alexander et al., 1991). Leithold (1993) compared facies preserved in the clinoform deposits of the Tropic Shale and Tununk Shale Member to these modern deposits. She suggested that at Big Water a vertical succession of facies records bottomset, foreset and perhaps distal topset deposits, while only bottomset facies are preserved at South Henry. Paleogeographic reconstructions by Zelt (1985), based on studies of both outcrops and well logs, suggest that the facies at Big Water accumulated between a distance of about 100 and a few tens of kilometers from the western shoreline of the seaway, whereas those at South Henry accumulated at least 100 km from the shoreline. Study of the interval between TT5 and TT6 in the Escalante core, which accumulated in a similar position relative to the shoreline as the interval at Big Water, provides a detailed picture of the processes that operated during construction of the bottomsets and foresets.

SEDIMENTOLOGICAL INDICATORS OF PROCESSES
AND RATES OF SEDIMENT ACCUMULATION

Evidence for Sedimentary Processes on the Clinoforms

Detailed study of sedimentary facies between bentonites TT5 and TT6 in the Escalante core reveals distinctive vertical changes in sediment composition, texture, and sedimentary structures. In general these changes are gradual, but it is useful for purposes of description and interpretation to divide the study interval into two units.

Lower unit.—

Description: The lower part of the study interval, from about 210 to 140 m (689 to 459 ft) in the Escalante core, is dominantly composed of marlstone and calcareous claystone with scattered white fecal pellets, foraminiferal tests, and fine shell debris. The intensity of bioturbation in this part of the stratigraphic section varies, but in general can be characterized as moderate to intense (Fig. 6). Preserved laminae are commonly defined by concentrations of fecal pellets and, more rarely, foramineral tests (Figs. 7, 8). The laminae typically range from < 1 mm to about 1 cm thick. Many laminae are partially disrupted by burrows. The bulk of the pellet-rich laminae in the lower interval have

regression occupied one-fourth of this time, and assuming 50% postdepositional compaction, the prodeltaic sediments at Big Water and in the Escalante core accumulated at an average rate of about 0.89 mm/y (89 cm/ky), while those at South Henry accumulated at an average rate of 0.12 mm/y (12 cm/ky). Given the evidence for numerous small-scale fluctuations in sea-level during that period (Leithold, 1994), and the presense of numerous marine flooding surfaces that represent times of little or no sediment accumulation during high-frequency, small-scale sea-level rises, these estimates can be considered minimum figures.

The geometry of the fine-grained strata between TT5 and TT6 suggests a clinoform morphology similar to that observed on modern subaqueous deltas (Nittrouer et al., 1986; Kuehl et al., 1989; Alexander et al., 1991; Leithold, 1993). The geometry of the ancient deposits suggests an average slope of about 0.10 degrees between Big Water and South Henry. For comparison, foresets on the modern Amazon subaqueous delta slope at an average angle of 0.19 degree (Kuehl et al., 1986a). On modern subaqueous deltas, such as those offshore of the Amazon and Huanghe river mouths, gently sloping topset deposits accumulate at low to moderate rates on the order of 1 mm or less per year when measured over 100-yr time scales (Kuehl et al., 1986a, 1986b; Nittrouer et al., 1986; Alexander et al., 1991). In con-

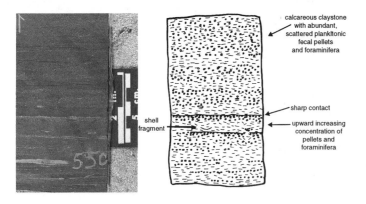

FIG. 8.—Photograph and sketch illustrating inversely graded beds in the lower part of the Escalante core study interval (168 m or 550 ft. in the core).

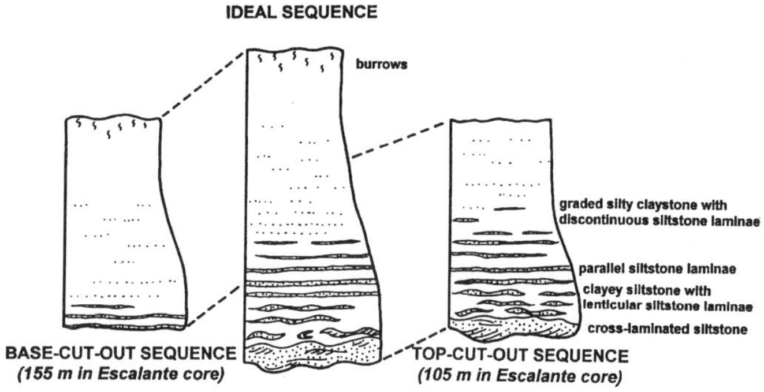

FIG. 9.—Ideal mud turbidite sequence and base cut-out and top cut-out variations in the Escalante Core (Stow et al., 1984).

diffuse, gradational lower and upper contacts with the surrounding marlstone and calcareous claystone and there is no obvious internal organization of pellets. Some of the laminae, however, show inverse grading, in which pellets become progressively more concentrated upwards, and in a few laminae become progressively more intermixed with foraminiferal tests (Fig. 8). These inversely graded laminae have gradational lower contacts and sharp upper contacts.

Thin, sharply based, calcareous siltstone and claystone beds that are devoid of pellets are interbedded with the pellet-rich marlstone and calcareous mudstone that dominates the lower unit. These calcareous siltstone and claystone beds typically are a few centimeters thick and have gradational tops. Some of the beds show normal grading in which the basal parts of the beds are composed of thinly interlaminated, coarse siltstone and claystone overlain by structureless calcareous claystone. Other beds are composed entirely of structureless, calcareous claystone with no obvious grading.

Interpretation: The pellet-rich marlstones and calcareous claystones that comprises the bulk of the lower unit record hemipelagic sedimentation on bottomsets and distal foresets where both terrigenous and biogenic sources of sediment were significant. Laminae rich in fecal pellets and foraminifera may record changes in the relative supplies of terrigenous and biogenic components, or changes in the intensities of physical processes at the seabed. Concentrations of fecal pellets and foraminifera in laminae with gradational bases and tops could represent, for example, times when rates of terrigenous sediment supply were diminished, perhaps during dry seasons or longer drought periods. These pellet concentrations could also represent times when biogenic sediment supplies increased, such as during seasonal plankton blooms, or times when fine, siliciclastic particles were winnowed from the seabed by bottom currents, leaving behind the larger biogenic aggregates.

Inversely graded, pellet-rich laminae in the lower unit are clearly the result of episodic reworking and winnowing of the seabed by bottom currents. The currents may have been associated with the deep thermohaline circulation of the seaway, and thus the deposits may be analogous to contourites described from the deep sea (e.g., Stow and Piper, 1984). The gradational bases and sharp tops of the laminae indicate that the currents episodically winnowed clay- and fine silt-sized grains from the bed and left behind pellets and, in some laminae, foraminiferal tests. Following these events the winnowed substrate was draped with

clay and silt deposited from suspension.

Sharply based claystone and siltstone beds in the lower unit are interpreted to be turbidites. Facies models for mud turbidites, analogous to the Bouma sequence for sandy turbidites, were discussed by Piper (1978), Stow and Shanmugam (1980), and Stow and Piper (1984) and variability in mud turbidites was described by Stow et al. (1984). The turbidites in the lower part of the study interval are missing the lower divisions of ideal mud turbidites (Fig. 9). Similar turbidites have been termed "base-cut-out" sequences by Stow et al. (1984) and are characteristic of distal settings, where relatively low-velocity, low-concentration turbidity currents deposit their loads. In the Tropic Shale, turbidites in the lower unit of the study interval probably accumulated on deltaic bottomsets. These beds probably represent deposition from the distal portions of turbidity currents initiated on the topsets or foresets.

Upper unit.—
Description: The upper part of the study interval, from a depth of about 140 to 104 m (459 to 341 ft) in the Escalante core, is primarily composed of silty, calcareous shale and mudstone. Carbonate content generally decreases upward within this unit (Fig. 3) and the intensity of bioturbation generally decreases from moderate to low levels (Fig. 6). The dominant sedimentary structures preserved in the unit are thin, normally graded beds with sharp bases. The graded beds average 1-2 cm thick but range up to 10 cm.

Some of the graded beds resemble those of the lower unit and are composed of thinly interlaminated siltstone and silty claystone that grades upward into claystone. More typically, however, the graded beds are coarser than similar beds in the lower unit. These graded beds have load casts at their bases, which indicates deposition on a soft, muddy seabed, and their basal parts are composed of ripple-cross-laminated, coarse siltstone (Fig. 10). Most commonly, the cross-lamination is unidirectional, which indicates deposition by migrating current ripples. In a few beds, however, bidirectional cross-laminae suggest the influence of waves. Both the current- and wave-rippled siltstones are overlain by alternating thin, lenticular to parallel laminae of siltstone and silty claystone.

The siltstone laminae become thinner and less distinct upward, and the laminated siltstone-silty claystone ultimately gives way to graded silty claystone and claystone. Some of the graded beds are bioturbated at their tops and are overlain by calcareous

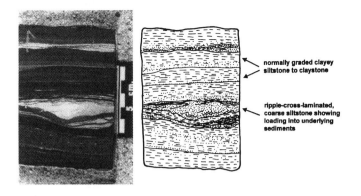

FIG. 10.—Photograph and sketch showing normally graded beds at 103 m (338 ft) in the Escalante core.

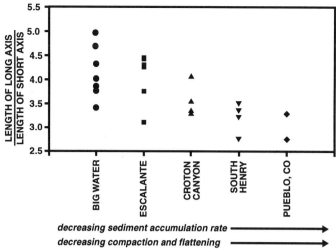

FIG. 11.—Change in pellet aspect along onshore-offshore transect from southern Utah to Colorado over a total distance of about 460 km.

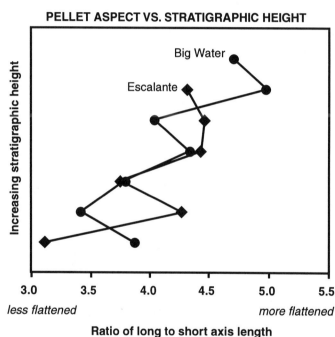

FIG. 12.—Upward change in pellet aspect in the Big Water section and Escalante core.

claystone with scattered fecal pellets. Most of the graded beds, however, are in direct, scoured contact with overlying graded beds. In general, the graded beds become thicker and coarser upward in the unit. The basal, ripple-cross-laminated parts of the beds, in particular, become thicker at the expense of the parallel-laminated siltstone and claystone. The graded claystone portions, in contrast, are thinner in graded beds higher in the unit.

Thin layers averaging 1-2 mm thick composed of shell fragments, fish scales, bones, and teeth are relatively common in the upper unit and become more common upward. Some of these layers are present at the bases of graded beds and others lie along laminae within the upper, generally fine-grained portions of the graded beds or within ungraded, calcareous mudstone.

Interpretation: The majority of the graded siltstone and claystone beds in the upper unit are interpreted as turbidites. In contrast to the turbidites in the lower unit, these beds commonly preserve ideal muddy turbidite sequences. A common variant, however, are beds that lack the fine-grained, upper part of the sequence and resemble the "top-cut-out" sequences of Stow et al. (1984) (Figs. 8, 9). The presence of more complete and coarser turbidites in the upper part of the study interval is consistent with its deposition in a more proximal setting such as within the foreset deposits of the prograding delta. In the upper unit, in contrast to the lower unit, non-turbiditic, hemipelagic sediment is relatively uncommon and evidence for biogenic reworking of turbidite deposits is rare. These observations, as well as abundant evidence for the loading of coarse siltstone beds in the basal parts of turbidites into the muddy sediment below, are consistent with relatively rapid sedimentation.

The presence of wave-rippled siltstone suggests that some graded beds in the upper unit were formed by storm reworking of the seabed rather than deposition from turbidity currents. The wave-rippled siltstone beds may represent extreme storms conditions when waves were large enough to mobilize sediment in many tens of meters water depth on the foresets. The turbidite beds in the succession also may owe their origins to storm processes. During storms, waves and currents may have suspended mud on the distal delta topsets, generating turbid layers that flowed down the sloping foresets under the influence of gravity. Similar modes of sediment dispersal were documented on the modern Huanghe delta by Wright et al. (1990) and on the Amazon subaqueous delta by Cacchione et al. (1995). In the Tropic Shale there is no evidence for channelization or slumping in

outcrop or in the Escalante core. On this ancient prodelta, as on the Huanghe and Amazon deltas, most of the foreset deposition probably resulted from widespread, "blanket-like" flows generated during wave events rather than from laterally concentrated or channelized flows.

Shell- and bone-rich layers in the upper unit probably have a variety of origins. The shell and bone concentrations at the bases of normally graded beds probably represent the coarsest particles that were entrained on the deltaic topsets and transported to the foresets in turbidity currents. These coarse particles would have been deposited rapidly as the turbidity currents decelerated. Shell and bone concentrations of this type would therefore have accumulated primarily on the most proximal foresets, which explains why they are increasingly found toward the top of the study interval. Where shell- and bone-rich layers are present within the graded beds or within ungraded, hemipelagic, calcareous mudstones, they may represent concen-

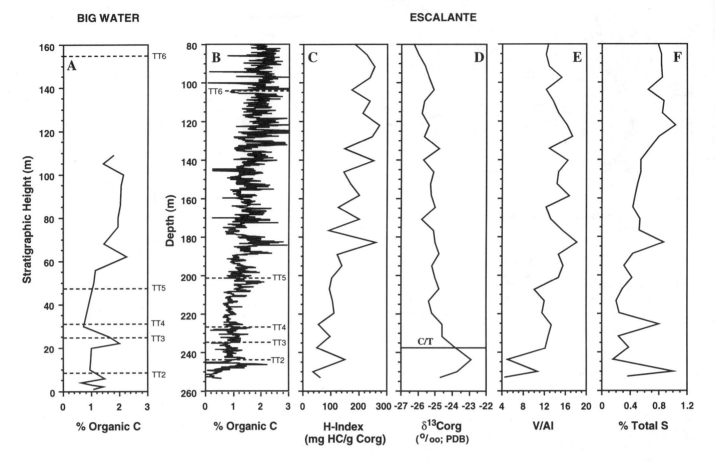

FIG. 13.—Vertical trends in (A) % OC in the Big Water section, and in (B) % OC, (C) H-Index, (D) δ^{13}C, (E) V/Al, and (F) % total S in the Escalante core. TT2-TT6 indicate the locations of the bentonites of Leithold (1994). C/T = Cenomanian-Turonian boundary.

trations of debris from benthic and nektonic macrofauna that were winnowed *in situ* by currents and waves on the foresets. Alternatively, some of the layers may represent episodic "die-offs" of benthic and nektonic macrofauna that occurred as the result of shifts in rates of sediment input and/or salinity. The increasing abundance of all of these types of shell and bone concentrations upward in the study interval attests to dynamic processes that operated on the upper foresets.

Evidence for Relative Rates of Sediment Accumulation

In the Escalante core, facies changes upward in the study interval reflect increasing proximity to the shoreline and increasing rates of terrigenous sediment input. Across southern Utah, moreover, the geometry of the interval between bentonites TT5 and TT6 indicates that sediment accumulation rates decreased from west to east. Data on pellet shapes, trends in the thicknesses of normally graded beds, and bioturbation provide additional evidence for these vertical and lateral changes.

Data on pellet shape from the Escalante core and from the outcrop sections reflect the general offshore decrease in sediment accumulation rates. The average ratio of long to short axis lengths of pellets generally decreases between Big Water and Pueblo, Colorado, ranging from about 5 to 2.8 (Fig. 11). In addition, vertical trends in pellet aspect ratios in the Big Water and Escalante samples (Fig. 12) show that pellets are more flattened upward in the clinoform deposits. In modern marine environ-

ments, calcareous fecal pellets may be partially to pervasively hardened by cementation or micritization soon after deposition (Bathhurst, 1971; Wanless et al., 1981). In the Tropic and Tununk shales, zooplankton fecal pellets were compacted in the upper part of the seabed before they were significantly hardened. The shapes of the pellets reflect the degree of compaction that occurred during this early stage of burial.

In modern clay-rich sediments, water content — and hence the cumulative amount of compaction — typically increases with sedimentation rate. Freshly deposited, clay-rich sediments typically have porosities of 70% to 90%, and they undergo compaction as they are buried (Berner, 1980). When clay-rich sediments accumulate rapidly they have less time to adjust to the accruing overburden, and they tend to have higher porosities and water contents than do sediments that accumulate more slowly. In the upper few meters of the seabed, porosity typically is reduced by 10 - 20% as aggregates of clay minerals collapse, pore volume is reduced, and water is expelled (Berner, 1980; Krone, 1984; Partheniades, 1984). With time and increased depth of burial, porosities of both rapidly and slowly deposited fine-grained sediments tend to converge (e.g., Einsele, 1992).

Pellet aspect ratios at the study sites reflect a seaward passage from rapidly accumulating foreset deposits in Utah to slowly accumulating bottomset deposits in Colorado. At the proximal sites, Big Water and Escalante, the increases in the aspect ratios upward in the sections also record the increase in sediment accumulation rates with time that accompanied the progradation of deltaic environments across the study area.

A general vertical increase in the thickness of graded beds in the Escalante core also reflects increasing sediment accumulation rates. This trend indicates that depositional events, either from settling from turbidity currents or storm resuspension, generally caused greater thicknesses of sediment to accumulate over the short term at more proximal localities on the foresets than on the more distal bottomsets. A notable exception to this general trend is a 9-cm-thick bed in the lower part of the study interval at about the 171 m (561.5 ft) in the core. This bed is composed of approximately 3 cm of lenticular and parallel laminated siltstone and mudstone overlain by about 6 cm of homogeneous claystone. The bed records an unusually large turbidity current, which transported a large amount of silt and clay to a bottomset environment in which sediment accumulated relatively slowly.

In the Escalante core, the intensity of bioturbation decreases upward in the study interval. Similar trends of upward decrease in intensity of bioturbation are found in outcrops of the Tropic and Tununk shales and are attributed an upward increase in sediment accumulation rate (Leithold, 1993; 1994). Bioturbated facies are overlain by laminated facies at Big Water, coeval deposits at South Henry were mostly bioturbated. Similar to modern subaqueous deltas (Nittrouer et al., 1986; Kuehl et al., 1986a, 1986b; Alexander et al., 1991), the thoroughly homogenized sedimentary deposits in the lower part of the study interval in the Escalante core and at Big Water and throughout the interval at South Henry accumulated slowly in bottomset and distal foreset environments. In these environments benthic infauna typically reworked sediments much more quickly than they accumulated. In contrast, the physically stratified sediments in the upper part of the interval in the Escalante core and at Big Water accumulated in more proximal foresets, where rates of sediment accumulation were at least episodically high enough to preclude biological reworking.

PATTERNS OF CARBON CONTENT AND TYPE

The interval between bentonites TT5 and TT6 in the Escalante core and outcrop at Big Water (Nix-Morris, 1996) is characterized by an upward-increasing organic carbon (OC) content (Fig. 13). Values of the hydrogen index (HI) also increase upward in the interval (Fig. 13), which suggests that a steadily higher proportion of lipid-rich, algal marine organic matter is preserved in the rocks in this interval. An upward increase in the relative proportion of marine organic matter in this interval is further indicated by a decrease in values of $\delta^{13}C$ of the organic matter (Fig. 13). Cretaceous marine organic carbon typically has $\delta^{13}C$ values of -27‰ to -29‰, while Cretaceous terrestrial organic matter (for example Cretaceous coals) typically has $\delta^{13}C$ values of -25‰ to -26‰ (Arthur et al., 1985a; Dean et al., 1986). The exception is marine organic matter deposited at the Cenomanian-Turonian oceanic anoxic event (OAE), which is up to 5‰ lower than that of older or younger Cretaceous marine organic matter (Schlanger et al., 1987; Arthur et al., 1988). In Cretaceous OC-rich sequences, other than Cenomanian-Turonian sequences, HI and $\delta^{13}C$ often are negatively correlated (as in Fig. 13) indicating that greater hydrogen richness of organic matter is the result of a greater contribution of marine organic matter (lower values of $\delta^{13}C$) rather than enhanced preservation of organic matter (Arthur and Dean, 1991).

With this background, we would interpret the profiles in Figure 13 as follows: The organic matter initially deposited in

the Tropic Shale as the Cretaceous shoreline transgressed over the shoreline sands of the Dakota Group was almost entirely terrestrial, with OC values <0.5%, HI values <50, and $\delta^{13}C$ values of about -25‰. The pulse of marine organic-matter production and preservation coincident with the global OAE beginning just below the Cenomanian-Turonian boundary is marked in the Escalante core by an increase in % OC, HI, and $\delta^{13}C$ (Fig. 13). For comparison, in the Bounds core in western Kansas, values of HI increase from 400 to 700 and values of $\delta^{13}C$ increase from -27‰ to -23‰ just below the Cenomanian-Turonian boundary (Dean et al., 1995). At Pueblo, in south-central Colorado, values of $\delta^{13}C$ increase from -28‰ to -24‰ just below the Cenomanian-Turonian boundary (Arthur et al., 1988; Pratt et al., 1993). The lower (more negative) values of $\delta^{13}C$ in Kansas and eastern Colorado before the Cenomanian-Turonian OAE reflect the more open marine conditions of these sites. After the Cenomanian-Turonian OAE, increasing transgression of the Greenhorn Sea resulted in more marine organic matter being produced and preserved at the site of the Escalante core as evidenced by higher values of % OC and HI, and lower, more negative values of $\delta^{13}C$ (Fig. 13). At most locations throughout the world, the Cenomanian-Turonian boundary usually is marked by a peak in OC as well as a positive excursion in the isotopic composition of inorganic and organic carbon. In the Escalante core, however, the peak in the amount (% OC) and marine character (higher values of HI) of the organic matter occur higher in the section. Therefore, the Tropic Shale in the Escalante core apparently recorded both global (OAE) and local events in the Western Interior Seaway.

There are several possible explanations for upward-increasing OC and HI values and decreasing values of $\delta^{13}C$ in the study interval. One possibility is that as the deltaic system prograded, primary productivity in the water column increased at Escalante and Big Water. In modern oceans, production of organic matter increases markedly between offshore ocean and coastal waters due to increased turbulent mixing and greater supplies of nutrients from rivers (Berger et al., 1989; Tyson, 1995). It is possible, therefore, that the study interval records an increase over time in productivity and hence in carbon flux to the seabed. Paleogeographic reconstructions (Zelt, 1985), however, suggest that the most proximal foreset deposits in the study interval accumulated within a few tens of kilometers of the shoreline where turbidity may have increasingly limited primary productivity. A decreasing abundance of fecal pellets upward in the interval (Nix-Morris, 1996) supports this idea, as does the sedimentology of the interval. Normally graded beds in the deposits suggest that frequent resuspension of fine-grained sediment on the topsets and proximal foresets led to frequent sediment gravity flow events, and imply high turbidity over the foresets. The hypothesis that such turbidity limited primary productivity is further supported by the carbonate content of the deposits in the study interval (Fig. 3). In the Escalante core, carbonate content in the bottomset and foreset facies between bentonites TT5 and TT6 decreases upward from a maximum of about 45% to a minimum of about 5%. If the productivity of calcareous phytoplankton was similar in the water column over both the foresets and the bottomsets, and assuming that preservation of biogenic carbonate remained constant throughout the deposition of the interval, an approximately sixteen-fold increase in the rate of terrigenous sediment input would be required to produce this change in carbonate content. The geometry of the clinoform deposits in the Tropic and Tununk shales in southern Utah, however, indicates only an eight-fold increase in sediment accumulation rates between the bottomsets and topsets. This result suggests that biogenic carbonate flux and perhaps primary productivity decreased during progradation.

Changes in oxygen levels in the Western Interior Seaway and, therefore, in preservation of organic matter, also possibly explain the observed trends in OC and HI values in the study interval. Glenister and Kauffman (1985) suggested that the seaway became progressively more restricted during the Greenhorn regression, and, consequently, oxygen levels in the water column could have fallen. For many years geologists have hypothesized a strong link between organic matter preservation and oxygen depletion (e.g., Demaison and Moore, 1980), but this relationship is the subject of much debate.

Several recent laboratory and field studies concluded that oxygen levels have little or no effect on degradation rates of organic matter (e.g., Kristensen and Blackburn, 1987; Cowie and Hedges, 1991, 1992; Lee, 1992). Several studies of modern marine sedimentary environments show, moreover, that there is little correlation between water column oxygen levels and organic carbon concentrations or burial efficiencies (e.g., Henrichs and Reeburgh, 1987; Betts and Holland, 1991; Pedersen and Calvert, 1990, 1991; Calvert et al., 1992).

Other workers maintain, however, that oxygen level must exert some control over organic matter degradation, perhaps by eliminating burrowing organisms (Lee, 1992; Hedges and Kiel, 1995; Savrda, this volume). The decrease in degree of bioturbation to low levels upward throughout the interval between TT5 and TT6 (Fig. 6) was interpreted by Leithold (1993) to reflect increasing rates of sediment accumulation, but it is also consistent with decreasing dissolved oxygen concentrations in bottom waters.

Oxygen-deficient bottom waters and increased sulfate reduction is further suggested by increasing concentrations of metals (see the V/Al ratio in Fig. 13) and sulfur upward in the study interval (Fig. 13). In the Bounds core in western Kansas, highest concentrations of metals and sulfur coincide with the Cenomanian-Turonian OAE (Dean et al., 1995), while in a core from south-central Colorado near Pueblo, as in the Escalante core, the highest concentrations occur well above the Cenomanian-Turonian boundary (Arthur et al., 1985b). This is further evidence that local circulation conditions as well as global events (OAE) may have controlled oxygen depletion in bottom waters of the Western Interior Seaway.

The effect of changes in oxygen abundance on preservation of marine carbon in the study interval thus remains unclear. What is clear from the data presented here, however, is that sediment accumulation rates increased throughout deposition of the interval. Organic carbon content is clearly related to sediment accumulation rates of both modern and ancient marine deposits (Heath et al., 1977; Müller and Suess, 1979; Ibach, 1982; Henrichs and Reeburgh, 1987; Betts and Holland, 1991; Hedges and Keil, 1995). In oxic marine environments, for example, organic carbon contents of sediments are positively correlated with sediment accumulation rates that range between about 10 and 100 cm/ky (Tyson, 1995). Where sediment accumulation rates exceed 100 cm/ky, however, dilution with inorganic mineral grains generally leads to a negative correlation between sediment accumulation rate and organic carbon content (Tyson, 1995). The average sediment accumulation rate of 89 cm/ky calculated for the study interval lies within the range over which a positive correlation is observed between accumulation rates and carbon

content in modern and ancient marine deposits.

Several mechanisms have been proposed to explain the relationship of organic content to sediment accumulation rates. A number of workers suggested that the efficiency of organic carbon preservation in marine environments increases with sediment accumulation rate (e.g., Henrichs and Reeburgh, 1987; Betts and Holland, 1991; Hedges and Keil, 1995). At high sediment accumulation rates, organic matter moves rapidly down through the upper zone of active diagenesis in the seabed, where bacterial and animal activities are highest, and where the supply of potent oxidants is greatest (Hedges and Keil, 1995). Other explanations for the positive correlation between organic carbon content and sediment accumulation include the interrelationship of carbon flux and biogenic sediment flux in deep sea sediments and the presence in rapidly deposited coastal sediments of abundant refractory terrigenous and reworked marine carbon (Hedges and Keil, 1995; Tyson, 1995). In the Tropic Shale, however, a general upward increase in hydrogen indices and decrease in values of $\delta^{13}C$ indicate that the burial rate of marine carbon during the progradation of the deltaic clinoforms increased more rapidly than the burial rate of terrigenous carbon.

DISCUSSION AND CONCLUSIONS

Marine environments near the western margin of the Cretaceous Western Interior Seaway were characterized by high rates of fine-grained sediment influx from the Sevier orogenic belt. Study of the Tropic and Tunupk shales in southern Utah and other fine-grained Cretaceous successions in the Western Interior reveals that during highstands, a large proportion of these sediments accumulated in subaqueous deltaic clinoforms that prograded eastward with time (Asquith, 1970; Winn et al., 1987; Leithold, 1993). The Tropic and Tunupk shales provide a record of sediment dispersal and accumulation on the clinoforms.

In the Escalante core, sedimentary structures in the Tropic Shale point to two dominant modes of fine-grained sediment dispersal from the shoreline to the basin. Bottomset deposits in the study interval accumulated primarily by suspension fallout, perhaps from river plumes. The abundance of fecal pellets in the bottomset deposits of indicates, however, that such plumes would have been dilute over the bottomsets. The pellets indicate that suspended sediment concentrations in surface waters were low enough to permit primary productivity and, hence, grazing by zooplankton.

Paleogeographic reconstructions by Zelt (1985) suggest that the bottomset deposits accumulated on the order of 100 km from the western shoreline of the seaway. On modern continental shelves, river plume waters may extend for tens to hundreds of kilometers from the shoreline (e.g., Barnes et al., 1972; Curtin and Legeckis, 1986; Jilan and Kangshan, 1989; Ortner et al., 1995). On these shelves suspended sediment concentrations in river plumes typically fall off rapidly with distance from river mouths due to aggregation of particles and rapid settling (e.g., Gibbs and Konwar, 1986). As a consequence, turbidity levels in surface waters typically decreases to levels required for primary production close to the shoreline. Suspended sediment concentrations in the plume of the Amazon, for example, become low enough to permit primary productivity within about 100 km of the river mouth, although this varies with discharge (DeMaster et al., 1986).

Turbidity currents were evidently a second important mode of fine-grained sediment dispersal along the western margin of the seaway (Winn et al., 1987). The absence of evidence for slumping or channeling in the Tropic and Tun150nk shales suggests that these might have been blanket-like flows similar to those documented by Wright et al. (1990) on the modern Huanghe delta and by Cacchione et al. (1995) on the Amazon subaqueous delta.

The Huanghe delta provides a particularly useful analogue to the Tropic and Tun150nk shales in that it prograded into a shallow epicontinental sea, the Gulf of Bohai. In this setting sediments rapidly deposited in front of the river mouth are frequently resuspended by tidal currents and storm waves and currents. Wright et al. (1990) showed that sediment suspensions on the Huanghe topsets evolve into turbidity currents that transport most of the sediment to the Huanghe delta foresets and topsets. The Western Interior Seaway is thought to have had a seasonal climate, and sediment transport is thought to have been strongly dominated by storm processes (Barron and Washington, 1982; Duke, 1985; Swift et al., 1987; Barron, 1989; Ericksen and Slingerland, 1990). Most of the fine-grained sediments emanating from river mouths during seasonal floods were likely initially deposited within a few tens of kilometers of the shoreline. On the ancient Tropic-Tun150nk deltaic topsets, as on the modern Huanghe delta, resuspension during subsequent storms may have transferred these sediments farther seaward in turbidity currents.

Turbidity currents were evidently particularly important to the accumulation of foreset deposits in the Tropic and Tun150nk shales. In the foreset environment, which according to paleogeographic reconstruction by Zelt (1985) was situated a least a few tens of kilometers from the shoreline, suspended sediment concentrations were probably high near the seabed. In the water column, however, suspended sediment concentrations were low enough to permit primary biogenic production. The preservation of marine carbon deposited on the foresets was enhanced by rapid burial and perhaps by oxygen depletion of bottom waters. Study of the Tropic and Tun150nk shales suggests that along the western margin of the seaway, proximal foreset facies may be important repositories of marine organic carbon.

ACKNOWLEDGMENTS

Work on the Tropic and Tun150nk shale by Leithold was supported by National Science Foundation grant EAR-9219840 and by the donors to the Petroleum Research Fund administered by the American Chemical Society. We are grateful to E. Kauffman for his collaboration on describing the Escalante core. We thank E. Merewether and M. Kirschbaum for helpful reviews of the manuscript.

REFERENCES

ALEXANDER, C. R., DEMASTER, D. J., AND NITTROUER, C. A., 1991, Sediment accumulation in modern epicontinental shelf setting: The Yellow sea: Marine Geology, v. 98, p. 51-72.

ARTHUR, M. A., AND DEAN , W. E., 1991, A holistic geochemical approach to cyclomania — Examples from Cretaceous pelagic limestone sequences, in Einsele, G., Ricken W., and Seilacher, A., eds., Cycles and Events in Stratigraphy: Berlin, Springer-Verlag, p. 126-166.

ARTHUR, M. A., DEAN, W. E., AND CLAYPOOL, G. E., 1985a, Anomalous ^{13}C enrichment in modern organic carbon: Nature, v. 315, p. 216-218.

ARTHUR, M. A., DEAN, W. E., AND PRATT, L. M., 1988, Geochemical and climatic effects of increased organic carbon burial at the Cenomanian/Turonian boundary: Nature, v. 335, p. 714-717.

ASQUITH, D. O., 1970, Depositional topography and major marine environments, late Cretaceous, Wyoming: American Association of Petroleum Geologists Bulletin, v. 54, p. 1184-1224.

BARNES, C. A., DUXBURY, A. C., AND MORSE, B.-A., 1972, Circulation and selected properties of the Columbia River effluent at sea, in Pruter, A. T., and Alverson, D. L., eds., The Columbia River Estuary and Adjacent Ocean Waters: Seattle, University of Washington Press, 868 p.

BARRON, E. J., 1989, Severe storms during Earth history: Geological Society of America Bulletin, v. 101, p. 601-612.

BARRON, E. J., AND WASHINGTON, W. M., 1982, Cretaceous climate: A comparison of atmospheric simulations with the geologic record: Palaeogeography, Palaeoclimatology, and Palaeoecology, v. 59, p. 3-29.

BATHHURST, R. G. C., 1971, Carbonate Sediments and Their Diagenesis: New York, Elsevier, 620 p.

BERGER, W. H., SMATACEK, V. S., AND WEFER, G., 1989, Oceanic productivity and paleoproductivity— an overview, in Berger, W. H., Smatacek, V. S., and Wefer, G., eds., Productivity of the Ocean: Present and Past: Chichester, Wiley, p. 1-34.

BERNER, R.A., 1980, Early Diagenesis: Princeton, N.J., Princeton University Press, 241 pp.

BETTS, J. N., AND HOLLAND, H. D., 1991, The oxygen content of ocean bottom waters, the burial efficiency of organic carbon, and the regulation of atmospheric oxygen: Palaeogeography, Palaeoclimatology, and Palaeoecology, v. 97, p. 5-18.

CACCHIONE, D. A., DRAKE, D. E., KAYEN, R. W., STERNBERG, R. W., KINEKE, G. C., AND TATE, G. B., 1995, Measurements in the bottom boundary layer on the Amazon subaqueous delta, in Nittrouer, C. A., and Kuehl, S. A., eds., Geological significance of sediment transport and accumulation on the Amazon continental shelf: Marine Geology, v. 125, p. 235-257.

CALVERT, S. E., BUSTIN, R. M., AND PEDERSEN, T. F., 1992, Lack of evidence for enhanced preservation of sedimentary organic matter in the oxygen minimum of the Gulf of California: Geology, v. 20, p. 757-760.

COWIE, G. L., AND HEDGES, J. I., 1991, Organic carbon and nitrogen geochemistry of Black Sea surface sediments from stations spanning the oxic: anoxic boundary, in Izdar, E., and Murray, J. W., eds., Black Sea Oceanography: Boston, Kluwer, p. 343-359.

COWIE, G. L., AND HEDGES, J. I., 1992, The role of anoxia in organic carbon preservation in coastal sediments: Relative stabilities of the major biochemicals under oxic and anoxic depositional conditions: Organic Geochemistry, v. 19, p. 229-234.

CURTIN, T. B., AND LEGECKIS, R. V., 1986, Physical observatons in the plume region of the Amazon River during peak discharge—I. Surface variability: Continental Shelf Research, v. 6, p. 31-51.

DEAN, W. E., ARTHUR, M. A., AND CLAYPOOL, G. E., 1986, Depletion of ^{13}C in Cretaceous marine organic matter: Source, diagenetic, or environmental signal?: Marine Geology, v. 70, p. 119-157.

DEAN, W. E., ARTHUR, M. A., SAGEMAN, B. B., AND LEWAN, M. D., 1995, Core descriptions and preliminary geochemical data for the Amoco Production Company Rebecca K. Bounds # 1 well, Greeley County, Kansas: United States Geological Survey Open-File Report 95-209, 243 p.

DEMAISON, G. J., AND MOORE, G. T., 1980, Anoxic environments and oil source bed genesis: American Association of Petroleum Geologists Bulletin, v. 64, p. 1179-1209.

DEMASTER, D. J., KUEHL, S. A., AND NITTROUER, C. A., 1986, Effect of suspended sediments on geochemical processes near the mouth of the Amazon River: examination of biological silica uptake and the fate of particle-reactive elements: Continental Shelf Research, v. 6, p. 107-125.

DROSER, M. L., AND BOTTJER, D. J., 1990, Trace fossils and ichnofabric in Ocean Drilling Program Leg 119 cores, in Barron, J., et al., Proceedings of the Ocean Drilling Program, Scientific Results, v. 119: Houston, Ocean Drilling Program, p. 635-641.

DUKE, W .L., 1985, Hummocky stratification, tropical hurricanes, and intense winter storms: Sedimentology, v. 32, p. 167-194.

EINSELE, G., 1992, Sedimentary Basins: Evolution, Facies, and Sediment Budget: Berlin, Springer-Verlag, 628 p.

ELDER, W. P., 1991, Molluscan paleoecology and sedimentation patterns of the Cenomanian-Turonian extinction interval in the southern Colorado Plateau region, in Nations, J. D., and Eaton, J. G., Stratigraphy, Depositional Environments, and Sedimentary Tectonics of the Western Margin, Cretaceous Western Interior Seaway: Boulder, Geological Society of America Special Paper 260, p. 113-137.

ENGLEMAN, E. E., JACKSON, L. L., NORTON, D. R., AND A. G. FISCHER, 1985, Determinations of carbonate carbon in geological materials by coulometric titration: Chemical Geology, v. 53, p. 125-128.

ERICKSEN, M. C. AND SLINGERLAND, R., 1990, Numerical simulations of tidal and wind-driven circulation in the Cretaceous Interior Seaway of North America: Geological Society of America Bulletin, v. 102, p. 1499-1516.

GIBBS, R. J., AND KONWAR, L., 1986, Coagulation and settling of Amazon River suspended sediment: Continental Shelf Research, v. 6, p. 127-149.

GLENISTER, L. M., AND KAUFFMAN, E. G., 1985, High resolution stratigraphy and depositional history of the Greenhorn regressive hemicyclothem, Rock Canyon Anticline, Pueblo, Colorado, in Pratt, L. M., Kauffman, E. G., and Zelt, F. B., eds., Fine-Grained Deposits and Biofacies of the Cretaceous Western Interior Seaway: Evidence of Cyclic Sedimentary Processes, Field Trip Guide Book 4: Tulsa, Society of Economic Paleontologists and Mineralogists, Tulsa, p. 170-183.

HATTIN, D. E., 1975, Petrology and origin of fecal pellets in Upper Cretaceous strata of Kansas and Saskatchewan: Journal of Sedimentary Petrology, v. 45, p. 686-696.

HEATH, G. R., MOORE, T. C., JR., AND DAUPHIN, J. P., 1977, Organic carbon in deep-sea sediments, in Andersen, N. R., and Malahoff, A., eds., The Fate of Fossil Fuel CO_2 in the Ocean: New York, Plenum, p. 605-625.

HEDGES, J. I., AND KEIL, R. G., 1995, Sedimentary organic matter preservation: An assessment and speculative synthesis: Marine Chemistry, v. 49, p. 81-115.

HENRICHS, S. M. AND REEBURGH, W. S., 1987, Anaerobic mineralization of marine sediment organic matter: Rates and the role of anaerobic processes in the oceanic carbon economy: Journal of Geomicrobiology, v. 5, p. 191-237.

IBACH, L. E. J., 1982, Relationship between sedimentation rate and total organic carbon content in ancient marine sediments: American Association of Petroleum Geologists Bulletin, v. 66, p. 170-183.

JILAN, S., AND KANGSHAN, W., 1989, Chanjiang river plume and suspended sediment transport in Hanzhou Bay: Continental Shelf Research, v. 9, p. 93-111.

KAUFFMAN, E. G., 1977, Upper Cretaceous cyclothems, biotas, and environments, Rock Canyon Anticline, Pueblo, Colorado, in Kauffman, E.G., ed., Cretacous Facies, Faunas, and Paleoenvironments Across the Western Interior Basin: Mountain Geologist, v. 14, p. 129-152.

KAUFFMAN, E. G., 1985, Cretaceous evolution of the Western Interior Basin of the United States, in Pratt, L. M., Kauffman, E. G., and Zelt, F. B., eds., Fine-Grained Deposits and Biofacies of the Cretaceous Western Interior Seaway: Evidence of Cyclic Sedimentary Processes, Field Trip Guide Book 4: Tulsa, Society of Economic Paleontologists and Mineralogists, p. IV-XIII.

KAUFFMAN, E. G., AND CALDWELL, W. G. E., 1993, The Western Interior basin in space and time, in Caldwell, W. G. E., and Kauffman, E. G., Evolution of the Western Interior Basin: St. John's, Geological Association of Canada Special Paper 39, p. 1-30.

KAUFFMAN, E. G., SAGEMAN, B. B., KIRKLAND, J. I., ELDER, W. P., HARRIES, P. J., AND VILLAMIL, T., 1993, Molluscan biostratigraphy of the Cretaceous Western Interior Basin, North America, in Caldwell, W. G. E., and Kauffman, E. G., Evolution of the Western Interior Basin: St. John's, Geological Association of Canada Special Paper 39, p. 397-434.

KRISTENSEN, E., AND BLACKBURN, T. H., 1987, The fate of organic carbon and nitrogen in experimental marine sediment systems: influence of bioturbation and anoxia: Journal of Marine Research, v. 45, p. 231-257.

KRONE, R. B., 1984, The significance of aggregate properties to transport processes: in Mehta, A. J., ed., Estuarine Cohesive Sediment Dynamics, Lecture Notes on Coastal and Estuarine Studies 14: Berlin, Springer-Verlag, p. 66-84.

KUEHL, S. A., DEMASTER, D. J., AND NITTROUER, C. A., 1986a, Nature of sediment accumulation on the Amazon continental shelf: Continental Shelf Research, v. 6, p. 209-225.

KUEHL, S. A., NITTROUER, C. A., AND DEMASTER, D. J., 1986b, Distribution of sedimentary structures in the Amazon subaqueous delta: Continental Shelf Research, v. 6, p. 311-336.

KUEHL, S. A., HARIU, T. M., AND MOORE, W. S., 1989, Shelf sedimentation off the Ganges-Bramaputra river system: Evidence for sediment bypassing to the Bengal fan: Geology, v. 17, p. 1132-1135.

LEE, C., 1992, Controls on organic carbon preservation: The use of stratified water bodies to compare intrinsic rates of decomposition in oxic and anoxic systems: Geochimica Cosmochimica Acta, v. 56, p. 3323-3335.

LEITHOLD, E. L., 1993, Preservation of laminated shale in ancient clinoforms; comparison to modern subaqueous deltas: Geology, v. 21, p. 359-362.

LEITHOLD, E. L., 1994, Stratigraphical architecture at the muddy margin of the Cretaceous Western Interior Seaway, southern Utah: Sedimentology, v. 41, p. 521-542.

MCKEE, B. A., NITTROUER, C. A., AND DEMASTER, D. J., 1993, Concepts of sediment deposition and accumulation applied to the continental shelf near the mouth of the Yangtze river: Geology, v. 11, p. 631-633.

MOORE, D. G., AND SCRUTON, P. C., 1957, Minor internal structures of some recent unconsolidated sediments: Bulletin of the American Association of Petroleum Geologists, v. 41, p. 2723-2751.

MÜLLER, P. J., AND SUESS, R., 1979, Productivity, sedimentation rate, and sedimentary organic matter in the oceans. 1., Organic carbon preservation: Deep-Sea Research, v. A26, p. 1347-1362.

NITTROUER, C. A., AND STERNBERG, R. W., 1981, The formation of sedimentary strata in an allochthonous shelf environment: The Washington continental shelf: Marine Geology, v. 42, p.201-232.

NITTROUER, C. A., DEMASTER, D. J., AND MCKEE, B.A., 1984, Fine-scale stratigraphy in proximal and distal deposits of sediment dispersal systems in the East China Sea: Marine Geology, v. 61, p. 13-24.

NITTROUER, C. A., KUEHL, S. A., DEMASTER, D. J., AND KOWSMANN, R. O., 1986, The deltaic nature of Amazon shelf sedimentation: Geological Society of America Bulletin, v. 97, p. 444-458.

NIX-MORRIS, W., 1996, Biodeposition and organic carbon burial within an ancient prodeltaic environment of the Turonian Greenhorn Sea: Unpublished M.S. Thesis, North Carolina State University, Raleigh, 39 p.

OBRADOVICH, J. D., 1993, A Cretaceous time scale: in Caldwell, W. G. E., and Kauffman, E. G., Evolution of the Western Interior Basin: St. John's, Geological Association of Canada Special Paper 39, p. 379-396.

ORTNER, P. B., LEE, T. N., AND JOHNSON, W. R., 1995, Mississippi River flood waters that reached the Gulf Stream: Journal of Geophysical Research, v. 100, p. 13595-13601.

PARTHENIADES, E., 1984, A fundamental framework for cohesive sediment dynamics, in Mehta, A. J., ed., Estuarine Cohesive Sediment Dynamics, Lecture Notes on Coastal and Estuarine Studies 14: Berlin, Springer-Verlag, p. 219-250.

PEDERSEN, T. F., AND CALVERT, S. E., 1990, Anoxia vs. productivity: What controls the formation of organic-carbon-rich sediments and sedimentary rocks?: American Association of Petroleum Geologists Bulletin, v. 74, p. 454-466.

PEDERSEN, T. F., AND CALVERT, S. E., 1991, Anoxia vs. productivity: What controls the formation of organic-carbon-rich sediments and sedimentary rocks?: Reply: American Association of Petroleum Geologists Bulletin, v. 75, p. 500-501.

PIPER, D. J. W., 1978, Turbidite muds and silts on deep-sea fans and abyssal plains, in Stanley, D. J., and Kelling, G., eds., Sedimentation in Submarine Canyons, Fans, and Trenches: Stroudsburg, Penn., Dowden, Hutchinson & Ross, p. 163-176.

POSAMENTIER, H. W., ALLEN, G. P., JAMES, D. P., AND TESSON, M., 1992, Forced regressions in a sequence stratigraphic framework: Concepts, examples, and exploration significance: American Association of Petroleum Geologists Bulletin, v. 76, p. 1687-1709.

PRATT, L. M., ARTHUR, M. A., DEAN, W. E., AND SCHOLLE, P. A., 1993, Paleo-oceanographic cycles and events during the Late Cretaceous in the Western Interior Seaway of North America, in Caldwell, W. G. E., and Kauffman, E. G., eds., Evolution of the Western Interior Basin: St. John's, Geological Association of Canada Special Paper 39, p. 333-354.

RHOADS, D. C., BOESCH, D. F., ZHICAN, T., FENGSHAN, X., LIQIANG, H., AND NILSEN, K. J., 1985, Macrobenthos and sedimentary facies on the Changjiang delta platform and adjacent continental shelf, East China Sea: Continental Shelf Research, v. 4, p. 189-213.

SCHLANGER, S. O., ARTHUR, M. A., JENKYNS, H. C., AND SCHOLLE, P. A., 1987, The Cenomanian-Turonian oceanic anoxic event, I. Stratigraphy and distribution of organic carbon-rich beds and the marine ^{13}C excursion, in Brooks, J., and Fleet, A., eds., Marine Petroleum Source Rocks: London, Geological Society of London, Special Publication, p. 371-399.

STOW, D. A. V., AND PIPER, D. A. V., 1984, Deep water fine-grained sediments: facies models, in Stow, D. A. V., and Piper, D. J. W., eds., Fine-Grained Sediments and Deep-Water Processes and Facies: Oxford, Blackwell Scientific Publications, p. 611-646.

STOW, D. A. V., AND SHANMUGAM, G., 1980, Sequence of structures in fine-grained turbidites: comparison of recent deep-sea and ancient flysch sediments: Sedimentary Geology, v. 25, p. 23-42.

STOW, D. A. V., ALAM, M., AND PIPER, D. J. W., 1984, Sedimentology of the Halifax Formation, Nova Scotia: Lower Paleozoic fine-grained turbidites, in Stow, D. A. V., and Piper, D J. W., eds., Fine-Grained Sediments and Deep-Water Processes and Facies: Oxford, Blackwell Scientific Publications, p. 127-144.

SWIFT, D. J. P., HUDELSON, P. L., BRENNER, R. L., AND THOMPSON, P., 1987, Shelf construction in a foreland basin: Storm beds, shelf sand bodies, and shelf-slope depositional sequences in the Upper Cretaceous Mesa Verde Group, Book Cliffs, Utah: Sedimentology, v. 34, p. 423-457.

TYSON, R. V., 1995, Sedimentary Organic Matter: London, Chapman & Hall, 615 p.

WANLESS, H. R., BURTON, E. A., AND DRAVIS, J., 1981, Hydrodynamics of carbonate fecal pellets: Journal of Sedimentary Petrology, v. 51, p. 27-36.

WINN, R. D., JR., BISHOP, M. G., AND GARDNER, P. S., 1987, Shallow-water and sub-storm-base deposition of Lewis Shale in Cretaceous Western Interior Seaway, south-central Wyoming: American Association of Petroleum Geologists Bulletin, v. 71, p. 859-881.

WRIGHT, L. D., WISEMAN, W. J., YANG, Z.-S., BORNHOLD, B. D., KELLER, G. H., PRIOR, D. B., AND SUHAYDA, J. N., 1990, Processes of marine dispersal and deposition of suspended silts off the modern mouth of the Huanghe (Yellow River): Continental Shelf Research, v. 10, p. 1-40.

ZELT, F. B., 1985, Natural gamma-ray spectrometry, lithofacies, and depositional environments of selected Upper Cretaceous marine mudrocks, western United States including Tropic Shale and Tununk Member of Mancos Shale, Unpublished Ph.D. Dissertation, Princeton University, Princeton, N.J., 334 p.

STABLE ISOTOPIC STUDIES OF CENOMANIAN-TURONIAN PROXIMAL MARINE FAUNA FROM THE U.S. WESTERN INTERIOR SEAWAY

MARK PAGANI AND MICHAEL A. ARTHUR

Department of Geosciences, Pennsylvania State University, University Park, Pennsylvania 16802

ABSTRACT: Macrofossils were sampled from 650 ft (198.1 m) of core material from the USGS Escalante No. 1 core, consisting mainly of proximal offshore marine calcareous mudstones of the Cretaceous Western Interior Seaway of North America. Samples range in age from Upper Cenomanian (*S. gracile* ammonite zone) through Upper Middle Turonian (*P. hyatti* ammonite zone) and include representatives of ammonites, inoceramid bivalves, and oysters. Diagenetically altered carbonates were identified through visual inspection and trace-element (Sr, Mn, Fe, S, K, Na) analysis of shell material. Carbon- and oxygen-isotopic measurements of carbonate shell material were performed on 170 individual shell samples.

In general, $\delta^{18}O$ values of inoceramids are more negative and $\delta^{13}C$ values are more positive than those of ammonites. Well-preserved oysters are significantly enriched in $\delta^{18}O$ in comparison with inoceramids. Inoceramid shell material is generally more enriched in $\delta^{13}C$ than the surrounding carbonate matrix. Inoceramid $\delta^{18}O$ is enriched relative to the surrounding matrix in the lower portion of the core, however, this relationship is reversed near the top of the core.

Isotopic compositions of inoceramids and ammonites reflect both variations in primary water column chemistry and biological effects. Overall, oxygen isotopic values appear to vary in concert with patterns of sea level variation reconstructed from basin-scale stratal patterns. Ammonite $\delta^{18}O$, in particular, appears to mirror regional short-term, base-level cyclicity. The amplitude of oxygen isotope variations in ammonites (2‰–3‰) suggests an interplay of at least two distinct water masses with very different isotopic compositions. The dominance of one water mass over the other may be related to sea level variation; a northern component water depleted in δw probably dominates during regressive intervals.

INTRODUCTION

During peak transgression in the early Turonian, the Cretaceous Western Interior Seaway of North America extended approximately 5000 km north to south (Caldwell, 1984), linking cool Arctic waters to a warm, southern Tethys Ocean (Kauffman, 1977, 1984). From the Cenomanian to Maestrichtian, this oceanic connection remained uninterrupted, which allowed the western interior to a act as a conduit through which at least two water masses with contrasting properties circulated (Kauffman, 1984).

Within and beyond the Western Interior Seaway, oceanographic conditions during the Cretaceous were significantly different than those of today (Saltzman and Barron, 1982; Barron, 1983; Arthur et al., 1985). Evidence for this difference includes abundant organic, carbon-rich (C_{org}) strata deposited on continental shelves and interiors, such as the western interior of North America (Kauffman, 1984; Sageman, 1985; Schlanger et al., 1987). How conditions favorable to C_{org} deposition developed has been widely debated; while several paleoceanographic reconstructions for the Cretaceous Western Interior Seaway consider the problem, the particulars of position and circulation of discreet water masses vary widely (see below).

The use of stable isotope ratio tracers has greatly enhanced water mass identification in modern oceans (Craig and Gordon, 1965). Similarly, stable isotopic analyses of sediment and shell material of the western interior are being used to infer paleoceanographic properties of the Cretaceous Western Interior Seaway (Tourtelot and Rye, 1969; Arthur et al., 1985; Wright, 1987; Kyser et al., 1993) as well as to evaluate global geochemical and climatic variations (Pratt, 1985; Arthur et al., 1987, 1988). This type of work must be performed on materials with original isotopic compositions intact. Seemingly well-preserved shell carbonate can be isotopically altered by diagenetic overgrowths, cement infilling, and dissolution and reprecipitation. Significant diagenetic overprinting can lead to erroneous interpretations of the chemical and physical properties of the water mass in which the skeletons grew. Thus, if one attempts to use $\delta^{18}O$ and $\delta^{13}C$ values to decipher the complex oceanographic history of an epieric

sea, such as the Cretaceous Western Interior Seaway, confidence that those values reflect primary water chemistry and temperature is critical.

There are no methods to quantitatively measure the extent of diagenetic overprinting. However, relative concentrations of specific trace elements incorporated into shell matrices can be used to qualitatively determine the extent of carbonate diagenesis (Veizer, 1983; Brand, 1987, 1990) as well as to infer chemical and physical characteristics of the sea water in which the organisms grew (Whittaker et al., 1987; Morrison, 1991). As a diagenetic tool, relative concentrations of Sr, Mg, Mn, Fe, and Na usually follow predictable trends with increasing diagenesis (Veizer, 1983).

CIRCULATION MODELS

Circulation models of the Cretaceous Western Interior Seaway were developed from faunal distribution (Gordon, 1973; Frush and Eicher, 1975; Eicher and Diner, 1985; Kauffman, 1984; Watkins, 1986; Fisher et al., 1994) and paleotemperature and salinity reconstructions using $\delta^{18}O$ and $\delta^{13}C$ values of shell material and inferred environmental tolerances (Wright, 1987). Numerical experiments using a wind-driven circulation model (Ericksen and Slingerland, 1990), qualitative analyses of atmospheric circulation models (Parrish et al., 1984) and/or general circulation models (e.g., Glancy et al., 1993), as well as modeling hypothetical distributions of temperature, salinity and calculated density differences among discreet water masses (Hay et al., 1993) have also been used to evaluate circulation within the basin. Although most of the models invoke a stratified basin (e.g., Jewell, 1993) to explain widespread organic carbon-rich strata, recent numerical simulation results suggest (for the early Turonian at least) the basin was well-mixed with a strong freshwater component along the basin shorelines due to high fluvial runoff (Slingerland et al., 1996).

Several working hypotheses explain the spatial relationships of major water masses within the basin: (1) a stratified seaway with warm, saline bottom waters derived from the south beneath

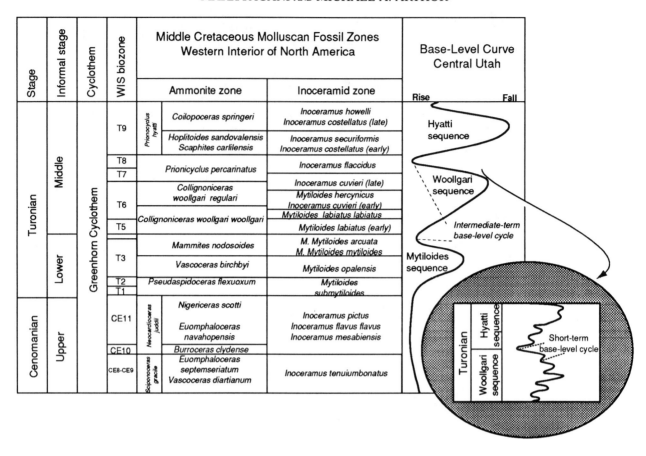

FIG. 1.—Intermediate- and short-term base-level curves for central Utah (Late Cenomanian through Turonian) adapted from Gardner and Cross (1994). Western Interior Seaway biozones from Kauffman et al. (1993).

cool, less saline northern waters freshened by fluvial runoff from the west (Watkins, 1986); (2) a stratified basin with cool, dense bottom waters and warm, low-salinity surface waters (Kauffman, 1977); (3) a seaway in which northern and southern water masses have a discrete contact and meet along a vertical to sloping front (Fisher et al., 1994; Hay et al., 1993). In the latter scenario, caballing may occur, forming a new deep-water mass that may subsequently be exported to global oceans (Hay et al., 1993); and (4) a well-mixed seaway, but one that retains several distinct water masses (Slingerland et al., 1996).

Models of circulation that have been proposed for the seaway include large-scale northern and southern gyres and cross-seaway circulation with bottom waters forming along the eastern margin of the basin (Wright, 1987), north-south propagation of water masses (Kauffman, 1977; Parrish et al., 1984; Watkins, 1986), storm-dominated circulation with resulting geostrophic currents running parallel to shoreline (Ericksen and Slingerland, 1990) as well as a combinations of these processes (Slingerland et al., 1996).

We evaluate water-mass characterisitics along the western margin of the Cretaceous Western Interior Seaway from the stable isotopic compositions of unusually well-preserved shell carbonate of macrofossils collected from the USGS Escalante No. 1 core, with the extent of diagenetic alteration qualitatively determined through evaluation of trace-element composition. These samples, which include inoceramid bivalves, oysters, and ammonites, range in age from late Cenomanian (*S. gracile* ammonite zone) through late middle Turonian (*P. hyatti* ammonite zone) and represent a period of maximum transgression within the basin (Fig. 1) (Kauffman, 1984; Leckie et al., 1991; E. G. Kauffman, pers. commun., 1995).

SAMPLE MATERIAL

Over 650 ft of core was sampled for well-preserved macrofossils from USGS Escalante No. 1 core. The set sampled included a mixed assemblage of inoceramid bivalves, ammonites, and oysters that ranged in age from late Cenomanian (*S. gracile*) through late middle Turonian (*P. hyatti*) (Fig. 1). Inoceramids represent the greatest number of samples in the set (n=106), followed by ammonites (n=43)/baculitids (n=6) and oysters (n=15) (Data Appendix 1). Of this set, 170 samples were analyzed for carbon and oxygen isotope composition and trace element content. Each sample was viewed under a binocular microscope at high magnification to determine quality, luster, and condition of original carbonate shell material. Shell material was removed and manually ground by mortar and pestle in preparation for analysis. Care was taken to exclude lithologic-matrix material from each hand-picked sample.

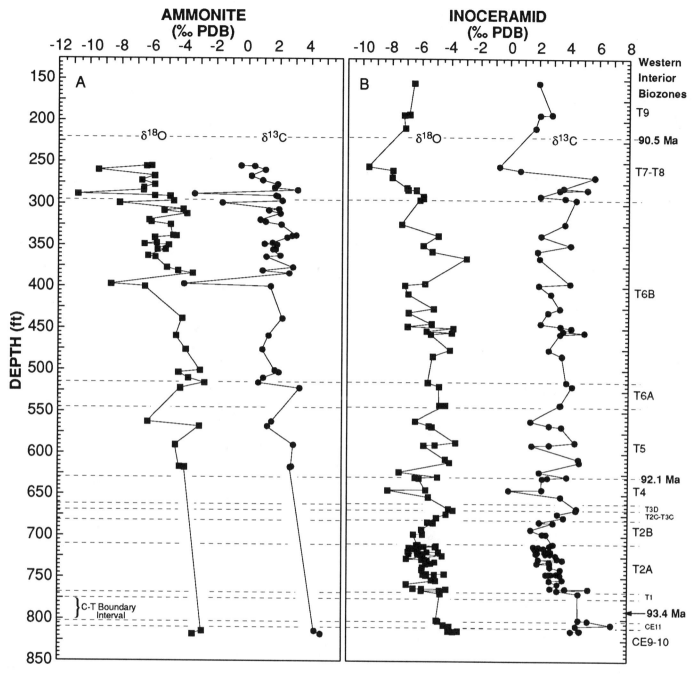

FIG. 2.—(A) Plot of δ¹⁸O (solid squares) and δ¹³C (solid circles) versus depth in ammonites. (B) Plot of δ¹⁸O (solid squares) and δ¹³C (solid circles) versus depth in inoceramids. Western Interior Seaway biozones defined by Kauffman et al. (1993). Associated radiometric ages from Obradovich (1993) and Kauffman et al. (1993). Arrow (93.4 Ma) indicates probable position of Cenomanian-Turonian boundary based on event stratigraphy (E. G. Kauffman, pers. commun., 1995).

ANALYTICAL METHODS

Isotopic measurements were performed on a Finnigan MAT 252 with an automated common acid bath system ("Fairbanks" device) in the Stable Isotope Biogeochemistry Laboratory at Penn State University. Each sample was reacted with 100% phosphoric acid at 90°C. Carbon and oxygen isotopic values are reported in (δ) notation as permil (‰) deviation from the Pee Dee belemnite (PDB) standard. NBS-19 was used as the calibration standard. Average precision of duplicate analyses was 0.034‰.

Concentrations of Fe, Mn, Ca, Sr, and S were measured on an inductively coupled plasma (ICP) spectrophotometer in the Penn State Materials Characterization Laboratory. Na and K were measured by atomic absorption spectrophotometry. Each sample was weighed, reacted with 5% HNO_3, and diluted with deionized water. Any insoluble residue was removed by vacuum filtration. Dried weights of insoluble residues were subtracted from total weights of original samples to calculate trace- and minor-element concentrations relative to 100% carbonate weight. Trace-metal concentrations are reported in ppm (Data Appendix 2) and/or recalculated relative to an element/Ca ratio.

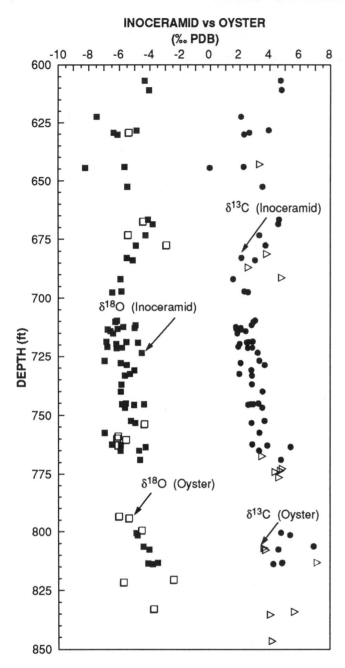

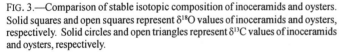

FIG. 3.—Comparison of stable isotopic composition of inoceramids and oysters. Solid squares and open squares represent $\delta^{18}O$ values of inoceramids and oysters, respectively. Solid circles and open triangles represent $\delta^{13}C$ values of inoceramids and oysters, respectively.

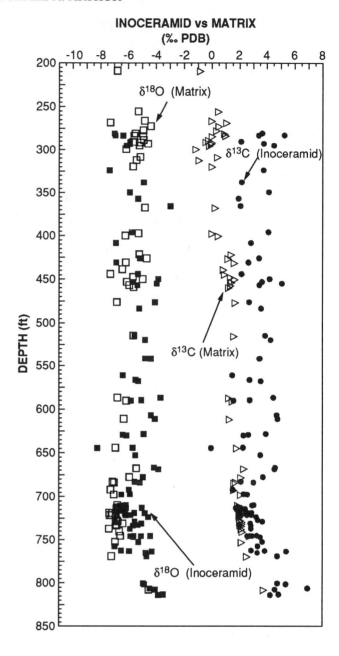

FIG. 4.—Comparison of stable isotopic composition of inoceramid and surrounding matrix. Solid squares and open squares represent $\delta^{18}O$ values of inoceramids and matrix, respectively. Solid circles and open triangles represent $\delta^{13}C$ values of inoceramids and matrix, respectively.

RESULTS

The full set of isotopic data is listed in Data Appendix 1 and plotted downcore (Figs. 2–4) against molluscan biostratigraphic assemblage zones and absolute ages (Kauffman et al., 1993). In general, $\delta^{18}O$ values of inoceramids are more negative and $\delta^{13}C$ are more positive than those of ammonites (Fig. 2) Isotopic values of oysters are very similar to those of inoceramids (Fig. 3); however, this relationship is complicated by diagenetic overprints. Where sampled, $\delta^{13}C$ of the surrounding matrix is generally lower than coeval inoceramid shell carbonate. Matrix $\delta^{18}O$ values indicate a more complex pattern, with a trend to-

ward more enriched values toward the top of the core (Fig. 4). At the base of the core inoceramid $\delta^{18}O$ is more enriched than the surrounding carbonate matrix; however, this relationship reverses at approximately 450 ft (137.2 m).

Before isotopic trends can be evaluated, diagenetic modification needs to be assessed. Our approach was similar to previous studies in which measured trace-element/Ca ratios were compared to elemental ranges associated with similar extant fauna or carbonate phases formed in equilibrium with modern seawater (Brand, 1990; Brand and Viezer, 1980; Morrison, 1991).

The trace elements Sr^{2+}, Mg^{2+}, Fe^{2+}, Mn^{2+}, and Na^+, are incorporated into shell carbonate via several pathways (Brand and

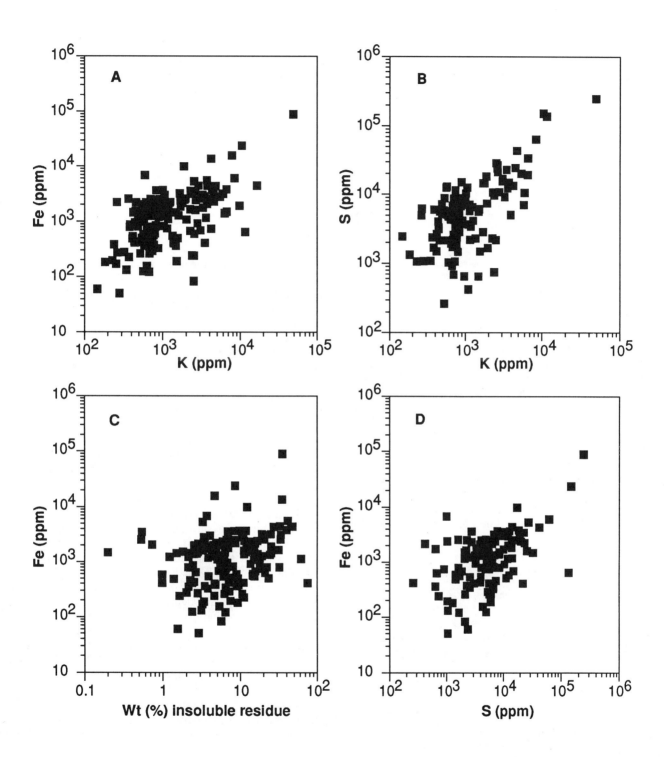

FIG. 5.—(A) Log/log cross plot of Fe versus K in all carbonate samples dissolved in 5% HNO₃. (B) Log/log cross plot of S versus K in all carbonate samples dissolved in 5% HNO₃. (C) Fe versus weight percent of insoluble residue collected through filtration of dissolved carbonate samples. (D) Log/log cross plot of Fe versus S in all carbonate samples dissolved in 5% HNO₃.

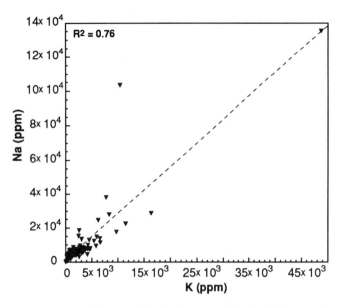

FIG. 6.—Cross plot of Na versus K in all carbonate samples. A significant positive correlation indicates insoluble residue as a probable source of Na contamination.

Veizer, 1980; Veizer, 1983). Most pathways, including occupation of interstitial positions between lattice planes, incorporation within the lattice due to structural defects, adsorption, and inclusion of impurities, are unquantifiable, but minor processes. The best understood process, substitution for Ca^{2+} in the $CaCO_3$ lattice, is believed to be the dominant pathway of trace-element incorporation (Veizer, 1983). In addition, specific biological fractionation may exert considerable control, especially for Sr^{2+} and Mg^{2+} (Brand and Veizer, 1980). At present, however, biological controls are poorly understood and difficult to ascertain for extinct species.

Concentration of trace elements within a $CaCO_3$ lattice formed in equilibrium with the fluid phase is a function of the distribution coefficient (D) defined as:

$$D = \frac{(m_i/m_c)_s}{(m_i/m_c)_l}$$

Where m_i and m_c stand for the molar concentration of the trace- and major-element (Ca^{2+}) respectively, and subscripts s and l define solid and liquid phases (McIntire, 1963). The distribution coefficient is also controlled by structural parameters of the solid phase, and thus varies between calcitic, aragonitic, and dolomitic phases (Veizer, 1983).

In general, one can predict trends in Sr/Ca, Na/Ca, Mg/Ca, Mn/Ca, and Fe/Ca ratios with increasing emplacement of diagenetic carbonate (Brand and Veizer, 1980; Veizer, 1983). These trends are caused by: (1) changes in the distribution coefficient between solid phases, such as would occur during alteration of aragonite or high-Mg calcite to diagenetic low-Mg calcite, and (2) changes in the relative concentrations of trace elements in diagenetic waters. Expected diagenetic trends would be decreasing Sr/Ca and Na/Ca ratios with concurrently increasing Mn/Ca and Fe/Ca ratios. Mg/Ca ratios could increase or decrease depending on whether the original carbonate was pre-

cipitated as high-Mg calcite (HMC) or low-Mg calcite (LMC) (Veizer, 1983).

Elements measured in this study included Ca, Mg, Mn, Fe, Na, Sr, S, and K (Data Appendix 2). Sulfur and potassium were analyzed to examine possible contamination by silicate and other detritus.

Use of HNO_3 in preparation for trace-element analysis, added an additional complication because of a potential contribution of Fe from nitric acid-soluble pyrite. This effect is more problematic for inoceramid chemistry because of an observed association with pyrite (intergrowths/replacements of shell material were common) in the core. We found, however, no significant relationship between concentrations of S, K, and Fe (Fig. 5). We conclude, therefore, that though there may have been additional Fe input in some samples, influence of pyrite-Fe on total Fe analyzed was negligible.

Trace Element Results

Ammonites.—
Prior to collection, the condition of aragonitic shell carbonate of ammonites and baculitids was visually assessed on the basis of luster, shell thickness, and overall quality (Data Appendix 1). A ranking of 3 indicates samples that appeared to preserve highly nacreous shell aragonite, 2 had nacreous aragonite typified by a low luster, 1 had slight to no luster, while those labeled with 0 had no luster and were in poor condition (thin, fragmented, or partly dissolved). Distributions of Sr/Ca, Na/Ca, Mg/Ca, Fe/Ca, and Mn/Ca for each category were compared against ranges found for the extant aragonitic cephalopod Nautilus (Fig. 7). Na/Ca ratios greatly elevated above modern ranges are believed to be a result of contamination (Fig. 6) and were subsequently disregarded as a diagenetic indicator.

In general, visual assessment correlated well with trace-element analysis; the bulk of diagenetically influenced material was derived from samples with a rating of 1 and 0 (Data Appendix 1).

Several studies use variations in $\delta^{18}O$ values in concert with trace-element data as an additional diagnostic tool for alteration. However, because this study utilizes oxygen-isotope values to interpret water mass distribution, the use of $\delta^{18}O$ as a diagenetic indicator is circular. Nonetheless, samples interpreted as diagenetically altered on the basis of trace-element trends are also dominantly more isotopically depleted than the average. The baculite group is an exception to this pattern. The baculite data also exhibited relatively anomalous trace-element trends when compared to those of the bulk of ammonites. All baculitids had rankings of 1 and 0 (Data Appendix 1), all had elevated Na/Ca, Mg/Ca, and Fe/Ca ratios, but all had negligible Mn content except for sample E500.85. Curiously, these baculite samples are also characterized by the most enriched $\delta^{18}O$ values. On the basis of available preservational criteria, we conclude that these samples have been altered from their primary values and are excluded from the set of unaltered isotope graphs.

Inoceramids.—
Inoceramids were visually ranked from 2, 1, or 0 (Data Appendix 1). A rank of 2 indicates thick, clean shells with a well-preserved prismatic outer layer (composed of LMC). Samples ranked 1 tended to be thinner and more poorly preserved, while those rated 0 were very thin and lacked prismatic character.

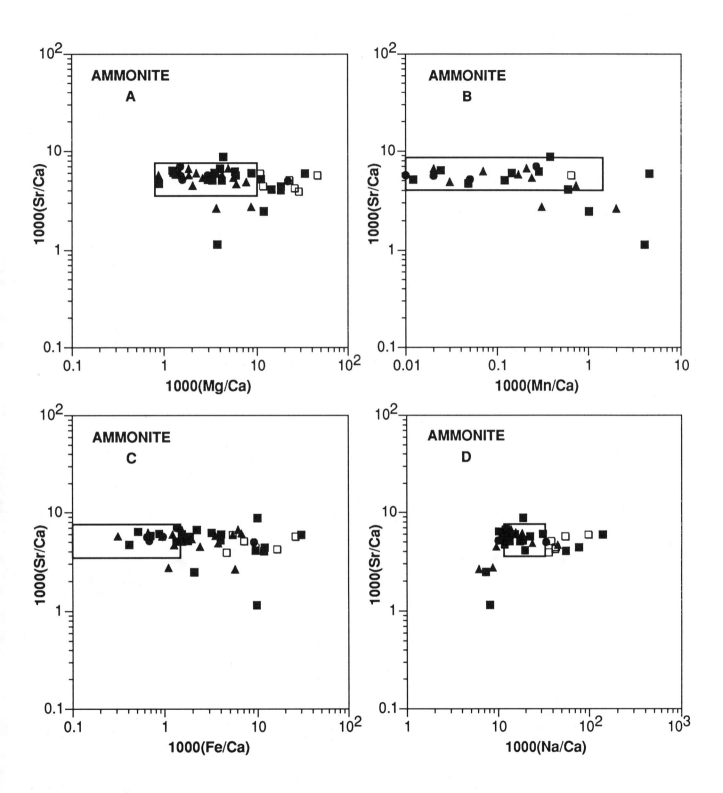

FIG. 7.—Diagenetic assessment of ammonites using trace-element cross plots. For all plots, the interior box brackets the chemical range for the aragonitic cephalopod *Nautilus* precipitated in equilibrium with modern seawater (Morrison and Brand, 1986). Solid circles represent best-preserved samples (ranked 3, in Data Appendix 2), solid triangles are ranked 2, solid squares are ranked 1, open squares represent poorly-preserved samples (ranked 0). (A) Log/log cross plot of Sr/Ca versus Mg/Ca. (B) Log/log cross plot of Sr/Ca versus Mn/Ca. (C) Log/log cross plot of Sr/Ca versus Fe/Ca. (D) Log/log cross plot of Sr/Ca versus Na/Ca.

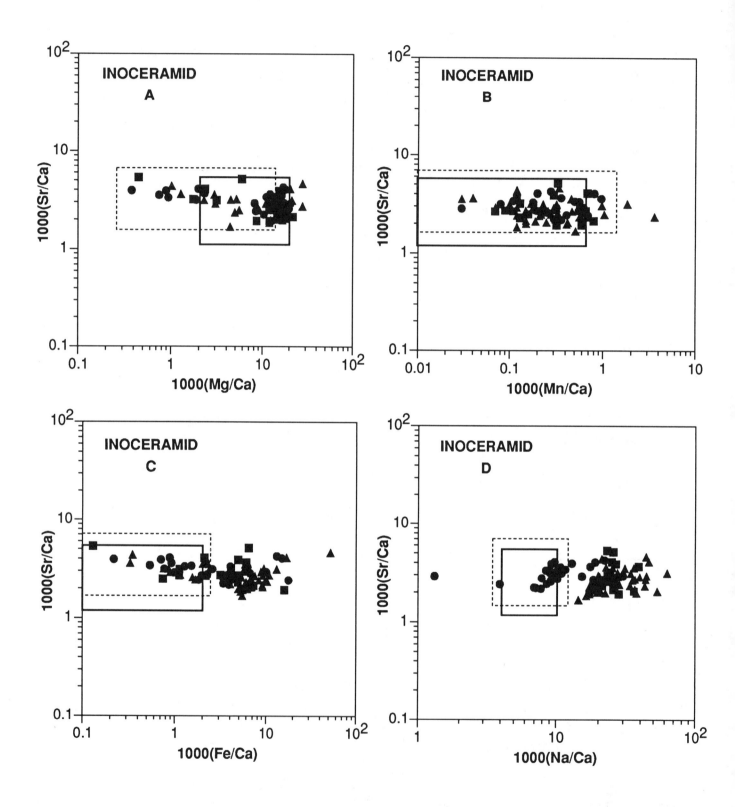

FIG. 8.—Diagenetic assessment of inoceramids using trace-element cross plots. For all plots, the interior, solid-line box brackets the chemical range for bivalves with low-Mg calcite shells precipitated in equilibrium with modern seawater (Milliman, 1974; Morrison and Brand, 1986). Dashed-line box extends the chemical range for bivalves with mixed mineralogy (aragonite-LMC) (Morrison and Brand, 1986). Solid circles represent best-preserved samples (ranked 2; in Data Appendix 2), solid triangles are ranked 1, solid squares are poorly-preserved samples (ranked 0). (A) Log/log cross plot of Sr/Ca versus Mg/Ca. (B) Log/log cross plot of Sr/Ca versus Mn/Ca. (C) Log/log cross plot of Sr/Ca versus Fe/Ca. (D) Log/log cross plot of Sr/Ca versus Na/Ca.

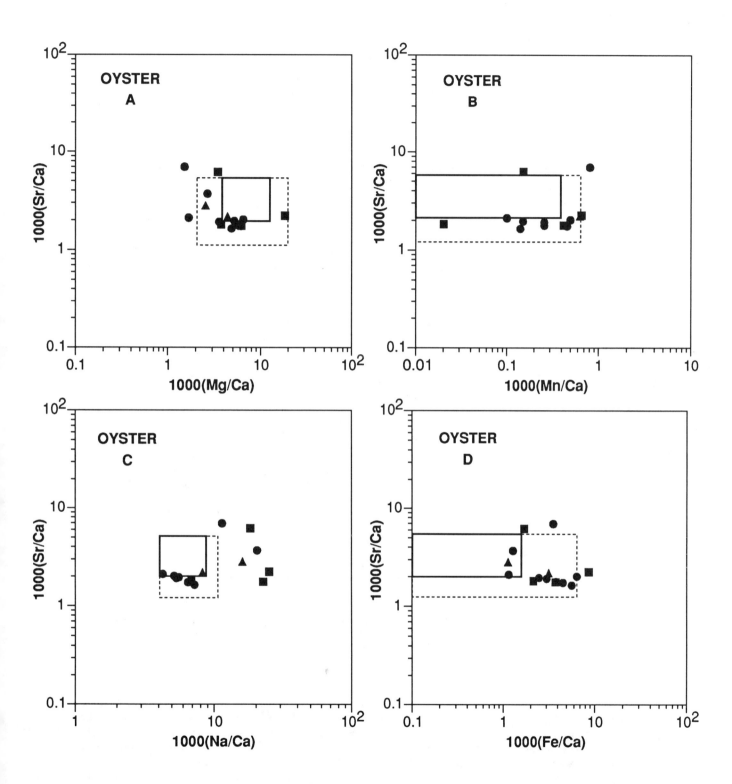

FIG. 9.—Diagenetic assessment of oysters using trace-element cross plots. For all plots, the interior solid-line box brackets the chemical range for oysters with calcite shells precipitated in equilibrium with modern seawater (Milliman, 1974; Morrison and Brand, 1986). Dashed-line box extends the chemical range for bivalves with low-Mg calcite (Morrison and Brand, 1986). Solid circles represent best-preserved samples (ranked 2; in Data Appendix 2), solid triangles are ranked 1, solid squares poorly-preserved samples (ranked 0). (A) Log/log cross plot of Sr/Ca versus Mg/Ca. (B) Log/log cross plot of Sr/Ca versus Mn/Ca. (C) Log/log cross plot of Sr/Ca versus Fe/Ca. (D) Log/log cross plot of Sr/Ca versus Na/Ca.

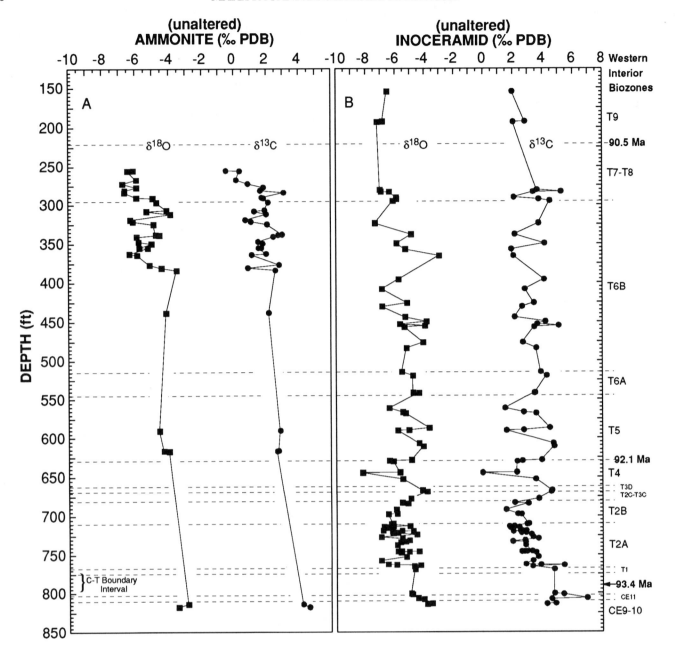

FIG. 10.—(A) Plot of δ¹⁸O (solid squares) and δ¹³C (solid circles) versus depth in the Escalante core for unaltered ammonites. (B) Plot of δ¹⁸O (solid squares) and δ¹³C (solid circles) versus depth in unaltered inoceramids in the Escalante core. Western Interior Seaway biozones are defined by Kauffman et al. (1993). Associated radiometric ages are from Obradovich (1993) and Kauffman et al. (1993). Arrow (93.4 Ma) indicates probable position of Cenomanian-Turonian boundary based on event stratigraphy (E. G. Kauffman, pers. commun., 1995).

Chemical distributions for each category were compared to both modern bivalves with LMC and those with a mixed mineralogy (aragonite and LMC) (Fig. 8).

Diagenetic interpretation proved more difficult for this set due to the lack of distinct trace-element trends. Samples assessed as altered were characterized by high Mg/Ca, Mn/Ca and Fe/Ca content with lower Sr/Ca and Na/Ca contents. Lower Sr/Ca and Na/Ca, however, were not used as strict alteration criteria. Na concentrations for some samples are much higher than modern, pure calcite values, which could suggest possible contamination by silicate material during preparation (Fig. 6). As indicated in Data Appendix 1, many of the samples designated as altered are also the most depleted in δ¹⁸O values.

Oysters.—

Like inoceramids, oysters were ranked from 2 to 0 on the basis of thickness and general quality of the shell material. Lower Sr and high Mn and Fe concentrations were primarily used to screen altered shells. Trace-element distributions were compared to both modern oysters and to bivalves with LMC (Fig. 9).

Though many oysters initially appeared well-preserved, trace-element analysis suggests differently. The majority of samples have lower Sr/Ca and higher Mn/Ca and Fe/Ca ratios than modern oysters. A trend toward increasing Mg/Ca ratio with decreasing Sr/Ca is characteristic.

In summary, it appears that the majority of oysters sampled were altered to differing degrees. Noteworthy is the relationship

TABLE. 1.—CALCULATED Δ¹⁸O(PDB) VALUES OF
CALCITE PRECIPITATED IN EQUILIBRIUM WITH
SEAWATER UNDER A VARIETY OF CONDITIONS.

$\delta^{18}O_c$(PDB)	Temp (°C)	Salinity (local)	Δfw
-4.24	30	34	
-3.22	25	34	
-4.53	30	32	5
-3.81	25	30	5
-4.83	30	32	10
-4.39	25	30	10
-5.12	30	32	15
-4.98	25	30	15
-5.41	30	32	20
-5.57	25	30	20
-5.71	30	32	25
-6.16	25	30	25
-6.00	30	32	30
-6.74	25	30	30

A global average salinity of 34 psu and δ¹⁸Ow of -1.2‰
(SMOW) is assumed. Δfw represents the difference between
the global average δ¹⁸O and the δ¹⁸O of freshwater input.

between well-preserved samples (that is, E820.54 and E677.62)
and relatively enriched δ¹⁸O values. Oyster δ¹⁸O values are usually lower than those measured in inoceramids, which suggests
the possibility of either greater diagenetic susceptibility for oyster carbonate or a difference in biologic isotope fractionation between inoceramids and oysters.

Stable Isotopic Results

Isotopic compositions for unaltered samples of inoceramids
and ammonites were plotted against core depth with associated
biozones and radiometric ages as compiled by Kauffman et al.
(1993) (Figs. 11–13).

Unaltered Ammonites.—

Oxygen isotope values average approximately -4.0‰ throughout the lower section of the core (Fig. 10). Though sampling
frequency begins to increase at approximately 385 ft (117.4 m),
there is evidence of two, rapid step-like transitions to more depleted δ¹⁸O values. The first, within T6B (*Collignoniceras
woollgari*) at 375 ft (114.3 m), represents a dramatic shift of average values (-4‰) to approximately -6‰. A return to heavier
values at the base of T7–T8 (*Prionocyclus percarinatus*) is followed by another step to even more depleted values (-6.5‰) at
290 ft (88.4 m).

Carbon isotopes decrease from +4.5‰ at the base of the core
to approximately +2‰ at 500 ft (152.4 m). Stable values persist
until 277 ft (84.4 m) within T7–T8 (*Prionocyclus percarinatus–
Prionocyclus percarinatus*), at which point a significant shift to
more negative values occurs.

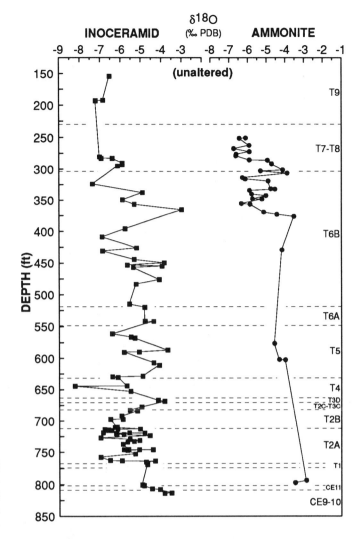

FIG. 11.—Plot of inoceramid δ¹⁸O (solid squares) and ammonite δ¹⁸O (solid circles)
versus depth in the Escalante core for unaltered samples. Western Interior Seaway
biozones are defined by Kauffman (1993).

Unaltered Inoceramids.—

δ¹⁸O rapidly decreases from relatively enriched values (average -3.7‰) at the base of the core to more depleted values averaging approximately -5.7‰ (Fig. 10). A 2‰ positive shift occurs at 700 ft (213.4 m) within biozone T2B through condensed
biozones T2C–T3C (*Pseudaspidoceras flexuoxum–Watinoceras-
coloradoense–Mammites nodosoides*), followed by lower and
relatively constant values averaging approximately -5.4‰. A
negative trend begins at 455 ft (138.7 m), within T6B
(*Collignoniceras woollgari*), reaching δ¹⁸O values averaging
-7‰ at the top of the core.

Carbon isotopes continually decrease from positive values
(+4.5‰) at the base of the core to less than +2‰ at 690 ft (210.3
m). A positive δ¹³C shift occurs in concert with increasing δ¹⁸O
values within biozone T2B, obtaining δ¹³C values as positive as
+4.6‰. This shift is followed by a decrease and stabilization of
values averaging +3.1‰. From the beginning of this positive
shift at 690 ft (210.3 m) to about 480 ft (146.3 m), carbon and
oxygen isotope variability closely track one another.

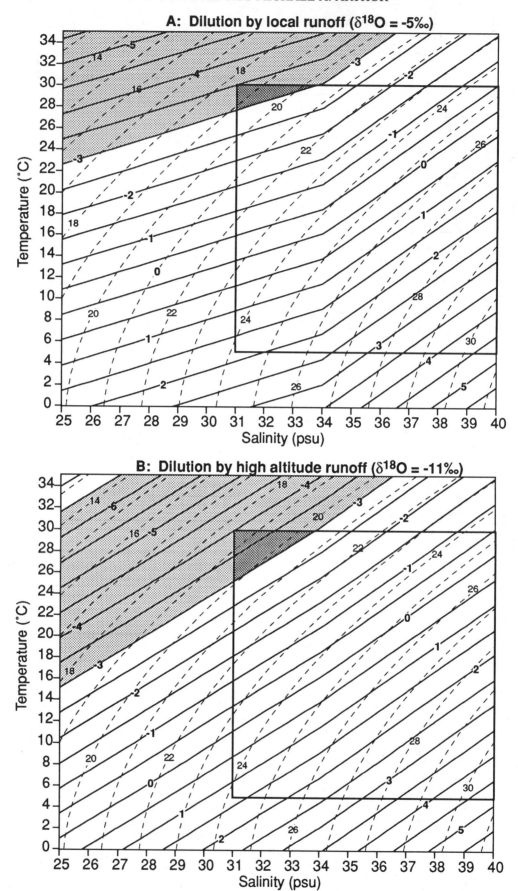

FIG. 12.—(A) Temperature-salinity diagram. Calculations assume δ^{18}O (SMOW) = -5‰ for freshwater runoff. (B) Temperature-salinity diagram. Calculations assume δ^{18}O (SMOW) = -11‰ for freshwater runoff. Density (dashed lines) of seawater expressed as σ$_t$. δ^{18}O (solid lines) values are for aragonite precipitated in equilibrium with seawater for a given salinity and temperature. All calculations assume a global average δ^{18}Ow for Cretaceous seawater = -1.2‰ with an average salinity of 34 psu. Interior solid box indicates environmental tolerances of planktonic foraminifera. Shaded region defines oxygen isotopic range for well-preserved ammonites.

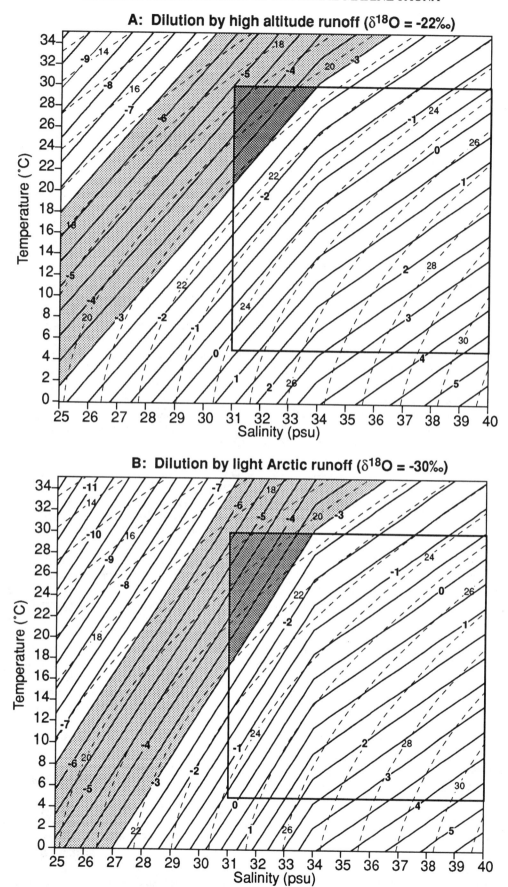

FIG. 13.—(A) Temperature-salinity diagram. Calculations assume δ¹⁸O (SMOW) = -22‰ for freshwater runoff. (B) Temperature-salinity diagram. Calculations assume δ¹⁸O (SMOW) = -30‰ for freshwater runoff. Density (dashed lines) of seawater expressed as σt. δ¹⁸O (solid lines) values are for aragonite precipitated in equilibrium with seawater for a given salinity and temperature. All calculations assume a global average δ¹⁸Ow for Cretaceous seawater = -1.2‰ with an average salinity of 34 psu. Interior solid box indicates environmental tolerances of planktonic foraminifera. Shaded region defines oxygen isotopic range for well-preserved ammonites.

Isotopic Composition of Inoceramids

The original $\delta^{18}O$ of calcite precipitated in equilibrium with seawater is a function of the temperature and $\delta^{18}O$ of the waters in which they precipitated (Epstein et al., 1951). The average oxygen isotopic composition of seawater may be modified by processes that effect local salinity, such as evaporation/fresh water runoff and mixing of water masses. There is some question, however, about whether or not isotopic values of epibenthic inoceramid bivalves directly reflect environmental signals. Although inoceramid $\delta^{18}O$ values are used to interpret water mass conditions during the Cretaceous (Saltzman and Barron, 1982), several researchers have questioned this appliction to the Cretaceous Western Interior Seaway (Tourtelot and Rye, 1969; Ludvigson et al., 1994). These authors attribute the common and anomalously low $\delta^{18}O$ and high $\delta^{13}C$ values to vital effects, that were possibly caused by chemosymbiosis with anaerobic bacteria (for example, as mentioned by Kauffman and Sageman, 1990; MacLeod and Hoppe, 1992). Although Pratt et al. (1993) have showed that the $\delta^{18}O$ values of inoceramids, for the same samples of Cretaceous Western Interior Seaway rocks, are consistently enriched relative to matrix and/or whole rock values, Fisher and Arthur (in prep.) showed that inoceramid $\delta^{18}O$ is consistently depleted relative to other benthic fauna and similar to well-preserved planktonic foraminifera for a suite of Cenomanian samples from the north-central Cretaceous Western Interior Seaway. Likewise, in the latter study, inoceramid $\delta^{13}C$ is substantially enriched relative to other benthic faunal elements (by 2–3‰), while it is similar to or more enriched than surface-dwelling foraminifers or carbonate fine-fraction. Unfortunately, a paucity of well-preserved oysters from our study prevents a definitive comparison against the isotopic composition of inoceramids. However, the oysters that are well-preserved have notably heavier $\delta^{18}O$ values with similarly heavy $\delta^{13}C$ (see E820.54 and E677.62B, Data Appendix 1).

Table 1 provides some calculated $\delta^{18}O$ values of calcite precipitated in equilibrium with seawater given a range of possible environmental conditions. Note that inoceramid $\delta^{18}O$ values are generally more depleted than one would expect for normal Cretaceous seawater (-4.2‰ at 30°C). These calculations indicate rather extreme and unrealistic conditions are necessary to account for the benthic calcite $\delta^{18}O$ values of ≤ -5‰ characteristic of inoceramids. Therefore, it appears that inoceramid bivalves from this site may not have precipitated in isotopic equilibrium with basinal waters and thus their isotopic compositions should not be interpreted directly as paleotemperatures. However, one could interpret the amplitude of relative changes in stable isotopic values in terms of paleoenvironmental events, particularly if these changes were also reflected in the stable isotopic composition of other organisims.

Paleoceanography

Support for the view that stable isotopic trends from inoceramids reflect temporal and spatial environmental variations within the water column can be found by identifying known oceanographic and isotopic events. For example, the positive $\delta^{18}O$ shift between 700 ft (213.4 m) and 670 ft (204.2 m), occurring within biozones T2A through T3C, corresponds with a well-documented incursion of Tethyan waters and maximum expansion of the Cretaceous Western Interior Seaway during Mammites nodosoides time (Leckie et al., 1991; Elder and Kirkland, 1994; Gardner and Cross, 1994). The condensed nature of T2C–T3C supports transgressive conditions during this interval. Increased circulation within the basin might have flushed-out dysaerobic bottom-waters with comparatively low $\delta^{13}C$ signatures and replaced them with waters having higher $\delta^{13}C$ more typical of the global isotopic average; this would account for the simultaneous positive shift of both $\delta^{13}C$ and $\delta^{18}O$.

The global expression of the positive $\delta^{13}C$ excursion near the Cenomanian–Turonian boundary is well-documented (Scholle and Arthur, 1980; Schlanger et al., 1987; Arthur et al., 1987). The excursion is inferred to reflect burial of isotopically-light organic-carbon during eustatic highstand (Arthur et al., 1987). The recovery from this shift to more negative $\delta^{13}C$ values within the basin is recorded by inoceramid $\delta^{13}C$ at the base of the core.

A trend toward lower $\delta^{18}O$ values for both inoceramids and ammonites, occurs within T6B (*Collignoniceras woollgari*) at 450 ft (137.2 m) and 380 ft (115.8 m), respectively (Fig. 11). Accompanying this trend at 450 ft (137.2 m) is a reduction in the abundance of planktonic foraminifera and an increase in quartz content (West et al., this volume). These factors point to a regression or shoreface progradation in conjunction with freshening and/or warming of the water column. In addition, ammonite $\delta^{18}O$, which reflects shallow to intermediate water column conditions, appears to be recording two rapid, step-like shifts, the first of which returns to more normal conditions prior to the second excursion.

Recent middle Cretaceous paleogeographic reconstructions for central Utah were developed by evaluating facies associations and stratal architecture, and subsequent recognition of stratigraphic sequences or cycles (Gardner and Cross, 1994). These stratigraphic cycles, viewed in context with biochronozones, define short-, intermediate-, and long-term base-level cycles. From the Turonian to early middle Coniacian, four intermediate-term cycles are defined, each composed of several short-term stratigraphic cycles. Of particular interest are both the informally named Woollgari sequence, which spans ammonite zones *Collignoniceras woollgari woollgari*, *Collignoniceras woollgari regulare* and *Prionocyclus percarinatus* (corresponding to T5–T6 and T7–T8 biozones) and the overlying Hyatti sequence (Fig. 1). The Woollgari sequence is comprised of five short-term cycles arranged in seaward-stepping, vertical-, and landward-stepping stacking patterns that contain shallow-marine to shoreface and marine-shelf facies associations. The correspondence of short-term cycles within the *Collignoniceras woollgari* zone with rapid, step-like $\delta^{18}O$ shifts recorded in ammonites strongly suggests that isotopic fluctuations were controlled by base-level cyclicity in some way. Thus, we contend that a rapid, negative $\delta^{18}O$ shift followed by increasingly positive values represents one short-term base-level fall and rise.

Water Mass Dynamics

Although the ammonite oxygen isotopic trend closely tracks base-level cyclicity, it is difficult to explain their negative $\delta^{18}O$ values in terms of simple freshening or warming of the water column. The presence of planktonic foraminifera constrains upper water-column salinity and temperature due to metabolic tolerances. Initial analyses of the Escalante sequence indicates that two foraminiferal morphotypes dominated: (1) populations with a biserial chamber arrangement (*Heterohelix*), and (2) populations with trochospiral chamber arrangement (*Hedbergella, Whiteinella*) (West et al., this volume). *Heterohelix* and *Hedbergella* are related to members of modern Globigerinida (Haynes, 1981). While most modern Globigerinida species require salinities of 35 to 36 practical salinity units (psu), several species can inhabit waters with salinities as low as 30.5 psu. Temperature constraints range from 30° to 5°C for most species (Haynes, 1981). Although a sharp reduction in absolute abundances of planktonic foraminifera occurs at and above approximately 340 ft (103.6 m), the presence of planktonic foraminifera throughout the rest of the core (West et al., this volume) argues against extreme temperature and salinity variations, particularly below 340 ft (103.6 m). However, if the water mass in which the ammonites lived mixed with freshwater that had comparatively low $\delta^{18}Ow$ values, reasonable temperature and salinity ranges are possible and still produce significantly depleted ammonite $\delta^{18}O_{aragonite}$.

Figures 12–13 represent a series of temperature–salinity–$\delta^{18}O$ diagrams originally developed and described by Railsback and Anderson (1989) and Railsback (1990). $\delta^{18}O$ values are for aragonite, calculated by inverting and applying the temperature–$\delta^{18}O_{aragonite}$ relationship [T(°C) = 21.8-4.69 ($\delta^{18}O_{aragonite}$-$\delta^{18}Ow$)] developed by Grossman and Ku (1986) and by assuming an average Cretaceous salinity of 34 psu and a global average $\delta^{18}Ow$ of -1.2‰ (SMOW) (standard mean ocean water) (Shackleton and Kennett, 1975). Two processes are represented in each chart: salinities below 34 psu suggest mixing of water masses with runoff and/or with constant $\delta^{18}Ow$, and salinities above 34 psu are a function of evaporation only. We assume an evaporative $\delta^{18}O$/salinity trend of 0.41‰/psu, similar to that of the modern Mediterranean Sea (Thunnell et al., 1987). Each figure has a different assumed $\delta^{18}Ow$ for waters that freshen average Tethyan seawater.

The global average $\delta^{18}O$ of precipitation is -4‰ to -5‰ lower than average seawater (Fig. 12) (Craig and Gordon, 1965). Runoff, if derived from high-altitude melt, can have significantly lower isotopic values considering typical modern gradients of 0.15‰–0.5‰ /100 m (Gat, 1980). Recently, Ludvigson et al. (1994) assessed the isotopic composition of meteoric waters on the eastern margin of the Cretaceous Western Interior Seaway by evaluating stable isotopic data from early diagenetic carbonates. They provide evidence for the presence of meteoric phreatic environments with $\delta^{18}O$ values of -7‰ (SMOW) Therefore, if we assume a baseline $\delta^{18}Ow$ value of -7‰ for mid-latitude, sea-level precipitation, a global $\delta^{18}Ow$ of -1.2‰, and altitudes for the mountains (Sevier orogeny) on the western flank of the seaway to be approximately 2700 m (Jordan, 1981), runoff from their eastern slopes would have a minimum value of -21.5‰ (SMOW) and a maximum of -11‰ (SMOW) (Fig. 12–13) (Glancy et al., 1993).

Figures 12–13 present several possible scenarios for basinal waters derived from mixing with local precipitation and/or discrete water masses with a $\delta^{18}Ow$ of -5‰, -11‰, -22‰, and -30‰, respectively. The box in each temperature–salinity–$\delta^{18}O$ diagram defines the salinity and upper temperature constraints warranted by modern planktonic foraminifera, and are assumed to apply to the Cretaceous as well. These diagrams show the difficulty of producing aragonite with low oxygen isotopic values (-5‰) with runoff typical of low- to mid-latitude precipitation (that is, > -10‰ (SMOW); Figs. 12). To produce these carbonate isotopic values, runoff $\delta^{18}Ow$ must be extraordinarily depleted, as is typical of high-altitude sources and those found

in the modern Arctic Ocean (Fig. 13) which have values of -30‰ (Spielhagen and Erlenkeuser, 1994).

Bice, Arthur and Marincovich (1996) suggested $\delta^{18}O$ values for late Maestrichtian runoff from the Alaskan north slope of -29‰ (SMOW). Dettman and Lohmann (1993) also found evidence for depleted Paleocene precipitation (-8‰ to -23‰ (SMOW)) in the northern region of the Western Interior Seaway. We contend that these relatively low $\delta^{18}O$ values of shell material in the Cretaceous Western Interior Seaway reflect the influence of a southward advected water mass that has been freshened by Arctic runoff with a highly negative $\delta^{18}Ow$.

This water mass, if in geostrophic balance, would flow southward and be closely associated with the western edge of the basin (Slingerland et al., 1996). During times of relative or eustatic sea-level rise, this northern water component moves westward in tandem with the transgressive shoreline. However, during sea-level fall and or/progradation of the shoreline, the water mass shifts progressively eastward and downward in the basin. Rapid base-level shifts are thus responsible for the isotopic "steps" in the ammonite $\delta^{18}O$ record.

SUMMARY

Stable isotopic compositions of well-preserved shell carbonate derived from inoceramids and ammonites of Upper Cenomanian through Middle Turonian age reflect temporal and spatial environmental variations within the water column. Inoceramid oxygen-isotopic compositions are probably influenced by vital effects and do not appear to directly record primary conditions. However, temporal trends in inoceramid isotopic values probably reflect relative environmental variations. Negative shifts in ammonite $\delta^{18}O$ coincide well with short-term, base-level cyclicity. The amplitude of these negative shifts reflects the influence of two distinct water masses. A freshened, northern component water with a depleted $\delta^{18}Ow$ signature is closely associated with the western boundary of the basin and dominates during regressive events and/or progradation of the shoreline. This agrees with results of previous studies of planktonic organisms (Leckie et al., 1991) as well as with recent numerical circulation models for the Cretaceous Western Interior Seaway (Slingerland et al., 1996).

ACKNOWLEDGMENTS

Macrofossil identification and biostratigraphic zonation were kindly performed by Bill Cobban (USGS) and Erle Kauffman, respectively. The authors thank reviewers Enriqueta Barrera and Thomas F. Anderson for their helpful comments and critical reviews. In addition, an informal review by Anthony Prave is much appreciated. The authors also thank Erle Kauffman for providing unpublished biostratigraphic data for the Escalante core and for his support, encouragement, and friendship throughout this and countless other endeavors.

Data Appendix 1: Stable isotope data and diagenetic rank assignments for carbonate macrofossils and matrix; all samples in ‰ relative to Pee Dee belemnite standard (calibration by NBS-19).

INOCERAMID (NOTE: Sample depth "E" refers to Escalante corehole)

Depth (ft)	$\delta^{13}C$ (‰PDB)	$\delta^{18}O$ (‰PDB)	Species	Rank	Diagenesis
E154.72	1.96	-6.54	Inoceramus sp.	2	
E192.45	2.82	-6.86	Inoceramus sp.	1	
E193.09	2.03	-7.22	Inoceramus sp.	2	
E208.82	1.74	-7.14	Inoceramus cuvieri (?)	2	
E255.29 (A)	-0.77	-9.60	Inoceramus sp.	1	√
E260.01 (B)	0.68	-7.98	Inoceramus sp.	1	√
E268.15	5.70	-8.02	Inoceramus sp.	0	√
E280.76	3.62	-7.03	Inoceramus sp.	1	
E282.93	5.22	-6.95	Inoceramus sp.	0	
E283.28	3.36	-6.40	Inoceramus sp.	2	
E290.50	2.07	-5.92	Inoceramus sp.	1	
E292.80	3.73	-5.91	Inoceramus sp.	1	
E295.01	4.47	-6.15	Inoceramus sp.	2	
E323.91	3.72	-7.37	Inoceramus cuvieri	2	
E337.88 (A)	2.12	-4.91	Inoceramus sp.	0	
E349.48 (B)	4.11	-5.91	Inoceramus cuvieri	1	
E356.88	1.89	-5.32	Inoceramus sp.	0	
E365.40	2.03	-2.99	Inoceramus sp.	0	
E395.96	4.08	-5.78	Inoceramus sp.	2	
E397.92	1.98	-7.14	Inoceramus sp.	1	√
E408.42	2.79	-6.91	Inoceramus sp.	0	
E421.54	3.47	-6.13	Inoceramus sp.	0	√
E425.99	3.40	-5.20	Inoceramus sp.	0	
E430.76	2.61	-6.89	Inoceramus sp.	2	
E443.99 (A)	2.11	-5.33	Inoceramus sp.	0	
E447.29	3.44	-6.95	Inoceramus sp.	1	√
E449.70	4.17	-3.86	Inoceramus sp.	1	
E453.00	3.62	-5.68	Inoceramus cuvieri	0	
E454.91	5.05	-3.97	Inoceramus sp.	0	
E456.70	3.42	-5.38	Inoceramus cuvieri	1	
E476.16 (A)	2.66	-4.10	Inoceramus sp.	0	
E483.39	3.55	-5.24	Inoceramus cuvieri	0	
E514.60	3.85	-5.57	Inoceramus sp.	0	
E519.47	4.24	-4.81	Inoceramus sp.	1	
E541.60	3.47	-4.80	Inoceramus sp.	1	
E541.75	3.42	-4.38	Mytiloides subhercynicus	1	
E561.17	1.45	-6.42	Mytiloides hercynicus	1	
E566.37	2.70	-5.50	Mytiloides subhercynicus	2	
E567.87	3.53	-5.30	Mytiloides subhercynicus	1	
E586.68	4.44	-3.69	Mytiloides subhercynicus	0	
E589.56	2.71	-5.08	Mytiloides subhercynicus	2	
E590.08	1.54	-5.84	Mytiloides subhercynicus	2	
E606.78	4.68	-4.38	Mytiloides subhercynicus	2	
E610.84	4.74	-4.10	Mytiloides subhercynicus	2	
E622.46	2.05	-7.51	Mytiloides sp.	2	
E628.37	3.90	-4.91	Mytiloides subhercynicus	1	
E629.42 (A)	2.61	-6.41	Mytiloides sp.	1	
E630.17	2.26	-6.15	Mytiloides hercynicus	0	
E644.07	2.23	-5.70	Mytiloides mytiloides	1	
E644.54	-0.07	-8.25	Mytiloides mytiloides	2	
E652.58	3.50	-5.51	Mytiloides mytiloides	2	
E666.77	4.58	-4.16	Mytiloides subhercynicus	2	
E668.61	4.53	-3.84	Mytiloides columbianus	1	
E673.32 (A)	3.29	-4.32	? Mytiloides Columbianus	0	√

Rank refers to visual assessment of diagenesis. A ranking of 2 represents relatively thick, clean shells with a well-preserved prismatic outer layer. Samples ranked 1 tended to be thinner and more poorly preserved, while those rated 0 were very thin and lacked prismatic character. A check mark in the last column indicates diagenesis on the basis of trace-element analysis.

DATA APPENDIX 1 (CONTINUED)

INOCERAMID

Depth (ft)	$\delta^{13}C$ (‰PDB)	$\delta^{18}O$ (‰PDB)	Species	Rank	Diagenesis
E677.62 (A)	3.70	-4.96	Mytiloides sp.	1	
E682.88	2.08	-5.55	Mytiloides columbianus	1	
E683.87	3.00	-5.17	Mytiloides columbianus	2	
E691.87	1.50	-5.96	Mytiloides columbianus	1	
E697.20	2.28	-5.90	Mytiloides columbianus	2	
E697.53	2.53	-6.51	Mytiloides sp.	0	
E709.73	3.02	-6.19	Mytiloides sp.	1	
E710.27	2.90	-6.30	Mytiloides opalensis	1	
E711.66	2.78	-4.98	Mytiloides sp.	1	√
E712.42	1.70	-5.79	Mytiloides sp.	1	√
E712.85	2.04	-5.04	Mytiloides sp.	1	
E713.26	1.72	-6.16	Mytiloides sp.	1	
E713.50	2.06	-6.80	Mytiloides sp.	1	
E714.24	2.38	-6.60	Mytiloides sp.	1	
E715.20	1.80	-6.45	Mytiloides sp.	1	
E718.65	2.86	-6.87	Mytiloides columbianus	1	
E718.83	2.60	-5.58	Mytiloides columbianus	0	
E719.09	2.45	-4.82	Mytiloides sp.	1	
E719.67	1.97	-6.24	Mytiloides opalensis	2	
E720.89	1.87	-6.81	Mytiloides columbianus	2	
E721.25	2.83	-5.86	Mytiloides columbianus	2	
E721.40	2.52	-6.21	Mytiloides opalensis	1	
E723.45	3.19	-4.57	Mytiloides columbianus	1	
E726.91	3.30	-6.99	Mytiloides columbianus	1	
E727.89	2.03	-5.95	Mytiloides sp.	1	√
E728.72	3.65	-5.56	Mytiloides columbianus	1	
E731.00	2.76	-5.07	Mytiloides sp.	1	
E732.60	1.94	-5.34	Mytiloides sp.	1	
E733.32	2.80	-5.68	Mytiloides columbianus	1	
E737.04	2.80	-5.90	Mytiloides columbianus	0	
E740.09	3.51	-5.95	Mytiloides opalensis	0	√
E745.20	3.24	-5.62	Mytiloides opalensis	0	
E745.46	2.89	-4.42	Mytiloides hattini	1	
E745.52	2.71	-5.86	Mytiloides opalensis	2	
E745.71	2.54	-5.09	Mytiloides opalensis	2	
E746.87	3.49	-5.67	Mytiloides hattini	2	
E752.45	3.64	-5.28	Mytiloides opalensis	2	
E753.33	2.77	-4.99	Mytiloides hattini	0	√
E757.50	3.30	-6.99	Mytiloides hattini	1	
E762.46	2.82	-6.52	Mytiloides opalensis	1	
E762.94 (A)	3.82	-5.95	Mytiloides sp.	?	
E763.55	5.37	-4.31	Mytiloides opalensis	1	
E765.02	3.27	-4.75	Mytiloides hattini	1	
E765.15	3.29	-5.97	Mytiloides hattini	1	√
E768.98	4.72	-4.69	Mytiloides opalensis	1	
E800.57	4.73	-4.94	Inoceramus pictus	2	
E801.54	5.34	-4.85	Inoceramus pictus	2	
E806.34	6.90	-4.47	Inoceramus pictus	2	
E807.68	4.55	-4.07	Inoceramus pictus	1	
E813.28	4.82	-3.51	Inoceramus pictus	1	
E813.46	4.79	-4.15	Inoceramus pictus	0	√
E813.86	4.22	-3.85	Inoceramus sp.	2	

DATA APPENDIX 1 (CONTINUED)

AMMONITE

Depth (ft)	$\delta^{13}C$ (‰PDB)	$\delta^{18}O$ (‰PDB)	Species	Rank	Diagenesis
E255 (A)	-0.50	-6.12	ammonite	3	
E255.46	0.36	-6.46	Collignoniceras woollgari	2	
E260.01 (A)	1.04	-9.50	Collignoniceras woollgari	0	√
E267.1	0.15	-5.93	Collignoniceras woollgari	2	
E272.3	0.87	-6.76	Collignoniceras woollgari	3	
E277.2	1.82	-5.92	Collignoniceras woollgari	3	
E281.3	1.63	-6.63	Collignoniceras woollgari	2	
E283.96	3.09	-6.65	Collignoniceras woollgari	1	
E288.45	-3.41	-10.79	Collignoniceras woollgari	2	√
E290.5 (B)	1.72	-5.92	Collignoniceras woollgari	2	
E291.15	1.82	-4.95	Collignoniceras woollgari	2	
E296.62	2.11	-4.73	Collignoniceras woollgari	2	
E298.96	-1.67	-8.16	Collignoniceras woollgari	2	√
E306.7	1.90	-4.13	Collignoniceras woollgari	1	
E308.19	1.26	-5.32	Collignoniceras woollgari	1	
E311.9	1.99	-3.89	Collignoniceras woollgari	3	
E319.34	0.72	-6.29	Collignoniceras woollgari	2	
E321.65	1.06	-6.14	Collignoniceras woollgari	1	
E324.97	2.05	-4.91	Collignoniceras woollgari	1	
E337.98 (B)	3.00	-4.77	Collignoniceras woollgari	1	
E338.6	2.75	-4.55	Collignoniceras woollgari	2	
E340.68	2.43	-5.91	ammonite	2	
E347.08	1.50	-5.80	Collignoniceras woollgari	1	
E348.37	0.99	-6.59	ammonite	1	√
E349.48 (A)	1.78	-5.03	Collignoniceras woollgari	1	
E355.08	1.70	-5.75	Collignoniceras woollgari	2	
E355.12	1.53	-5.24	Collignoniceras woollgari	2	
E362.59	1.99	-6.35	Collignoniceras woollgari	1	
E363.97	1.11	-5.88	Collignoniceras woollgari	1	
E376.85	2.80	-5.15	Collignoniceras woollgari	1	
E380.78	0.87	-4.44	ammonite	1	
E384.04	2.56	-3.53	Collignoniceras woollgari	1	
E397.13	-4.06	-8.69	Collignoniceras woollgari	1	√
E399.89	1.40	-6.54	ammonite	1	√
E438.52	2.14	-4.17	Collignoniceras woollgari	1	
E459.1	1.27	-4.55	Collignoniceras woollgari	?	
E476.16 (B)	0.88	-3.94	Baculites sp.	1	√
E500.85	1.67	-3.03	Baculites yokoyamai	1	√
E503.28	1.92	-4.39	Baculites yokoyamai	1	
E509.92	0.94	-3.78	Baculites yokoyamai	0	√
E515.64	0.63	-2.76	Baculites yokoyamai	0	√
E522.04	3.25	-4.28	Collignoniceras woollgari	0	√
E562.32	1.47	-6.34	Collignoniceras woollgari	0	√
E567.87 (B)	1.21	-3.07	Baculites yokoyamai	0	√
E590.22	2.86	-4.58	Collignoniceras woollgari	2	
E616.52	2.75	-4.32	Collignoniceras woollgari	2	
E617.09	2.66	-4.00	Collignoniceras woollgari	2	
E813.96	4.24	-2.88	Anisoceras coloradonense	1	
E817.66	4.63	-3.47	Scipnoceras gracile	2	

Rank refers to visual assessment of diagenesis. A ranking of 3 represents samples that appeared to preserve highly nacreous shell aragonite, 2 had nacreous aragonite typified by a low luster, 1 had slight to no luster, while those labeled with 0 had no luster and were in relatively poor condition (thin, fragmented, or partly dissolved). A check mark in the final column indicates diagenesis as determined by trace-element chemistry.

DATA APPENDIX 1 (CONTINUED)

OYSTER

Depth (ft)	$\delta^{13}C$ (‰PDB)	$\delta^{18}O$ (‰PDB)	Species	Rank	Diagenesis
E629.42 (B)	3.35	-5.39	Pseudoperna sp.	2	√
E667.61	3.82	-4.45	Pseudoperna sp.	2	√
E677.62 (B)	4.79	-2.92	Pseudoperna bentonensis	1	
E712.36	1.51	-5.66	Pseudoperna sp.	0	√
E753.84	3.51	-4.36	Psuedoperna sp.	0	√
E759.15	4.85	-6.08	Pseudoperna bentonensis	2	
E759.98	4.72	-6.15	Pseudoperna bentonensis	1	slight
E760.57	4.37	-5.57	Pseudoperna bentonensis	2	√
E762.94 (B)	4.62	-6.08	Pseudoperna sp.	0	√
E793.54	3.66	-6.03	Pycnodonte sp.	2	√
E794.35	3.76	-5.38	Pycnodonte sp.	2	√
E799.59	7.17	-4.53	Phelopteria sp.	0	√
E820.54	5.67	-2.44	Pycnodonte sp.	2	
E821.71	4.07	-5.73	Pycnodonte sp.	2	√
E832.96	4.18	-3.74	Crassostrea soleniseus	2	

Rank refers to visual assessment of diagenesis. A ranking of 2 represents relatively thick, clean well-preserved shells. Samples ranked 1 tended to be thinner and more poorly preserved, while those rated 0 were very thin and poorly preserved. A check mark in the last column indicates diagenesis on the basis of trace-element analysis.

DATA APPENDIX 1 (CONTINUED)

MATRIX

Depth (ft)	$\delta^{13}C$ (‰PDB)	$\delta^{18}O$ (‰PDB)	Material
E503.29	1.34	-6.99	Matrix
E514.60	1.59	-5.63	Matrix
E541.90	0.76	-6.53	Matrix
E586.68	1.23	-6.76	Matrix
E590.22	1.48	-6.14	Matrix
E610.84	1.27	-6.31	Matrix
E644.07	1.79	-6.91	Matrix
E667.61	2.32	-5.40	Matrix
E677.62	2.12	-5.90	Matrix
E682.88	1.74	-7.11	Matrix
E683.87	1.62	-7.00	Matrix
E691.87	1.60	-7.25	Matrix
E697.53	2.10	-7.02	Matrix
E709.73	1.89	-6.76	Matrix
E711.66	2.03	-6.50	Matrix
E712.36	1.97	-6.76	Matrix
E712.42	1.97	-6.76	Matrix
E712.85	2.02	-6.80	Matrix
E713.26	2.08	-6.63	Matrix
E713.50	2.03	-6.58	Matrix
E714.29	2.09	-6.83	Matrix
E715.20	2.13	-6.70	Matrix
E718.65	2.02	-7.35	Matrix
E718.83	2.02	-7.14	Matrix
E719.09	2.16	-6.91	Matrix
E721.25	2.10	-7.19	Matrix
E721.40	1.98	-7.33	Matrix
E723.45	2.01	-6.86	Matrix
E726.91	2.07	-6.80	Matrix
E727.89	2.03	-6.77	Matrix
E728.72	2.18	-6.71	Matrix
E731.00	2.18	-6.38	Matrix
E732.60	2.05	-6.89	Matrix
E733.32	2.16	-6.82	Matrix
E737.04	2.14	-7.37	Matrix
E740.09	2.20	-6.63	Matrix
E745.20	2.73	-6.55	Matrix
E752.45	2.15	-6.92	Matrix
E768.98	2.53	-7.19	Matrix
E807.68	3.77	-4.51	Matrix

Matrix represents material surrounding individual macrofossils.

Data Appendix 2: Trace element data for carbonate macrofossils expressed as ppm of sample dry weight.

INOCERAMID

Depth	Ca (ppm)	Fe (ppm)	Mg (ppm)	Mn (ppm)	Sr (ppm)	Na (ppm)	K (ppm)	S (ppm)
E154.72	0	522	3809	174	1143	3102	414	2612
E192.45	68411	59	155	16	212	771	146	2392
E193.09	0	1563	20638	1628	5990	18294	2604	26042
E208.82	370735	369	3007	231	1060	490	239	0
E255 a	373152	253	591	17	1896	3749	515	5686
E260.01b	329120	3099	3781	601	1031	4695	1643	0
E268.15	346162	717	803	236	1389	4272	4154	0
E280.76	370334	128	378	45	1590	3576	343	1055
E282.93	332161	1347	579	43	1047	3937	844	3656
E283.28	384319	339	762	76	1532	3752	669	6689
E290.50a	367313	417	3166	114	991	3250	432	3955
E292.8	344939	826	1762	60	1082	4275	815	6520
E295.01	358136	773	270	45	1238	3748	378	1644
E323.91	342770	184	4773	37	1148	3996	1500	5949
E337.88a	379417	280	4479	62	935	3545	732	5310
E349.48b	364725	120	469	16	1294	3909	562	4968
E356.88	359341	398	4786	77	956	3592	582	5548
E365.40	374778	49	169	0	1964	4114	280	1050
E395.96	337437	261	630	26	1046	3052	264	6029
E397.92	351141	1478	1921	360	857	2692	692	8462
E408.42	367115	2335	2150	120	1857	4700	723	1033
E421.54	348483	2513	7511	274	733	4450	1942	1618
E425.99	269690	1609	4406	0	972	6473	2232	2232
E430.76	364913	1468	5587	203	1194	3096	588	4618
E443.99	373081	803	4581	41	984	4709	1027	6849
E447.29	361855	1951	1578	184	598	1977	659	0
E449.70	339058	348	1021	24	969	4835	1463	636
E453.00	354905	1721	5405	106	1355	4844	1169	3341
E454.91	437916	489	1358	0	1348	4864	535	0
E456.70	370176	1487	3987	111	1054	3387	451	2097
E476.16a	348657	1080	4717	33	933	4582	1162	6308
E483.39	377163	1567	5921	92	996	4228	587	5456
E514.60	3395832	15496	42013	235	8905	37864	7767	0
E519.47	273480	2510	5074	0	727	7843	3922	4902
E541.60	120758	1054	1718	0	349	3538	2005	0
E541.75	30449	176	362	0	105	519	184	1297
E567.87a	254205	2756	7025	37	681	5574	3378	0
E589.56	368755	2283	5460	68	1191	3814	1680	2767
E590.08	339119	3443	6948	109	967	5179	2260	10358
E606.78	373605	267	334	43	1426	3469	321	0
E610.84	369115	472	346	0	1206	3772	409	1462
E622.46	3755423	6687	3140	258	894	1479	590	1000
E628.37	249583	4232	5084	0	1008	7750	4750	42500
E629.42a	258084	13442	7156	88	1173	7500	4167	0
E644.07	364969	2441	4542	54	907	3332	697	4183
E644.54	370980	2442	5810	11	1038	3578	912	4102
E652.58	375818	1221	5522	94	979	3452	702	6429
E666.77	334914	724	771	122	1203	5856	1126	5631

DATA APPENDIX 2 (CONTINUED)

INOCERAMID

Depth	Ca (ppm)	Fe (ppm)	Mg (ppm)	Mn (ppm)	Sr (ppm)	Na (ppm)	K (ppm)	S (ppm)
E668.61	337204	675	989	9	1178	7188	3035	15176
E673.32a	323315	5150	3833	191	615	4735	2507	27855
E677.62a	374372	1512	3456	82	912	0	0	0
E682.88	345323	2335	5044	112	979	8496	2637	20508
E683.87	374019	842	5847	39	1022	2956	384	5845
E691.87	377534	2353	6041	174	1323	4765	1893	7180
E697.20	323960	4277	5526	91	1349	7158	3734	14523
E709.73	359582	1868	4286	42	655	2367	682	0
E710.27	330666	2200	5933	139	670	12716	4526	23707
E711.66	350362	1400	5292	338	1054	3811	915	0
E712.42	358935	2126	5594	237	825	2993	455	5508
E712.85	371017	1804	4560	153	720	3906	1065	0
E713.26	275186	2060	5287	130	721	5634	1761	14085
E713.50	319401	2914	4834	80	649	2987	1051	12168
E714.29	328849	4310	6985	105	1012	6455	2869	0
E715.20	324251	1987	5401	0	628	7353	3677	0
E718.65	349579	2412	6718	41	831	2945	704	5762
E718.83	352028	1697	5901	47	793	4604	959	639
E719.09	318027	529	4734	0	768	3046	736	4016
E719.67	309778	3090	5397	0	894	9247	5822	6849
E720.89	358618	5410	6477	291	1423	6787	3670	0
E721.25	355944	1368	5350	112	775	2545	525	8547
E721.4	383088	1931	5988	72	797	3141	833	0
E723.45	359867	1436	5071	111	781	3119	743	4455
E726.91	359780	1304	6490	102	878	3179	729	3146
E727.89	359636	2143	5844	200	827	2588	260	4860
E728.72	366853	567	6011	0	933	3380	563	12676
E731.00	365958	1962	5447	119	920	3215	554	12195
E732.6	368921	2175	5298	55	729	3200	1247	4067
E733.32	375208	2069	5890	69	1090	3318	813	3893
E737.04	252194	1718	3643	0	513	5357	3274	0
E740.09	365179	1977	3106	117	694	2596	890	0
E745.20	342242	1158	4984	46	784	3333	885	14583
E745.52	373576	1554	7396	97	950	3442	634	4227
E745.71	357537	1343	6166	68	962	3655	800	8182
E746.87	382692	1279	3945	115	841	2664	760	4153
E752.45	364780	1408	5702	133	782	2840	420	0
E753.33	355871	1785	6227	232	945	4582	1685	14151
E757.50	331221	2728	6018	69	784	4933	3027	0
E762.46	332142	2998	6225	49	705	7229	4518	0
E763.55	306025	232	1356	30	953	15083	2479	0
E765.02	367353	1739	6293	94	947	3859	829	5208
E765.15	347309	3578	5865	202	799	9758	4275	13011
E800.57	368354	940	5846	67	1132	4041	475	5644
E801.54	367162	664	5708	149	882	3149	637	1893
E806.34	362966	81	139	0	1402	4703	2475	2122
E807.68	355530	315	5175	44	1022	4913	873	6550
E813.46	372352	1409	5604	234	947	4119	1141	2140
E813.86	372861	726	6261	205	987	6851	759	4848

AMMONITE

Depth	Ca (ppm)	Fe (ppm)	Mg (ppm)	Mn (ppm)	Sr (ppm)	Na (ppm)	K (ppm)	S (ppm)
E255 a	373152	253	591	17	1896	3749	515	5685
E255.46	342485	233	501	23	2115	5406	2359	726
E260.01a	244953	136	2663	0	1431	24395	6250	0
E267.10	333096	210	1630	70	2206	4675	1085.14	417
E272.30	368900	500	555	100	2535	4521	734	11206
E277.20	375971	363	1116	5	2103	4375	767	2289
E281.30	358411	724	1981	86	1900	4666	662	903
E283.96	353900	782	1414	0	2323	4311	793	8533
E288.45	347552	3384	1291	1159	393	2852	2581	23098
E290.50a	367314	417	3166	114	991	3250	433	3955
E291.15	381971	479	847	6	2257	5222	1535	1451
E296.62	373200	118	325	0	2111	4452	668	1670
E298.96	363605	2143	1325	720	951	2261	610	5384
E306.70	363776	150	319	14	1689	4316	625	4336
E308.19	405642	539	1339	39	2030	5489	761	1730
E311.90	339074	218	516	7	1876	4274	369.05	0.00
E319.34	376485	933	763	275	1670	3649	440	3271
E321.65	362434	254	483	0	2076	4361	673	2142
E337.98b	472009	777	1484	0	2454	6128	817	4085
E338.60	370957	585	682	7	2426	4274	790	1437
E340.68	360300	349	666	0	2037	4532	579	0
E347.08	380995	333	482	0	2285	4308	753	2589
E348.37	373496	776	4413	312	912	2779	483	7030
E349.48a	356928	1138	2053	86	2192	4751	571	2105
E355.08	292595	1136	2230	10	1410	7051	3571	11905
E355.12	372462	587	969	0	1970	4786	632	0
E362.59	313434	583	1845	0	1750	7127	2522	0
E363.97	314952	1249	3475	0	1621	5925	1590	0
E376.85	368811	190	455	6	2320	3822	228	1036
E380.78	342186	527	1427	0	1704	5980	1329	0
E384.04	343666	612	1016	3	1747.09	6502	695.07	672.65
E397.13	353863	3526	1528	113	3062	6735	832	2772
E399.89	320495	3039	4614	161	1298	6391	2350	9399
E438.5	350165	1405	3053	43	2065	5432	1339	2232
E459.10	405052	3700	8834	0	1985	13750	6563	18750
E476.16b	363371	4320	6633	0	1577	28448	16379	0
E500.85	197339	5898	6611	741	1153	27778	8333	62500
E503.29	258121	3036	4688	0	1031	14549	5943	10246
E509.92	23899	405	618	0	100	1036	518	259
E515.64	303759	1456	8734	0	1169	11225	6633	33163
E522.04	336075	2481	7601	0	1688	13208	3066	16509
E562.32	6135	165	283	4	34	343	254	0
E567.87b	276014	2717	3220	0	1207	12174	5435	19565
E590.22	332339	2288	1925	0	2010	6172	1778	0
E616.52	331796	1838	1231	0	1940	4380	855	0
E617.09	482889	632	2884	0	2228	22297	11487	135135
E813.96	258623	397	904	0	1535	8168	3465	22277
E817.66	363657	1316	1524	61	2081	6629	1121	6884

DATA APPENDIX 2 (CONTINUED)

OYSTERS

Depth	Ca (ppm)	Fe (ppm)	Mg (ppm)	Mn (ppm)	Sr (ppm)	Na (ppm)	K (ppm)	S (ppm)
E629.42b	388820	967	2018	58	748	2179	573	0
E667.61	369669	2104	1790	52	596	2675	1216	6323
E677.62b	345627	389	872	0	963	5517	1379	0
E712.36	338659	2928	6115	220	746	8408	3251	8969
E753.84	364238	780	1351	6	651	2500	625	0
E759.15	396403	1204	1417	101	745	2133	687	2531
E759.98	374965	1201	1627	235	808	3086	926	0
E760.57	375858	1440	2155	97	655	2518	1079	3597
E762.94b	374746	1417	2286	157	650	8456	1471	5515
E793.54	381235	2468	2472	191	755	1978	364	2306
E794.35	366852	1676	2260	167	629	2386	1008	10081
E799.59	388648	655	1327	57	2373	7069	2241	0
E820.54	313152	402	836	0	1137	6386	1359	0
E821.71	308697	1108	470	250	2110	3518	484	5057
E832.96	387440	444	651	40	803	1664	698	7395

REFERENCES

ARTHUR, M. A., DEAN, W. E., AND PRATT, L. M., 1988, Geochemical and climatic effects of increased marine organic carbon burial at the Cenomanian/Turonian boundary, Nature, v. 335, p. 714-717.

ARTHUR, M. A., DEAN, W. E., AND SCHOLLE, P. A., 1985, Comparative geochemical and mineralogical studies of two cyclic trangressive pelagic limestone units, Cretaceous Western Interior basin, U.S., in Pratt, L. M., Kauffman, E. G., and Zelt, F. B., eds., Fine-grained Deposits and Biofacies of the Cretaceous Western Interior Seaway: Evidence of Cyclic Sedimentary Processes, Field Trip Guidebook No. 4: Tulsa, Society for Economic Paleontologists and Mineralogists, p. 157-169.

ARTHUR, M. A., SCHLANGER, S. O., AND JENKYNS, H. C., 1987, The Cenomanian-Turonian oceanic anoxic event, II. Palaeoceanographic controls on organic matter production and preservation, in Brooks, J. and Fleet, A. J., eds., Marine Petroleum Source Rocks: Cambridge, Geological Society Special Publication No. 26, p. 401-420.

BARRON, E. J., 1983, A warm, equable Cretaceous: The nature of the problem, Earth Science Review, v. 29, p. 305-338.

BICE, K., ARTHUR, M. A., AND MARINCOVICH JR., L., 1996, Late Paleocene Artic Ocean shallow-marine temperatures from mollusc stable isotopes, Paleoceanography, v. 11, p. 241-249.

BRAND, U., 1987, Depositional analysis of the Breathitt formation's marine horizons, Kentucky, U.S.A.: Trace elements and stable isotopes, Chemical Geology, v. 65, p. 117-136.

BRAND, U., 1990, Chemical diagenesis and dolomitization of Paleozoic high-Mg calcite crinoids, Carbonates and Evaporites, v. 5, p. 179-195.

BRAND, U., AND VEIZER, J., 1980, Chemical diagenesis of a muticomponent carbonate system-1: trace elements, Journal of Sedimentary Petrology, v. 50, p. 1219-1236.

CRAIG, H., AND GORDON, L. I., 1965, Deuterium and oxygen 18 variation in the ocean and the marine atmosphere, in Tongiorgi, E., ed., Stable Isotopes in Oceanographic Studies and Paleotemperatures: Pisa, Consiglio Nationale delle Ricerche Laboratorio di Geologia Nucleare, p. 161-182.

DETTMAN, D., AND LOHMANN, K. C., 1993, Seasonal change in Paleogene surface water $\delta^{18}O$; fresh-water bivalves of western North America, in Swart, P. K., Lohmann, K. C., McKenzie, J. A., and Savin, S. eds.: Climate Change in Continental Isotopic Records: Washington, D.C. American Geophysical Union, Geophysical-Monograph, v. 78, p. 153-163.

EICHER, D. L., AND DINER, R., 1985, Foraminifera as indicators of water mass in the Cretaceous Greenhorn sea, Western Interior., in Pratt, L. M., Kauffman, E. G., and Zelt F. B., eds., Fine-grained Deposits and Biofacies of the Cretaceous Western Interior Seaway: Evidence of Cyclic Sedimentary Processes, Field Trip Guidebook No. 4: Tulsa, Society for Economic Paleontologists and Mineralogists, p. 60-71.

ELDER, W. P., AND KIRKLAND, J. I., 1994, Cretaceous paleogeography of the southern Western Interior region, in Caputo, M. V., Peterson, J. A., and Franczyk, K. J., eds., Mesozoic Systems of the Rocky Mountain Region, USA, Tulsa, Society

for Economic Paleontologists and Mineralogists, The Rocky Mountain Section, p. 415-440.

EPSTEIN, S., BUCHSBAUM, R., LOWENSTAM, H., AND UREY, H. C., 1951, Carbonate-water isotopic temperature scale, Bulletin of the Geological Society of America, v. 62, p. 417-426.

ERICKSEN, M. C., AND SLINGERLAND, R., 1990, Numerical simulations of tidal and wind-driven circulation in the Cretaceous Interior Seaway of North America, Geological Society of America Bulletin, v. 102, p. 1499-1516.

FISHER, C. G., HAY, W. W., AND EICHER, D. L., 1994, Oceanic front in the Greenhorn Sea (late middle through late Cenomanian), Paleoceanography, v. 9, p. 879-892.

FRUSH, M. P., AND EICHER, D. L., 1975, Cenomanian and Turonian foraminifera and paleoenvironments in the Big Bend region of Texas and Mexico, in Caldwell, W.G.E., ed., The Cretaceous System in North America: St John's, Geological Association of Canada Special Paper 13, p. 277-301.

GARDNER, M. H., AND CROSS, T. A., 1994, Middle Cretaceous paleogeography of Utah, in Caputo, M. V., Peterson, J. A., and Franczyk, K. J., eds., Mesozoic Systems of the Rocky Mountain Region, USA, Tulsa, Society for Economic Paleontologists and Mineralogists, The Rocky Mountain Section, p. 471-502.

GAT, J. R., 1980, The isotopes of hydrogen and oxygen in precipitation, in Fritz, P., and Fontes, J., eds., Handbook of Environmental Isotope Geochemistry: New York, Elsevier, p. 21-47.

GLANCY, T. J., ARTHUR, M. A., BARRON, E. J., AND KAUFFMAN, E. G., 1993, A paleoclimate model for the North American Cretaceous (Cenomanian-Turonian) epicontinental sea, in Caldwell, W. G. E., and Kauffman, E. G., eds., Evolution of the Western Interior Basin: St. John's, Geological Association of Canada Special Paper 39, p. 219-241.

GORDON, W. A., 1973, Marine life and ocean surface currents in the Cretaceous, Journal of Geology, v. 81, p. 269-284.

GROSSMAN, E. L., AND KU, T., 1986, Oxygen and carbon isotope fractionation in biogenic aragonite: Temperature effects, Chemical Geology, v. 59, p. 59-74.

HAY, W. W., EICHER, D. L., AND DINER, R., 1993, Physical oceanography and water masses in the Cretaceous Western Interior Seaway, in Caldwell, W. G. E., and Kauffman, E. G., eds., Evolution of the Western Interior Basin: St. John's, Geological Association of Canada Special Paper 39, p. 297-318.

HAYNES, J. R., 1981, Foraminifera: New York, John Wiley and Sons, p. 433.

JEWELL, P. W., 1993, Water-column stability, residence times, and anoxia in the Cretaceous North American seaway, Geology, v. 21, p. 579-582.

JORDAN, T. E., 1981, Thrust loads and foreland basin evolution, Cretaceous, Western United States, The American Association of Petroleum Geologists Bulletin, v. 65, p. 2506-2520.

KAUFFMAN, E. G., 1977, Geological and biological overview: Western Interior Cretaceous Basin, The Mountain Geologist, v. 14, p. 75-99.

KAUFFMAN, E. G., 1984, Paleobiogeography and evolutionary response dynamic in the Cretaceous Western Interior Seaway of North America, in Westermann, G. E. G., ed., Jurassic-Cretaceous Biochronology and Paleogeography of North America: St. John's, Geological Association of Canada Special Paper 27, p. 273-306.

KAUFFMAN, E. G., AND SAGEMAN, B. B., 1990, Biological sensing of benthic environments in dark shales and related oxygen-restricted facies, *in* Ginsburg, R. N., and Beaudoin, B., eds., Cretaceous Resources, Events and Rhythms: Dordrecht, Kluwer Academic Publishers, NATO Advanced Science Institute Series, p. 121-138.

KAUFFMAN, E. G., SAGEMAN, B. B., KIRKLAND, J. I., ELDER, W. P., HARRIES, P. J., AND VILLAMIL, T., 1993, Molluscan biostratigraphy of the Cretaceous Western Interior Basin, North America, *in* Caldwell, W. G. E., and Kauffman, E. G., eds., Evolution of the Western Interior Basin: St. John's, Geological Association of Canada Special Paper 39, p. 397-434.

KYSER, T. K., CALDWELL, W. G. E., WHITTAKER, S. G., AND CADRIN, A. J., 1993, Paleoenvironment and geochemistry of the northern portion of the Western Interior Seaway during Late Cretaceous time, *in* Caldwell, W. G. E., and Kauffman, E. G., eds., Evolution of the Western Interior Basin: St. John's, Geological Association of Canada Special Paper 39, p. 355-378.

LECKIE, R. M., SCHIMDT, M. G., FINKELSTEIN, D., AND YURETICH, R., 1991, Paleoceanographic and paleoclimatic interpretations of the Mancos Shale (Upper Cretaceous), Black Mesa Basin, Arizona: Boulder, Geological Society of America Special Paper 260, p. 139-152.

LUDVIGSON, G. A., WITZKE, B. J., GONZALEZ, L. A., HAMMOND, R. H., AND PLOCHER, O. W., 1994, Sedimentology and carbonate geochemistry of concretions from the Greenhorn marine cycle (Cenomanian-Turonian), eastern margin of the Western Interior Seaway, *in* Shurr, G. W., Ludvigson, G. A., and Hammond, R. H., eds., Perspectives on the Eastern Margin of the Cretaceous Western Interior Basin: Boulder, Geological Society of America Special Paper 287, p. 145-173.

MACLEOD, K. G. AND HOPPE, K. A., 1992, Evidence that Inoceramid bivalves were benthic and harbored chemosynthetic symbionts, Geology, v. 20, p. 117-120.

McINTIRE, W. L., 1963, Trace element partition coefficients: A review of theory and applications to geology, Geochimica et Cosmochimica Acta, v. 27, p. 1209-1264.

MILLIMAN, J. D., 1974, Recent Sedimentary Carbonates: Part 1. Marine Carbonates: New York, Springer-Verlag, p. 375.

MORRISON, J. O., 1991, Cretaceous marine invertebrates: A geochemical perspective, Ph.D. dissertation, University of Ottawa, Ottawa, Ontario, Canada, p. 352.

MORRISON, J. O., AND BRAND, U., 1986, Geochemistry of recent marine invertebrates, Geoscience Canada, v. 13, p. 237-254.

OBRADOVITCH, J., 1993, A Cretaceous time scale, *in* Caldwell, W. G. E., and Kauffman, E. G., eds., Evolution of the Western Interior Basin: St. John's, Geological Association of Canada Special Paper 39, p. 379-396.

PARRISH, J. T., GAYNOR, G. C., AND SWIFT, D. J. P., 1984, Circulation in the Cretaceous Western Interior Seaway of North America, a review, *in* Stott, D. F., and Glass, D. J., eds., The Mesozoic of Middle North America: Calgary, Canadian Society of Petroleum Geologists Memoir 9, p. 221-231.

PRATT, L. M., 1985, Isotopic studies of organic matter and carbonate in rocks of the Greenhorn marine cycle, *in* Pratt, L. M., Kauffman, E. G., and Zelt F. B., eds., Fine-grained Deposits and Biofacies of the Cretaceous Western Interior Seaway: Evidence of Cyclic Sedimentary Processes, Field Trip Guidebook No. 4: Tulsa, Society for Economic Paleontologists and Mineralogists, p. 122-134.

PRATT, L. M., ARTHUR, M. A., DEAN, W. E., AND SCHOLLE, P.A., 1993, Paleoceanographic cycles and events during the late Cretaceous in the Western Interior Seaway of North America, *in* Caldwell, W. G. E., and Kauffman, E. G., eds., Evolution of the Western Interior Basin: Geological Association of Canada Special Paper 39, p. 333-353.

RAILSBACK, L. B., 1990, Influence of changing deep ocean circulation on the Phanerozoic oxygen isotopic record, Geochimica et Cosmochimica Acta, v. 54, p. 1501-1509.

RAILSBACK, L. B., AND ANDERSON, T. F., 1989, Paleoceanographic modeling of temperature-salinity profiles from stable isotopic data, Paleoceanography, v. 4, p. 585-591.

SAGEMAN, B. B., 1985, High-resolution stratigraphy and paleobiology of the Hartland Shale Member: Analysis of an oxygen-deficient epicontinental sea, *in* Pratt, L. M., Kauffman, E. G., and Zelt F. B., eds., Fine-grained Deposits and Biofacies of the Cretaceous Western Interior Seaway: Evidence of Cyclic Sedimentary Processes, Field Trip Guidebook No. 4: Tulsa, Society for Economic Paleontologists and Mineralogists, p. 110-121.

SALTZMAN, E. S., AND BARRON, E. J., 1982, Deep circulation in the Late Cretaceous: Oxygen isotope paleotemperatures from Inoceramus remains in D.S.D.P. cores, Palaeogeography, Palaeoclimatology, Palaeoecology, v. 40, p. 167-181.

SCHLANGER, S. O., ARTHUR, M. A., JENKYNS, H. C., AND SCHOLLE, P. A., 1987, The Cenomanian-Turonian oceanic anoxic event, I. Stratigraphy and distribution of organic carbon-rich beds and the marine $\delta^{13}C$ excursion, *in* Brooks, J., and Fleet, A. J., eds., Marine Petroleum Source Rocks: Cambridge, Geological Society Special Publication No. 26, p. 371-399.

SCHOLLE, P. A., AND ARTHUR, M. A., 1980, Carbon isotope fluctuations in Cretaceous pelagic limestones: Potential stratigraphic and petroleum exploration tool, Bulletin American Association of Petrology Geologists, v. 64, p. 67-87.

SHACKLETON, N. J., AND KENNETT, J. P., 1975, Paleotemperature history of the Cenozoic and the initiation of Antarctic glaciation: Oxygen and carbon isotope analyses in DSDP sites 277, 279, and 281, Initial Reports of the Deep Sea Drilling Program; v. 29, p. 743-755.

SLINGERLAND, R., KUMP, L. R., ARTHUR, M. A., FAWCETT, P. J., SAGEMAN, B. B., AND BARRON, E. J., 1996, Estuarine circulation in the Turonian Western Interior seaway of North America, Geological Society of America Bulletin, v. 108, p. 941-952.

SPIELHAGEN, R. F., AND ERLENKEUSER, H., 1994, Stable oxygen and carbon isotopes in planktic foraminifers from Arctic Ocean surface sediments: Reflection of the low salinity surface water layer, Marine Geology, v. 119, p. 227-250.

THUNNELL, R. C., WILLIAMS, D. F., AND HOWELL, M., 1987, Atlantic-Mediterranean water-exchange during the Late Neogene, Paleoceanography, v. 2, p. 661-678.

TOURTELOT, H. A., AND RYE, R. O., 1969, Distribution of oxygen and carbon isotopes in fossils of Late Cretaceous age, Western Interior region of North America, Geological Society of America Bulletin, v. 80, p. 1903-1922.

VEIZER, J., 1983, Chemical diagenesis of carbonates: Theory and application of trace element technique, *in* Arthur, M. A., and Andersen, T. F., eds., Stable Isotopes in Sedimentary Geology: Tulsa, Society for Economic Paleontologists and Mineralogists Short Course No. 10, p. 1-100.

WATKINS, D. K., 1986, Calcareous nannofossil paleoceanography of the Cretaceous Greenhorn Sea, Geological Society of America Bulletin, v. 97, p. 1239-1249.

WHITTAKER, S. G., KYSER, T. K., AND CALDWELL, W. G. E., 1987, Paleoenvironmental geochemistry of the Claggett marine cyclothem in south-central Saskatchewan, Canadian Journal of Earth Science, v. 24, p. 967-984.

WRIGHT, E. K., 1987, Stratification and paleocirculation of the Late Cretaceous Western Interior Seaway of North America, Geological lSociety of America Bulletin, v. 99, p. 480-490.

GEOCHEMICAL EXPRESSIONS OF CYCLICITY IN CRETACEOUS PELAGIC LIMESTONE SEQUENCES: NIOBRARA FORMATION, WESTERN INTERIOR SEAWAY

WALTER E. DEAN[1] AND MICHAEL A. ARTHUR[2]

[1]*U.S. Geological Survey, Federal Center, Denver, Colorado 80225*
[2]*Department of Geosciences, Pennsylvania State University, University Park, Pennsylvania 16802*

ABSTRACT: Marked cycles in color, carbonate and organic-carbon concentrations, geophysical log characteristics (for example, gamma-ray and sonic velocity), degree of bioturbation, and composition of benthic faunal assemblages characterize the pelagic carbonate sequences of the Upper Cretaceous Niobrara Formation in the Cretaceous Western Interior Seaway of North America and seem to be attributable to Milankovitch orbital forcing (periodicities of ca. 20-400 ky). A hierarchy of cycles is present, starting with decimeter- to meter-scale bedding cycles with nominal 20 ky periodicity that are bundled into larger cycles of perhaps 100 ky, which are, in turn, encompassed by much longer cycles of >1 My. The longer cycle may not be Milankovitch in origin but instead may be related to variations in sea level and tectonism in the Western Interior Seaway instead. Geochemical data constrain interpretations of the depositional mechanisms that led to the cyclicity in that subtle variations in mineralogy and redox conditions are easily seen in major- and minor-element geochemical data. Carbon-isotope and Rock-Eval pyrolysis data on the organic fraction across the cycles suggest that cyclic variations in lithology are in part caused by changes in surface-water biotic productivity and preservation of organic matter at the seafloor.

All cycles in the Niobrara Formation apparently are the product of a combination of changes in dilution of biogenic components by terrigenous detritus, carbonate production in surface waters, and degree of oxygenation of bottom-water. All of these factors are interrelated and ultimately controlled by fresh-water runoff to the seaway from uplifted highlands to the west, and possibly from the low-relief eastern margin, as well as by large-scale transgressive-regressive cycles. Geochemically, the cycles can be expressed in terms of a three-component system of $CaCO_3$, clastic material (represented by aluminum, Al), and organic matter (represented by organic carbon, OC). Within the bedding cycles, the clastic-rich marlstone beds almost always contain the highest concentrations of OC. However, on the scale of the entire formation, variations in clastic material (Al) are not always in phase with those of organic matter (OC), which suggests that factors controlling the production and preservation of organic matter were not always related to factors controlling the deposition of detrital clastic material.

Q-mode factor analyses show that most major, minor, and trace elements (especially Ti, Mg, Na, K, Ce, Li, Nd, Sc, and Y) are associated with Al in the clastic fraction. Strontium is mainly in the carbonate fraction, and most other trace elements (especially Cd, Cu, Mo, Ni, V, and Zn) are associated with the organic fraction. Most iron and sulfur are present as pyrite (FeS_2) in these rocks which, like most carbonate-rich rocks and sediments, are iron-limited with respect to pyrite formation. Concentrations of manganese (Mn), which is very sensitive to redox conditions, are highest in the limestone beds of the Fort Hays Limestone Member, which were deposited under oxic bottom-water conditions, and are very low in the Smoky Hill Chalk Member, which was deposited under suboxic to anoxic bottom-water conditions.

INTRODUCTION

The Cretaceous formations that were deposited in the Western Interior Seaway (WIS) of North America provide remarkable examples of sedimentary cycles on scales of tens of thousands of years to tens of millions of years (Gilbert, 1895; Hattin, 1971; Kauffman, 1977; Fischer, 1986; Pratt, 1981; Arthur et al., 1984; Barron et al., 1985; Barlow and Kauffman, 1985; Arthur and Dean, 1991; Pratt et al., 1993). The basin is one of the most studied in the world because of the striking expressions of these cycles in the subsurface and in outcrop and the abundance of well-exposed sections. The basin may have been particularly sensitive to eustatic, orbital, and climatic cycle forcing because of its unique setting as a partly restricted, north-south arm of the east-west Tethys ocean and its large latitudinal span. Transgressive-regressive cycles on scales of 6-10 My, the second order cycles of Kauffman (1977), are well documented in the WIS (Kauffman and Caldwell, 1993). Nine of these second-order cycles are recognized. They reached maximum transgressions during cycles five (Greenhorn) and six (Niobrara), when the seaway was wide and deep enough to limit detrital clastic influx to the basin center to allow relatively pure carbonate sediments to accumulate. Superimposed on these second-order cycles are Kauffman's third-order cycles that have periodicities of 1-3 My and are particularly well-expressed in the Smoky Hill Chalk (Kansas) or Limestone (Colorado) Member of the Niobrara Formation. These second-order cycles were used by Scott and Cobban (1964) to subdivide that member into informal marlstone-limestone units. Fourth-order cycles with periodicities of hundreds

of thousands of years can often be identified on geochemical and mechanical well logs. We show and discuss some examples of fourth-order cycles based on geochemical data.

Higher-frequency cycles have periodicities of tens of thousands of years that are in the range of solar-terrestrial orbital cycles, the so-called Milankovitch band. Perhaps the most striking of the Milankovitch-band cycles, particularly in outcrop, are the distinctive cyclic interbeds of different lithologies, which are mostly controlled by variation in amounts of carbonate and organic matter. These cycles are particularly well developed in pelagic carbonate units of the Niobrara and Greenhorn Formations of the WIS and sequences of the Tethyan realm (e.g., Fischer et al., 1985; Bottjer et al., 1986; ROCC Group, 1986; Arthur and Dean, 1991; Pratt et al., 1993; Sageman et al., this volume). Most of these carbonate sequences are characterized by cyclically interbedded chalks/limestones and shales/marlstones with estimated periodicities in the 20-40 ky range. Some carbonate cycles may be explained by diagenetic redistribution of carbonate (e.g., Hallam, 1964, 1986; Ricken, 1985, 1986; Raiswell, 1988), but the variations in carbonate, organic carbon, and inorganic geochemical characteristics within limestone/marlstone cycles of many epicontinental and deep-sea depositional environments most likely reflect cyclic climatic variations accompanied by changes in surface-water productivity, bottom-water energy and oxygenation, and flux and composition of detrital clastic material (Arthur and Dean, 1991). G. K. Gilbert (1895) first proposed that solar-terrestrial orbital cycles might be forcing mechanisms for the striking limestone/marlstone bedding cycles that are so prominent in outcrops of the Niobrara and Greenhorn

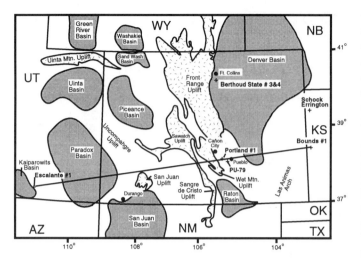

FIG. 1.–Map showing locations of major basins and uplifts in Colorado and adjacent states, and locations of cores that have been analyzed as part of the Cretaceous Western Interior Continental Scientific Drilling Project.

Formations. More recent studies suggest that the cycles in the carbonate units of the Western Interior do indeed lie within the Milankovitch band and have strengthened Gilbert's hypothesis (e.g., Hattin, 1986; Laferriere et al., 1987; Pratt et al., 1993; Elder et al., 1994; Kauffman, 1995; Sageman et al., 1997 and this volume).

We examine the nature of cyclicity in the Niobrara Formation in cores recovered in the Coquina Oil, Berthoud State No. 4 and No. 3 wells near Fort Collins, Colorado, the Amoco No. 1 Rebecca K. Bounds well in western Kansas, and the USGS No. 1 Portland well near Cañon City, Colorado (Fig. 1). To emphasize some points made using several geochemical characteristics, data from the Niobrara Formation in these wells will be supplemented with data from the Niobrara Formation in outcrops and quarries (Fig. 1), and, for comparison, from the Bridge Creek Limestone Member of the Greenhorn Formation. In particular, we outline how geochemistry constrains the interpretation of changes in paleoceanographic and paleoclimatic conditions; primary depositional processes; production and preservation of organic matter and other biogenic components; sources of detritus; and diagenesis. There is a marked contrast in the geochemistry of interbedded lithologies within the Niobrara Formation and other cyclic pelagic limestone sequences (e.g., Arthur and Dean, 1991), and these geochemical variations and implied mineralogical variations primarily reflect changes in: (1) the relative amounts of different components that reach the seafloor, (2) redox conditions on the sea floor and in the sediments that result from changes in the flux of organic matter and dissolved oxygen, (3) diagenetic recrystallization and cementation of biogenic components, and (4) the composition of detrital minerals delivered to the depositional sites, probably as the result of variations in weathering in the source regions and/or flux of volcanic detritus to the depositional site.

ANALYTICAL METHODS

Total carbonate and inorganic carbon were determined by coulometry (Engleman et al., 1985). Carbonate in the untreated sample is reacted with perchloric or phosphoric acid to liberate CO_2, which is then titrated in a coulometer cell to measure car-

bonate carbon. Total carbon is measured by by combusting an untreated sample at 960° C to liberate CO_2, and titrating the CO_2 in a coulometer cell. Values of organic carbon (OC) were determined by difference between total carbon and carbonate carbon. The coulometer technique has a precision of better than ±5% for both carbonate and total carbon. Percent $CaCO_3$ was calculated by dividing percent carbonate carbon by 0.12, which is the fraction of carbon in $CaCO_3$.

Pyrolysis of organic matter was performed using a Rock-Eval pyrolysis instrument, for comparison with similar analyses commonly used in the study of ancient black shales. The Rock-Eval pyrolysis method provides a rapid estimate of the hydrocarbon type and abundance in sediments and sedimentary rocks (Espitalie et al., 1977; Peters, 1986). Programmed heating of a sample in a helium atmosphere generates hydrocarbons and CO_2. The hydrocarbons generated are measured with a flame ionization detector, and the CO_2 yield is monitored by a thermal conductivity detector. The yield of free or adsorbed hydrocarbons are determined by heating the sample in flowing helium to 250°C and is recorded as the first characteristic peak on a pyrogram (S1; mg HC/g sample). The S1 peak is roughly proportional to the content of organic matter that can be extracted from the rock or sediment with organic solvents. The second peak is composed of pyrolytic hydrocarbons generated by thermal breakdown of kerogen as the sample is heated from 250°C to 550°C (S2, mg HC/g sample). CO_2 also is generated by kerogen degradation and is retained during the heating interval from 250°C to 390°C and analyzed as the integral of the third peak on the pyrogram (S3, mg CO_2/g sample). Our data are expressed in terms of a hydrogen index (HI), in which the S2 peak is normalized to the OC content of the sample (mg HC/g OC), and oxygen index (OI), which is the S3 peak normalized to the OC content of the sample (mg CO_2/g OC).

Stable-carbon isotope ratios were determined using standard techniques (Pratt and Threlkeld, 1984; Dean et al., 1986). Powdered samples for carbon-isotope determinations of organic carbon were oven dried at 40°C and reacted with an excess of 0.5N HCl for 24 hours to dissolve carbonate minerals. The washed residue from each sample was combusted at 1,000°C under oxygen pressure in an induction furnace. The resulting CO_2 was dehydrated and purified in a high-vacuum, gas-transfer system. Whole-rock samples for determining carbon and oxygen isotope ratios in carbonate were reacted with 100% phosphoric acid, and the evolved CO_2 was dehydrated and purified in a high-vacuum gas-transfer system. All isotope ratios in CO_2 were determined using an isotope-ratio mass spectrometer. Results are reported in the standard per mil (‰) δ-notation relative to the University of Chicago Pee Dee belemnite (PDB) marine-carbonate standard,

$$\delta^{13}C ‰ = [(R_{sample}/R_{PDB})-1] \times 10^3,$$

where R is the ratio ($^{13}C:^{12}C$) or ($^{18}O:^{16}O$). Analytical precision of these analyses is ±0.2‰.

For inorganic geochemical analysis, splits of 171 powdered samples of the Niobrara Formation in the Bounds core used for carbon analyses were analyzed for concentrations of 10 major elements (Si, Al, Fe, Mg, Ca, Na, K, Ti, P, and Mn) by wavelength-dispersive X-ray fluorescence spectrometry (XRF; Baedecker, 1987). The same samples were analyzed for 40 major, minor, and trace elements (Al, Fe, Mg, Ca, Na, K, Ti, P, Mn, Ag, As, Au, Ba, Be, Bi, Cd, Ce, Co, Cr, Cu, Eu, Ga, Ho, La, Li, Mo, Nb, Nd, Ni, Pb, Sc, Sn, Sr, Ta, Th, U, V, Y, Yb, and Zn) by induction-coupled, argon-plasma emission spectrometry (ICP; Baedecker, 1987). Splits of 112 samples of the Niobrara Formation in the Berthoud State No. 4 core and 48 samples of the Niobrara Formation in the Berthoud State No. 3 core were analyzed by ICP only. Concentrations of eight elements (Ag, Au, Bi, Eu, Ho, Sn, Ta, and U) were below detection limits of the ICP method in all samples analyzed. For many samples (usually those high in CaCO3), concentrations of several other trace elements also were below detection limits. As a measure of precision, 10 splits of one sample of the Tropic Shale, equivalent, in part, to the Greenhorn Formation at the western edge of the basin, in the USGS Escalante core (Fig. 1) were analyzed along with the samples from the Bounds core. The mean, standard deviation, and coefficient of variation (standard deviation as a percent of the mean) for each element in these 10 splits is given in Table 1.

Multivariate statistical analyses on the geochemical data for the Niobrara Formation in the Bounds and Berthoud State cores were conducted using a modified version of the extended CABFAC program described by Klovan and Miesch (1976). Prior to the analyses, concentrations of all elements were transformed to proportions of their range of concentrations because of the widely differing ranges of values from fractions of a percent to thousands of parts per million. The transformed values, therefore, all ranged from 0 to 1. Several iterative runs through the program were made for each of the data sets trying a principal component solution, a varimax solution, and an oblique solution, and using the correlation coefficient and cosine theta values as measures of similarity (Klovan and Miesch, 1976). For all three data sets, using cosine theta values and an oblique axis solution with five axes (factors) seemed to produce the most interpretable results. The 5-factor models explained 85% of the variance in the transformed variables in the Berthoud State No. 4 core, 88% in the Berthoud state No. 3 core, and 91% in the Bounds core.

The geochemical data described in this paper are available in digital form at the World Data Center-A for Paleoclimatology, NOAA/NGDC 325 Broadway, Boulder, CO 80303 (phone: 303-497-6280; fax: 303-497-6513).

CYCLES IN THE NIOBRARA FORMATION

Pelagic limestone beds of the Niobrara Formation were deposited during an overall major transgressive episode of the WIS (e.g., Kauffman, 1977; Arthur et al., 1985; Kauffman and Caldwell, 1993) in the late Turonian (ca. 88.8 Ma) through early Campanian (ca. 82 Ma; Fig. 2). Deposition occurred in a rapidly subsiding basin (Cross and Pilger, 1978; Jordan, 1981; Beaumont et al., 1993) characterized by substantial clastic sediment input (Ryer, 1993) from uplifted tectonic terranes of the Sevier orogenic belt to the west (Weimer, 1983; Armstrong and Ward, 1993). Concomitant volcanic activity in the same region supplied abundant windblown ash (Kauffman, 1977; Elder, 1988) that generated more than 400 bentonite beds that are traceable over 2,000 km in Cretaceous sequences of the WIS (Obradovich and Cobban, 1975).

Manifestation of Cycles

We sampled the entire Smoky Hill Member in only one core, the Berthoud State No. 4, where it is about 80 m thick (Fig. 2). Percent CaCO3 in the 1-m samples from this core clearly define the third-order cycles that Scott and Cobban (1964) used to sub-

TABLE 1.—*Mean, standard deviation, and coefficient of variation (standard deviation as a percent of the mean) of 10 analyses of a single sample.*

Element	Mean	Std. Dev.	Coef. Var.
Al	7.81	0.057	0.73
Ca	1.81	0.032	1.75
Fe	2.90	0.000	0.00
K	2.14	0.052	2.41
Mg	1.40	0.000	0.00
Na	0.78	0.005	0.62
P	0.07	0.000	0.00
Ti	0.33	0.012	3.55
Mn	100	6.9	6.90
As	11	1.0	9.54
Ba	347	4.8	1.39
Be	2	0.0	0.00
Ce	64	1.5	2.33
Co	11	0.5	4.87
Cr	76	0.4	0.56
Cu	19	1.1	6.01
Ga	18	0.7	3.98
La	35	0.7	2.02
Li	53	0.0	0.00
Nb	17	0.7	4.24
Nd	28	1.1	3.91
Ni	15	0.7	4.79
Pb	15	0.8	5.45
Sc	11	0.3	2.85
Sr	190	0.0	0.00
Th	13	1.0	7.24
V	111	3.2	2.85
Y	21	0.7	3.40
Zn	81	1.7	2.05
		Av. coef. var. =	2.88
Oxide			
SiO_2	61.1	0.09	0.15
Al_2O_3	14.8	0.04	0.28
Fe_2O_3	4.15	0.037	0.89
MgO	2.30	0.011	0.48
CaO	2.45	0.014	0.57
Na_2O	0.96	0.015	1.56
K_2O	2.66	0.007	0.26
TiO_2	0.63	0.006	0.95
P_2O_5	0.16	0.005	3.13
		Av. coef. var. =	0.92

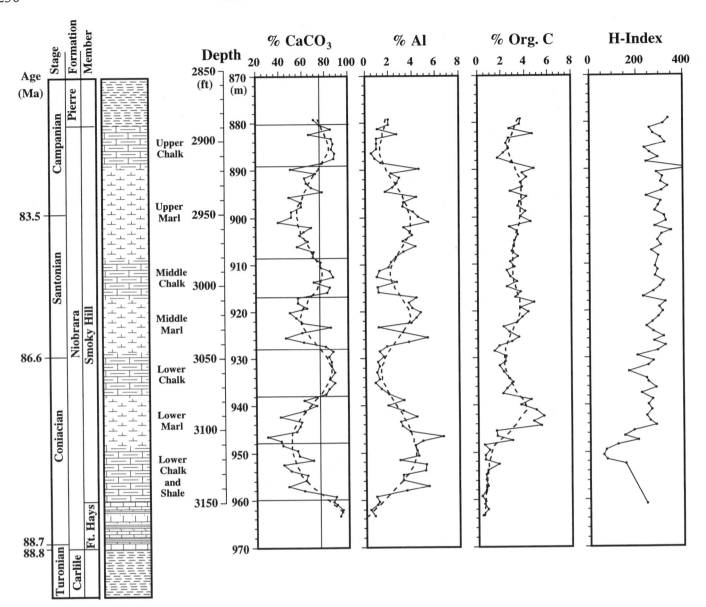

FIG. 2.–Ages, stratigraphic nomenclature, lithology, and down-core plots versus depth of % CaCO₃, % aluminum (Al), % organic carbon (Org C), and Rock-Eval pyrolysis hydrogen index (H-Index) in 1-m samples from the Berthoud State No. 4 core. Dashed lines are weighted moving-average smoothed curves. Ages of stage boundaries are from Obradovich, (1993).

divide the Smoky Hill into informal marl/chalk units (named to the right of the lithology column in Fig. 2). In the Berthoud State No. 4 core, 75% CaCO₃ (vertical line on the plot of % CaCO₃) provides a convenient cut-off for the informal marl/chalk subunits (horizontal lines on the plot of % CaCO₃). Because the Niobrara Formation is composed of essentially two lithologic components, clay and biogenic calcite, there is a distinct antithetic relation between variations in % Al and % CaCO₃, the geochemical proxies for those components (Fig. 2). This can be seen at the scale of the 10- to 20-m cycles that subdivide the Smoky Hill Member, and smaller-scale cycles (several meters) that are manifested as minima below the smoothed curve for % CaCO₃, and as maxima above the smoothed curve for %Al (Fig. 2).

Bedding cycles that are tens of centimeters to several meters in thick occur in cores of the Niobrara Formation and are even

more pronounced in outcrop where differential weathering related to carbonate content has emphasized these small-scale lithologic alternations. For example note the change from laminated marlstone to bioturbated limestone through several bedding cycles in the Smoky Hill Member of the Niobrara Formation in the Berthoud State No. 4 core (Fig. 3). The examples of laminated marlstone and bioturbated limestone illustrated in Figure 3 are fairly extreme in terms of variations in CaCO₃, and not all cycles in the Smoky Hill are composed of such high-amplitude variations, but rather are manifested as subtle variations in color and/or degree of bioturbation (Savrda, this volume). Geochemical data for three of the cycles in Figure 12 from 3041-3043.5 ft (927-928 m) are shown in Figure 4. At the scale of bedding cycles, clay-rich marlstone beds usually have high OC concentrations relative to the limestone beds (Fig. 4). At the scale of the entire formation, however, variations in OC are not always

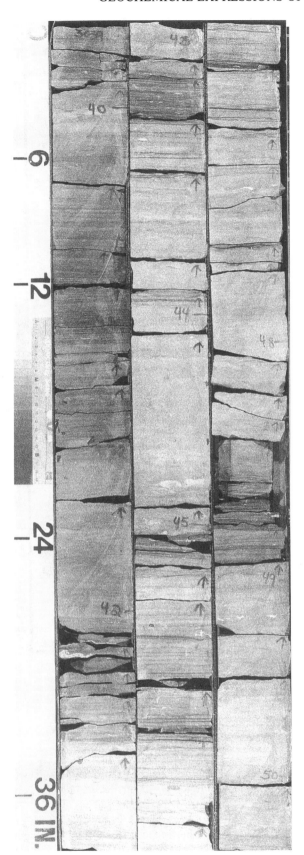

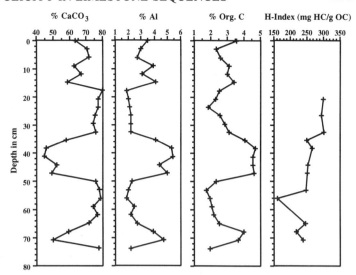

Fig. 4.–Down-core plots versus depth of % CaCO₃, % aluminum (Al), % organic carbon (Org C), and Rock-Eval pyrolysis hydrogen index (H-Index) in 3-cm samples through three limestone/marlstone cycles in the Berthoud State No. 4 core (926.9-927.7 m in Figure 2; 3041-3943.5 ft in Figure 3).

in phase with those of clay (% Al) (Fig. 2). Overall, the Smoky Hill Member is characterized by relatively high concentrations of hydrogen-rich organic matter (OC >2%; HI >200 mgHC/g OC) (Fig.2), and variable CaCO₃ contents.

Because coring in the Amoco Bounds well began at 530 ft (161.5 m), only the lower 20 m of the Smoky Hill was cored (Fig. 5). However, essentially 100% of the Fort Hays Member was recovered. The upper Fort Hays and lower Smoky Hill Members in the Bounds core was sampled at about 10-cm intervals, which allowed some of the bedding cycles to be detected by geochemical analyses. These can be seen in Figure 6 by the distinct minima below the smoothed curve for % CaCO₃, and as maxima above the smoothed curve for % Al, especially between 177 m and 181 m in the Smoky Hill and between 183.5 m and 185.5 m in the Fort Hays. As in the Berthoud State No. 4 core, the variations in OC are not always in phase with those of clay (% Al). This can be seen by the high Al content and relatively low OC content above 176 m, and the high OC content and low Al content between 182 m and 183 m. At a smaller scale, not all peaks in % OC correspond to peaks in % Al. Part of the gamma-ray log from Figure 5 is repeated in Figure 6 to show the similarities between gamma-ray intensity and % OC. Detailed studies of gamma-ray intensity and geochemical facies of Cenomanian-Turonian rocks of the Western Interior Basin were conducted by Zelt (1985a, b), who found that the main contributor to total gamma-ray intensity was [238]U and its daughter products, which are associated with organic matter. Therefore, the variations in the gamma-ray logs of the Niobrara (Figs. 5, 6) probably reflect, in part, cyclic variations in concentrations of organic matter and thus provide a continuous record of OC concentration. The three peaks in the gamma-ray log in Figure 5 are unique in that the rest of the gamma-ray log above 157 m shows little variation with an average value of about 60 API units. This suggests that the Smoky Hill cored in the Bounds well was the richest in OC, and probably corresponds to the OC-rich lower marl subunit in the Berthoud State No. 4 core (Fig. 2).

Much of the Smoky Hill Member is dark and finely laminated or characterized by subtle cycles of laminated marlstone alternating with bioturbated limestone (Fig. 3) (Savrda and

Fig. 3.–Photograph of 10 feet (3 m) of the Smoky Hill Limestone Member of the Niobrara Formation between 3040 ft and 3050 ft in the Berthoud State No. 4 core showing interbedded cycles of bioturbated limestone and laminated marlstone. Geochemical data for three of the cycles between 3041 ft and 3043.5 ft are plotted in Figure 4.

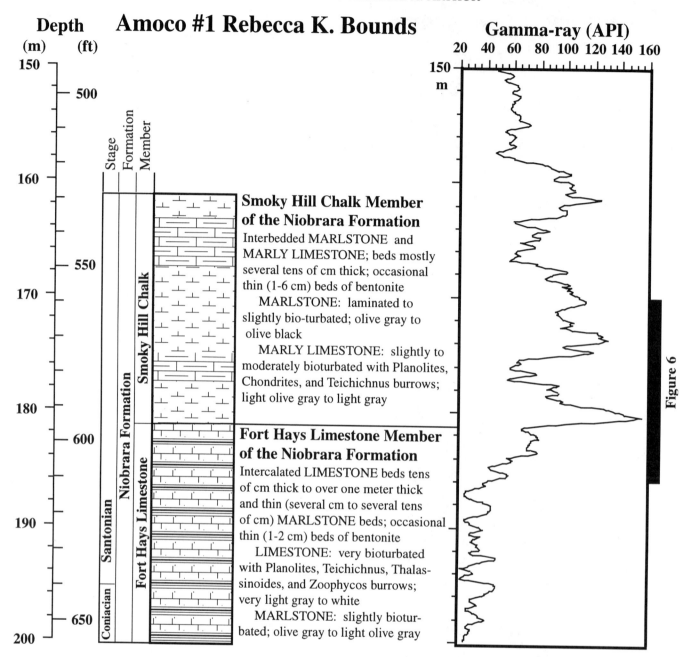

FIG. 5.–Depth, stratigraphic nomenclature, lithology, and gamma-ray intensity for the Niobrara Formation (part) recovered in the Bounds core.

Bottjer, 1986; Savrda, this volume), the Fort Hays Member is generally bioturbated throughout (Fig. 7) and has much greater lithologic and color contrasts between bedding couplets. These contrasts are particularly striking in outcrop, but can be seen very well in cores. In the Bounds core, there are 63 white to light gray, bioturbated limestone beds in 24.4 m of section (Fig. 8) separated by olive-gray marlstone beds that are moderately to well bioturbated (Dean et al., 1995). The Fort Hays Member in the Portland core, collected just west of Pueblo (Fig. 1), contains 33 bedding cycles in 11 m of section (Fig. 9). Many of the bedding cycles in the Fort Hays Member in the Portland well can be seen in the gamma-ray and, particularly, in the neutron porosity logs (Fig. 10).

Age and Duration of Small-Scale Cycles

The Niobrara Formation ranges in age from late Turonian (ca. 88.8 Ma) to early Campanian (ca. 82 Ma) and includes all of the Coniacian and Santonian (Kauffman et al., 1993). If we apply the time scale of Obradovich (1993), which is based on $^{40}Ar/^{39}Ar$ ages of bentonites in the WIS, to interpolate ages for the tops and bottoms of the members of the Niobrara Formation, we obtain an estimate of about 0.3 My for the duration of the Fort Hays Member and about 6.5 My for the duration of the Smoky Hill Member. The Smoky Hill Member is about 80 m thick in the Berthoud State No. 4 core, which gives an estimated sedimentation rate of about 12.3 m/My uncorrected for compaction.

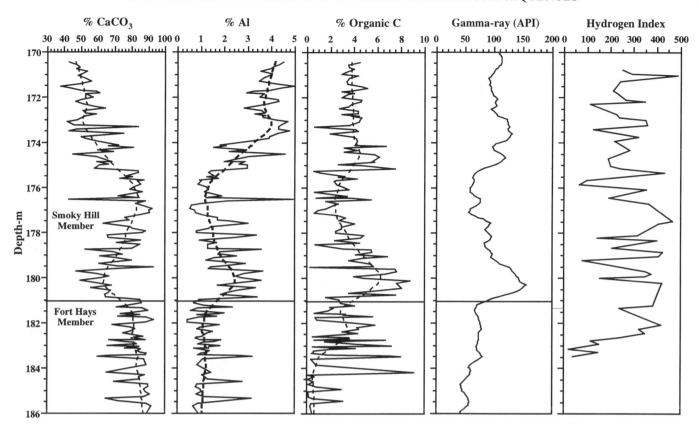

FIG. 6.–Down-core plots versus depth of % CaCO₃, % aluminum (Al), % organic C, gamma-ray intensity, and Rock-Eval pyrolysis hydrogen index (H-Index) in samples from the Bounds core. Dashed lines are weighted moving-average smoothed curves.

At this rate, the 30-m cycles in % CaCO₃ and % Al (third-order cycles; dashed smoothed curves in Fig. 2) have an average duration of about 2.5 My. The 3-4-m cycles in % CaCO₃ and % Al (fourth-order cycles) in the Berthoud State No. 4 core at a sedimentation rate of 12.3 m/My have periodicities of several hundred thousands of years. Fourth-order cycles in the Niobrara Formation in the Bounds core are manifested as the three cycles in gamma-ray intensity in Figure 5, which average about 8 m between 158 m and 181 m, and the ca. 6-m cycles in % CaCO₃, % Al, and % OC (dashed smoothed curves between 174 and 181 cm in Fig. 6).

Bedding cycles in the Smoky Hill Member in the Berthoud State No. 4 core average about 30-40 cm in thickness (Fig. 4) which, at 13 m/My gives an average periodicity of about 23-30 ky. Near Pueblo, Colorado (Fig. 1), the bedding cycles in the Smoky Hill are thicker due to a greater proportion of detrital clastic material with an average uncorrected sedimentation rate of about 49 m/My. In the Pueblo section, the limestone/marlstone bedding couplets in the Smoky Hill Member average about 1 m in thickness for an average periodicity of about 20 ky.

Barlow and Kauffman (1985) report that the Fort Hays Member at Pueblo contains about 40 bedding cycles in 12.5 m of section, an average of 31 cm per cycle. Assuming a duration of 0.3 My for the Fort Hays (Kauffman et al., 1993; Obradovich, 1993), the average periodicity of the bedding cycles is about 7.4 ky. The Fort Hays Member in the Portland core contains 33 bedding cycles in 11 m of section (Fig. 9), an average of 33 cm per cycle for an average periodicity of 9.1 ky. The Fort Hays Member in the Bounds core contains 63 bedding cycles in 24.4 m of section (Fig. 8), an average of 39 cm per cycle for an average periodicity

of 4.8 ky. According Scott et al. (this volume), the base of the Fort Hays Limestone Member in the Bounds core is marked by an unconformity; the upper Turonian and lower Coniacian sections are missing. Therefore, more than 63 bedding cycles were deposited in the Fort Hays at the locality of the Bounds core, which would make the average periodicity even less than 4.8 ky. The 0.3 My duration for the Fort Hays estimated from the Obradovich (1993) ⁴⁰Ar/³⁹Ar time scale seems too low. Because of the poor resolution of the Obradovich data at the Turonian/Coniacian boundary (there are no data points for the upper Turonian and lower Coniacian parts of the section), the age span of the Fort Hays could be considerably greater than the estimated 0.3 My, and, therefore, the estimated periodicities would be greater. Fischer et al. (1985) and Fischer (1986, 1993) suggested that the bedding cycles in the Niobrara Formation have an average periodicity of 21 ky, and correspond to the Earth's cycle of axial precession. The calculated periodicities for the bedding cycles in the Smoky Hill Member are very close to 21 ky, and it is possible that with better age data the periodicities of the Fort Hays bedding cycles also would be close to 21 ky.

Bundling of bedding cycles in the Niobrara Formation into sets of about five commonly are observed in outcrops where weathering has accentuated the bedding cycles (e.g., Fischer, 1980,1993; Barlow and Kauffman, 1985; Laferriere et al., 1987; Laferriere and Hattin, 1989; Scott et al., this volume), which suggests the possible presence of some influence by the ca. 100-yr cycle of the Earth's orbital eccentricity (Fischer, 1980, 1986, 1993). This would also support a 21-ky periodicity for the small-scale cycles. There is some evidence for bundling of bedding cycles in the Fort Hays Member in the Bounds core in that some

Fig. 7.–Photographs of bioturbated limestones in the Fort Hays Limestone Member of the Niobrara Formation in the Bounds core showing (A) bioturbated limestone overlying an oyster-shell island, and (B) bioturbated limestone with large *Chondrites* burrows at the base and large *Teichichnus* burrows at the top.

thinner limestone beds are grouped between thicker limestone beds (e.g., limestone beds 29-31, 34-38, 52-54, and 57-61 in Fig. 8).

Mechanisms for Cycle Development

Direct depositional mechanisms that might produce cyclic fluctuations in pelagic carbonate and detrital clastic material (clay) have been variously attributed to cyclic fluctuations in carbonate production (e.g., Eicher and Diner, 1989), carbonate dissolution, and the influx of diluting clay (e.g., Einsele, 1982; Arthur et al., 1984; Fischer et al., 1985; ROCC Group, 1986; Bottjer et al., 1986; Dean and Gardner, 1986; Elder et al., 1994). However, it is usually difficult to determine which of these mechanisms or combinations of mechanisms produced the cycles in any particular sedimentary sequence, even in Quaternary sequences. If a sequence was deposited close to the carbonate compensation depth (CCD), then carbonate dissolution might be suspected as the primary mechanism and this might be manifested by fragmentation of foraminifera or corrosion of nannofossils. If a site is close to a climatically sensitive land mass, then fluctuations in influx of wind- or river-borne clastic material might be suspected as the main mechanism. If there are similar cyclic variations in abundance of other biogenic components such as biogenic silica, then fluctuations in productivity may have produced the cycles. Regardless of the direct mechanism, the strik-

ing similarities in periodicities of pelagic carbonate cycles from a variety of depositional settings suggests that the overall forcing mechanism was global.

In the presumed ice-free world of the Cretaceous, however, it is more difficult to establish links between orbital cycles, climate change, and carbonate sedimentation. Computer climate-model simulations by Barron et al. (1985), Glancy et al. (1986, 1993), and Park and Oglesby (1991, 1994) suggest that orbital variations in the Cretaceous, through their effect on solar insolation, would have produced regions of intense precipitation with a distinct seasonality. Changes in precipitation in the Sevier volcanic highland to the west of the Western Interior Seaway should have had a marked effect on runoff and delivery of detrital clastic material at all scales. Paleotemperature and salinity profiles of the seaway, reconstructed from oxygen and carbon isotopic signatures of mollusc shells in the Pierre Shale by Wright (1987), suggested the presence of three water masses: a more saline, warm bottom layer; a cooler intermediate layer; and a low-salinity surface layer. Wright proposed that the low-salinity surface layer was caused by runoff from the Sevier highland to the west, and the warm, saline bottom water was produced by westward return flow of surface waters that were heated and evaporated along the eastern margin of the seaway. Circulation and climate modeling of the early Turonian Western Interior Seaway by Slingerland et al. (1996) suggested that runoff from the eastern margin of the seaway formed a northward-flowing coastal jet whereas runoff from the western margin contributed to strong, southward-flowing coastal jet. These coastal jets established a strong counterclockwise gyre with significant export of water from the basin that drew in both Tethyan and boreal surface waters. During this transgressive episode, sufficient nutrient-rich Tethyan subthermocline water was drawn into the basin to stimulated upwelling and high organic productivity associated with the Greenhorn cycle. Similar upwelling conditions probably also existed during the Niobrara transgressive cycle.

Kauffman (1988) suggested that stratification may have been produced by warm surface waters from the proto-Gulf of Mexico over cold, deeper water from the Arctic Ocean. This stratification would have been enhanced by freshwater runoff and resulted in episodic oxygen depletion in mid-water to bottom-water masses. Hay et al. (1993) argued that the polar and subtropical waters that entered the Western Interior Seaway would certainly have very different temperature and salinity characteristics, but the two water masses might have had the same densities, and the front where these two surface-water masses met may have produced a third, denser water mass that became the bottom water of the seaway and perhaps a significant source of intermediate water to the world ocean. In their model, because of abrupt environmental changes at the front, plankton from both surface-water masses would have downwelled at the convergence, which would impose a large biological oxygen demand on the descending third water mass and hence oxygen-depleted bottom waters in the seaway. This effect presumably would have been greatest at peak transgressions when the organic-carbon-rich hydrocarbon source rocks of the Greenhorn and Niobrara Formations were deposited. Fisher et al. (1994) used the oceanic-front model to explain the ecologic exclusion of benthic foraminifera in Cenomanian/Turonian Greenhorn and equivalent strata north of about the Wyoming-Montana border. Based on presence and absence of benthic foraminifers, they determined that the width of the front ranged from 5.75 to 17.5 km and moved back and forth across the approximate position of the Wyoming-Montana

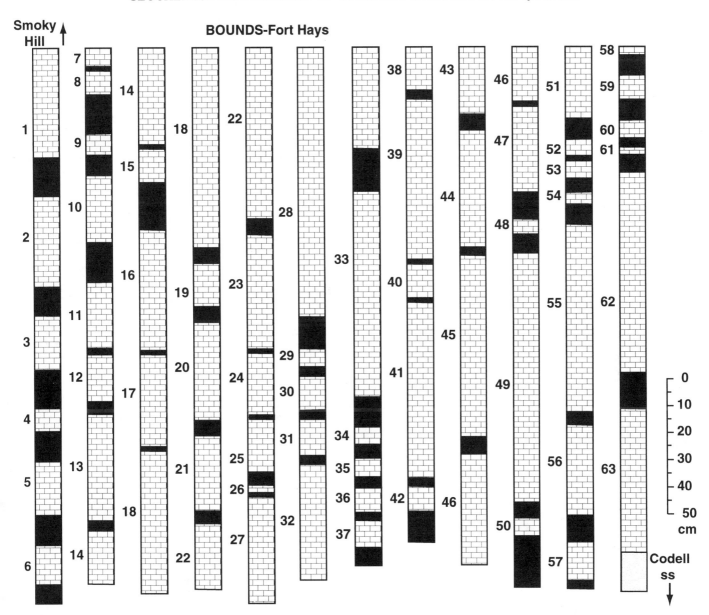

Fɪɢ. 8.–Thicknesses of 63 limestone beds and interbedded marlstone beds in the Fort Hays Member of the Niobrara Formation in the Bounds core.

border 25 times during a 1-My period in middle to late Cenomanian.

Although the above models were used to explain the circulation characteristics of the water masses of the seaway on scales of millions to tens of millions of years, shorter-term climatically influenced modifications of these water masses, whatever they actually were, would have produced cyclic variations within the Milankovitch band. Detailed clay mineralogy studies by Pratt (1981, 1984) showed that the clay-rich, clay-poor bedding cycles in the Greenhorn Formation were the result of varying supply of detrital terrigenous clastic material from the western highland. She suggested that the clay- and organic-carbon-rich beds in the Greenhorn were deposited during wetter time periods when density stratification of the water column caused by buoyant plumes of brackish water and suspended clay was greatest. This model was extended to include the limestone/marlstone bedding cycles in the Niobrara Formation (Pratt et al., 1993). Density stratification would also result in oxygen-deficient bottom waters that

would enhance the preservation of autochthonous marine organic matter which is more abundant and better preserved in the clay-rich marlstone half of a bedding couplet (Fig. 4). The bioturbated limestone beds of the Niobrara and Greenhorn probably were deposited during drier time intervals when less diluting detrital clastic material was transported into the basin and the water column was thoroughly mixed.

The links that emerge, therefore, between the Earth's orbital cycles and cyclic fluctuations in carbonate in the Greenhorn and Niobrara cycles, particularly the bedding cycles, are that changes in axial precession and obliquity with periodicities of about 20 ky and 40-50 ky, respectively (Fischer, 1993; Sageman et al., 1997), produced changes in insolation that altered patterns of precipitation in the Sevier highland, which, in turn, affected the delivery of terrigenous clay and the fresher-water surface layer of the seaway. Correlations of basinal carbonate cycles in the Greenhorn Formation to nearshore parasequences in time-equivalent strata along the western margin of the seaway by El-

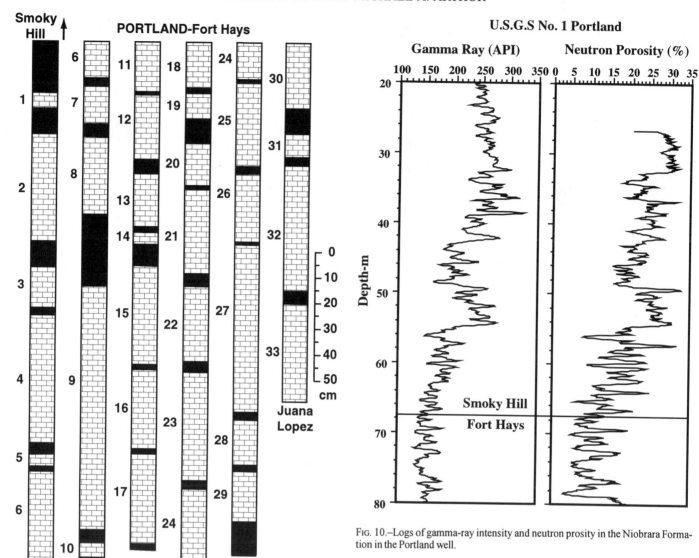

FIG. 9.–Thicknesses of 33 limestone beds and interbedded marlstone beds in the Fort Hays Member of the Niobrara Formation in the Portland core.

FIG. 10.–Logs of gamma-ray intensity and neutron prosity in the Niobrara Formation in the Portland well.

der et al. (1994) argue that clastic dilution produced the cycles. Elder et al. (1994) observed that the progradational phases of the nearshore parasequences correlate with clay-rich units in the basin center, and that the transgressive phases are characterized by reduced clastic influx from the west and deposition of carbonate-rich sediment. Although we believe that variation in terrigenous dilution was the main cause of the carbonate cycles in both the Niobrara and Greenhorn Formations, we cannot rule out variations in carbonate production as a cause for at least some of the variation in carbonate content as has been argued by Eicher and Diner (1989) and Arthur and Dean (1991). However, we can rule out dissolution in an undersaturated water column because the seaway was only several hundred meters deep (e.g., Eicher and Diner, 1989).

CHEMICAL COMPONENT ASSOCIATIONS:
RESULTS OF Q-MODE FACTOR ANALYSES

The Q-mode factor analyses for each of the three data sets essentially reduced the 29 or so analytical variables to five composite chemical variables called factor loadings. To determine which elements had the most influence in each factor, the factor loadings were correlated with the concentration of each element. The results of this correlation analysis for the Berthoud State No. 4 core are given as correlation coefficients in Tables 2-4.

Berthoud State No. 4 Core

Factor 1 for the Berthoud State No. 4 core is a detrital clastic (aluminosilicate) factor, and the elements that contributed most to the composition of this factor, in order of decreasing correlation coefficients in Table 2, are Mg, Ti, Al, K, Li, Na, Sc, Ce, and Nd. Down-core variations in Factor 1 loadings and in % Mg (Fig. 11) are similar to variations in % Al (Fig. 2) with high concentrations of Mg and Al defining the upper marl, middle marl, lower marl, and lower limestone and shale subunits of the Smoky Hill Member (Scott and Cobban, 1964). Factor 1 loadings also more clearly delineate four cyclic limestone/marlstone units between 940 and 960 m that are 3-4 m thick and are the basis for the name of the lower limestone (or chalk) and shale subunit.

The aluminosilicate elements that contributed to the composition of Factor 1 loadings, except for Mg and Li, also contributed to Factor 3 loadings (Table 2), but not as much as concentrations of OC and many redox-sensitive trace metals commonly

TABLE 2 – *Correlation coefficients between Q-mode factor loadings and concentrations of elements in the Niobrara Formation, Berthoud State No. 4 core.*

Variable	Factor 1	Factor 2	Factor 3	Factor 4	Factor 5
Al	.66	-.78	.62	-.31	.15
Ca	-.58	.80	-.63	.35	-.23
Fe	-.06	-.78	.58	-.25	.89
K	.65	-.80	.58	-.26	.18
Mg	.74	-.67	.29	-.13	.09
Na	.56	-.73	.61	-.21	.08
P	.26	-.55	.40	.34	.07
Ti	.70	-.77	.58	-.30	.13
Mn	.05	-.09	-.33	.91	-.17
As	-.26	-.18	.32	-.09	.37
Ba	.31	-.24	.39	-.74	.30
Cd	-.50	-.25	.80	-.18	.35
Ce	.51	-.85	.64	-.14	.27
Co	-.06	-.75	.57	-.13	.79
Cr	.26	-.32	.48	-.45	.12
Cu	-.13	-.45	.92	-.46	.34
Li	.58	-.43	.27	-.39	.08
Mo	-.38	-.41	.59	-.33	.70
Nd	.50	-.77	.60	-.13	.24
Ni	-.31	-.55	.75	-.35	.75
Pb	.33	-.80	.69	-.27	.42
Sc	.53	-.85	.66	-.24	.32
Sr	-.39	.74	-.54	.03	-.08
V	-.32	-.31	.91	-.42	.29
Y	.20	-.74	.62	.15	.31
Zn	-.40	-.25	.86	-.30	.32
Total S	-.21	-.66	.60	-.32	.87
Organic C	-.45	-.13	.75	-.50	.32
Carb. C	-.55	.83	-.68	.34	-.29

TABLE 3 – *Correlation coefficients between Q-mode factor loadings and concentrations of elements in the Niobrara Formation, Berthoud State No. 3 core.*

Variable	Factor 1	Factor 2	Factor 3	Factor 4	Factor 5
Al	.81	-.50	.46	-.75	-.76
Ca	-.59	.54	-.42	.65	-.06
Fe	.21	-.72	.68	-.36	.50
K	.84	-.51	.48	-.77	-.09
Mg	.86	-.35	.20	-.43	-.49
Na	.56	-.41	-.20	-.48	-.03
P	.21	-.44	.14	-.28	.52
Ti	.79	-.51	.45	-.74	-.05
Mn	.05	-.31	-.28	.66	-.26
As	.06	-.64	.63	-.28	.60
Ba	-.08	-.18	-.06	-.07	.20
Cd	.23	-.55	.82	-.49	.41
Ce	.65	-.52	.32	-.62	.08
Co	.31	-.70	.75	-.46	.46
Cr	.64	-.52	.73	-.79	.18
Cu	.41	-.46	.84	-.72	.35
Ga	.85	-.52	.47	-.74	-.07
La	.70	-.59	.36	-.61	.09
Li	.57	-.32	.28	-.60	-.15
Mo	.05	-.53	.79	-.41	.48
Nd	.53	-.57	.27	-.50	.28
Ni	.23	-.60	.82	-.52	.51
Pb	.69	-.68	.62	-.69	.23
Sc	.76	-.25	.40	-.73	.06
Sr	-.63	.70	-.39	.40	.14
Th	.58	-.48	.18	-.48	.08
V	.25	-.38	.82	-.65	.39
Y	.68	.54	.38	-.69	.25
Yb	.76	-.57	.26	-.64	.42
Zn	.35	-.50	.85	-.64	.41

associated with high concentrations of OC. The elements that contributed most to the composition of Factor 3 loadings, in order of decreasing correlation coefficients in Table 2, are Cu, V, Zn, Cd, OC, Ni, Pb, Sc, Ce, Y, Al, Na, Nd, S, Mo, Fe, K, Ti, Co, Cr and, P. The close association of detrital clastic material (aluminosilicates) and organic matter, emphasized by Factor 3 element associations, is particularly well developed in the decimeter-scale bedding cycles in the Smoky Hill Member (three of which are illustrated by the data shown in Fig. 4), but, in general, the association applies to the entire formation (Fig. 2). Down-core variations in Factor 3 loadings and in the concentration of V (one of the strongest Factor 3 elements; Fig. 11) are similar to variations in % OC (Fig. 2). Notice that although stratigraphic variations in % Al, % OC, and Factor 1 and 3 loadings are generally similar throughout the Smoky Hill Member, there are some distinct differences. Highest values of % OC and Factor 3 loadings occur in the marl subunits as do highest values of % Al and Factor 1 loadings, but some of the maxima are somewhat offset. For example, highest concentrations of OC and Factor 3 loadings in the lower part of the Smoky Hill occur in the upper part of the lower marl subunit whereas highest concentrations of Al and highest Factor 1 loadings occur in the lower part of the lower

marl and in the underlying lower limestone and shale subunit.

Factor 2 in the Berthoud State No. 4 core is a carbonate factor with CC, Ca, and Sr contributing to the loadings. Samples with Factor 2 loadings >1 (Fig. 11) and with CaCO$_3$ concentrations >75% (Figs. 11, 2) clearly define the upper, middle, and lower limestone subunits of the Smoky Hill Member (Scott and Cobban, 1964) and the Fort Hays Limestone Member. Based on Factor 2 characteristics, the Fort Hays Limestone is similar in composition to the limestone subunits of the Smoky Hill Member. However, the Fort Hays is clearly separated from the Smoky Hill limestone units by Factor 4 which is most influenced by its higher concentrations of manganese (Table 2; Fig. 11). Concentrations of CC and Ca had some influences on Factor 4 loadings, which reflects the generally purer CaCO$_3$ composition of the Fort Hays relative to the limestones of the Smoky Hill, as did concentrations of phosphorus (P) and yttrium (Y) (Table 2).

Factor 5 in the Berthoud State No. 4 core is a sulfide (pyrite) factor with strongest contributions from Fe, S, Co, Ni, and Mo, in order of decreasing correlation coefficients with Factor 5 loadings in Table 2. Highest loadings for Factor 5 occur in the upper marl and upper limestone subunits of the Smoky Hill Member

TABLE 4.– *Correlation coefficients between Q-mode factor loadings and concentrations of elements in the Niobrara Formation, Bounds core.*

Variable	Factor 1	Factor 2	Factor 3	Factor 4	Factor 5
Si	.96	-.63	-.18	-.12	.19
Al	.96	-.67	-.11	-.31	.23
Ca	-.92	.74	-.04	.28	-.25
Fe	.78	-.81	.31	-.39	.25
K	.95	-.66	-.13	-.16	.24
Mg	.91	-.52	-.32	.08	.10
Na	.92	-.65	-.09	-.23	.24
P	.29	-.58	-.29	.24	.73
Ti	.95	-.67	-.07	-.25	.24
Mn	-.32	.06	-.31	.91	-.01
As	.57	-.75	.51	-.49	-.23
Ba	.91	-.60	-.17	-.22	.27
Ce	.87	-.68	-.10	-.29	.41
Co	.40	-.73	-.40	-.21	.30
Cr	.87	-.72	.18	-.44	.25
Cu	.36	.79	.47	-.48	.51
Ga	.94	-.66	-.07	-.14	.14
La	.80	-.68	-.25	-.10	.52
Li	.95	-.68	-.11	-.17	.22
Mo	.15	-.70	.72	-.53	.42
Nb	.88	.70	.12	-.42	.28
Ni	.38	-.84	.68	-.47	.38
Sc	.95	-.66	-.10	-.20	.22
Sr	-.77	.82	.02	.03	-.38
Th	.86	-.62	-.08	-.15	.17
V	.67	-.65	.39	-.67	.31
Y	.51	-.54	-.42	.13	.68
Zn	.38	-.54	.69	-.38	-.12
Organic C	.26	-.74	.73	-.58	.37

(Fig. 11). The 2-3-m cycles in Factor 5 loadings (Fig. 11) correspond to 2-3-m cycles in OC between 880 and 900 m (Fig. 2) but have much larger amplitudes.

Berthoud State No. 3 Core

As expected, the five factors defined by the geochemistry for the Berthoud State No. 3 core are similar to those for the Berthoud State No. 4 core. By far the most dominant factor is a detrital clastic (aluminosilicate) Factor 1 with similar element associations (Table 3). The element that most strongly contributed to Factor 1 loadings in both cores was Mg (Fig. 12), followed in the Berthoud State No. 3 core by Ga, K, Al, Ti, Yb, Sc, La, Pb, Y, Ce, Cr, Th, Li, Na, and Nd, in decreasing order of correlation coefficients with Factor 1 in Table 3. Variations in Factor 1 loadings and associated elements (Fig. 12) generally are antithetic to those in carbonate Factor 2 (Fig. 12, Table 3), and these two factors define the marl and limestone subunits of the Smoky Hill Member (Fig. 12). Factor 3 loadings are defined by redox-sensitive elements, principally Zn, Cu, V, Cd, Ni, Mo, Co, Cr,

Fe, As, and Pb, in order of decreasing correlation coefficients with Factor 3 loadings in Table 3. We do not have analyses of OC for samples from the Berthoud State No. 3 core, but OC undoubtedly would be associated with Factor 3 as it was in the Berthoud State No. 4 core. Also, we do not have analyses of S for samples from the Berthoud State No. 3 core, but, by comparison with the Berthoud State No. 3 core, S probably would be associated with Factor 5, because it is influenced mainly by As, P, Ni, Fe, and Mo. As in the Berthoud state No. 4 core, Factor 5 loadings are highest at the top of the Smoky Hill Member (Fig. 12). As in the Berthoud State No. 4 core, Factor 4 is dominated by the association of Mn with the relatively pure, highly bioturbated, limestones of the Fort Hays Member (Fig. 12).

Amoco Bounds Core

The first four factors in the Bounds core essentially grouped the same elements as were grouped in the first four factors in the Berthoud State cores, namely a detrital-clastic Factor 1, a carbonate Factor 2, an organic-metal Factor 3, and a Mn-carbonate Factor 4 in the Fort Hays Member (Table 4; Fig. 13). The antithetic relation between Factor 1 and Factor 2 loadings is very pronounced (Fig. 13) because in terms of major components, the Niobrara Formation is essentially a two component system, detrital clastic material and CaCO3 (see also plots of % Al and % CaCO3 in Fig. 6). The limestone/marlstone cycles in the Bounds core appear to be more pronounced with higher amplitudes (Fig. 6) than in the Berthoud State No. 4 core (Fig. 2), but this is an artifact of higher resolution sampling of the Bounds core. Whereas the Berthoud State No. 4 core contains all of the Smoky Hill Member in 80 m of section, the Bounds core only contains the lower 11 m of the Smoky Hill which is 171-189 m thick in western Kansas (Hattin, 1982). Down-core variations in the Berthoud State No. 4 core (Fig. 2) are defined with a 1-m sampling interval over the 100 m of section of the entire formation. The variations in the Bounds core (Fig. 13) are defined with about a 10-cm sampling interval over about 16 m of section including the upper part of the Fort Hays Member and the lower part of the Smoky Hill Member. At that scale of sampling we are able to define decimeter-scale bedding cycles that were only defined by close-interval sampling for three cycles in the Berthoud State No. 4 core (Fig. 4). At this 10-cm scale of resolution, both the Bounds and Berthoud State No. 4 cores show variations in % CaCO3 of 20-30%.

In general, concentrations of OC and redox-sensitive metals are similar to those of the clastic Factor 1 elements, illustrated in Figures 6 and 13 by Al, but, like the Berthoud State No. 4 core, variations in OC and Al are not always in phase. For example, in Figure 6, note the large spikes in % OC (up to 6-8%) in the Fort Hays Member that do not always coincide with peaks in % Al, and a maximum in % OC at 175 m is followed by a maximum in % Al at 173 m. The main differences between the Bounds factor analysis and those of the Berthoud State cores is that Factor 5 in the Bounds core is a P-rare-earth-element factor determined by P, Y, La, Cu, Mo, and Ce, in decreasing order of correlation coefficients with Factor 5 loadings in Table 4. Note in Figure 13 that highest Factor 5 loadings and highest concentrations of P tend to occur in carbonate-poor horizons in the Fort Hays Member and at the base of the Smoky Hill Member. The elements that defined a sulfide association (Factor 5; Fe, Co, Ni, and Mo) in the Berthoud State No. 4 core were grouped with Factor 1 and Factor 3 in the Bounds factor analysis.

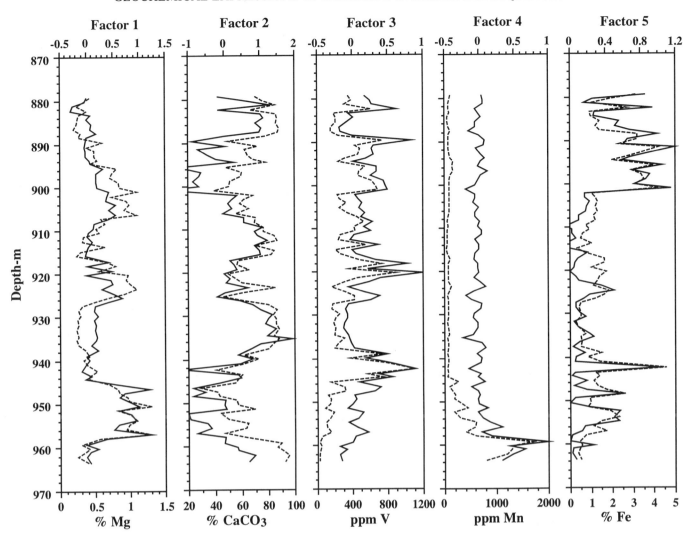

FIG. 11.—Down-core plots versus depth of loadings for the five Q-mode factors (solid curves, top scales), and for % Mg, % CaCO3, ppm V, ppm Mn, and % Fe (dashed curves, bottom scales) in the Berthoud State No. 4 core.

GEOCHEMICAL INDICATORS OF CLASTIC SOURCE VARIATIONS

Concentrations of many major and trace elements vary inversely with CaCO3 content in the Niobrara Formation, particularly those elements that are commonly associated with the non-carbonate (detrital-clastic) fraction (e.g. Al, Na, K, Mg, Ti, Li, Sc, and Ce). Because subtle compositional variations often are obscured by fluctuations in CaCO3 that cause variable dilution of the clastic fraction, we calculated element/element ratios for some of the more important elements and plotted them versus depth in core (Figs. 14 and 15).

The Niobrara and Greenhorn Formations were deposited during maximum transgression of the Western Interior Seaway of North America. The seaway was essentially a northwestern arm of the Tethys Sea and provided a connection between Tethys and the Arctic Ocean. There were four major sediment sources that competed against one another during deposition of the Niobrara and Greenhorn: biogenic carbonate, windblown volcanic ash, terrigenous detritus from the Sevier highland to the west, and terrigenous detritus from the low-relief eastern margin. We cannot assume that any of these inputs were constant; in fact, all probably varied in response to environmental changes. It is generally assumed that the low-relief eastern margin of the basin was not a significant source of sediment, but recent circulation modeling results of the Western Interior Seaway by Slingerland et al. (1996) suggest that there was runoff from the eastern margin of the seaway and, therefore, sediment supply. The Precambrian Transcontinental arch in the northern Great Plains was a positive tectonic feature that undoubtedly influenced Niobrara deposition (Shurr, 1984). The lithologic variations discussed above, and oxygen isotope data (Pratt et al., 1993) suggest that continental runoff to the basin changed through time, probably related to changes in sea level, and an eastern sediment source may have been part of these changes. Therefore, the flux of terrigenous detritus probably varied significantly and could have produced most of the observed variations in carbonate content.

Pratt (1981, 1984) demonstrated some changes in the relative proportions of detrital clay minerals (discrete illite) and quartz superimposed on a background of mixed-layer illite/smectite, which was produced by altered volcanic ash, in the Bridge Creek Limestone in the PU-79 core from near Pueblo, Colorado (Fig. 1). The concentrations of discrete illite and clay-sized quartz, relative to total clay minerals, are highest in the marlstones and calcareous shales from which Pratt (1981, 1984) concluded that

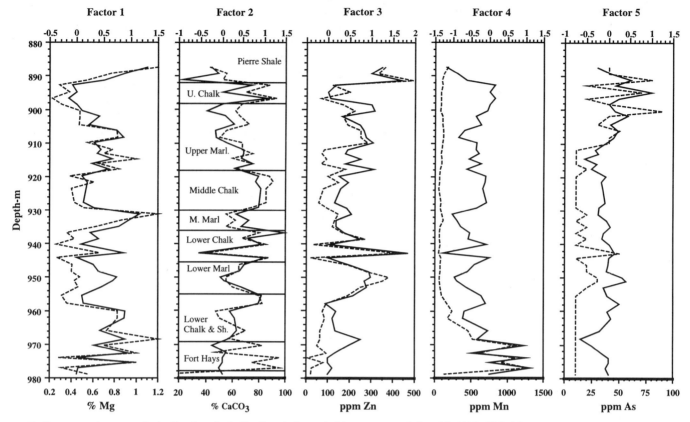

FIG. 12.–Down-core plots versus depth of loadings for the five Q-mode factors (solid curves, top scales), and for % Mg, % CaCO3, ppm Zn, ppm Mn, and ppm As (dashed curves, bottom scales) in the Berthoud State No. 3 core.

the influx of river-borne detritus was high during deposition of the marlstone and calcareous shale units of the Bridge Creek Limestone, and even higher in the underlying Hartland Shale Member. The abundance of mixed illite/smectite, on the other hand, is controlled by dispersal of volcanic ash by both rivers and wind, and its concentration, relative to total clay minerals, is highest in the limestone beds of the Bridge Creek Limestone. This suggests greater importance of eolian activity during deposition of the limestone beds, which reflect deposition under dry conditions. A similar situation, with higher concentrations of mixed-layer illite/smectite in limestone beds, was reported by Arthur et al. (1985) for the Niobrara Formation in the Ideal Cement Quarry near Fort Collins, Colorado, and by Pollastro and Martinez (1985) for the Niobrara Formation in outcrops near Pueblo, Colorado. They concluded that whereas mixed illite/smectite was the product of diagenesis of volcanic ash, the concentrations of discrete illite and chlorite are interpreted as detrital phases that diluted the background flux of volcanic ash during deposition of the marlstone units. Interpretation of drier, windier conditions during deposition of the limestone beds is in line with the oxygen isotope evidence suggesting that the limestones represent the more arid portion of each climatic-depositional cycle in that unit (e.g. Pratt et al., 1993).

The ratio K/Fe+Mg also is a measure of the relative contributions of detrital and volcanic clastic material in that most of the K probably is in discrete illite and most Fe and Mg are in smectite. If the above interpretation is correct, then limestone beds should contain more Fe+Mg and, therefore, lower K/Fe+Mg ratios. Figure 16D shows that there is a distinct negative corre-

lation between K/Fe+Mg and % CaCO3 in the Bridge Creek Limestone in both the Bounds and PU-79 cores; this indicates that the limestone beds do indeed contain more mixed-layer illite/smectite derived from altered volcanic ash (Fe+Mg) than the marlstone beds.

Figure 16B shows that, in general, there is a positive correlation between Si/Al and % CaCO3 for the Bridge Creek Limestone in the Bounds and PU-79 cores in that beds with <60% CaCO3 have low Si/Al ratios whereas the highest ratios occur in some, but not all, limestone beds. Certainly the greatest variability in Si/Al ratios in both cores, but particularly in the Bounds core, are in beds with >60% CaCO3. This silica may be detrital quartz or biogenic silica. The detrital Si may be river-borne or eolian quartz. Pratt (1981; 1984) concluded that the influx of river-borne quartz was greatest during deposition of the marlstone units of the Bridge Creek, which is the opposite of what we observed based on Si/Al ratios. Because the higher values of Fe+Mg (illite/smectite) suggest that there was a higher relative influx of volcanogenic material during deposition of the limestone beds, we conclude that most of the variation in Si/Al seen in Figure 16B probably is due to variation in influx of eolian quartz, although we cannot rule out contributions from biogenic silica as a cause for at least some of the variability.

Unfortunately, relations between K/Fe+Mg and Si/Al ratios and marlstone/limestone beds in the Niobrara Formation are less clear than those in the Bridge Creek. In general, the Si/Al ratio is not related to carbonate content in the Bounds core (the only one for which we have Si data). The K/Fe+Mg ratio in the Niobrara in the Berthoud State No. 4 core shows the most dis-

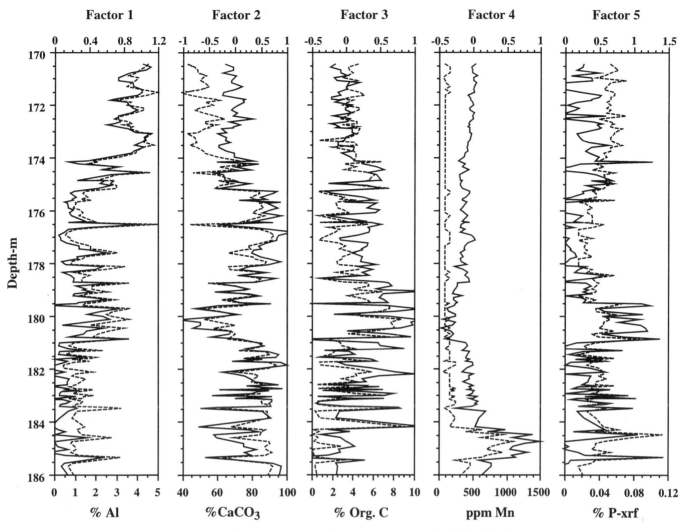

FIG. 13.–Down-core plots versus depth of loadings for the five Q-mode factors (solid curves, top scales), and for % Al, % CaCO₃, % total organic carbon (TOC), ppm Mn, and % P (dashed curves, bottom scales) in the Bounds core.

tinctive inverse relation with % CaCO₃ (Fig. 16C), perhaps reflecting the closer proximity of the Berthoud State site to the western shoreline. The limestones in the Bounds core show a slight tendency to have higher K/Fe+Mg ratios than marlstones, which suggests that this site received more detrital illite during deposition of the limestones than during deposition of the marlstones. Unless more illite is being brought in from an eastern source, there is no explanation for this.

Arthur and Dean (1991) showed that cyclic variations in the Na/K ratio in the Bridge Creek Limestone are related to changes in the importance of discrete illite (high potassium) and other detrital phases versus mixed-layer clays derived from altered volcanic material. There are excellent correlations between Na/K and % CaCO₃ in all three data sets for the Bridge Creek (Fig. 17). The average Na/K ratio in bentonites in the Pierre Shale is about 3.0, which represents a pure, mixed-layer-clay (formed, for example, by alteration of volcanic detritus) end member (Schultz et al., 1980). The fact that Na/K ratios are highest in bioturbated limestone beds (higher % CaCO₃) (Fig. 17), suggests that the proportion of mixed-layer clays is highest in those beds, and the proportion of K-bearing detrital illite is lowest. Another possibility is that the Na/K ratio may reflect

available moisture and hence decomposition of feldspars in the source area. High Na/K ratios in limestones deposited during dry periods may be the result of less decomposition of Na-plagioclase. Similarly, low Na/K ratios in marlstones deposited during wet periods would be due to more complete decomposition of Na-plagioclase. Both processes discussed above would result in higher Na/K ratios in the limestones, and are consistent with inferred environmental conditions.

Arthur and Dean (1991) found that the Al/Ti ratio in Cretaceous pelagic limestones in the North Atlantic generally increased with increasing Na/K ratio. Based on averages of analyses of numerous samples of the Pierre Shale from throughout the Western Interior basin by Schultz et al. (1980), the average Al/Ti ratio in bentonites is about 40, whereas the average in marine shales and siltstones is about 24. For comparison, the Al/Ti ratio in average shale (Turekian, 1972) is 19. Most of the values plotted in Figs. 14, 15, and 17 are between these shale and bentonite values, which suggests a mixture of detrital and volcanic clastic sources. We reasoned that perhaps the Al/Ti ratio in the Western Interior carbonate units might allow us to distinguish between the detrital-source and feldspar-decomposition options for the high Na/K ratios in the limestone beds of the Bridge Creek.

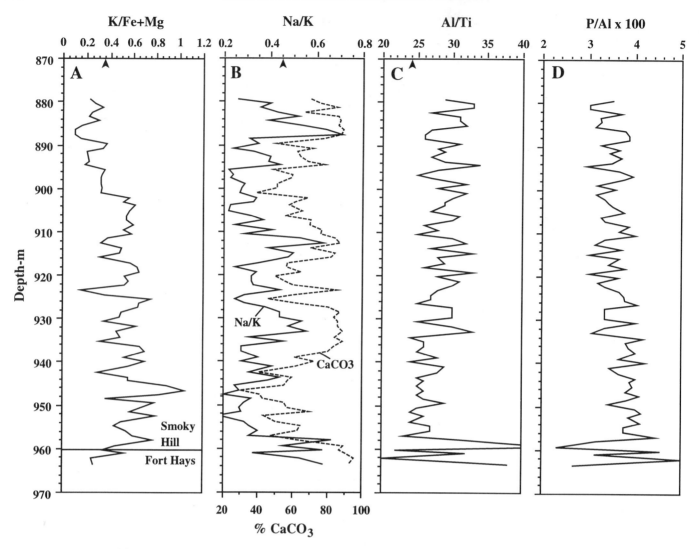

Fig. 14.–Down-core plots of element ratios versus depth for the Niobrara Formation in the Berthoud State No. 4 core for (A) K/Fe+Mg, (B) Na/K and CaCO3, (C) Al/Ti, and (D) P/Al. Arrows indicate values of ratios in average marine shale and siltstone in the Pierre Shale (Schultz at al., 1980).

If the higher Na/K ratios in limestone beds were due to more volcanic ash, then there should also be a positive correlation between % CaCO3 and Al/Ti. However, as seen in Figs. 17 H-J, there is no correlation between Al/Ti and % CaCO3, which suggests that feldspar decomposition, and, therefore, available moisture, is the main factor controlling the Na/K ratio. Because of the clay mineralogy and geochemical data, which show that the concentration of discrete illite and, therefore, the K/Fe+Mg ratio (Fig. 16D) are distinctly lower in the limestone beds of the Bridge Creek (Pratt, 1984), the relative abundances of detrital versus volcanic clastic influx also must play a role in determining the geochemistry of the acid-insoluble fraction of the Bridge Creek.

The relative proportion of K-rich detrital clay (illite) is significantly higher in the Niobrara Formation than in the Bridge Creek Limestone (Arthur et al., 1985) resulting in an Na/K ratio that is overall 3-4 times lower (Fig. 17), which suggests greater influence of fluvial sediment source regions to the west during much of Niobrara deposition. As discussed above, we cannot rule out the possibility that greater available moisture during deposition of the Niobrara may also be partly responsible for the lower Na/K ratio in the Niobrara relative to the Bridge Creek. Although there is no apparent correlation between Na/K and %

CaCO3 in the Niobrara in the basin-center Bounds core (Fig. 17B), the limestone beds tend to have a higher Na/K ratio. A positive correlation between Na/K and % CaCO3 is better seen in the Niobrara in the Berthoud State No. 4 core (Figs. 14B and 17A), where pulses of high Na/K occur in the Fort Hays Limestone Member and in the lower, middle, and upper limestone subunits of the Smoky Hill Member (Fig. 14B).

To summarize what we have learned from these element ratios, we have plotted them against each other for the Bridge Creek Limestone in the Bounds core (Fig. 18). We chose that unit in that core because the ratios show the most distinct differences between limestones and marlstones (Fig. 16 and 17), and because the core contains bentonites A, B, and C of Elder (1985; 1988). Samples from these bentonites, and several bentonitic samples adjacent to them, were excluded from previous plots but we include them in Figure 18 to illustrate the distinctive geochemical characteristics of the volcanic clastic component. Figure 18A shows that the K/Fe+Mg ratio by itself is not a particularly distinctive indicator of the volcanic fraction, but the Al/Ti ratio is. Earlier we stated that the average Al/Ti ratio in bentonites in the Pierre Shale analyzed by Schultz et al. (1980) was about 40, in marine shales and siltstones it was about 24, and in

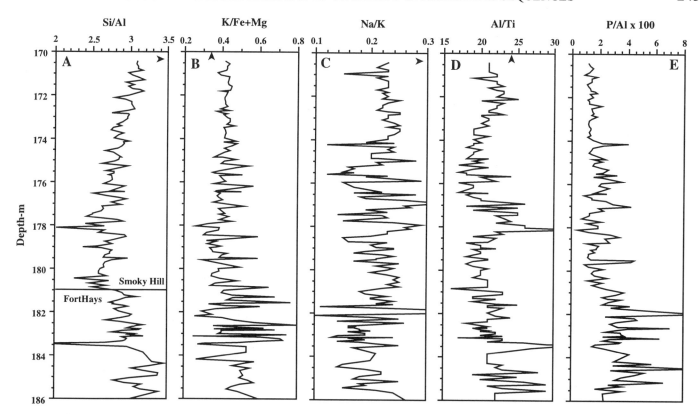

FIG. 15.–Down-core plots of element ratios versus depth for the Niobrara Formation in the Bounds core for (A) Si/Al, (B) K/Fe+Mg, (C) Na/K, (D) Al/Ti, and (E) P/Al. Arrows indicate values of ratios in average marine shale and siltstone in the Pierre Shale (Schultz at al., 1980).

World average shale it was 19. The bentonites in the Bridge Creek in the Bounds core have distinctly higher Al/Ti ratios. Given these extreme values of Al/Ti in the bentonites and the generally low values of Al/Ti in the Bridge Creek (close to World shale average), we conclude that volcanic clastic material was relatively minor compared to detrital clastic material. Figs. 17H and J show that some of the limestone beds in the Bridge Creek do have Al/Ti values that approach those of the bentonites, particularly in the PU-79 core, which suggests that volcanic clastics were important components in those beds.

Figures. 17B and C show that the Na/K ratio also is distinctively high in the bentonites. The average Na/K in bentonites analyzed by Schultz et al. (1980) in the Pierre Shale is about 3, and this is similar to those in the Bridge Creek. The extreme values of Na/K and Al/Ti combined (Fig. 18C), together with generally lower values of K/Mg+Fe in the Bridge Creek relative to those in the Niobrara (Figs. 16C and 16 D), further emphasize the relatively minor contributions of volcanic clastics to the Bridge Creek.

Finally, several lines of geochemical evidence suggest that the Bridge Creek Limestone in the Bounds core may have received detrital clastic material from an eastern source. The K/Fe+Mg ratio is lower in limestone beds in the Bridge Creek in the PU-79 and Bounds cores (Fig. 16D), which suggests that there is more illite/smectite (high Fe+Mg) than discrete illite (high K). The opposite is true in the Bounds core, which suggests that this site received more detrital illite than the site of the PU-79 core during deposition of the limestone beds. A high proportion of illite relative to smectite in the Greenhorn Formation in central Kansas was reported by Doveton (1991) based on

high potassium contents from spectral gamma-ray logs.

DIAGENETIC GEOCHEMICAL SIGNALS

Geochemical Indicators of Redox Cycles

The Rock-Eval hydrogen index (HI) and oxygen index (OI) provide measures of the type of organic matter in sediments and sedimentary rocks (e.g. Tissot and Welte, 1984; Peters, 1986). A plot of OI versus HI (Figs. 19A) is a conventional way of displaying results of Rock-Eval pyrolysis and defining types of kerogen in rocks. Three types of kerogen (I, II, and III) and their thermal maturation pathways are shown by the curves on Figure 19A. Type I kerogen consists of extremely lipid-rich, sapropelic, algal organic compounds typified by Eocene Green River oil shale. Type III kerogen is the other extreme and consists of H-poor organic compounds, such as those found in terrestrial organic matter. Type II kerogen also consists of H-rich compounds such as those found in autochthonous marine organic matter. Marine hydrocarbon source rocks typically contain type II kerogen. Type III kerogen also may be formed by oxidation of type II kerogen, which lowers the HI and increases the OI. Therefore, the extreme range of HI and OI values shown for the Niobrara Formation in Figure 19A could result from a mixture of marine and terrestrial organic matter, from oxidation of marine organic matter in the depositional environment and hydrocarbon generation on deep burial. The HI and OI values for the Niobrara Formation in the Bounds core plotted in Figure 19A are for the same samples plotted in Figure 6; the points labeled BS#4 1-m are for samples of the Smoky Hill Member collected at 1-m intervals and plotted

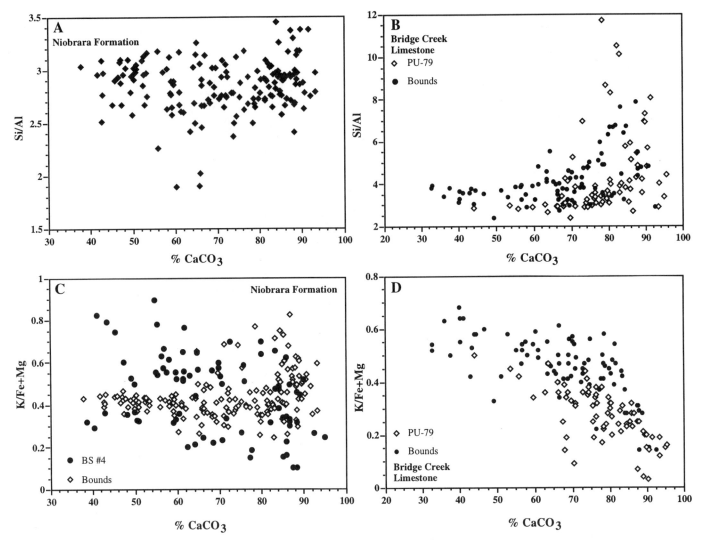

FIG. 16.–Cross plots of % CaCO₃ versus Si/Al for (A) the Niobrara Formation in the Bounds core, (B) the Bridge Creek Limestone in the Bounds and PU-79 cores; and of K/Fe+Mg for (C) the Niobrara Formation in the Bounds and Berthoud State No. 4 cores, (D) Bridge Creek Limestone in the Bounds and PU-79 cores.

in Figure 2; the points labeled BS#4 C-cycles are samples from bedding cycles plotted in Figure 4.

The results plotted in Figure 19A can be interpreted in terms of thermal maturation or oxidation of an original type II kerogen. In theory, samples of bioturbated limestones, which indicate well-oxygenated bottom waters, should contain more degraded organic matter than samples of laminated marlstone. The relation between organic matter degradation and selective loss of hydrogen as a function of degree of oxidation was clearly demonstrated by Pratt (1984) and Pratt et al. (1986) for the interbedded laminated to microburrowed (marginally oxic to anoxic), OC-rich marlstone units versus the bioturbated (oxic), OC-poor limestone units of the Bridge Creek Limestone in the PU-79 core (Fig. 1) from near Pueblo, Colorado.

An HI-OI plot of the Bridge Creek is similar to that shown for the Niobrara in Figure 19A, but detailed study of the organic matter in the Bridge Creek by Pratt (1984) demonstrated that the lowering of HI is due to increased oxidation related to bioturbation, and not by changes in the abundance of terrestrial organic matter. Consequently, there is an inverse correlation between HI and % CaCO₃ in the Bridge Creek – that is, the

bioturbated limestone beds have lower values of HI. However, for the samples from the Niobrara Formation plotted in Figure 19A there is no correlation between concentration of CaCO₃ and HI (insert), and values of HI in the Smoky Hill Member in the Berthoud State No. 4 core are moderately high with little variation (Fig. 2). The Niobrara Formation in the Berthoud State cores was buried much deeper (perhaps 2500 m) than the Niobrara in the Bounds core (perhaps 500 m). The Berthoud State cores are from the Berthoud oil field in northeastern Colorado and the Niobrara Formation is a probable source for that oil. Therefore, we would expect the Niobrara Formation in the Berthoud State cores to be well along the thermal maturation pathway, and, in fact, fluid inclusion studies of samples from the Berthoud State No. 4 core indicate that a peak paleotemperature of 100° C was reached at 70 Ma and 2438 m depth (Crysdale and Barker, 1990). However, the uniformity of HI values above 200 mgHC/g OC for the Smoky Hill in the Berthoud State No. 4 core (Fig. 2) indicates the good preservation of marine organic matter under generally oxygen-depleted bottom waters throughout deposition of that unit. Because values of HI do not vary cyclically through OC-rich and -poor units, we conclude that cyclic variations in

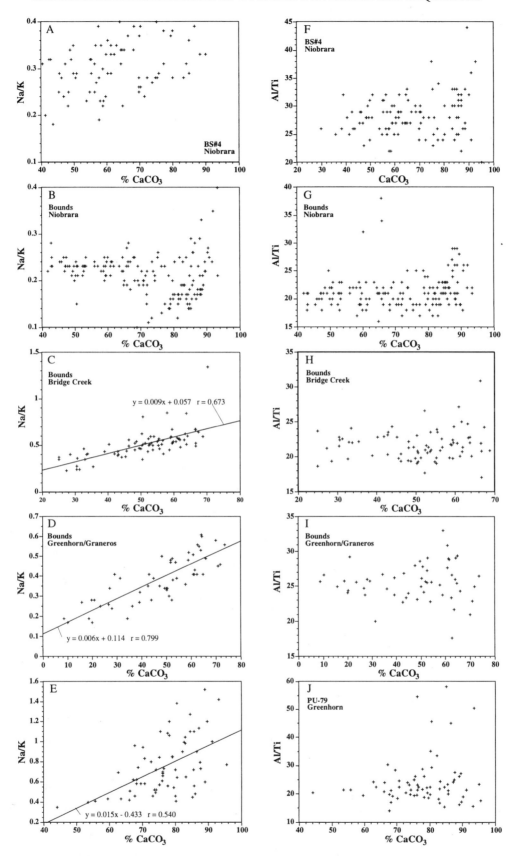

FIG. 17.–Cross plots of % CaCO₃ versus Na/K for (A) the Niobrara Formation in the Berthoud State No. 4 core, (B) the Niobrara Formation in the Bounds core, (C) the Bridge Creek Limestone in the Bounds core (D) the Greenhorn Formation and Graneros Shale in the Bounds core (Dean et al., 1995), and (E) the Greenhorn Formation in the PU-79 core; and of CaCO₃ versus Al/Ti for (F) the Niobrara Formation in the Berthoud State No. 4 core, (G) the Niobrara Formation in the Bounds core, (H) the Bridge Creek Limestone in the Bounds core (I) the Greenhorn Formation and Graneros shale in the Bounds core (Dean et al., 1995), and (J) the Greenhorn Formation in the PU-79 core.

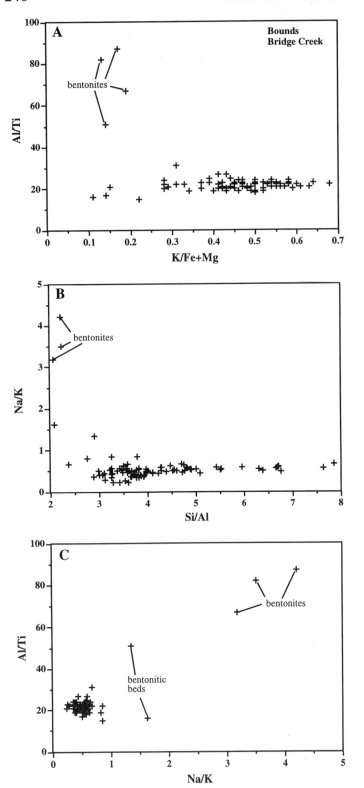

FIG. 18.–Cross plots of (A) K/Fe+Mg versus Al/Ti, (B) Si/Al versus Na/K, and (C) Na/K versus Al/Ti for the Bridge Creek Limestone in the Bounds core.

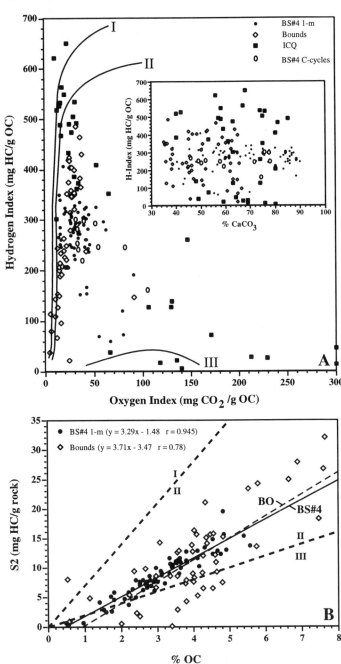

FIG. 19.–(A) A plot of Rock-Eval pyrolysis oxygen index versus hydrogen index for samples of the Niobrara Formation from the Berthoud State No. 4 core, the Bounds core, and the Ideal Cement Quarry (ICQ) near Fort Collins, Colorado. Insert shows a cross plot of % CaCO₃ versus hydrogen index for samples from the Berthoud State No. 4 core, Bounds core, and Ideal Cement Quarry. Curves labeled with Roman numerals refer to thermal maturation pathways for types I, II, and III kerogen. (B) Scatter plot of % OC versus Rock-Eval pyrolysis S2 for samples of the Niobrara Formation from the Berthoud State No. 4 and Bounds cores. Linear regression equations are given for the best fit lines for the Berthoud State No. 4 data (solid line) and the Bounds data (thin dashed line). The heavy, dashed lines labeled I/II and II/III represent the boundaries between pyrolitic hydrocarbon yields of type I and type II kerogen, and type II and type III kerogen, respectively, as defined by Langford and Blanc-Valleron (1990) (see text for method of determination).

amount of organic matter (% OC) are due mainly to variable OC-flux and dilution and not to preservation alone. The HI values in the Berthoud State No. 4 core probably are somewhat lower than expected for such laminated OC-rich strata because the organic matter is mature as the result of the high paleo-burial depth. We were surprised, however, that the samples from the Bounds core had some of the lowest values of HI because it was not deeply buried and is further from any presumed sources of terrestrial organic matter.

Because the HI is derived from S2 by normalizing to % OC (HI = [S2/%OC] x 100)], Langford and Blanc-Valleron (1990)

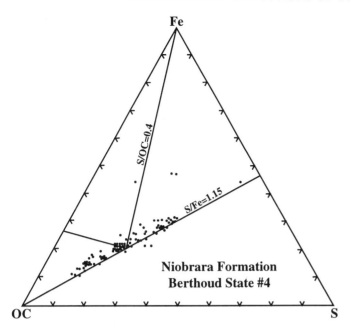

FIG. 20.–Triangular diagram of Fe-S-OC in the Niobrara Formation in the Berthoud State No. 4 core. Line marked S/OC=0.4 represents a constant S/OC ratio of 0.4, the average ratio in normal marine sediments (Goldhaber and Kaplan, 1974; Berner and Raiswell, 1983). Line marked S/Fe=1.15 represents the stoicheometric ratio of Fe and S in pyrite.

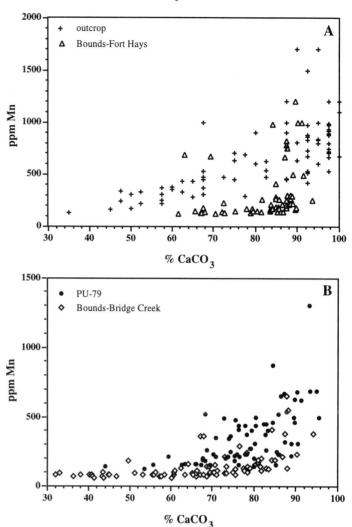

FIG. 21.–Scatter plots of % CaCO3 versus ppm Mn in (A) samples of the Fort Hays Limestone from the Bounds core and from outcrops in Colorado and New Mexico, and (B) samples of the Bridge Creek Limestone in the Bounds and PU-79 cores.

suggested that the average value for HI for a particular data set could be obtained from the slope of a regression equation for a graph of % OC versus S2 (Fig. 19B). The data plotted for samples of the Niobrara Formation from the Berthoud State No. 4 and Bounds cores show that they have similar regression equations with slopes of 3.29 and 3.71, respectively. Multiplying these slope coefficients by 100 (Langford and Blanc-Valleron, 1990) yields average values of HI of 330 and 370, respectively. Average percentages of pyrolizable hydrocarbons can also be obtained from the slope coefficients by multiplying by 10 (to convert mg HC per 1000 mg OC to mg HC per 100 mg OC, or wt. % HC), yielding average values of 33% and 37% pyrolizable hydrocarbons, respectively. Langford and Blanc-Valleron (1990) further suggest that lines with zero intercept and slopes of 7 (HI=700; 70 wt. % HC)) and 2 (HI=200; 20 wt. % HC) separate the pyrolitic hydrocarbon yields of type I and type II, and type II and type III kerogen, respectively (heavy dashed lines in Fig. 19B). From the HI-OI plot (Fig. 19A), one has the impression that, although there is considerable scatter, the organic matter in both cores contains more type II kerogen than type III. Using the criteria of Langford and Blanc-Valleron (1990), however, Figure 19B indicates that most of the organic matter is type II but closer to type III, probably because of oxidation of type II organic matter.

Diagenetic Fe-S-OC relations in the Niobrara and Bridge Creek Limestones, and in Cretaceous shales of the Western Interior, were discussed by Dean and Arthur (1989). Most modern marine sediments accumulating in continental margin settings with oxygenated bottom waters, and with sufficient organic matter to promote sulfate reduction below the sediment-water interface, have a positive correlation between OC and sulfide-sulfur with a S/OC ratio of about 0.4 (Goldhaber and Kaplan, 1974; Berner and Raiswell, 1983). Data from these sediments fall along a line of constant S/OC on a Fe-S-OC ternary plot (line marked 0.4 in Fig. 20). Samples from the Niobrara have variable S/OC

ratios indicated by scatter of points on either side of the 0.4 ratio line in Figure 20. This suggests either that seawater in the Cretaceous Western Interior basin contained less sulfate than modern seawater, or, more likely, that sulfide escaped before being fixed as iron sulfide, probably due to limitation by reactive iron. Most samples from the Niobrara have S/Fe ratios close to that of stoichiometric FeS_2 (S/Fe=1.15 line in Fig. 20), which suggests that most, if not all, reactive Fe was consumed during bacterial sulfate reduction in the OC-rich units. Iron limitation on pyrite formation in carbonate-rich rocks is common and explains the typical linear Fe/S and nonlinear S/OC relations in many such rocks. Rocks that were not iron-limited during early diagenesis – for example, most black shales – plot along a line of constant S/OC ratio such as the modern marine line (line marked 0.4 in Figure 20).

Under variable redox conditions, iron concentrations of original sediment tend to be preserved because, once reduced, iron is fixed in one of several ferrous-iron minerals, such as pyrite. Manganese concentrations in sediments, on the other hand, are highly variable and almost entirely depend on redox conditions because minerals containing reduced Mn (for example, rhodochrosite, $MnCO_3$) are rare in marine sediments and rocks. Con-

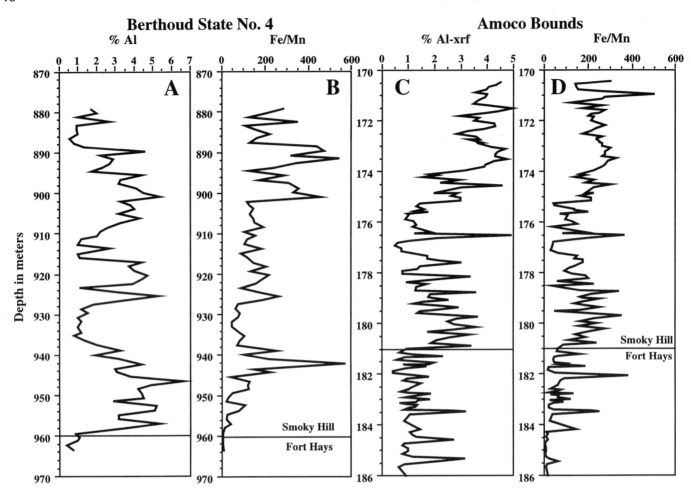

FIG. 22.–Down-core plots of % Al and Fe/Mn versus depth for the Niobrara Formation in the Berthoud State No. 4 core (A and B) and in the Bounds core (C and D).

sequently, concentrations of Mn greater than a few hundred ppm usually indicate more oxidizing conditions. Higher concentrations of Mn (up to 0.3%, 3000 ppm) in the Fort Hays were reported by Yamamoto et al. (1979) in an outcrop section in west-central Kansas. They suggested that the high Mn concentrations might be contributed by volcanic ash. However, the clear, positive correlation between Mn concentration and % CaCO$_3$, shown in Figure 21A for the Fort Hays in samples from the Bounds core and in outcrop samples from Colorado and New Mexico, indicates that most of the Mn is in the carbonate, probably substituting for Ca in calcite lattice of cements in the limestone beds. Figure 21B shows that Mn and % CaCO$_3$ also correlate well in the Bridge Creek Limestone, although values of Mn are somewhat lower than those in the Fort Hays.

Redox fluctuations are well illustrated in profiles of Mn concentration (Figs. 11, 12) or Fe/Mn ratio (Fig. 22) for the Niobrara Formation. In general, the basal Niobrara was deposited in a more oxic environment, as indicated by higher degree of bioturbation (Savrda, this volume) and lower OC concentrations (Figs. 11, 12). Consequently, the concentrations of Mn are considerably higher as shown by the very low Fe/Mn ratios in the Fort Hays Member in Figure 22. The low Mn concentrations (Figs. 11, 12) and high Fe/Mn ratios (Fig. 22) throughout the Smoky Hill Member suggest that the sediments remained reduced throughout most of Smoky Hill deposition. As discussed earlier, iron is mostly associated with the sulfide factor (Factor

5, Fig. 11) so that variations in Fe/Mn mainly reflect variations in sulfide content. In the Bounds core, however, the iron is associated with the detrital clastic fraction as indicated by the high correlation coefficient between Factor 1 loadings and % Fe in Table 4. The variations in Fe/Mn in the Bounds core, therefore probably mostly reflect variation in detrital clastic material (high iron) and associated more reducing conditions (low manganese). This combination has resulted in some striking cyclic variations in Fe/Mn in the Bounds core. Figure 22D shows that there are several longer cycles in the Bounds core with wavelengths of about 6 m that probably represent fourth order cycles with periodicities of several hundred thousand years and coincide with cycles in % Al (Fig. 22C), % OC, and gamma-ray intensity (Fig. 11). The 6-m, fourth order cycles have 40-50-cm cycles superimposed on them that represent the Milankovitch bedding cycles; they are particularly well developed between 177 and 182 m (Fig. 22D). The Fort Hays Member was deposited under more oxic conditions than the Smoky Hill Member, as evidenced by the predominance of highly bioturbated limestone in the Fort Hays and predominance of laminated marlstone in the Smoky Hill. Therefore, the Mn remained in the Fort Hays strata during early diagenesis, which resulted in lower Fe/Mn ratios. Under more reducing conditions that existed during deposition of the Smoky Hill, however, reduced Mn diffused out of the sediments and accumulated in the anoxic to dysoxic bottom waters, which increased Fe/Mn ratios.

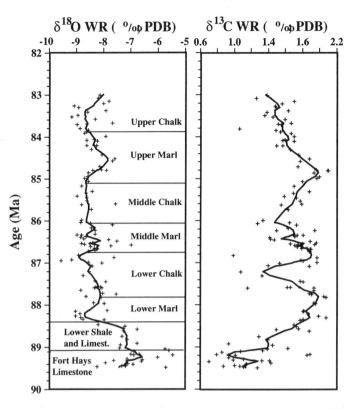

FIG. 23.–Down-core plots versus age for whole-rock (WR) carbonate values of $\delta^{18}O$ and $\delta^{13}C$ for samples of the Fort Hays and Smoky Hill Members of the Niobrara Formation in the Berthoud State No. 3 core.. Raw data points are shown by plot symbols, and the solid lines are smoothed curves. The data are plotted versus age using Decade of North American Geology time scale (Palmer, 1983).

Geochemical Indicators of Burial Diagenesis

The light stable-isotope ratios of carbon and oxygen in biogenic components are used in paleoenvironmental studies as indicators of dynamics of carbon cycling, salinity, and temperature (e.g. Arthur et al., 1985). In a similar way, minor-element data from biogenic carbonates are commonly thought to indicate the effects of seawater chemical composition and temperature. However, such data commonly are collected and interpreted without regard for potential diagenetic overprints. Such diagenetic problems may typify the pelagic sequences we analyzed because all are highly lithified. Burial depths of sequences examined in Deep Sea Drilling Project and Ocean Drilling Project sites typically are <1000 m but may be as much as 2500 m for the Western Interior sequences. The initial carbonate mineralogy of these units was most likely stable, low-magnesium calcite. Because of extensive cementation and low porosities (<10%), we would expect, for example, a diagenetic signal characterized by isotopically light $\delta^{18}O$ and relatively low Sr/Ca values in the limestones (e.g., Schlanger and Douglas, 1974; Scholle, 1977).

As discussed by Pratt et al. (1993) the $\delta^{18}O$ values for the deeply buried Niobrara Formation (Fig. 23) are more negative than -6‰, which is at least 3‰ more depleted in ^{18}O than expected for carbonate precipitated in isotopic equilibrium with Cretaceous seawater (Scholle and Arthur, 1980). The oxygen-isotope values in the more carbonate-rich intervals are depleted by 1-2‰ relative to those from marlstones. These data suggest that burial diagenetic effects might be important. Pratt et al.

(1993) summarize data showing that, on average, the most deeply buried Smoky Hill sections (in cores from Colorado) are about 3.5‰ depleted in ^{18}O in comparison with the shallowest buried sequences (e.g., Kansas outcrops), which they attribute to greater burial diagenesis in the Colorado sections. This decrease in $\delta^{18}O$ with increasing burial probably is the result of alteration of volcanic materials and neoformation of clay minerals that decrease the $\delta^{18}O$ of pore waters which, in turn, leads to isotopically light carbonate cements in pelagic carbonate sediments (e.g., Lawrence et al., 1979). Higher burial temperatures also produce calcite cements that are relatively depleted in ^{18}O (e.g., Scholle, 1977). Values of $\delta^{18}O$ in the Fort Hays Member are about the same, ca. -7‰, in Kansas and Colorado (Pratt el al., 1993), which suggests that the Fort Hays was not influenced by burial diagenesis. The values of $\delta^{18}O$ in the Fort Hays Member are still 4-5‰ lower than values in the open-marine Austin Chalk, which typically are -2‰ to -3‰ (Czerniakowski et al., 1984). This corresponds to about the expected range for seawater of normal salinity and at a temperature of 20° C. The lower values of $\delta^{18}O$ in the Niobrara Formation, even after accounting for burial diagenesis, suggest that the Western Interior Seaway was more depleted in ^{18}O than the open ocean, probably because of the influence of freshwater runoff from the Sevier highland on the western margin of the seaway (e.g., Pagani and Arthur, this volume).

Cyclic variations in whole-rock $\delta^{13}C$ values also occur in the Niobrara Formation (Fig. 23) and may provide additional clues to diagenetic history. In general, lower values of $\delta^{13}C$ correspond to intervals of higher carbonate content, such as the Fort Hays Limestone and the chalk subunits of the Smoky Hill. Carbon-isotope values in carbonates can be interpreted in terms of primary processes, such as changes in water-mass stratification and consequent supply of isotopically light CO_2 from decomposition of isotopically light organic matter in bottom waters (e.g., Berger and Vincent, 1986). However, $\delta^{13}C$ values of lithified pelagic carbonates that contain significant amounts of organic matter most likely reflect addition of carbonate cements precipitated in equilibrium with pore-waters that contained dissolved CO_2 that is substantially more ^{13}C-depleted than CO_2 in bottom waters (Irwin et al., 1977; Scholle and Arthur, 1980). The heavier carbon-isotope values of beds with lower carbonate concentrations probably represent values that are closer to the original primary biogenic components because of less cementation during diagenesis (Arthur et al., 1985).

Pratt et al. (1993) showed that whole-rock samples from the Fort Hays Member of the Niobrara have values of $\delta^{13}C$ (Fig. 23) that are similar to those in whole-rock samples of OC-lean pelagic chalks from the United Kingdom, which were deposited in open-ocean conditions (Scholle and Arthur, 1980). Values of $\delta^{13}C$ in the Smoky Hill Member, however, are lower (<2.0‰;) than those in UK chalks (2.0-2.5‰). This difference between the carbon-isotope compositions of Smoky Hill and UK carbonates is considered to indicate the extent of burial diagenetic overprinting of whole-rock carbon isotope values in the OC-rich Smoky Hill. The difference is greatest (ca. 2 ‰) in the upper parts of the section (Pratt et al., 1993).

The concentrations of Mg and Sr also provide a measure of the degree of carbonate diagenesis because of the expected loss of Sr and Mg during recrystallization and cementation (e.g., Matter et al., 1975; Brand and Veizer, 1980; Baker et al., 1982; Renard, 1986). Because the Niobrara Formation consists mostly of impure limestones that average about 32% insoluble residue, whole-rock Mg and Sr contents may be influenced by detrital

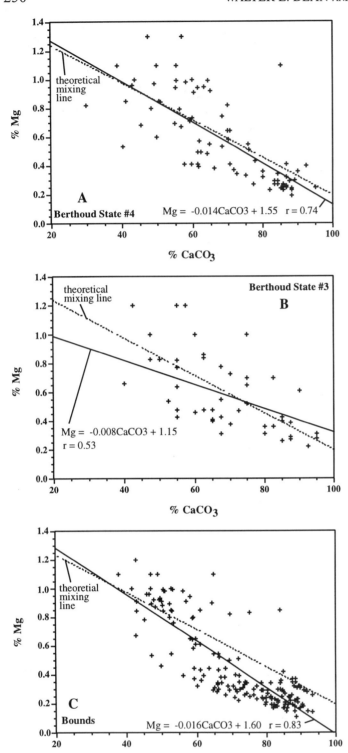

FIG. 24.–Cross plots of % CaCO₃ versus % Mg in the Niobrara Formation in the (A) Berthoud State No. 4, (B) Berthoud State No. 3, and (C) Bounds cores. Solid lines represent linear regressions. Dashed lines represent theoretical mixing lines of clay containing 1.5% Mg and pelagic carbonate containing 0.2% Mn.

clastic material and volcanic ash. The purer limestone beds in the Niobrara Formation contain 0.3-0.4% Mg. The Mg content is inversely proportional to the CaCO₃ content (Fig. 24), and marlstone beds usually contain 0.7-0.8% Mg. All of this variation can be explained by dilution of the carbonate with clastic

material and volcanic ash. We do not have data on the composition of acid-insoluble residues from the Niobrara Formation, but the mineralogy of insoluble residues is very similar to that of the overlying Pierre Shale (Pollasto and Martinez, 1985; Pollasto and Scholle, 1987). Average marine shale and bentonite from the Pierre Shale have Mg concentrations of 1.5% and 1.8%, respectively (Schultz et al., 1980). Assuming dilution of a pure pelagic carbonate containing 0.2% Mg (Scholle, 1977) with clay containing 1.5% Mg, we can construct a theoretical mixing line between these two end members. Figure 24 shows that the observed values in the Niobrara Formation from the two Berthoud State cores and the Bounds core are very close to this mixing line, which implies that the observed variations in Mg concentration in the Niobrara Formation are indeed due to dilution of pelagic carbonate with terrigenous clay.

The Sr concentrations in the Niobrara Formation are higher than expected for pure pelagic carbonates that lost Sr during diagenetic dissolution-reprecipitation reactions (e.g., Scholle, 1977; Brand and Veizer, 1980). For example, modern pelagic, low-magnesium calcites contain about 1600 ppm Sr , whereas relatively pure British chalks contain an average of about 500 ppm Sr (Scholle, 1977). Concentrations of Sr in shale and bentonite in the Campanian Pierre Shale are about 250 ppm (Schultz et al., 1980). Assuming that the insoluble residues in the Niobrara Formation have similar Sr concentrations, we can construct a theoretical mixing line between 250 ppm Sr for clay and 1600 ppm for pure pelagic carbonate. Figure 25 shows that, like Mg, observed concentrations of Sr in the Niobrara Formation are remarkably close to those predicted by the theoretical mixing line, which suggests that the carbonate in the Niobrara did not lose Sr through diagenesis. If the carbonates of the Niobrara had lost Sr through diagenesis, then the initial Sr concentration of the pelagic carbonate must have been considerably higher than 1600 ppm, the concentration of Sr in modern pelagic carbonates. A higher concentration of Sr in the surface waters of the Western Interior Seaway relative to European shelf environments might have occurred due to weathering of Sr-rich volcanic rocks in the Sevier highland west of the seaway. However, the remarkable approximation of the theoretical mixing line to the observed Sr values in the Niobrara suggests that the initial Sr concentration in pure pelagic carbonate was close to 1600 ppm and little has been lost through diagenesis. Also, if Sr was lost through diagenesis, we would expect the loss to be much greater in the more deeply buried Berthoud State cores than in the shallow-buried Bounds core, but the Sr concentrations are the same in all three cores.

BIOGENIC PRODUCTIVITY AND CYCLES

As discussed above, there are several possibilities for the origin of the cyclic interbeds of calcareous and clay-rich sediment (Fig. 3) other than diagenetic carbonate redistribution (e.g., Eder, 1982; Ricken, 1986; Raiswell, 1988), namely variation in production of biogenic calcite by coccoliths, variation in rate of carbonate dissolution, and variation in supply of terrigenous sediment. Each of these factors may play a role, particularly in deeper-water environments. Knowledge of accumulation rates of components across individual cycles would help to discriminate between the various possibilities, but it is difficult to perform these calculations without precise age control. We estimated the average periodicity of the bedding cycles at between 20-40 ky, but in to calculate relative changes in element accumulation rates across

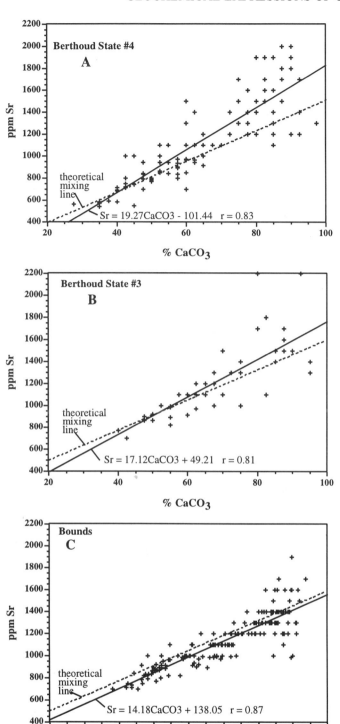

FIG. 25.–Cross plots of % CaCO3 versus ppm Sr in the Niobrara Formation in the (A) Berthoud State No. 4, (B) Berthoud State No. 3, and (C) Bounds cores. Solid lines represent linear regressions. Dashed lines represent theoretical mixing lines of clay containing 250 ppm Sr and pelagic carbonate containing 1600 ppm Sr.

high- and low-carbonate parts of cycles, we need to make assumptions about the amount of time that each lithologic type (hemicycle) represents. We suspect that changes in primary productivity played a role in the origin of the cyclicity in carbonate and OC contents, but supporting evidence for this hypothesis is

not easily obtained. Some geochemical features of cyclic strata might help to assess the role of varying primary production and carbon flux in the formation of cyclic sequences. Most promising in this regard are variations in concentrations and relative fluxes of silica and phosphorus.

As discussed above, variation in silica in the carbonate units of the Western Interior is complicated by variations in detrital and eolian quartz concentrations, and by the no-analogue Cretaceous ocean in which diatoms were not a major component of the phytoplankton, in contrast to modern high-productivity settings (e.g. Thierstein, 1989). High Si/Al ratios resulting from high concentrations of radiolaria in sediments may indicate initial enhanced preservation of biogenic opal, but do not necessarily indicate high primary productivity. Ratios of Si/Al >8 in some limestone beds of the Bridge Creek (Fig. 16B), can be considered "excess" silica over the Si/Al ratio in normal background terrigenous material (ca. 3-4). Because of the association of higher Si/Al ratios with other indicators of increased eolian influx during the deposition of the limestone units, we interpret these high ratios as being due mainly to increased eolian quartz. Eicher and Diner (1989) suggested that the high values of Si/Al in limestone, relative to those in marlstone or shale interbeds, represent relative increases in proportion of biogenic opal in the limestone beds; we cannot rule this out as a cause of at least some of the excess silica in the limestone beds. Eicher and Diner argue further that limestone beds represent deposition during times of highest overall productivity, not just carbonate productivity. We do not agree with this argument because modern coccoliths, the main source of CaCO3 in the WIS limestones, do not compete well with other organisms, for example, diatoms and dinoflagellates, in more eutrophic environments (e.g. Watkins, 1989).

The P/Al ratio could indicate variations in biologic productivity in that the flux of phosphatic fish debris to the seafloor might be proportional to surface-water primary production. However, the interpretation of phosphate concentrations in sediments is complicated by other possible phosphorus sources, such as that associated with organic matter, sorption on manganese- and iron-oxyhydroxides (e.g., Froelich et al., 1982) and coatings on planktonic foraminifera and other calcareous components (e.g., Sherwood et al., 1987). The factor analyses for the Niobrara Formation showed that phosphorus and associated light rare-earth elements tend to be associated with carbonate. In the Bounds core phosphate was a separate factor with highest loadings in the Fort Hays Limestone Member and the lower carbonate units of the Smoky Hill Member (Fig. 13). Highest loadings for Factor 5 and highest concentrations of phosphorus in the Bounds core (Fig. 13) are not in the limestone beds, but instead in the carbonate-poor layers, which suggests remobilization and reprecipitation of phosphate in certain horizons, or selective winnowing and concentration of phosphatic lags (e.g., fish debris) in these horizons. Figure 26 shows that there is a pattern of increasing phosphorus enrichment with increasing CaCO3 in both the Niobrara Formation and Bridge Creek Limestone. The shapes of the curves suggest mixing of essentially two end-members, a carbonate-rich component with a high P/Al ratio, and a terrigenous component with a lower P/Al ratio. Although increasing P/Al could reflect higher relative productivity – that is, flux of phosphatic fish debris and phosphorus-bearing organic matter – in the carbonate-rich beds, we believe that relative phosphorus enrichment in the limestone beds is most likely due to the processes of sorption suggested by Sherwood et al. (1987).

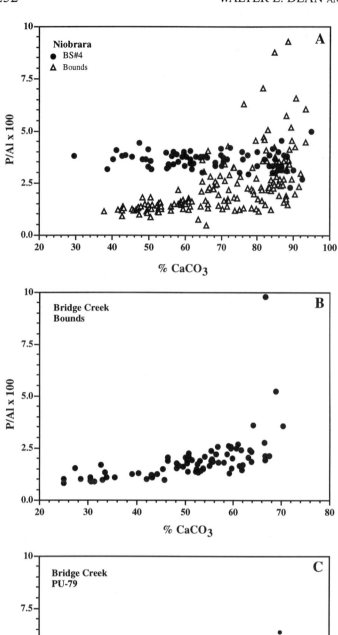

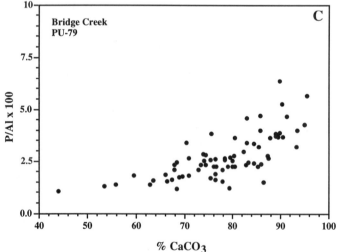

FIG. 26.—Cross plots of % CaCO₃ versus P/Al x 100 for (A) the Niobrara Formation in the Bounds and Berthoud State No. 4 cores, (B) the Bridge Creek Limestone in the Bounds core, and (C) the Bridge Creek Limestone in the PU-79 core.

CONCLUSIONS

Cyclic variations on scales of decimeters to tens of meters in concentrations of carbonate, detrital clastic material, organic carbon, and in degree of bioturbation characterize the cyclic pelagic limestone sequences of the Niobrara Formation. The periodicities of the dominant small-scale (bedding) cycles are estimated at 20-40 ky/cycle and probably are attributable to Milankovitch forcing. Cycles on the scale of several meters are prominent in plots of concentrations of many elements and in gamma-ray and sonic-velocity logs and represent estimated periodicities of several hundred thousands of years. Cycles in carbonate and clay that were used to subdivide the Smoky Hill Member into informal marlstone and limestone units have periodicities of 1-3 My.

Deposition of cyclic sedimentary sequences were the result of (1) cyclic changes in production of biogenic material in surface waters and flux to sediments that resulted from variations in upwelling or salinity, (2) cyclic changes in climate, weathering, runoff, and rate of supply of terrigenous detrital material that diluted the fluxes of biogenic components such as pelagic carbonate, and (3) fluctuations in water column stability and bottom-water production, circulation and oxygenation that affected preservation and accumulation of organic matter as well as carbonate. The Niobrara Formation cycles apparently are the product of all three of these mechanisms, which were interrelated and controlled by fresh-water runoff to the Cretaceous Western Interior Seaway of North America from uplifted highlands to the west and the low-relief eastern margin.

Although the cycles in the Niobrara Formation on all scales can adequately be represented as a three-component system of carbonate, clay (represented in our analyses by either Al or Mg), and organic carbon, multivariate Q-mode factor analyses of elemental geochemical data identified other elements that are associated with these three components and also defined several other element associations that reflect more subtle geochemical differences in the Niobrara. The detrital clastic (aluminosilicate) association is dominated by most major elements and several trace elements. An organic-carbon-trace-element association is dominated by organic carbon and redox-sensitive trace elements such as V, Cu, Zn, Cd, and Ni. As expected, Sr is the main element other than Ca associated with the carbonate component, although the limestones of the Fort Hays Member are characterized by unusually high concentrations of Mn that reflect more oxidizing conditions during deposition of the Fort Hays Member. In cores from northeastern Colorado there is a sulfide (pyrite) association in the upper part of the Smoky Hill Member dominated by Fe, S, Co, Ni, and Mo. In a core from western Kansas there is an additional P-rare-earth-element association in the more clay-rich beds of the Fort Hays Member and lower Smoky Hill Member.

The main sources of sediment that competed during deposition of the carbonate units of the Niobrara and Greenhorn Formations were biogenic carbonate, windblown volcanic ash, and terrigenous detritus transported by winds, rivers, and surface and bottom currents. We used the ratios K/Fe+Mg, Si/Al, Na/K, and Al/Ti to try to detect relative changes in influx of detrital illite, volcanic smectite, and eolian quartz. We found that the K/Fe+Mg ratio by itself is not a particularly distinctive indicator of the volcanic fraction, but Al/Ti and Na/K are. The Na/K ratio is complicated because it may also respond to available moisture as it affects selective decomposition of Na plagioclase. These indicators suggest that volcanic clastic material was a minor contributor to the total clastic influx except in some of the limestone beds of the Bridge Creek Limestone. In the Niobrara Formation, there is little difference in the composition of the clastic faction that can be related to limestone/marlstone cycles and, therefore, dry/wet climatic cycles.

Evidence for influx of detrital-clastic material from an eastern-margin source diluting background carbonate deposition at the basin center site of the Bounds core include (1) modeling results which suggest that runoff from the eastern margin of the seaway exited as a northward-directed coastal jet; (2) more bedding couplets in the Niobrara and Bridge Creek Limestones in the Bounds core relative to the Portland core; and (3) higher K/Fe+Mg in the Bridge Creek Limestone in the Bounds core than in the PU-79 core from Pueblo, Colorado, suggesting a greater influx of detrital illite.

The organic matter in the Smoky Hill Member of the Niobrara Formation in northeastern Colorado is uniformly enriched in Type II marine kerogen (hydrogen indices >200) regardless of whether it is in an OC-rich marlstone or an OC-poor limestone. This is in contrast to the Bridge Creek Limestone Member of the Greenhorn Formation where the organic matter in bioturbated limestone beds is highly oxidized. Because hydrogen indices (HI) in the Niobrara do not vary cyclically through OC-rich and OC-poor units, we conclude that the cyclic variations in % OC are due mainly to variable OC-flux and dilution and not to preservation.

Values of $\delta^{18}O$ in bulk carbonate are about 2‰ lower in the Smoky Hill Member in more deeply buried sections in eastern Colorado than in shallower buried sections in western Kansas. We attribute most of this difference to greater burial diagenesis in Colorado with formation of isotopically light carbonate cements. Even when corrected for burial diagenesis, the values of $\delta^{18}O$ are lower than those in open-marine carbonates of the same age. We view this as further evidence for seawater of lower than normal salinity in the Western Interior Seaway.

Values of Sr/Ca and Mg/Ca are often used as measures of carbonate diagenesis because of the expected loss of Sr and Mg during recrystallization and cementation. In the Niobrara Formation, however, the observed concentrations of Sr and Mg are remarkably close to those predicted by theoretical mixing of pure pelagic biogenic carbonate and Cretaceous clay, which suggests that the carbonates in the Niobrara have not lost Sr or Mg through diagenesis.

We conclude that the dominant process controlling relative variations in carbonate and clay contents of the Niobrara and Greenhorn Formations was fluctuations in river-borne detrital clastic material from the western Sevier highlands. However, there is some biological (e.g.,Eicher and Diner, 1989) and geochemical evidence for higher productivity by calcareous and possibly siliceous organisms during deposition of the more carbonate-rich units. Geochemical evidence includes higher Si/Al and P/Al ratios in the carbonate-rich units.

ACKNOWLEDGMENTS

We gratefully acknowledge support from the USGS Energy Resources Program and DOE Grant OE-FG02-92ER14251. We are grateful to the many analysts in the USGS laboratories in Denver who skillfully performed the many geochemical analyses. Ted Daws (U.S.G.S.) provided the Rock-Eval data on samples from the Berthoud State cores, and Tim White (PSU) provided Rock-Eval and coulometer data for samples from the Bounds core. Al Miesch introduced us to the wonderful world of Q-mode factor analysis. Pete Scholle provided the carbon and oxygen isotope analyses of samples from the Berthoud State No. 3 core, as well as a lot of inspiration on the origin of Cretaceous cycles. Al Fischer has provided endless inspiration, guidance, and enthusiasm over the years in the pursuit of his beloved "Gilbert cycles," most recently in his review of this paper. Brad Sageman also provided a thoughtful review of the paper, and taught us a few modern methods in cycle analysis (Sageman et al., 1997, and this volume).

REFERENCES

ARMSTRONG, R. L., AND WARD, P., 1993, Latest Triassic to earliest Eocene magmatism in the North American Cordilleran: Implications for the Western Interior Basin, in Caldwell, W. G. E., and Kauffman, E. G., eds., Evolution of the Western Interior Basin: Geological Association of Canada Special Paper 39, p. 49-72.

ARTHUR, M. A., AND DEAN, W. E., 1991, A holistic geochemical approach to cyclomania–Examples from Cretaceous pelagic limestone sequences, in Einsele, G., Ricken ,W., and Seilacher, A., eds., Cycles and Events in Stratigraphy: Berlin, Springer-Verlag, p. 126-166.

ARTHUR, M. A., DEAN, W. E., BOTTJER, D., AND SCHOLLE, P. A., 1984, Rhythmic bedding in Mesozoic-Cenozoic pelagic carbonate sequences: The primary and diagenetic origin of Milankovitch-like cycles, in Berger, A., Imbrie, J., Hays, J. D., et al., eds., Milankovitch and Climate, Part 1: Amsterdam, Riedel Publishing Company, p. 191-222.

ARTHUR, M. A., DEAN, W. E., AND CLAYPOOL, G. E., 1985a, Anomalous ^{13}C enrichment in modern organic carbon: Nature, v. 315, p. 216-218.

ARTHUR, M. A., DEAN, W. E., POLLASTRO, R. M., CLAYPOOL, G. E., AND SCHOLLE, P. A., 1985b, Comparative geochemical and mineralogical studies of two cyclic transgressive pelagic limestone units, Cretaceous Western Interior Basin, U.S., in Pratt, L. M., Kauffman, E. G., and Zelt, F. B., eds., Fine-grained Deposits and Biofacies of the Western Interior Seaway: Evidence of Cyclic Sedimentary Processes, Field Trip Guidebook No. 4: Tulsa, Society of Economic Paleontologists and Mineralogists, p. 16-27.

ARTHUR, M. A., DEAN, W. E., AND PRATT, L. M., 1988, Geochemical and climatic effects of increased organic carbon burial at the Cenomanian/Turonian boundary: Nature, v. 335, p. 714-717.

BAEDECKER, P. A., ed., 1987, Geochemical Methods of Analysis: Washington D.C., United States Geological Survey Bulletin 1770, 129 p.

BAKER, P. A., GIESKES, J. M., AND ELDERFIELD, H., 1982, Diagenesis of carbonates in deep-sea sediments–Evidence from Sr/Ca ratios and interstitial dissolved Sr^{2+} data: Journal of Sedimentary Petrology, v. 52, p. 71-82.

BARLOW, L. K., AND KAUFFMAN, E. G., 1985, Depositional cycles in the Niobrara Formation, Colorado Front Range, in Pratt, L. M., Kauffman, E. G., and Zelt, F. B., eds., Fine-grained Deposits and Biofacies of the Western Interior Seaway: Evidence of Cyclic Sedimentary Processes, Field Trip Guidebook No. 4: Tulsa, Society of Economic Paleontologists and Mineralogists, p. 184-198.

BARRON, E. J., ARTHUR, M. A., AND KAUFFMAN, E. G., 1985, Cretaceous rhythmic bedding sequences: A plausible link between orbital variations and climate: Earth and Planetary Science Letters, v. 72, p. 327-340.

BEAUMONT, C., QUINLAN, G. M., AND STOCKMAL, G. S., 1993, The evolution of the Western Interior Basin: Causes, consequences and unsolved problems, in Caldwell, W. G. E. and Kauffman, E. G., eds., Evolution of the Western Interior Basin: St. John's, Geological Association of Canada, Special Paper 39, p. 97-117.

BERGER, W. H. AND VINCENT, E., 1986, Deep-Sea Carbonates: Reading the carbon isotope signal, Geol. Rundschau, v. 75: p. 249-269.

BERNER, R. A., AND RAISWELL, R., 1983, Burial of organic carbon and pyrite sulfur in sediments over Phanerozoic time: A new theory: Geochimica et Cosmochimica Acta, v. 47, p. 855-862.

BOTTJER, D. J., ARTHUR, M. A., DEAN, W. E., HATTIN, D. E., AND SAVRDA, C. E., 1986, Rhythmic bedding produced in Cretaceous pelagic carbonate environments: Sensitive recorder of climatic cycles: Paleoceanography, v.1, p. 467-481.

BRAND, U., AND VEIZER, J., 1980, Chemical diagenesis of a multicomponent carbonate system–1: Trace elements: Journal of Sedimentary Petrology, v. 50, p. 1219-1236.

CROSS, T. A., AND PILGER, R. H., Jr., 1978, Tectonic controls of Late Cretaceous sedimentation, Western Interior: Nature, v. 274, p. 865-903.

CRYSDALE, B. L., AND BARKER, C. E., 1990, Thermal and fluid migration history in the Niobrara Formation, Berthoud oil field, Denver Basin, Colorado, in Nuccio, V. F., and Barker, C. E., eds. Applications of Thermal Maturity Studies to Energy Exploration: Denver, Rocky Mountain Section, Society of Economic Paleontologists and Mineralogists, p. 153-160.

CZERNIAKOWSKI, L. A., LOHMAN, K. C., AND WILSON, J. L., 1984, Closed-system marine burial diagenesis–Isotopic data from the Austin Chalk and its components: Sedimentology, v. 31, p. 863-877.

DEAN, W. E., AND ARTHUR, M. A., 1989, Iron-sulfur-carbon relationships in organic-carbon-rich sequences I: Cretaceous Western Interior Seaway: American Journal of Science, v. 289, p. 708-743.

DEAN, W. E., AND GARDNER, J. V., 1986, Milankovitch cycles in Neogene deep-sea sediment: Paleoceanography, v. 1, p. 539-553.

DEAN, W. E., ARTHUR, M. A., AND CLAYPOOL, G. E., 1986, Depletion of ^{13}C in Cretaceous marine organic matter: Source, diagenetic, or environmental signal?: Marine Geology, v. 70, p. 119-157.

DEAN, W. E., ARTHUR, M. A., SAGEMAN, B. B., AND LEWAN, M. D., 1995, Core Descriptions and Preliminary Geochemical Data for the Amoco Production Company Rebecca K. Bounds # 1 Well, Greeley County, Kansas: Washington, D.C., United States Geological Survey Open-File Report 95-209, 243 p.

DOVETON, J. H., 1991, Lithofacies and geochemical facies profiles from nuclear wireline logs: New subsurface templates for sedimentary modeling, in Franseen, E. K., Watney, W. L., Kendall, C. G. St. C., and Ross, W., eds., Sedimentary Modeling: Computer Simulations and Methods for Improved Parmeter Definition: Lawrence, Kansas, Kansas Geological Survey Bulletin 233, p. 101-110.

EDER, F. W., 1982, Diagenetic redistribution of carbonate, a process in forming limestone-marl alternations (Devonian and Carboniferous, Rheinisches Schiefegebirge, W. Germany). in Einsele, G., and Seilacher, A., eds., Cyclic and Event Stratification: Berlin, Springer, p. 98-112.

EICHER, D. L., AND DINER, R., 1989, Origin of Cretaceous Bridge Creek cycles in the Western Interior, United States: Palaeogeography, Palaeoclimatology, and Palaeoecology, v. 74, p. 127-146.

EINSELE G., 1982, Limestone-marl cycles (periodites): Diagnosis, significance, causes - A review, in Einsele, G., and Seilacher, A., eds., Cyclic and Event Stratification: Berlin, Springer-Verlag, p. 8-53.

ELDER, W. P., 1985, Biotic patterns across the Cenomanian-Turonian boundary interval, Greenhorn Formation, Rock Canyon Anticline, Pueblo, Colorado, in Pratt, L. M., Kauffman, E. G., and Zelt, F. B., eds., Fine-grained Deposits and Biofacies of the Western Interior Seaway: Evidence of Cyclic Sedimentary Processes, Field Trip Guidebook No. 4: Tulsa, Society of Economic Paleontologists and Mineralogists, p. 157-169.

ELDER, W. P., 1988, Geometry of Upper Cretaceous bentonite beds–Implications about volcanic source areas and paleowind patterns, Western Interior, United States: Geology, v. 16, p. 835-838.

ELDER, W. P., GUSTASON, E. R., AND SAGEMAN, B. B., 1994, Correlation of basinal carbonate cycles to nearshore parasequences in the Late Cretaceous Greenhorn seaway, Western Interior, U.S.A.: Geological Society of America Bulletin, v. 106, p. 892-902.

ENGLEMAN, E. E., JACKSON, L. L., NORTON, D. R., AND FISCHER, A. G., 1985., Determinations of carbonate carbon in geological materials by coulometric titration: Chemical Geology, v. 53, p. 125-128.

ERICKSEN, M. C., AND SLINGERLAND, R., 1990, Numerical simulations of tidal and wind-driven circulation in the Cretaceous Interior Seaway of North America: Geological Society of America Bulletin, v. 102, p. 1499-1516.

FISCHER, A. G., 1980, Gilbert-bedding rhythms and geochronology, in Yochelson, E. I., ed., The Scientific Ideas of G. K. Gilbert: Boulder, Geological Society of America Special Paper 183, p. 93-104.

FISCHER, A. G., 1986, Climatic rhythms recorded in strata: Annual Reviews of Earth and Planetary Sciences, v. 14, p. 351-376.

FISCHER, A. G., 1993, Cyclostratigraphy of Cretaceous chalk-marl sequences, in Caldwell, W. G. E., and Kauffman, E. G., eds., Evolution of the Western Interior Basin: St. John's, Geological Association of Canada Special Paper 39, p. 283-296.

FISCHER, A. G., HERBERT, T., AND PREMOLI-SILVA, I., 1985, Carbonate bedding cycles in Cretaceous pelagic and hemipelagic sediments, in Pratt, L. M., Kauffman, E. G., and Zelt, F. B., eds., Fine-grained Deposits and Biofacies of the Western Interior Seaway: Evidence of Cyclic Sedimentary Processes, Field Trip Guidebook No. 4: Tulsa, Society of Economic Paleontologists and Mineralogists, p. 1-10.

FISHER, C. G., HAY, W. W., AND EICHER, D. L., 1994, Oceanic front in the Greenhorn Sea (late middle through late Cenomanian): Paleoceanography, v. 9, p. 879-892.

FROELICH, P.N., BENDER, M. L., LUEDTKE, N. A., HEATH, G. R., AND DEVRIES, J., 1982, The marine phosphorus cycle, American Journal of Science, v. 282, p. 474-511.

GILBERT, G. K., 1895, Sedimentary measurement of geologic time: Journal of Geology, v. 3, p. 121-127.

GLANCY, T. J., BARRON, E. J., AND ARTHUR, M. A., An initial study of the sensitivity of modeled Cretaceous climate to cyclical insolation forcing: Paleoceanography, v. 1, p. 523-537.

GLANCY, T. J., JR., ARTHUR, M. A., BARRON, E. J., AND KAUFFMAN, E. G., 1993, A paleoclimate model for the North American Cretaceous (Cenomanian-Turonian) epicontinental sea, in Caldwell, W. G. E. and Kauffman, E. G., eds., Evolution of

the Western Interior Basin: Geological Association of Canada, Special Paper 39, p. 219-241

GOLDHABER, M. B., AND KAPLAN, I. R., 1974, The sulfur cycle, in Goldberg, E. D., ed., The Sea, v. 5, Marine Chemistry: New York, John Wiley and Sons, Inc., p. 569-655.

HALLAM, A., 1964, Origin of the limestone-shale rhythms in the Blue Lias of England: A composite theory: Journal of Geology, v. 72, p. 157-168.

HALLAM, A., 1986, Origin of minor limestone-shale cycles: Climatically induced or diagenetic?: Geology, v. 14, p. 609-612.

HATTIN, D. E., 1971, Widespread synchronously deposited burrow-mottled limestone beds in the Greenhorn Limestone (Upper Cretaceous) of Kansas and southeastern Colorado: American Association of Petroleum Geologists Bulletin, v. 55, p. 412-431.

HATTIN, D. E., 1982, Stratigraphy and depositional environment of Smoky Hill Chalk Member, Niobrara Chalk (Upper Cretaceous) of the type area, western Kansas: Lawrence, Kansas Geological Survey Bulletin, 225, 108 p.

HAY, W. W., EICHER, D. L., AND DINER, R., 1993, Physical oceanography and water masses in the Cretaceous Western Interior Seaway, in Caldwell, W. G. E., and Kauffman, E. G., eds., Evolution of the Western Interior Basin: St. John's, Geological Association of Canada Special Paper 39, p. 297-318.

IRWIN, H., CURTIS, C., AND COLEMAN, M., 1977, Isotopic evidence for source of diagenetic carbonates formed during burial of organic-rich sediments: Nature, v. 269, p. 209-213.

JORDAN, T. E., 1981, Thrust loads and foreland basin evolution, Cretaceous western United States: American Association of Petroleum Geoogists Bulletin, v. 64, p. 2506-2520.

KAUFFMAN, E. G., 1977, Geological and biological overview: Western Interior Cretaceous Basin: The Mountain Geologist, v. 14, p. 75-100.

KAUFFMAN, E. G., 1988, Concepts and methods of high-resolution event stratigraphy, in Annual Review of Earth and Planetary Sciences, v. 16, p. 605-654.

KAUFFMAN, E. G., 1995, Global change leading to biodiversity crisis in a greenhouse world: The Cenomanian-Turonian (Cretaceous) mass extinction, in Effects of Past Global Change on Life: Washington, D. C., National Academy Press, p. 47-71.

KAUFFMAN, E. G., AND CALDWELL, W. G. E., 1993, The Western Inteior Basin in space and time, in Caldwell, W. G. E., and Kauffman, E. G., eds., Evolution of the Western Interior Basin: St. John's, Geological Association of Canada, Special Paper 39, p. 1-30.

KAUFFMAN, E. G., SAGEMAN, B. B., KIRKLAND, J. I., ELDER, W, P., HARRIES, P. J., and VILLAMIL, T., 1993, Molluscan biostratigraphy of the Cretaceous Western Interior Basin, North America, in Caldwell, W. G. E. and Kauffman, E. G., eds., Evolution of the Western Interior Basin: Geological Association of Canada Special Paper 39, p.397-434. .

KLOVAN, E., AND MIESCH, A. T., 1976, Extended CABFAC and QMODEL computer programs for Q-mode factor analysis of compositional data: Computers and Geosciences, v. 1, p.161-178.

LAFERRIERE, A. P., AND HATTIN, D. E., 1989, Use of rhythmic bedding patterns for locating stuctural features, Niobrara Formation, United States Western Interior: American Association of Petroleum Geologists Bulletin, v. 73, p. 630-640.

LAFERRIERE, A. P., HATTIN, D. E., AND ARCHER, A. W., 1987, Effects of climate, tectonics, and sea-level changes on rhythmic bedding patterns in the Niobrara Formation (Upper Cretaceous), Western Interior: Geology, v. 15, p. 233-236.

LANGFORD, F. F., AND BLANC-VALLERON, 1990, Interpreting Rock-Eval pyrolysis data using graphs of pyrolizable hydrocarbons vs. total organic carbon: American Association of Petroleum Geologists Bulletin, v. 74, p. 799-804.

LAWRENCE, J. R., DREYER, J. I., ANDERSON, T. F., AND BRUECKNER, H. K., 1979, Importance of alteration of volcanic material in the sediments of Deep Sea Drilling Site 323: Chemistry, $^{18}O/^{16}O$ and $^{87}Sr/^{86}Sr$: Geochimica et Cosmochimica Acta, v. 47, p. 573-588.

MATTER, A., DOUGLAS, R. G., AND PERCH-NIELSEN, K., 1975, Fossil preservation, geochemisty, and diagenesis of pelagic carbonates from Shatsky Rise, northwest Pacific, in Larson, R. L., Moberly, R., et al., Initial Reports of the Deep Sea Drilling Project, v. 32: Washington, D.C., United States Government Printing Office, p. 891-921.

OBRADOVICH, J. D., 1973, A Cretaceous time scale, in Caldwell, W. G. E., and Kauffman, E. G., eds., Evolution of the Western Interior Basin: St. John's, Geological Association of Canada Special Paper 39, p. 379-396.

OBRADOVICH, J. D., AND COBBAN, W. A., 1975, A time-scale for the Late Cretaceous of the Western Interior of North America: St. John's, Geological Association of Canada Special Paper 13, p. 31-45.

PALMER, A. R., 1983, The Decade of North American Geology 1983 geologic time scale: Geology, v. 11, p. 503-504.

PARK, J. AND OGLESBY, R. J., 1991, Milankovitch rhythms in the Cretaceous: A GCM modeling study: Palaeogeography, Palaeoclimatology, and Palaeoecology (Global and Planetary Change), v. 90, p. 329-355.

PARK, J. AND OGLESBY, R. J., 1994, The effect of orbital cycles on Late and Middle

Cretaceous climate: A comparative general circulation model study: London, International Association of Sedimentology Special Publication 19, p. 509-529.

PETERS, K. E., 1986, Guidelines for evaluating petroleum source rock using programmed pyrolysis: American Association of Petroleum Geologists Bulletin, v. 70, p. 318-329.

POLLASTRO, R. M., AND MARTINEZ, C. J., 1985, Whole-rock, insoluble residue, and clay mineralogies of marl, chalk, and bentonite, Smoky Hill Shale Member, Niobrara Formation near Pueblo, Colorado - Depositional and diagenetic implications, in Pratt, L. M., Kauffman, E. G., and Zelt, F. B., eds., Fine-grained Deposits and Biofacies of the Western Interior Seaway: Evidence of Cyclic Sedimentary Processes, Field Trip Guidebook No. 4: Tulsa, Society of Economic Paleontologists and Mineralogists, p. 215-222.

POLLASTO, R. M., AND SCHOLLE, P. A., 1987, Diagenetic relationships in a hydrocarbon-productive chalk–The Cretaceous Niobrara Formation, in Mumpton, F. A., ed., Studies in Diagenesis: Washington, D.C., United States Geological Survey Bulletin 1578, p. 219-236.

PRATT, L. M., 1981, A paleo-oceanographic interpretation of the sedimentary structures, clay minerals, and organic matter in a core of the Middle Cretaceous Greenhorn Formation near Pueblo, Colorado: Unpublished Ph.D. Dissertation, Princeton University, Princeton, New Jersey, 176 p.

PRATT, L. M., 1984, Influence of paleoenvironmental factors on preservation of organic matter in Middle Cretaceous Greenhorn Formation, Pueblo, Colorado: American Association of Petroleum Geologists Bulletin, v. 68, p. 1146-1159.

PRATT, L. M., 1985, Isotopic studies of organic matter and carbonate in rocks of the Greenhorn marine cycle, in Pratt, L. M., Kauffman, E. G., and Zelt, F. B., eds., Fine-grained Deposits and Biofacies of the Western Interior Seaway: Evidence of Cyclic Sedimentary Processes, Field Trip Guidebook No. 4: Tulsa, Society of Economic Paleontologists and Mineralogists, p. 38-48.

PRATT, L. M., AND THRELKELD, C. N., 1984, Stratigraphic significance of $^{13}C/^{12}C$ ratios in mid-Cretaceous rocks of the western interior, U.S.A., in Stott, D. F., and Glass, D. J., eds., Mesozoic of Middle North America, Calgary, Canadian Society of Petroleum Geology Memoir 9, p. 305-312.

PRATT, L. M., CLAYPOOL, G. E., AND KING, J. D., 1986, Geochemical imprint of depositional conditions on organic matter in laminated-bioturbated interbeds from fine-grained marine sequences: Marine Geology, v. 70, p. 67-84.

PRATT, L. M., ARTHUR, M. A., DEAN, W. E., AND SCHOLLE, P. A., 1993, Paleo-oceanographic cycles and events during the Late Cretaceous in the Western Interior Seaway of North Ameica, in Caldwell, W. G. E., and Kauffman, E. G., eds., Evolution of the Western Interior Basin: St. John's, Geological Association of Canada Special Paper 39, p. 333-354.

RAISWELL, R., 1988, Chemical model for the origin of minor limestone-shale cycles by anaeobic methane oxidation: Geology, v. 16, p. 641-644.

RENARD, M., 1986, Pelagic carbonate chemostratigraphy (Sr, Mg, ^{18}O, ^{13}C): Marine Micropaleontology, v. 10, p. 117-164.

RICKEN, W., 1985, Epicontinental marl-limestone alternations: Event deposition and diagenetic bedding (Upper Jurassic, southwest Germany), in Bayer, U., and Seilacher, A., eds., Sedimentary and evolutionary cycles: New York, Springer-Verlag, Lecture Notes in Earth Sciences, v. 1, p. 127-162.

RICKEN, W., 1986, Diagenetic bedding: A model for marl-limestone alternations: New York, Springer-Verlag, Lecture Notes in Earth Sciences, v. 6, 210p.

ROCC (RESEARCH ON CRETACEOUS CYCLES) GROUP, 1986, Rhythmic bedding in Upper Cretaceous pelagic carbonate sequences - Verying sedimentary response to climatic forcing: Geology, v. 14, p. 153-156.

RYER, T. A., 1993, Speculations on the origins of Mid-Cretaceous clastic wedges, central Rocky Mountain region, United States, in Caldwell, W. G. E., and Kauffman, E. G., eds., Evolution of the Western Interior Basin: St. John's, Geological Association of Canada Special Paper 39, p. 189-198.

SAGEMAN, B. B., RICH, J., ARTHUR, M. A., BIRCHFIELD, G. E., AND DEAN, W. E., 1997, Evidence for Milankovitch periodicities in Cenomanian-Turonian lithologic and geochemical cycles, Western Inteior, U.S.A.: Journal of Sedimentary Research, v. 67, p. 286-302.

SAVRDA, C.E. AND BOTTJER, D. J., 1986, Trace-fossil model for reconstruction of paleo-oxygenation in bottom waters: Geology, v. 14 p. 3-6.

SCHLANGER, S.O., AND DOUGLAS, R. G., 1974, The pelagic ooze-chalk-limestone transition and its implications for marine stratigraphy, in Hsu, K. J. and H. C. Jenkins, eds., Pelagic Sediments—On Land and Under the Sea, London: International Association of Sedimentologists Special Publication 1, p. 117-148.

SCHOLLE, P. A., 1977, Chalk diagenesis and its relation to petroleum exploration–Oil from chalks, a modern miracle?: American Association of Petroleum Geologists Bulletin, v. 61, p. 982-1009.

SCHOLLE, P. A., AND ARTHUR, M. A., 1980, Carbon isotopic fluctuations in Cretaceous pelagic limestones–Potential stratigraphic and petroleum exploration tool: American Association of Petroleum Geologists Bulletin, v. 64, p. 67-87.

SCHULTZ, L. G., TOURTELOT, H. A., GILL, J. R., AND BOERNGEN, J. G., 1980, Composition of the Pierre Shale and equivalent rocks, Northern Great Plains Region: Washington, D. C., United States Geological Survey Professional Paper 1064-B, 114 p.

SCOTT, G. R., 1969, General and engineering geology of the northern part of Pueblo, Colorado: Washington, D. C., United States Geological Survey Bulletin 1262, 131 pp.

SCOTT, G. R., AND COBBAN, W. A., 1964, Stratigraphy of the Niobrara Formation at Pueblo, Colorado: Washington, D. C., United States Geological Survey Professional Paper 454-L, 30p.

SHERWOOD, B. A., SAGER, S. L., AND HOLLAND, H. D., 1987, Phosphorus in foraminiferal sediments from North Atlantic Ridge cores and in pure limestones: Geochimica et Cosmochimica Acta, v. 51, p. 1861-1866.

SHURR, G. W., 1984, Regional setting of Niobrara Formation in northern Great Plains: American Association of Petroleum Geologists Bulletin, v. 68, p. 598-609.

SLINGERLAND, R., KUMP, L. R., ARTHUR, M. A., FAWCETT, P. J., SAGEMAN, B. B., AND BARRON, E. J., 1996, Estuarine circulation in the Turonian Western Interior seaway of North America: Geological Society of America Bulletin, v. 108, p. 941-952.

THIERSTEIN, H. R., 1989, Inventory of paleoproductivity records: The mid-Cretaceous enigma, in Berger, W. H., Smetacek, V. S., and Wefer, G., eds., Productivity in the ocean: Past and present: New York, John Wiley and Sons, p. 355-375.

TISSOT, B. P., AND WELTE, D. H., 1984, Petroleum Formation and Occurrence, 2nd ed.: New York, Springer-Verlag, 538 p.

TUREKIAN, K. K., 1972, Chemistry of the Earth: New York, Holt, Rinehart, and Winston, 131 p.

WATKINS, D. K., 1989, Nannoplankton productivity fluctuations and rhythmically-bedded pelagic carbonates of the Greenhorn Limestone (Upper Cretaceous): Palaeogeography, Palaeoclimatology, and Palaeoecology, v. 74, p. 75-86.

WEIMER, R. J., 1983, Relationships of unconformities, tectonics, and sea level changes, Cretaceous of the Denver Basin and adjacent areas, in Reynolds, M. W., and Dolly, E. D., eds., Mesozoic paleogeography of the west-central United States: Denver, Rocky Mountain Section, Society of Economic Paleontologists and Mineralogists, p. 359-376.

WRIGHT, E. K., 1987, Stratification and paleocirculation of the Late Cretaceous Western Interior Seaway of North America: Geological Society of America Bulletin, v. 99, p. 480-490.

YAMAMOTO, S., HONJO, S., AND MERRIAM, D. F., 1979, Quantitative chemical stratigraphy of the Niobrara Chalk (Cretaceous) in western Kansas, in Gill, D., and Merriam, D. F., eds., Geochemical and Petrophysical Studies in Sedimentology: Oxford, Pergamon Press, p. 235-244.

ZELT, F. B., 1985a, Natural gamma-ray spectrometry, lithofacies, and depositional environments of selected Upper Cretaceous marine mudrocks, western United States, including Tropic Shale and Tununk Member of the Mancos Shale: Unpublished Ph.D. Dissertation, Princeton University, Princeton, New Jersey, 340 p.

ZELT, F. B., 1985b, Paleoceanographic events and lithologic/geochemical facies of the Greenhorn marine cycle (Upper Cretaceous) examined using natural gamma-ray spectrometry, in Pratt, L. M., Kauffman, E. G., and Zelt, F. B., eds., Fine-grained Deposits and Biofacies of the Western Interior Seaway: Evidence of Cyclic Sedimentary Processes, Field Trip Guidebook No. 4: Tulsa, Society of Economic Paleontologists and Mineralogists, p. 49-59.